RF Circuit Design:

Theory and Applications

Reinhold Ludwig

Worcester Polytechnic Institute

Gene Bogdanov

Worcester Polytechnic Institute

Upper Saddle River, NJ 07458

Library of Congress Cataloging-in-Publication Data

Ludwig, Reinhold.
  RF circuit design : theory and applications / Reinhold Ludwig, Gene Bogdanov.
    p. cm.
  Includes bibliographical references and index.
  ISBN 0-13-147137-6 (hardcover)
  1. Radio circuits--Design and construction. 2. Radio frequency. I. Bogdanov, Gene. II. Title.
  TK6553.L823 2007
  621.384'12--dc22
                          2007040036

Vice President and Editorial Director, ECS: **Marcia J. Horton**
Associate Editor: **Alice Dworkin**
Editorial Assistant: **William Opaluch**
Marketing Manager: **Tim Galligan**
Managing Editor: **Scott Disanno**
Production Liaison: **Rose Kernan**
Senior Operations Supervisor: **Alexis Heydt-Long**
Operations Specialist: **Lisa McDowell**
Cover Design: **Jayne Conte**
Director, Image Resource Center: **Melinda Patelli**
Manager, Rights and Permissions: **Zina Arabia**
Manager, Visual Research: **Beth Brenzel**
Manager, Cover Visual Research & Permissions: **Karen Sanatar**
Image Permission Coordinator: **John Ferreri**
Composition/Full-Service Project Management: **Laserwords**
Printer/Binder: **Courier Westford**

MATLAB and SIMULINK are registered trademarks of The MathWorks, 3 Apple Hill Drive, Natick, MA

Copyright © 2009 by Pearson Education, Inc., Upper Saddle River, New Jersey, 07458. Pearson Prentice Hall. All rights reserved. Printed in the United States of America. This publication is protected by Copyright and permission should be obtained from the publisher prior to any prohibited reproduction, storage in a retrieval system, or transmission in any form or by any means, electronic, mechanical, photocopying, recording, or likewise. For information regarding permission(s), write to: Rights and Permissions Department.

Pearson Prentice Hall™ is a trademark of Pearson Education, Inc.
Pearson® is a registered trademark of Pearson plc
Prentice Hall® is a registered trademark of Pearson Education, Inc.

Pearson Education Ltd., London
Pearson Education Singapore, Pte. Ltd.
Pearson Education, Canada, Inc.
Pearson Education–Japan
Pearson Education Australia PTY, Limited
Pearson Education North Asia, Ltd., Hong Kong
Pearson Educación de Mexico, S.A. de C.V.
Pearson Education Malaysia, Pte. Ltd.
Pearson Education, Upper Saddle River, New Jersey

10 9 8 7 6 5 4 3 2 1

ISBN-13: 978-0-13-147137-5
ISBN-10:   0-13-147137-6

DEDICATION

To our families and the memory of my father F. Ludwig

Contents

Preface — xiii

***Chapter 1.* Introduction** — 1
   1.1 Importance of Radio Frequency Design — 2
   1.2 Dimensions and Units — 6
   1.3 Frequency Spectrum — 9
   1.4 RF Behavior of Passive Components — 10
      1.4.1 Resistors at High Frequency — 16
      1.4.2 Capacitors at High Frequency — 19
      1.4.3 Inductors at High Frequency — 22
   1.5 Chip Components and Circuit Board Considerations — 25
      1.5.1 Chip Resistors — 25
      1.5.2 Chip Capacitors — 27
      1.5.3 Surface-Mounted Inductors — 28
   1.6 RF Circuit Manufacturing Processes — 29
   1.7 Summary — 32

***Chapter 2.* Transmission Line Analysis** — 41
   2.1 Why Transmission Line Theory? — 42
   2.2 Examples of Transmission Lines — 45
      2.2.1 Two-Wire Lines — 45
      2.2.2 Coaxial Line — 46
      2.2.3 Microstrip Lines — 47
   2.3 Equivalent Circuit Representation — 49

2.4 Theoretical Foundation ... 52
   2.4.1 Basic Laws ... 52
2.5 Circuit Parameters for a Parallel-Plate Transmission Line ... 57
2.6 Summary of Different Line Configurations ... 61
2.7 General Transmission Line Equation ... 62
   2.7.1 Kirchhoff Voltage and Current Law Representations ... 62
   2.7.2 Traveling Voltage and Current Waves ... 66
   2.7.3 Characteristic Impedance ... 67
   2.7.4 Lossless Transmission Line Model ... 67
2.8 Microstrip Transmission Lines ... 68
2.9 Terminated Lossless Transmission Line ... 73
   2.9.1 Voltage Reflection Coefficient ... 73
   2.9.2 Propagation Constant and Phase Velocity ... 74
   2.9.3 Standing Waves ... 75
2.10 Special Termination Conditions ... 78
   2.10.1 Input Impedance of Terminated Lossless Line ... 78
   2.10.2 Short-Circuit Terminated Transmission Line ... 79
   2.10.3 Open-Circuited Transmission Line ... 82
   2.10.4 Quarter-Wave Transmission Line ... 83
2.11 Sourced and Loaded Transmission Line ... 87
   2.11.1 Phasor Representation of Source ... 87
   2.11.2 Power Considerations for a Transmission Line ... 88
   2.11.3 Input Impedance Matching ... 91
   2.11.4 Return Loss and Insertion Loss ... 92
2.12 Summary ... 95

Chapter 3. The Smith Chart ... 103
3.1 From Reflection Coefficient to Load Impedance ... 104
   3.1.1 Reflection Coefficient in Phasor Form ... 104
   3.1.2 Normalized Impedance Equation ... 106
   3.1.3 Parametric Reflection Coefficient Equation ... 108
   3.1.4 Graphical Representation ... 110
3.2 Impedance Transformation ... 112
   3.2.1 Impedance Transformation for General Load ... 112
   3.2.2 Standing Wave Ratio ... 113
   3.2.3 Special Transformation Conditions ... 116
   3.2.4 Computer Simulations ... 120

3.3	Admittance Transformation	123
	3.3.1 Parametric Admittance Equation	123
	3.3.2 Additional Graphical Displays	126
3.4	Parallel and Series Connections	127
	3.4.1 Parallel Connection of R and L Elements	127
	3.4.2 Parallel Connection of R and C Elements	128
	3.4.3 Series Connection of R and L Elements	129
	3.4.4 Series Connection of R and C Elements	129
	3.4.5 Example of a T-Network	131
3.5	Summary	135

Chapter 4. Single- and Multiport Networks — 145

4.1	Basic Definitions	146
4.2	Interconnecting Networks	154
	4.2.1 Series Connection of Networks	154
	4.2.2 Parallel Connection of Networks	156
	4.2.3 Cascading Networks	157
	4.2.4 Summary of $ABCD$ Network Representations	158
4.3	Network Properties and Applications	163
	4.3.1 Interrelations between Parameter Sets	163
	4.3.2 Analysis of Microwave Amplifier	166
4.4	Scattering Parameters	169
	4.4.1 Definition of Scattering Parameters	169
	4.4.2 Meaning of S-Parameters	172
	4.4.3 Chain Scattering Matrix	175
	4.4.4 Conversion between Z- and S-Parameters	177
	4.4.5 Signal Flowgraph Modeling	178
	4.4.6 Generalization of S-Parameters	184
	4.4.7 Practical Measurements of S-Parameters	188
4.5	Summary	195

Chapter 5. An Overview of RF Filter Design — 205

5.1	Basic Resonator and Filter Configurations	206
	5.1.1 Filter Types and Parameters	206
	5.1.2 Low-Pass Filter	210
	5.1.3 High-Pass Filter	213
	5.1.4 Bandpass and Bandstop Filters	214
	5.1.5 Insertion Loss	221

5.2 Special Filter Realizations 224
   5.2.1 Butterworth-Type Filters 225
   5.2.2 Chebyshev-Type Filters 228
   5.2.3 Denormalization of Standard Low-Pass Design . . . 236
5.3 Filter Implementation 245
   5.3.1 Unit Elements 247
   5.3.2 Kuroda's Identities 247
   5.3.3 Examples of Microstrip Filter Design 249
5.4 Coupled Filter 257
   5.4.1 Odd and Even Mode Excitation 257
   5.4.2 Bandpass Filter Section 260
   5.4.3 Cascading Bandpass Filter Elements 262
   5.4.4 Design Example 264
5.5 Summary 268

Chapter 6. Active RF Components — 277

6.1 Semiconductor Basics 278
   6.1.1 Physical Properties of Semiconductors 278
   6.1.2 The pn-Junction 285
   6.1.3 Schottky Contact 295
6.2 RF Diodes 298
   6.2.1 Schottky Diode 299
   6.2.2 PIN Diode 301
   6.2.3 Varactor Diode 307
   6.2.4 IMPATT Diode 310
   6.2.5 Tunnel Diode 312
   6.2.6 TRAPATT, BARRITT, and Gunn Diodes 313
6.3 Bipolar-Junction Transistor 313
   6.3.1 Construction 314
   6.3.2 Functionality 316
   6.3.3 Frequency Response 322
   6.3.4 Temperature Behavior 324
   6.3.5 Limiting Values 328
   6.3.6 Noise Performance 329
6.4 RF Field Effect Transistors 330
   6.4.1 Construction 330
   6.4.2 Functionality 331

	6.4.3 Frequency Response	338
	6.4.4 Limiting Values	339
6.5	Metal Oxide Semiconductor Transistors	339
	6.5.1 Construction	340
	6.5.2 Functionality	341
6.6	High Electron Mobility Transistors	342
	6.6.1 Construction	343
	6.6.2 Functionality	343
	6.6.3 Frequency Response	346
6.7	Semiconductor Technology Trends	347
6.8	Summary	352

Chapter 7. Active RF Component Modeling — 361

7.1	Diode Models	362
	7.1.1 Nonlinear Diode Model	362
	7.1.2 Linear Diode Model	364
7.2	Transistor Models	367
	7.2.1 Large-Signal BJT Models	367
	7.2.2 Small-Signal BJT Models	376
	7.2.3 Large-Signal FET Models	388
	7.2.4 Small-Signal FET Models	391
	7.2.5 Transistor Amplifier Topologies	395
7.3	Measurement of Active Devices	397
	7.3.1 DC Characterization of Bipolar Transistor	397
	7.3.2 Measurements of AC Parameters of Bipolar Transistors	398
	7.3.3 Measurements of Field Effect Transistor Parameters	403
7.4	Scattering Parameter Device Characterization	404
7.5	Summary	413

Chapter 8. Matching and Biasing Networks — 421

8.1	Impedance Matching Using Discrete Components	422
	8.1.1 Two-Component Matching Networks	422
	8.1.2 Forbidden Regions, Frequency Response, and Quality Factor	431
	8.1.3 T and Pi Matching Networks	442
8.2	Microstrip Line Matching Networks	446
	8.2.1 From Discrete Components to Microstrip Lines	446
	8.2.2 Single-Stub Matching Networks	450
	8.2.3 Double-Stub Matching Networks	454

8.3		Amplifier Classes of Operation and Biasing Networks	458
	8.3.1	Classes of Operation and Efficiency of Amplifiers	458
	8.3.2	Bipolar Transistor Biasing Networks	463
	8.3.3	Field Effect Transistor Biasing Networks	469
8.4		Summary	478

Chapter 9. **RF Transistor Amplifier Design** — 485

9.1		Characteristics of Amplifiers	486
9.2		Amplifier Power Relations	487
	9.2.1	RF Source	487
	9.2.2	Transducer Power Gain	488
	9.2.3	Additional Power Relations	489
9.3		Stability Considerations	492
	9.3.1	Stability Circles	492
	9.3.2	Unconditional Stability	494
	9.3.3	Stabilization Methods	501
9.4		Constant Gain	504
	9.4.1	Unilateral Design	504
	9.4.2	Unilateral Figure of Merit	510
	9.4.3	Bilateral Design	512
	9.4.4	Operating and Available Power Gain Circles	515
9.5		Noise Figure Circles	521
9.6		Constant VSWR Circles	525
9.7		Broadband, High-Power, and Multistage Amplifiers	529
	9.7.1	Broadband Amplifiers	529
	9.7.2	High-Power Amplifiers	540
	9.7.3	Multistage Amplifiers	543
9.8		Summary	550

Chapter 10. **Oscillators and Mixers** — 559

10.1		Basic Oscillator Models	560
	10.1.1	Feedback Oscillator	560
	10.1.2	Negative Resistance Oscillator	562
	10.1.3	Oscillator Phase Noise	574
	10.1.4	Feedback Oscillator Design	578
	10.1.5	Design Steps	581
	10.1.6	Quartz Oscillators	585

10.2	High-Frequency Oscillator Configuration	587
	10.2.1 Fixed-Frequency Oscillators	591
	10.2.2 Dielectric Resonator Oscillators	598
	10.2.3 YIG-Tuned Oscillator	603
	10.2.4 Voltage-Controlled Oscillator	604
	10.2.5 Gunn Element Oscillator	608
10.3	Basic Characteristics of Mixers	609
	10.3.1 Basic Concepts	610
	10.3.2 Frequency Domain Considerations	612
	10.3.3 Single-Ended Mixer Design	614
	10.3.4 Single-Balanced Mixer	622
	10.3.5 Double-Balanced Mixer	623
	10.3.6 Integrated Active Mixers	624
	10.3.7 Image Reject Mixer	628
10.4	Summary	641

Appendix A. **Useful Physical Quantities and Units** — **647**

Appendix B. **Skin Equation for a Cylindrical Conductor** — **653**

Appendix C. **Complex Numbers** — **657**

C.1	Basic Definition	657
C.2	Magnitude Computations	657
C.3	Circle Equation	658

Appendix D. **Matrix Conversions** — **659**

Appendix E. **Physical Parameters of Semiconductors** — **663**

Appendix F. **Long and Short Diode Models** — **665**

F.1	Long Diode	666
F.2	Short Diode	666

Appendix G. **Couplers** — **669**

G.1	Wilkinson Divider	669
G.2	Branch Line Coupler	672
G.3	Lange Coupler	676

Appendix H. **Noise Analysis** — **677**

H.1	Basic Definitions	677
H.2	Noisy Two-Port Networks	679

H.3	Noise Figure for Two-Port Network	682
H.4	Noise Figure for Cascaded Multiport Network	685

Appendix I. Introduction to MATLAB — 689

- I.1 Background — 689
- I.2 Brief Example of Stability Evaluation — 691
- I.3 Simulation Software — 693
  - I.3.1 Overview — 693
  - I.3.2 File Organization — 693

Index — 695

Preface

High-frequency circuit design continues to enjoy significant industrial attention, triggered by a host of radio-frequency (RF) and microwave (MW) products. Improved semiconductor devices, new board materials, and advanced fabrication technologies have made possible a proliferation of high-speed digital and analog systems that profoundly influence wireless communication, global positioning, radar, remote sensing, and related electrical and computer engineering disciplines. As a consequence, this interest has translated into market demands for trained engineers and professionals with knowledge of high-frequency circuit design principles. Since the publication of the first edition of this textbook in January, 2000, the need for well-educated RF professionals has surged, making a text that teaches the fundamentals of high-frequency circuits even timelier.

The objective of this second edition remains the same: to present the fundamental RF design aspects and the underlying distributed circuit theory with minimal emphasis on electromagnetics. We have written this book in a manner that requires no EM background beyond a first year undergraduate physics course in fields and waves. Students and practicing engineers equipped with rudimentary exposure to circuit theory and/or microelectronics can read this book and grasp the entire spectrum of high-frequency circuit principles involving passive and active discrete devices, transmission lines, filters, amplifiers, mixers, oscillators and their design procedures. Lengthy mathematical derivations are either relegated to the appendices or placed in examples, thereby separating dry theoretical details from the main text. Although de-emphasizing theory creates a certain loss in precision, it promotes readability and focus on the underlying circuit concepts.

What has changed from the first edition? Besides our obvious attempt to eliminate typos and inconsistencies, the second edition was improved in several important ways. First, we have added *Practically Speaking* sections at the end of each chapter. In these sections, key design concepts and measurement procedures are discussed in detail. Topics such as the construction of an attenuator, a microstrip filter, or the simulation of a low noise RF amplifier with bias and matching networks, are presented similarly to a lab component that accompanies the lectures. Equipped with the right instrumentation and software simulator, the reader can easily replicate the circuits. Second, topics of interest, helpful definitions, and noteworthy observations are placed on the *margins* and offset from the main text. In addition to highlighting their importance, this approach allows us to emphasize and better explain items that do not directly fit into the flow of the main text. For example, the coverage of a Phase Lock Loop (PLL) system would exceed the scope of this book. However, a brief explanation of a PLL provides context and extra motivation for the underlying high-frequency circuits. It furthermore inspires the readers to explore these topics on their own. Third, more emphasis is placed on nonlinear design principles, specifically in regard to oscillators and their associated resonator circuits.

Accepting the challenge to deliver a high degree of linear and nonlinear design experience, we have included a number of examples that analyze in considerable depth, often extending over several pages, the philosophy and the intricacies of various modeling approaches. While linear scattering parameter simulations are adequate under certain conditions, nonlinear simulations, for instance the harmonic balance analysis, are required for more sophisticated designs. Oscillator and mixer, as well as amplifier designs can greatly benefit from a nonlinear circuit simulation. Naturally, the use of appropriate simulation tools creates problems in terms of their capabilities, accuracies, speeds, and not least costs. The availability of circuit simulators and RF software tools has steadily increased over the years. Indeed, the authors are routinely contacted about simulators that offer "exceptional" performances under particular constraints. It is not our goal to render an assessment or endorsement of a specific simulator (the authors have no commercial, nor professional ties with any vendor). In general, professional high-frequency simulators are expensive and require familiarity to use them effectively. Several years ago, the ECE department at WPI decided after an extensive review to adopt Advanced Design Systems (ADS) of Agilent Technologies as the default high-frequency circuit simulator for its undergraduate and graduate electrical and computer engineering students. For this reason, and because of its wide-spread industrial use, we rely on ADS simulations for most of our circuits. However, for readers without access to commercial simulators, we created a number of standard Matlab M-files that can be downloaded from our website listed in Appendix G. Because Matlab is a popular and relatively inexpensive mathematical tool, many examples discussed in this book can be executed and the results graphically displayed in a matter of seconds. Specifically, the various Smith Chart computations of impedance transformations should appeal to the reader.

Since our goal focuses on circuits, the textbook purposely omitted high-speed digital circuits as well as coding and modulation aspects. Although important, these topics would require too many additional pages and would move the book too far away from its original intent of providing a fundamental, one- or two-semester introduction to RF circuit design. In the ECE department at WPI, this does not constitute a disadvantage, as most of these topics are taught in specialized communication systems engineering courses.

The organization of this text is as follows: **Chapter 1** presents a general explanation of why basic circuit theory needs to be modified as the operating frequency is increased to a level where the wavelength becomes comparable with circuit dimensions. **Chapter 2** then develops the fundamental concepts of distributed circuit theory. **Chapter 3** introduces the Smith Chart as a generic tool for dealing with the periodic impedance behavior on the basis of the reflection coefficient. **Chapter 4** presents networks and flow-graph representations, and how the terminal conditions can be described with so-called scattering parameters. The network models and their scattering parameter descriptions are utilized in **Chapter 5** to develop passive RF filter configurations. To address active devices, **Chapter 6** provides a review of key semiconductor fundamentals, followed by their circuit models representation in **Chapter 7**. The impedance matching and biasing of bipolar and field effect transistors is taken up in **Chapter 8**. **Chapter 9** focuses on a number of key high-frequency amplifier configurations and their design intricacies, ranging from low noise to high power applications. Finally, **Chapter 10** introduces the reader to nonlinear systems and their design, covering oscillator and mixer circuits.

This book is used in the ECE department at WPI as a required text for its standard 7-week (5 lecture hours per week) course in RF circuit design (ECE 3113, *Introduction to RF Circuit Design*). The course has primarily attracted an audience of 3rd and 4th year undergraduate students with a background in microelectronics. The course does not include a separate laboratory, although a total of six practical circuits (all part of the Practically Speaking sections) are presented to the students who are then instructed to conduct their own measurements with a network analyzer. In addition, ADS simulations are incorporated as part of the regular lectures. Each chapter is self-contained, with the goal of providing wide flexibility in organizing the course material. At WPI, the content of approximately one three semester hour course is compressed into a 7-week period (consisting of a total of 28–29 lectures). The topics covered in ECE 3113 are shown in the table below.

EE 3113, Introduction to RF Circuit Design

Chapter 1, Introduction	Sections 1.1–1.6
Chapter 2, Transmission Line Analysis	Sections 2.1–2.12
Chapter 3, Smith Chart	Sections 3.1–3.5
Chapter 4, Single- and Multi-Port Networks	Sections 4.1–4.5
Chapter 7, Active RF Component Modeling	Sections 7.1–7.2
Chapter 8, Matching and Biasing Networks	Sections 8.1–8.4
Chapter 9, RF Transistor Amplifier Designs	Sections 9.1–9.4

The remaining material is targeted for a second (7-week) term covering more advanced topics such as microwave filters, equivalent circuit models, oscillators and mixers. An organizational plan is provided below.

Advanced Principles of RF Circuit Design

Chapter 5, A Brief Overview of RF Filter Design	Sections 5.1–5.5
Chapter 6, Active RF Components	Sections 6.1–6.6
Chapter 7, Active RF Component Modeling	Sections 7.3–7.5
Chapter 9, RF Transistor Amplifier Designs	Sections 9.5–9.8
Chapter 10, Oscillators and Mixers	Sections 10.1–10.4

Obviously, the entire course organization remains subject to change depending on total classroom time, student background, and interface requirements with related courses. At the writing of this 2nd edition, a new graduate course is being designed that combines the advanced RF circuit topics of Chapters 5–10 with a classical graduate-level electromagnetics text.

ACKNOWLEDGEMENTS

The authors are grateful to a number of colleagues, students, and practicing engineers. Prof. Fred Looft, head of the ECE department, was instrumental in providing departmental funding for the networked ADS simulator resources and the recently acquired network analyzers. Our thanks go to Korné Vennema and Scott Blum of NXP (formerly Philips Semiconductors) for providing technical RF expertise, sponsoring student projects, and making available measurement equipment. Professor Sergey N. Makarov added assistance through technical discussions. Brian Foley, Peter Serano, Shaileshkumar Raval, Dr. Rostislav Lemdiasov, Aghogho Obi, Souheil Benzerrouk, Dr. Funan Shi are current and former graduate students who provided insight, sometimes a fresh view, and always much appreciated ambience and support in the Center for Imaging and Sensing (CIS) at WPI. The authors are particularly grateful to Prof. Diran Apelian, director of the Metal Processing Institute at WPI, and Scott Biederman of GM for introducing them to the importance of microwave imaging and RF principles in material processing. R. L. would like to acknowledge his former co-author Dr. Pavel Bretchko; his brilliant effort and hard work helped shape the original text and laid the foundation of this second edition. Tom Robbins, the publisher of the first edition, is thanked for his constant support and editorial insight over the past 7 years. It is professionals like Mr. Robbins to whom the academic publishing industry owes its existence.

The staff of Prentice-Hall, specifically Alice Dworkin, Rose Kernan, and G. Muthukumar, Senior Project Manager, Laserwords Private Limited, Chennai, India, are thanked for their support in making this book project a reality.

1

Introduction

1.1 Importance of Radio Frequency Design 2
1.2 Dimensions and Units 6
1.3 Frequency Spectrum 9
1.4 RF Behavior of Passive Components 10
    1.4.1 Resistors at High Frequency 16
    1.4.2 Capacitors at High Frequency 19
    1.4.3 Inductors at High Frequency 22
1.5 Chip Components and Circuit Board Considerations 25
    1.5.1 Chip Resistors 25
    1.5.2 Chip Capacitors 27
    1.5.3 Surface-Mounted Inductors 28
1.6 RF Circuit Manufacturing Processes 29
1.7 Summary 32

For the past several years, analog and digital design engineers have continually been developing and refining circuits for increasingly higher operating frequencies. Analog circuits for wireless communication in the low to high gigahertz (GHz) range and concomitantly the rapid improvement of clock speeds of microprocessors, memory chips, and peripheral units in high-performance mainframes, workstations, and personal computers exemplify this trend. Global positioning systems require carrier frequencies in the range of 1227.60 and 1575.42 MHz, wireless local area networks and HiperLAN operate at 2.4 GHz, and optical communication channels can transport data of up to 40 gigabits per second (Gbps). The low-noise amplifier in a personal communication system (PCS) may operate at 1.9 GHz and fit on a circuit board smaller in size than a dime. Satellite broadcasting in the C-band involves 4 GHz uplink and 6 GHz downlink systems. In general, due to the rapid expansion of wireless communication, more compact amplifier, filter, oscillator, and mixer circuits are being designed and placed in service at frequencies generally above 1 GHz. There is little doubt that this trend will continue unabated, resulting not only in engineering systems with unique capabilities, but also special design challenges not encountered in conventional low-frequency systems.

This chapter reviews the implications as one migrates from low- to high-frequency circuit operation. It motivates and provides the physical rationales that have resulted in the need for new engineering approaches to design and optimize these circuits. The example of a mobile phone circuit, components of which will be analyzed in more detail in later chapters, serves as a vehicle to outline the goals and objectives of this textbook, and its organization.

The chapter begins with a brief historical discussion explaining the transition from direct current (DC) to high-frequency modes of operation. As the frequency increases and the associated wavelengths of the electromagnetic waves becomes comparable to the dimensions of the discrete circuit components such as resistors, capacitors, and inductors, these components start to deviate in their electric responses from the ideal frequency behavior. It is the purpose of this chapter to provide the reader with an appreciation and understanding of high-frequency passive component characteristics. In particular, due to the availability of sophisticated measurement equipment, the design engineer must know exactly why and how the high-frequency behavior of his or her circuit differs from the low-frequency realization. Without this knowledge, it will be impossible to develop and understand the special requirements of high-performance systems.

LUMPED THEORY

Circuit elements are assumed to have zero spatial extent (point form).

DISTRIBUTED THEORY

Circuit elements are modeled as having finite size relative to the wavelength.

1.1 Importance of Radio Frequency Design

The beginning of electrical circuit design is most likely traced back to the late eighteenth and early nineteenth centuries when the first reliable batteries became available. Named after their inventor A. Volta (1745–1827), the Voltaic cells permitted the supply of reliable DC energy to power the first crude circuits. However, it soon became apparent that low-frequency alternating current (AC) power sources can transport electricity more efficiently and with less electric loss when transmitted over some distance, and that distributing the electric energy could be facilitated through transformers that operate in accordance with Faraday's induction law. Due to pioneering work by such eminent engineers as Charles Steinmetz, Thomas Edison, Werner Siemens, and Nikola Tesla, the power generation and distribution industry quickly gained entry into our everyday life. It was James Maxwell (1831–1879) who, in a paper first read in 1864 to the Royal Society in London, postulated the mutual coupling of the electric and magnetic fields; their linkage through space gives rise to wave propagation. In 1887 Heinrich Hertz experimentally proved the radiation and reception of electromagnetic energy through air. This discovery heralded the rapidly expanding field of wireless communication, from radio and TV transmissions in the 1920s and 1930s to mobile phones and Global Positioning Systems (GPS) in the 1980s and 1990s. With the advent of third generation (3G) wireless systems and high-speed optical communication in the new millennium we can anticipate faster proliferation of high- and ultrahigh-frequency components, modules, and systems. Unfortunately, the design and development of enabling high-frequency circuits for today's information technology applications is not so straightforward. As will be discussed in detail, conventional Kirchhoff-based voltage and current law analyses, as presented to first- and second-year undergraduate electrical engineering students, apply only to DC and low-frequency lumped parameter systems consisting of networks of resistors, capacitors, and inductors. They fail when applied to circuits governed by electromagnetic wave propagation.

The main purpose of this textbook is to provide the reader with theoretical and practical aspects of analog circuit design when the frequency of operation extends into the **radio frequency** (RF) and **microwave** (MW) domains. In these domains, typically starting at several hundred kHz, the wavelengths of the electric signals shorten to a point where they compare with generic dimensions of the circuit, affecting its functionality. Here, conventional circuit analysis principles on the basis of Kirchhoff's theory fail. From a practical point of view the following questions face the design engineer:

- At what upper frequency does conventional circuit analysis require adjustment?
- What characteristics make the high-frequency behavior of electronic components so different from their low-frequency behavior?
- What "new" circuit theory replaces the classical Kirchhoff theory?
- How is this theory applied to the practical design of high-frequency analog circuits?

> **MOBILE PHONE ELEMENTS**
>
> Key components of a mobile phone include:
>
> (a) Antenna
> (b) RF switch
> (c) Power amplifier (PA)
> (d) Low-noise amplifier (LNA)
> (e) Mixer
> (f) Voltage-controlled oscillator (VCO)
> (g) Filter (bandpass, low-pass)
> (h) Converter (ADC, DAC)
> (i) Digital base band processor

This book intends to provide comprehensive answers to these questions by developing not only the theoretical background, but by also delivering the practical applications through a host of examples and design projects.

To identify more clearly the issues that we will address, let us examine the generic RF system shown in Figure 1-1.

Typical applications of this configuration are **mobile phones** and **wireless local area networks** (WLANs). The entire block diagram in Figure 1-1 can be called a **transceiver**, since it incorporates both transmitter and receiver circuits and uses a single antenna for communication. In this configuration the input signal (either a voice or a digital signal from a computer) is first digitally processed. If the input signal is a voice signal, as is the case in mobile phones, it is first converted into digital form, then compressed to reduce the time of transmission, and finally appropriately coded to suppress noise and communication errors.

After the input signal has been digitally preprocessed, it is converted back to analog form via a **digital-to-analog converter** (DAC). This low-frequency signal is mixed with a high-frequency carrier signal provided by a local oscillator. The combined signal is subsequently amplified through a **power amplifier** (PA) and then routed to the antenna, whose task is to radiate the encoded information as electromagnetic waves into free space.

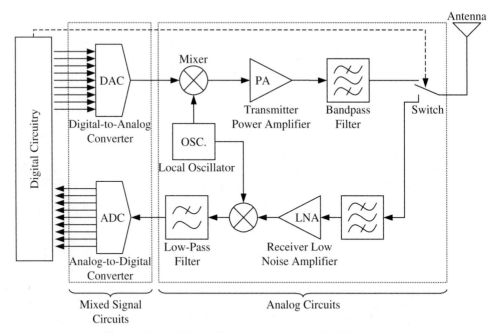

Figure 1-1 Block diagram of a generic RF system.

In the block diagram of Figure 1-1, let us focus on the transmitter PA. This could be a 2 GHz PA for mobile phones that may be implemented as a dual-stage amplifier. Details of the circuit diagram for the first stage PA are shown in Figure 1-2(a).

We notice that the input signal is fed through a DC blocking capacitor into an **input matching network**, needed to match the input impedance of the transistor (type BFG425W of NXP), operated in common-emitter configuration, to the output impedance of the mixer that precedes the PA. The matching is needed to ensure optimal power transfer and eliminate performance degrading reflections. The **interstage matching network** must then match the output impedance of the transistor to the input impedance of the second stage of the PA. Key components in the matching networks are microstrip lines shown by the shaded rectangles in Figure 1-2(a). At high frequency, these distributed elements exhibit unique electric properties that differ significantly from low-frequency lumped circuit elements. We also notice additional networks to bias the input and output ports of the transistor. The separation of high-frequency signals from the DC bias is achieved through two RF blocking networks that feature so-called **radio frequency chokes** (RFCs).

The actual dual-stage circuit board implementation is given in Figure 1-2(b), which shows the microstrip lines as copper traces of specific lengths and widths. Attached to the microstrip lines are chip capacitors, resistors, and inductors.

> **SINGLE CHIP MOBILE PHONE TRANSCEIVER**
>
> Despite intense research and development efforts, single chip system-on-chip (SoC) solutions remain elusive. Key problem areas are digital and analog mixed-signal integration on a single wafer as well as the incorporation of high-quality filter elements.

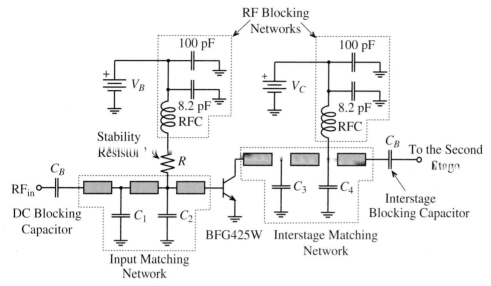

Figure 1-2(a) Simplified circuit diagram of the first stage of a 2 GHz power amplifier for a mobile phone.

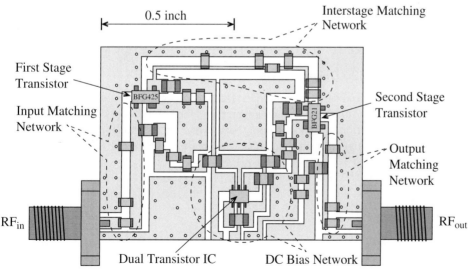

Figure 1-2(b) Printed circuit board layout of the power amplifier.

To understand, analyze, and ultimately build such a PA circuit requires knowledge of a number of crucial RF topics discussed in this textbook:

- Microstrip line impedance behavior is discussed in Chapter 2, Transmission Line Analysis, and its quantitative evaluation is considered in Chapter 3, The Smith Chart.
- The ability to reduce a complicated circuit into simpler constituents whose input-output is described through a two-port network representation. This is discussed in Chapter 4, Single- and Multiport Networks.
- Strategies of generically developing particular impedance versus frequency responses, as encountered in filter design. Chapter 5, A Brief Overview of RF Filter Design, outlines the basic discrete and distributed filter theories, and Chapter 8, Matching Networks, delves into a detailed circuit implementation as related to Figure 1-2(b).
- High-frequency bipolar-junction and field-effect transistors as well as RF diodes are investigated in Chapter 6, Active RF Components, in terms of their physical basis, followed by Chapter 7, Active Circuit Device Models, where large-signal and small-signal circuit models are analyzed.
- The overall amplification requirements, as related to gain, linearity, noise, and stability, are the basis of Chapter 9, RF Transistor Amplifier Design.
- In addition to amplifiers, Chapter 10, Oscillators and Mixers, focuses on important RF circuit systems, as shown in Figure 1-1.

PHASORS

Harmonic signals can be represented as the real part of an exponential form such as

$$E_x = \text{Re}\left(E_{0x} e^{-j\beta z} e^{j\omega t}\right)$$

where, unlike conventional circuit theory, the phasor $E_{0x} e^{-j\beta z}$ now contains a space factor.

A successful RF design engineer is aware of these concepts and applies them in the design, construction, and testing of particular RF circuits. As the preceding example implies, our concern in this textbook is mostly geared toward analog RF circuit theory and applications. We purposely neglect mixed and digital signals and the associated modulation and coding since their treatment would exceed the size and scope of this textbook.

1.2 Dimensions and Units

To understand the upper frequency limit, beyond which conventional circuit theory can no longer be applied, we should recall the representation of an electromagnetic wave. In free space, plane electromagnetic (EM) wave propagation in the positive z-direction is typically written in sinusoidal form:

$$E_x = E_{0x} \cos(\omega t - \beta z) \quad (1.1\text{a})$$

$$H_y = H_{0y} \cos(\omega t - \beta z) \quad (1.1\text{b})$$

where E_x and H_y are the x-directed electric and the y-directed magnetic field vector components, as shown qualitatively in Figure 1-3. Here E_{0x} and H_{0y} represent constant amplitude factors in units of V/m and A/m.

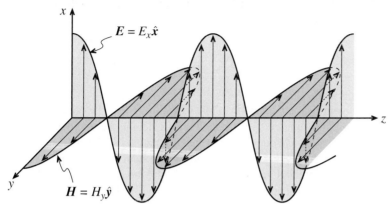

Figure 1-3 Electromagnetic wave propagation in free space. The electric and magnetic fields are shown at a fixed instance in time as a function of space ($\hat{x}$, $\hat{y}$ are unit vectors in x- and y-direction).

These waves possess an angular frequency ω, and a propagation constant β that defines the spatial extent in terms of the **wavelength** λ, such that $\beta = 2\pi/\lambda$. Classical field theory based on Maxwell's equations reveals that the ratio between electric and magnetic field components is defined in terms of the so-called **intrinsic impedance** Z_0

$$\frac{E_x}{H_y} = Z_0 = \sqrt{\mu/\varepsilon} = \sqrt{(\mu_0\mu_r)/(\varepsilon_0\varepsilon_r)} = 377 \ \Omega \sqrt{\mu_r/\varepsilon_r} \quad (1.2)$$

based on the material-dependent permeability $\mu = \mu_0\mu_r$ and permittivity $\varepsilon = \varepsilon_0\varepsilon_r$, with μ_0 and ε_0 being absolute permeability and permittivity of free space (their values are listed in Appendix A) and μ_r and ε_r denoting relative values.

We also point out that the field components are orthogonal to each other and both are orthogonal to the direction of propagation. This is known as **transverse electromagnetic mode** (TEM) and, since we deal exclusively with RF, it is the only mode that is considered in this text. TEM wave propagation is in stark contrast to the various **transverse electric** (TE) and **transverse magnetic** (TM) wave modes, which are the underlying principles of MW and optical communication. In these cases the field vectors are no longer perpendicular to the direction of propagation.

TEM

These modes occur in waves traveling in free space and along microwave transmission lines. The velocity of propagation depends on the dielectric properties (example: coaxial cable) and in some cases geometry (example: microstrip). In the latter case, the waves are classified as **quasi-TEM** (technically not TEM, but exhibiting similar behavior).

TE and TM

These modes occur in microwave waveguides and optical fibers. These modes can be thought of as waves repeatedly reflecting off the walls of the waveguide. TE and TM modes do not propagate below their cutoff frequencies. TEM transmission lines also support TE and TM modes, but their cutoff is usually much higher than the operating frequency.

The phase velocity v_p of the TEM wave in a nonmagnetic medium ($\mu_r = 1$) can be found via

$$v_p = \frac{\omega}{\beta} = \frac{1}{\sqrt{\varepsilon\mu}} = \frac{1}{\sqrt{\varepsilon_0\mu_0}} \cdot \frac{1}{\sqrt{\varepsilon_r}} = \frac{c}{\sqrt{\varepsilon_r}} \qquad (1.3)$$

where c is the speed of light.

Relevant quantities, units, and symbols used throughout the book are summarized in Tables A-1 and A-2 in Appendix A. Although we are dealing here with rather abstract concepts of electromagnetic wave quantities, we can immediately relate (1.1) to circuit parameters by observing that the electric field E, as the unit of V/m already implies, can intuitively be understood as a normalized voltage wave. Similarly, the magnetic field H, given in units of A/m, can be regarded as a normalized current wave.

Example 1-1. Intrinsic wave impedance, phase velocity, and wavelengths

Compute the intrinsic wave impedance, phase velocity, and wavelengths of an electromagnetic wave in free space and a printed circuit board (PCB) material whose dielectric constant is 4.6 for the frequencies $f = 30$ MHz and 3 GHz.

Solution. Relative permeability and permittivity of free space are equal to unity. Therefore, from (1.2) we determine that the intrinsic impedance in this case is equal to

$$Z_0 = \sqrt{\frac{\mu}{\varepsilon}} = \sqrt{\frac{\mu_0}{\varepsilon_0}} = \sqrt{\frac{4\pi \times 10^{-7}}{8.85 \times 10^{-12}}} = 377 \; \Omega.$$

The phase velocity in free space according to (1.3) is equal to

$$v_p = \frac{1}{\sqrt{\varepsilon\mu}} = \frac{1}{\sqrt{\varepsilon_0\mu_0}} = 2.999 \times 10^8 \; \text{m/s}$$

which happens to be the speed of light $v_p = c$. For an $\varepsilon_r = 4.6$ circuit board, we find $v_p = c/(\sqrt{\varepsilon_r}) = 1.4 \times 10^8$ m/s. The wavelength is evaluated by the following expressions:

$$\lambda = \frac{2\pi}{\beta} = \frac{2\pi v_p}{\omega} = \frac{v_p}{f} \qquad (1.4)$$

PRINTED CIRCUIT BOARD MATERIALS

Board materials that are glass based can be used up to 150 – 250°C. Selected dielectric materials are:

FR4 with $\varepsilon_r = 4.6$,
epoxy with $\varepsilon_r = 3.9$, and
polyimide with $\varepsilon_r = 4.5$.

Alternatively, polytetrafluoroethylene (PTFE) based materials have the lowest dielectric constants and can be operated beyond 300°C. They include:

PTFE with $\varepsilon_r = 2.1$,
thermoset PTFE with $\varepsilon_r = 2.8$, and glass-based PTFE with $\varepsilon_r = 2.4$.

where f is the operating frequency. Using equation (1.4), we find that the wavelengths for the electromagnetic wave propagating in free space at a frequency of 30 MHz is equal to $\lambda = 10$ m, and at 3 GHz is a minute $\lambda = 10$ cm. In the dielectric medium, this reduces to 4.67 m and 4.67 cm for 30 MHz and 3 GHz, respectively.

This example conveys an appreciation of how the wavelength changes as a function of frequency. As the frequency increases, the wavelength reduces to dimensions comparable to the size of circuit boards or even individual discrete components. The implication of this fact will be analyzed in Chapter 2.

> Modern PCB fabrication techniques can stack 40 or more layers with interconnections between the layers established through so-called **vias**.
>
> Unfortunately, the thermal conductivity of most boards is typically less than $0.5\,\text{W/m}\,°\text{C}$, very poor.

1.3 Frequency Spectrum

Because of the vast scope of applications, engineers have to deal with a broad range of frequencies of circuit operation. Over the years several efforts have been made to classify the frequency spectrum allocation. The first designations for industrial and government organizations were introduced in the United States by the Department of Defense during and shortly after World War II. However, the most common frequency spectrum classification in use today was created by the **Institute of Electrical and Electronics Engineers** (IEEE). In the U.S. the **Federal Communications Commission** (FCC) allocates and regulates the frequency bands for all private and commercial applications. A selected list of frequency bands and their typical applications is provided in Table 1-1.

Based on Table 1-1 and calculations carried out in Example 1-1, we note that the VHF/UHF band, as typically encountered in television sets, constitutes the point at which the wavelength first may reach dimensions equivalent to the physical extent of the electronic system. It is this region where we need to begin to take into account the wave nature of current and voltage signals. Clearly, the situation becomes even more critical in the SHF and the X to Ka bands. Without being able to assign exact limits, the RF frequency range customarily extends from the VHF to the SHF bands, whereas the MW frequency range traditionally has been associated with radar systems operating in the X band and above.

> **FREQUENCY REUSE FOR MOBILE SYSTEMS**
>
> The operating frequency is of primary interest to the antenna designer who has to deal with electromagnetic radiation effects and directivity patterns.
>
> To be useful for communication, information must first be modulated onto a single frequency carrier or a frequency band.
>
> In mobile systems the frequency is reused to enable multiple user access. Three access techniques are in widespread use:
>
> TDMA (time division multiple access),
> FDMA (frequency division multiple access), and
> CDMA (code division multiple access).

Table 1-1 Frequency bands and their applications

Frequency Band	Frequency	Typical Application
VHF (Very High Frequency)	88 – 108 MHz	FM broadcasting
UHF (Ultrahigh Frequency)	824 – 894 MHz 810 – 956 MHz	CDMA mobile phone service GSM mobile phone service
UHF (Ultrahigh Frequency)	2,400 MHz	WLAN
SHF (Superhigh Frequency)	5,000 – 5,850 MHz	Unlicensed National Information Infrastructure
SHF (Superhigh Frequency)	6,425 – 6,523 MHz	Cable Television Relay
SHF (Superhigh Frequency)	3,700 – 4,200 MHz	Geostationary fixed satellite service
X Band	8 – 12.5 GHz	Marine and airborne radar
Ku Band	12.5 – 18 GHz	Remote sensing radar
K Band	18 – 26.5 GHz	Radar
Ka Band	26.5 – 40 GHz	Remote sensing radar

1.4 RF Behavior of Passive Components

From conventional AC circuit analysis we know that a resistance R is frequency independent and that an ideal capacitor C and an ideal inductor L can simply be described by their reactances X_C and X_L as follows:

$$X_C = -\frac{1}{\omega C} \qquad (1.5a)$$

$$X_L = \omega L. \qquad (1.5b)$$

The implications of (1.5) are such that a capacitor of $C = 1$ pF and an inductor of $L = 1$ nH at low frequencies represent, respectively, either an open- or short-circuit condition. This can be easily proved; for instance at 60 Hz we obtain

$$|X_C(60 \text{ Hz})| = \frac{1}{2\pi \cdot 60 \cdot 10^{-12}} = 2.65 \times 10^9 \ \Omega \approx \infty \qquad (1.6a)$$

$$|X_L(60 \text{ Hz})| = 2\pi \cdot 60 \cdot 10^{-9} = 3.77 \times 10^{-7} \ \Omega \approx 0. \qquad (1.6b)$$

It is important to point out that resistances, inductances, and capacitances are not only created by wires, coils, and plates as typically encountered in conventional low-frequency electronics. Even a single

---

REACTANCES

The reactance of a capacitor follows from basic voltage-current relations:

$$v_c = \frac{1}{C}\int i(t)dt$$

or written as a phasor:

$$V_C = \frac{1}{C}\left(\frac{I}{j\omega}\right) = jX_C I.$$

Equally, for the inductor:

$$v_L = L\left(\frac{di}{dt}\right)$$

which yields as a phasor:

$$V_L = Lj\omega I = jX_L I.$$

Note that a capacitor has negative reactance, while an inductor has positive reactance.

straight wire or a copper trace on a **printed circuit board** (PCB) layout possesses frequency-dependent resistance and inductance. For instance, a cylindrical copper conductor of radius a, length l, and conductivity σ_{cond} has a DC resistance of

$$R_{DC} = \frac{l}{\pi a^2 \sigma_{cond}}. \qquad (1.7)$$

Under DC conditions, current flows uniformly distributed over the entire conductor cross-sectional area. With AC, the situation is complicated by the fact that the alternating charge carrier flow establishes a magnetic field that induces an electric field (in accordance with Faraday's induction law) whose associated current density opposes the initial current flow. The effect is strongest at the center $r = 0$, where the impedance is significantly increased. The result is a current flow that tends to reside at the outer perimeter with increasing frequency. As derived in Appendix B, the z-directed current density J_z can be approximated by

$$J_z \cong \frac{pI}{2\pi a j \sqrt{r}} \exp\left(-(1+j)\frac{a-r}{\delta}\right) \qquad (1.8)$$

where $p^2 = -j\omega\mu\sigma_{cond}$, and I is the total current flow in the conductor. Most important in the exponent of (1.8) is the factor δ, the so-called **skin depth**

$$\delta = \frac{1}{\sqrt{\pi f \mu \sigma_{cond}}} \qquad (1.9)$$

which describes the spatial drop-off in current density as a function of frequency f, permeability μ, and conductivity σ_{cond}. The exponential decrease of current density with depth applies to most conductor geometries. Its equivalent representation is a layer of uniform current density with a thickness δ (skin layer), which is convenient for resistance computations. Further calculations reveal that the normalized resistance R and the internal inductance L_{in} under high-frequency conditions (typically $f \geq 500$ MHz) can be cast in the form

$$R/R_{DC} \cong a/(2\delta) \qquad (1.10)$$

and

$$(\omega L_{in})/R_{DC} \cong a/(2\delta). \qquad (1.11)$$

As the name implies, the internal inductance is due to the magnetic field internal to the wire. For the equations (1.10) and (1.11) to be valid, it is assumed that $\delta \ll a$. In most cases, the relative permeability of the conductor is equal to unity (i.e., $\mu_r = 1$). Because of the inverse square root frequency behavior, the skin depth is large for low frequencies and decreases for increasing frequencies. Figure 1-4 illustrates the skin depth behavior as a function of frequency for material conductivities of copper, aluminum, gold, and a typical Pb-Sn solder.

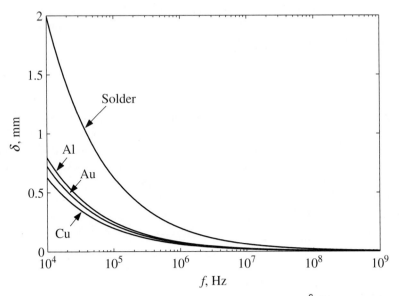

Figure 1-4 Skin depth behavior of copper $\sigma_{Cu} = 64.5 \times 10^6$ S/m, aluminum $\sigma_{Al} = 40.0 \times 10^6$ S/m, gold $\sigma_{Au} = 48.5 \times 10^6$ S/m, and typical solder $\sigma_{solder} = 6.38 \times 10^6$ S/m.

If we consider the conductivity of copper, we can plot the AC current density (1.8) normalized with respect to the DC current density $J_{z0} = I/(\pi a^2)$, as schematically shown for the axisymmetric wire in Figure 1-5(a).

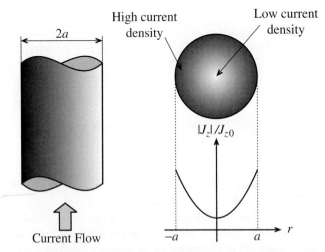

Figure 1-5(a) Schematic cross-sectional AC current density representation normalized to DC current density.

For a fixed wire radius of, let us say, $a = 1$ mm, we can now plot $|J_z|/J_{z0}$ as a function of radius r for various frequencies, as given in Figure 1-5(b). This plot actually uses the exact theoretical result derived in Appendix B, which is accurate even when $\delta > a$.

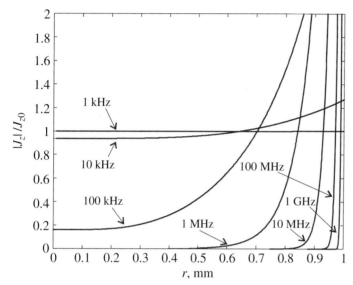

Figure 1-5(b) Frequency behavior of normalized AC current density for a copper wire of radius $a = 1$ mm.

We notice the significant increase in current flow at the outer perimeter of the wire even for moderate frequencies of less than 1 MHz. At frequencies around 1 GHz, the current flow is almost completely confined to the surface of the wire with negligible radial penetration. With reference to (1.8), and as depicted in Figure 1-5(a), the skin depth δ has a simple physical meaning. It denotes the reduction in the current density to the e^{-1} factor (approximately 37%) of its value at the surface. If we rewrite (1.10) slightly, we find

$$R = R_{DC}\frac{a}{2\delta} = R_{DC}\frac{\pi a^2}{2\pi a \delta}. \quad (1.12)$$

This equation shows that the resistance increases inverse proportionally with the cross-sectional skin area, seen as dotted lines in Figure 1-6. Also depicted in Figure 1-6 is the resistance behavior based on the more elaborate skin equation for a cylindrical conductor, derived in Appendix B.

Although the internal inductance can attain significant values, in most circuit designs it is the external inductance L_{ex} that dominates the inductive behavior of the component or circuit. To find L_{ex} we must determine the magnetic field exterior to the wire that is responsible for the inductance. For a cylindrical wire of radius a and length l, a useful approximate formula can be developed

$$L_{ex} \cong \frac{\mu_0 l}{2\pi}\left[\ln\left(\frac{2l}{a}\right) - 1\right]. \tag{1.13}$$

The details are found in the referenced literature at the end of this Chapter. To appreciate their relative magnitudes, an approximate comparison between the influence of internal and external inductances is given in Example 1-2. To keep the analysis simple, we assume L_{ex}, as given in (1.13), does not appreciably change with frequency.

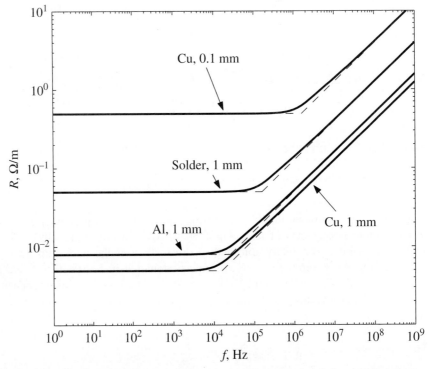

Figure 1-6 The exact theoretical per-unit-length resistance as a function of frequency for round wires of varying materials and radii. The dashed lines represent the DC and skin depth based resistance approximations.

Example 1-2. Comparison between internal and external inductance

A copper wire of AWG 26 is 2 cm long. For the frequencies of 100 MHz, 2 GHz, and 5 GHz compare the internal and external inductances.

Solution. From Appendix B, we find for AWG 26 a diameter $d = 16$ mils. Therefore, the radius in mm is

$$a = 8 \text{ mil} = 8 \times (2.54 \times 10^{-5} \text{m}) = 0.2032 \text{ mm}.$$

We next compute the internal inductance according to (1.11)

$$L_{in} \cong \frac{a}{2\delta} \frac{R_{DC}}{\omega}$$

with $R_{DC} = l/(\pi a^2 \sigma_{cond}) = 2.39 \times 10^{-3} \Omega$, and the external inductance according to (1.13)

$$L_{ex} = \frac{\mu_0 l}{2\pi}\left[\ln\left(\frac{2l}{a}\right) - 1\right] = 17.1 \text{ nH}.$$

Results of the inductance calculations are summarized in Table 1-2.

INDUCTANCE DEFINITIONS

The internal inductance L_{in} is associated with the magnetic field buildup within the current carrying conductor. The external inductance L_{ex} takes into account the magnetic field outside the current-carrying conductor. Added together, they determine the total inductance.

In most practical circuit configurations, one has to consider multiconductor coupling where the magnetic field of one conductor links with one or more adjacent conductors. In this case, we have to use the concept of mutual inductance where the current of one conductor induces currents in neighboring conductors. Mutual inductance plays an important role in crosstalk analyses and signal integrity investigations.

Table 1-2 External vs. internal inductance at various frequencies

f, GHz	δ, μm	L_{ex}, nH	L_{in}, nH	L_{in}/L_{ex}
0.1	6.266	17.1	0.0617	3.60 10^{-3}
2	1.40	17.1	0.0138	8.05 10^{-4}
5	0.886	17.1	0.00872	5.09 10^{-4}

The example makes clear that the external inductance can often exceed the internal inductance by more than two orders of magnitude.

1.4.1 Resistors at High Frequency

Perhaps the most common circuit element in low-frequency electronics is a resistor whose purpose is simply to produce a voltage drop by converting some of the electric energy into heat. Considering them as discrete elements, we can differentiate among several types of resistors:

- Carbon-composition resistors of high-density dielectric granules
- Wire-wound resistors of nickel or other winding material
- Metal-film resistors of temperature-stable materials
- Thin-film chip resistors of aluminum- or beryllium-based materials.

Of these types, mainly the thin-film chip resistors find application nowadays in RF and MW circuits as **surface-mounted devices** (SMDs). This is due to the fact that they can be produced in extremely small sizes with good RF performance, as Figure 1-7 shows.

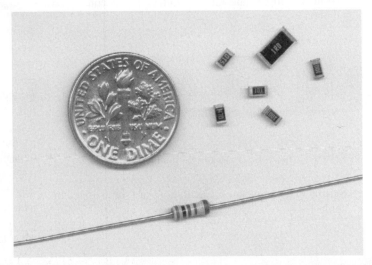

Figure 1-7 One- and quarter-watt thin-film chip resistors in comparison with a conventional quarter-watt resistor.

As the previous section has shown, even a straight wire possesses inductance. Consequently, the electric equivalent circuit representation of a high-frequency resistor of nominal value R is more complicated and has to be modified so that the finite lead dimensions as well as parasitic capacitances are taken into account. This situation is depicted in Figure 1-8.

Here, the two inductances L model the leads, while the capacitances are used to account for the physical wire arrangement, with charge separation effects modeled by capacitance C_a and interlead capacitance C_b. The lead resistance is generally neglected

RF Behavior of Passive Components

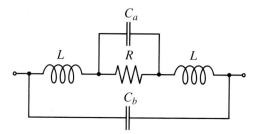

Figure 1-8 Electric equivalent circuit representation of a high frequency resistor.

when compared with the nominal resistance R. For a wire-wound resistor the model is more complex, as Figure 1-9 shows.

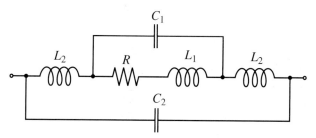

Figure 1-9 Electric equivalent circuit representation of a wire-wound resistor at high frequency.

Here, in addition to the lead inductances L_2 and the contact capacitance, we need to include the inductance L_1 of the wire-wound resistor, which acts as a coil, and the stray capacitance C_1 established between the windings. The interlead capacitance C_2 (or C_b in Figure 1-8) is usually much smaller than the internal or stray capacitance, and in many cases can be safely neglected.

> **HOW PRECISE ARE CIRCUIT MODELS?**
>
> As we learned from the previous discussions, resistances are frequency dependent. Consequently, even though models like Figure 1-8 and Figure 1-9 are widely used, they may require additional refinement, for instance $R = R(f)$.
>
> Depending on the test conditions and available instruments, a host of sophisticated circuit models have been devised by a number of component manufactures.

RF&MW→

Example 1-3. RF impedance response of metal-film resistors

Find the high-frequency impedance behavior of a 2 kΩ metal-film resistor (see Figure 1-8) with 2.5 cm copper wire connections of AWG 26 and a stray capacitance C_a of 5 pF. Assume the conductivity of copper to be $\sigma_{Cu} = 64.5 \times 10^6 \times \Omega^{-1} \cdot m^{-1}$.

Solution. In Example 1-2 we have determined that the radius of an AWG 26 wire is $a = 2.032 \times 10^{-4}$ m. According to (1.13) the external inductance of the straight wire at high frequency is approximately equal to

$$L_{ex} = \frac{\mu_0 l}{2\pi}\left[\ln\left(\frac{2l}{a}\right) - 1\right] = 52.0 \text{ nH}$$

where the $l = 2 \times$ (length of single lead) to account for two connections. It should be noted that the preceding formula for the computation of the lead inductance is applicable only for frequencies where the skin depth is smaller than the radius of the wire, implying that $\delta = (\pi f \mu \sigma)^{-1/2} \ll a$ or explicitly expressed in terms of frequency, $f \gg 1/(\pi \mu \sigma_{Cu} a^2) = 95$ kHz.

Knowing the inductance of the leads, we can now compute the impedance of the entire circuit as

$$Z = j\omega L_{ex} + \frac{1}{j\omega C_a + 1/R}.$$

The result of the computation is presented in Figure 1-10, where the absolute value of the impedance of the resistor is plotted versus frequency.

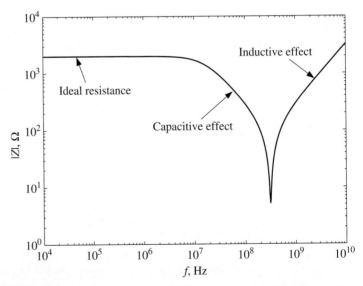

Figure 1-10 Absolute impedance value of a 2000-Ω thin-film resistor as a function of frequency.

As seen, at low frequencies the impedance of the resistor is equal to R. However, as the frequency increases and exceeds 5 MHz, the effect of the stray capacitance becomes dominant, which causes the impedance of the resistor to decrease. Beyond the resonance at approximately 300 MHz, the total impedance increases due to the lead inductance.

This example underscores the care that is required when dealing with the ubiquitous, seemingly frequency-independent resistors. Not all resistors exhibit exactly the same response as shown in Figure 1-10. However, it is typical to see single, often multiple, resonance points that occur when the frequency reaches into the GHz range.

1.4.2 Capacitors at High Frequency

In most RF circuits, chip capacitors find widespread application for the tuning of filters and matching networks as well as in biasing circuits for active components such as transistors. It is therefore important to understand their high-frequency behavior. Elementary circuit analysis defines capacitance for a parallel-plate capacitor whose plate dimensions are large compared to their separation as follows:

$$C = \frac{\varepsilon A}{d} = \varepsilon_0 \varepsilon_r \frac{A}{d} \qquad (1.14)$$

where A is the plate surface area, and d denotes the plate separation. Ideally, there is no current flow between the plates. However, at high frequencies the dielectric materials become lossy (i.e., there is a conduction current flow). The impedance of a capacitor must thus be written as a parallel combination of conductance G_e and susceptance ωC:

$$Z = \frac{1}{G_e + j\omega C}. \qquad (1.15)$$

In this expression, the current flow at DC is due to the conductance $G_e = \sigma_{diel} A/d$, with σ_{diel} being the conductivity of the dielectric. It is now customary to introduce the **loss tangent** $\tan\Delta = \sigma_{diel}/(\omega\varepsilon)$ and insert it into the expression for G_e to yield

$$G_e = \frac{\sigma_{diel} A}{d} = \frac{\omega \varepsilon A}{d}\tan\Delta = \omega C \tan\Delta. \qquad (1.16)$$

Some practical values for the loss tangent are summarized in Table A-3. The corresponding electric equivalent circuit with parasitic lead inductance L, series resistance R_s describing losses in the lead conductors, and dielectric loss resistance $R_e = 1/G_e$, is shown in Figure 1-11.

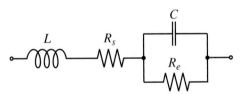

Figure 1-11 Electric equivalent circuit of a capacitor at high frequency.

Example 1-4. RF impedance response of capacitor

Compute the high frequency impedance of a 47 pF capacitor whose dielectric medium consists of an aluminum oxide (Al_2O_3) possessing a loss tangent of 10^{-4} (assumed to be frequency independent) and whose leads are 1.25 cm AWG 26 copper wires ($\sigma_{Cu} = 64.516 \times 10^6 \; \Omega^{-1} \cdot m^{-1}$).

Solution. Similar to Example 1-3, the inductance associated with the leads is given by (1.13)

$$L_{ex} = \frac{\mu_0 l}{2\pi}\left[\ln\left(\frac{2l}{a}\right) - 1\right] = 22.5 \; nH$$

where $l = 2 \times 1.25 \, cm$ accounts for the lengths of both leads and where $a = 2.032 \times 10^{-4}$ m. The series resistance of the leads is computed from (1.12) to be

$$R_s = R_{DC}\frac{a}{2\delta} = \frac{1}{2\pi a \sigma_{Cu}}\sqrt{\pi f \mu_0 \sigma_{Cu}} = \frac{l}{2a}\sqrt{\frac{\mu_0 f}{\pi \sigma_{Cu}}}$$

$$= 4.84\sqrt{\frac{f}{Hz}} \; \mu\Omega.$$

Finally, in accordance with (1.16), the parallel leakage resistance is equal to

$$R_e = \frac{1}{G_e} = \frac{1}{2\pi f C \tan\Delta} = \frac{3.39 \times 10^7}{f/Hz} \; M\Omega.$$

The frequency response of the magnitude of the impedance based on the circuit model in Figure 1-11 for the capacitor is shown in Figure 1-12.

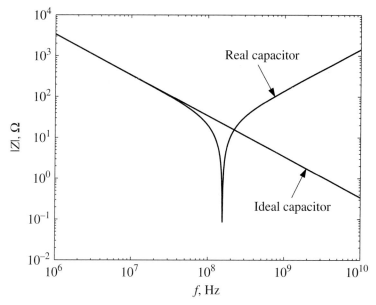

Figure 1-12 Absolute value of the capacitor impedance as a function of frequency.

In computing the parallel leakage resistance R_e, we have assumed the loss tangent $\tan\Delta$ to be frequency independent. In reality, however, this factor may significantly depend upon the operating frequency. Unfortunately, data sheets often report the loss tangent at relatively low frequencies, which poorly represents the RF performance.

Since the loss tangent can also be defined as the ratio of an **equivalent series resistance** (ESR) to the capacitor's reactance, many data sheets list ESR instead of $\tan\Delta$. The ESR value is thus given as

$$\text{ESR} = \frac{\tan\Delta}{\omega C}.$$

This indicates that $\text{ESR} \to 0$ as $\tan\Delta \to 0$.

As with the resistor in Example 1-3, the capacitor reveals a resonance behavior due to the presence of dielectric losses and finite lead wires.

The construction of a surface-mounted ceramic capacitor is shown in Figure 1-13. The component is a rectangular block of a ceramic dielectric into which a number of interleaved metal electrodes are sandwiched. The purpose of this type of packaging is to provide a high capacitance per unit volume by maximizing the electrode surface area. Capacitance values range from 0.47 pF to 100 nF with operating voltage ranging from 16 V to 63 V. The loss tangent is usually listed by the manufacturer as $\tan\Delta \leq 10^{-3}$ at a 1 MHz test frequency. Again, this loss tangent can significantly increase as the frequency reaches into the GHz range.

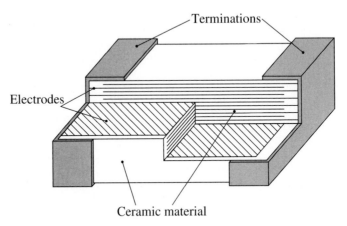

Figure 1-13 Actual construction of a surface-mounted ceramic multilayer capacitor.

Besides capacitance and loss tangent, manufacturers list a nominal voltage that cannot be exceeded at a particular operating temperature range (for instance, $T \leq 85\,°C$). Furthermore, the capacitance is temperature dependent, as further discussed in the problem section of this chapter.

1.4.3 Inductors at High Frequency

Although not employed as often as resistors and capacitors, inductors generally are used in transistor biasing networks, for instance as **RF chokes** (RFCs) to short circuit the device to DC voltage conditions. Since a coil is generally formed by winding a straight wire on a cylindrical former, we know from our previous discussion that the windings represent an inductance in addition to the frequency-dependent wire resistance. Moreover, adjacent wires constitute separated moving charges, thus giving rise to a parasitic capacitance effect as shown in Figure 1-14.

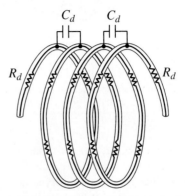

Figure 1-14 Distributed capacitance and series resistance in the inductor coil.

RF Behavior of Passive Components

The equivalent circuit model of the inductor is shown in Figure 1-15. The parasitic shunt capacitance C_s and series resistance R_s represent composite effects of distributed capacitance C_d and resistance R_d, respectively.

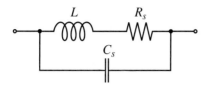

Figure 1-15 Equivalent circuit of the high-frequency inductor.

Example 1-5. RF impedance response of an RFC

Estimate the frequency response of an RFC formed by $N = 3.5$ turns of AWG 36 copper wire on a 0.1 inch air core. Assume that the length of the coil is 0.05 inch. The parasitic shunt capacitance in this RFC is approximately 0.3 pF.

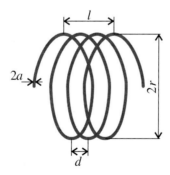

Figure 1-16 Inductor dimensions of an air-core coil.

Solution. The dimensions of the coil are shown in Figure 1-16. From Table A-4 in Appendix A, we find that the radius of the AWG 36 wire is $a = 2.5$ mils $= 63.5$ μm. The radius of the coil core is $r = 50$ mils $= 1.27$ mm. The length of the coil is $l = 50$ mils $= 1.27$ mm.

To estimate the inductance of the coil, we will not use the well known formula $L = \pi r^2 \mu_0 N^2 / l$ for a long solenoidal coil because

$r \ll l$ does not apply. Instead, and since $l > 0.8r$, we can employ the formula for a short air-core solenoid:

$$L = \frac{10\pi r^2 \mu_0 N^2}{9r + 10l}. \tag{1.17}$$

Equation (1.17) will not give an exact value for the inductance, but a rather good approximation. Substituting the given values into (1.17), we obtain $L = 32.3$ nH. We neglect the skin effect and compute the series resistance R_s as the DC resistance of the wire:

$$R_s = \frac{l_{\text{wire}}}{\sigma_{\text{Cu}} \pi a^2} = \frac{2\pi r N}{\sigma_{\text{Cu}} \pi a^2} = 0.034 \ \Omega.$$

The frequency response of the RFC impedance is shown in Figure 1-17.

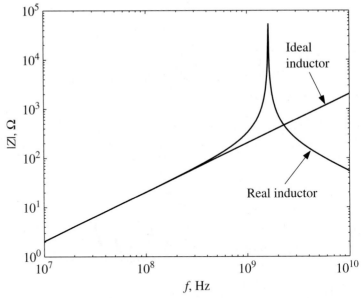

Figure 1-17 Frequency response of the impedance of an RFC.

RFCs find widespread use for biasing RF circuits. However, as Figure 1-17 shows, the frequency dependency can form complicated resonance conditions with additional elements in an RF system. Indeed, certain matching circuits rely on the RFCs as tuning elements.

As can be seen from Figure 1-17, the behavior of the RFC deviates from the expected behavior of an ideal inductance at high frequencies. First, the impedance of the RFC increases more rapidly as the frequency approaches resonance. Second, as the frequency continues to increase, the influence of the parasitic capacitance C_s becomes dominant and the impedance of the coil decreases.

If the RFC had zero series resistance, then the overall impedance behavior at resonance would reach infinity, but due to the nonzero contribution of R_s, the maximum value is finite. To characterize the effect of the coil resistance, the quality factor Q is commonly used:

$$Q = \frac{|X|}{R_s} \tag{1.18}$$

where X is the reactance and R_s is the series resistance of the coil. This quality factor definition applies precisely only for simple lumped component systems. The important concept of the quality factor will be treated more extensively in Chapter 5. The quality factor characterizes the resistive loss in this passive circuit, and for tuning purposes it is desirable that this factor is as high as possible. The inductor Q generally increases with frequency, levels off, and then drops close to the self-resonance. Manufacturers of high-quality inductors typically supply measured Q versus f curves in their datasheets.

1.5 Chip Components and Circuit Board Considerations

The practical realization of passive and active components on printed circuit boards (PCBs) is primarily accomplished in chip form and placed on specially fabricated plastic or ceramic board materials. In the early days of PCB fabrication, the word "printed" referred to the process of applying patterns of insulation material onto a thin base layer of copper, followed by etching away the uncovered copper material and thereby leaving behind the circuit traces. Today, a sophisticated photolithographic processes is employed to produce the often extremely narrow traces with tight tolerances.

In the following section we examine the three most common passive chip elements in terms of their sizes and electric characteristics.

1.5.1 Chip Resistors

The size of chip resistors can be as small as 40 by 20 mils (where 1 mil = 0.001 inch = 0.0254 mm) for 0.05 W power ratings and up to 1 by 1 inch for 500 W ratings in RF power amplifiers. The chip resistor sizes that are most commonly used in circuits operating up to several watts are summarized in Table 1-3.

A general rule of thumb in determining the size of the chip components from the known size code is as follows: the first two digits in the code denote the length L in terms of tens of mils, and the last two digits denote the width W of the component. The thickness of the chip resistors is not standardized and depends on the particular component type.

Table 1-3 Standard sizes of chip resistors

Geometry	Size Code	Length L, mils	Width W, mils
	0402	40	20
	0603	60	30
	0805	80	50
	1206	120	60
	1812	120	180

The resistance values range from 1/10 Ω up to several MΩ. Higher values are difficult to manufacture and result in high tolerances. Typical resistor tolerance values range from ±5% to ±0.01%. Another difficulty that arises with high-value resistors is that they are prone to produce parasitic fields, adversely affecting the linearity of the resistance versus frequency behavior. A conventional chip resistor realization is shown in Figure 1-18.

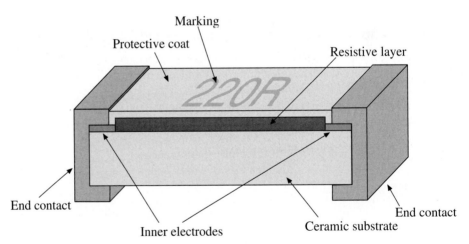

Figure 1-18 Cross-sectional view of a typical chip resistor.

A metal film (usually nichrome) layer is deposited on a ceramic body (usually aluminum oxide). This resistive layer is trimmed to the desired nominal value by reducing its length and inserting inner electrodes. Contacts are made on both ends of the resistor that allow the component to be soldered to the board. The resistive film is coated with a protective layer to prevent environmental interferences.

1.5.2 Chip Capacitors

The chip capacitors are implemented either as a conventional single-plate configuration, as shown in Figure 1-19, or a multiple-layer design (see Figure 1-13).

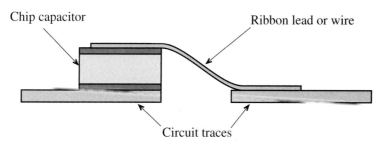

Figure 1-19 Cross section of a typical single-plate capacitor connected to the board.

Frequently, single-plate capacitors are combined in clusters of two or four elements sharing a single dielectric material and a common electrode, as shown in Figure 1-20.

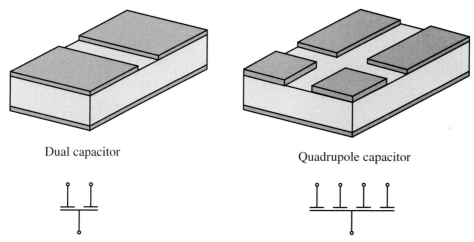

Figure 1-20 Clusters of single-plate capacitors sharing a common dielectric material.

The standard sizes of capacitors range from a minimum of 15 mils square in a single layer configuration to 400 by 425 mils at higher values. Typical values for commercial capacitors range from 0.1 pF to several μF. The tolerances vary from ±2% to ±50%. For small capacitances, tolerances are usually expressed in terms of pF instead of percent; for example, we often encounter capacitors with the nominal values given as (0.5 ± 0.25) pF.

1.5.3 Surface-Mounted Inductors

The most common implementation of surface-mounted inductors is still the wire-wound coil. A typical example of such an inductor with air core is shown in Figure 1-21. Modern manufacturing technology allows us to fabricate these inductors with extremely small footprints. Their dimensions are comparable to those of chip resistors and capacitors. Typical sizes of surface-mounted wire-wound inductors range from 60 by 30 mils to 180 by 120 mils. The inductance values cover the range from 1 nH to 1000 µH.

Figure 1-21 Typical size of an RF wire-wound air-core inductor (courtesy of Coilcraft, Inc.).

When thickness constraints of the circuit play a major role, flat inductors are often employed that can be integrated with microstrip transmission lines. A generic configuration of a flat coil is shown in Figure 1-22. Although such thin-wire coils have relatively low inductances on the order of 1 to 500 nH, it is the frequency in the GHz range that helps push the reactance beyond 1 kΩ. The physical construction can be as small as 2 mm by 2 mm.

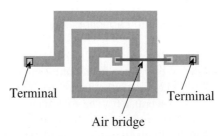

Figure 1-22 Flat coil configuration. An air bridge is made by using either a wire or a conductive ribbon.

1.6 RF Circuit Manufacturing Processes

Flat coils are used in both integrated and **hybrid circuits**. Hybrid circuits differ from ordinary circuits in that discrete semiconductor elements are placed on the dielectric substrate in die form (without case) and are connected to the conductors on the board using bond wires. After the entire circuit is assembled, it is then placed into a single case to protect it from environmental interferences. Resistors and capacitors for hybrid circuits can directly be implemented on the board by metal-film deposition. This approach permits significant reduction in the size of the circuit.

1.6 RF Circuit Manufacturing Processes

Despite much progress in circuit fabrication technology, many RF circuits are still implemented on conventional printed circuit boards because of their low cost and readily available layout tools. Substrates made out of **fire retardent** (FR) epoxy-based glass materials like **FR4** are the basis of single- or multilayer substrates. Fine-line resolution of 5 mil and spacings of approximately the same dimensions can be achieved. Unfortunately, high loss tangents (0.03 at 1 GHz) typically preclude the use of FR4 above 2 GHz.

More sophisticated hybrid **microwave integrated circuits** (MICs) allow operation at much higher frequencies (up to 20 GHz) due to the use of ceramics such as alumina (Al_2O_3), aluminum nitride (AlN), and beryllia (BeO) with loss tangents well below 0.001. In addition, higher dielectric strength, often exceeding 25 kV/mm, and low thermal expansion coefficients, typically less than $5 \times 10^{-6} \, °C^{-1}$ facilitate high-density circuit designs. Active and passive circuit elements are either wire-bonded to the ceramic substrate or directly attached to the metal traces. Line widths of less than 1 mil at spacings of similar dimensions can be achieved.

One of the most actively pursued MIC manufacturing processes is the so-called cofired ceramics technology. Specifically, **low-temperature cofired ceramics** (LTCCs) and **high-temperature cofired ceramics** (HTCCs) utilize unfired, so-called green, ceramic tapes that can be punched with via holes and screen-printed with passive components (R, L, C). Multiple tapes are then collated and stacked on top of each other before firing at around 900°C for LTCC, or 1500°C for HTCC. As a result, three-dimensional structures with more than 20 layers can be produced; they are also known as **Multichip Modules - Ceramic** (MCMs-C). Because of the differences in melting temperatures, traces made out of gold, copper, or silver as metallization can only be fired at 900°C whereas tungsten (also known as wolfram) with a much higher melting point can be fired at 1500°C. Figure 1-23 illustrates the generic concept of creating a 3D module composed of multiple layers of ceramics and a metallic base material acting as a heat sink.

The highest degree of integration can be accomplished when using a **Microwave Monolithic Integrated Circuit** (MMIC) design. Here, the device placements and interconnections are directly carried out on the wafer level. Starting with a Si, GaAs, or InP wafer and

> **WAFER FABRICATION**
>
> Cylindrical ingots several feet long of semiconductor materials like Si and GaAs are ground to diameters of 200 mm (6 inch) or 300 mm (8 inch) and sliced into micrometer thick layers, the wafers.
>
> These circular wafers are subjected to a sequence of chemical processes that etch the circuit configuration as patterned on a set of masks. On top of the circuit, layers of metal interconnect are created to form pathways between the circuit elements.
>
> After processing in a specialized factory called wafer fab, the wafer is diced into individual die that are wire-bonded, packaged, and assembled into chips. The number of fully functioning die determines the wafer yield.

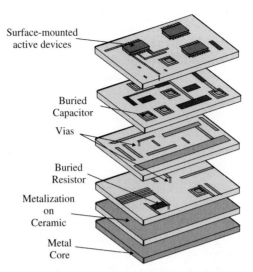

Figure 1-23 Construction of a three-dimensional LTCC/HTCC module made out of individual layers of ceramic tape that are collated, stacked, and fired (courtesy of Lamina Ceramics Inc.).

employing semiconductor fabrication steps, transistors, diodes, resistors, capacitors, and inductors are fabricated through ion implantation or epitaxial growth in conjunction with photolithographic processes involving often dozens of mask layers. Clearly, MMICs achieve extremely high operating frequencies at low noise levels. Owing to the highly sophisticated and expensive microelectronic fabrication processes, only high-volume circuit realizations can normally justify the associated manufacturing cost.

Practically Speaking.   Measuring the impedance of a coil

The electric behavior of resistors, capacitors, and inductors can be measured with a conventional LCR meter if the maximum frequency does not become too high, typically less than 25 MHz.

In Figure 1-24, impedance and quality factor measurements of a plastic core torroid inductor are shown with an LCR meter (HP 4192A). As one observes: at low frequency the finite wire resistance dominates, followed by the linear inductive (ωL) behavior in the range from 5 kHz to 1 MHz. At around 2 MHz the imaginary part approaches zero

RF Circuit Manufacturing Processes

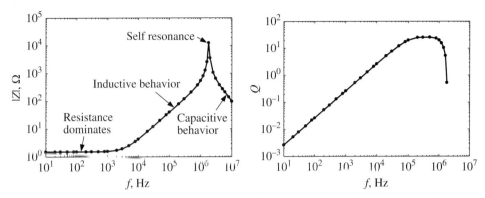

Figure 1-24 Impedance and quality factor behavior of a real, non-magnetic core inductor as measured by the HP 4192A LCR meter.

and self resonance occurs (the total reactance is zero). This is then succeeded by a sharp decline in impedance due to the predominant influence of the coil's capacitive effect. The quality factor ($Q = \omega L/R$) increases almost linearly up to the self resonance point, attaining a value of approximately 30, before dropping sharply as the frequency exceeds 1 MHz. The instrument and the toroidal coil are shown in Figure 1-25. For a test frequency of 100 kHz (right LCD display) the instrument records an inductance of 63.58 µH (left LCD display) and a quality factor of 20.0 (center LCD display).

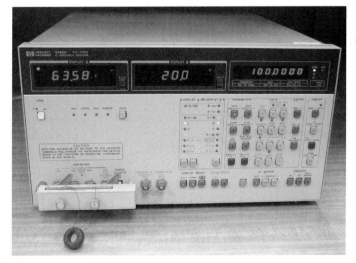

Figure 1-25 LCR meter with a plastic core toroidal inductor connected to the test fixture and a measurement taken at 100 kHz.

> *Specifically designed air core inductors for RF applications in the range 1 nH to 60 nH can reach self-resonance frequencies well into the 3–5 GHz range and quality factors as high as 300.*

1.7 Summary

In this chapter, the evolution from low- to high-frequency systems is discussed and placed in a historical context. A key concept when dealing with high-frequency applications is the fact that the electromagnetic wave nature begins to dominate over Kirchhoff's current and voltage laws. Issues such as propagation constant and phase velocity,

$$\beta = 2\pi/\lambda \text{ and } v_p = \frac{\omega}{\beta} = \frac{1}{\sqrt{\varepsilon\mu}} = \frac{c}{\sqrt{\varepsilon_r}}$$

gain importance.

A consequence of the electromagnetic wave nature is the skin effect, which forces the current to flow close to the surface of the conducting structures. The depth of field penetration from the surface can be determined via the skin depth equation:

$$\delta = \frac{1}{\sqrt{\pi f \mu \sigma}}.$$

With the skin depth we can approximately characterize the frequency-dependent resistance and reactance of components at RF frequency. As an example, the simple cylindrical lead wires exhibit resistances and reactances that become a function of frequency

$$R \approx R_{DC}\frac{a}{2\delta} \text{ and } X = \omega L_{in} \approx R_{DC}\frac{a}{2\delta}.$$

These wires, in conjunction with the respective R, C, and internal L_{in} as well as external L_{ex} elements, form electric equivalent circuits whose performance markedly deviates from the ideal element behavior. As a consequence, we find that the constant resistance at low frequency is no longer constant, but displays a second-order system response with a resonance dip. The dielectric material in a capacitor becomes lossy at high frequencies (i.e., allows the flow of a small conduction current). The degree of loss is quantified by the loss tangent, which is tabulated for a range of engineering materials. Therefore, a capacitor exhibits an impedance behavior that follows an inverse frequency response only at low frequencies. Finally, inductors represent an impedance response that follows a linear increase at low frequencies before deviating from the ideal behavior by reaching a resonance peak and then turning capacitive.

An important performance characteristic is the quality factor Q of an inductor or capacitor. Directly related is the loss tangent, or dissipation factor, which is inversely proportional, that is, the higher the Q-factor, the lower the loss tangent. The inverse relationship can be established through simple circuit considerations. For instance, for a capacitor with plate area A and separation d we know from elementary circuit theory that the capacitance is $C = \varepsilon A/d$ and the shunt resistance is $R_p = d/(\sigma A) = 1/G_p$. Consequently, for a parallel circuit it is seen that

$$Q = \frac{B}{G_p} = \frac{\omega C}{G_p} = \frac{\omega \varepsilon}{\sigma} = \frac{1}{\tan \Delta}.$$

Through a similar argument one can show that an identical inverse relationship is valid for inductors.

Passive RF component vendors will always attempt to keep the physical dimensions of resistors, capacitors, and inductors as small as possible. This is desirable since the wavelengths of high-frequency voltage and current waves become ever smaller, eventually reaching the characteristic sizes of the circuit components. For this reason, and to reduce footprint and power, new circuit board fabrication strategies, like hybrid MIC and MMIC, are pursued in an effort to realize the smallest components. As discussed in depth in subsequent chapters, when the wavelength is comparable in size with the discrete electronic components, basic circuit analysis no longer applies.

Further Reading

J. Israelsohn, "The ABCs of Integrated Ls and Cs," *EDN Magazine*, July 11, 2002, pp. 51–60.

B. Beker, C. Cokkinides, and M. Sechrest, "Field, Circuit, and Visualization based Simulation Methodology of Passive Electronic Components," *IEEE Proceedings of 33rd Annual Simulation Symposium*, 2000, pp. 157–164.

V. F. Perna, "A Guide to Judging Microwave Capacitors," *Microwaves*, Vol. 9, 1970, pp. 40–42.

R. G. Arnold and D. J. Pedder, "Microwave Lines and Spiral Inductors in MCM-D Technology," *IEEE Trans. on Components, Hybrids, Manufact. Tech.*, Vol. 15, 2001, pp. 1038–1043.

S. Chaki, S. Andoh, Y. Sasaki, N. Tanino, and O. Ishihara, "Experimental Study on Spiral Inductors," *IEEE MTT-S Digest*, 1995, pp. 753–756.

F. Zandman, P.-R. Simon, and J. Szwarc, *Resistor Theory and Technology*, Scitech Publishing, Park Ridge, NJ, 2001.

I. Bohl and P. Bhartia, *Microwave Solid State Design*, John Wiley, New York, 1988.

C. Bowick, *RF Circuit Design,* Newmes, Newton, MA, 1982.

D. K Chen, *Fundamentals of Engineering Electromagnetics,* Addison-Wesley, Reading, MA, 1993.

R. A. Chipman, *Transmission Lines,* Schaum Outline Series, McGraw-Hill, New York, 1968.

L. N. Dworsky, *Modern Transmission Line Theory and Applications,* Robert E. Krieger, Malabar, FL, 1988.

M. F. Iskander, *Electromagnetic Fields and Waves,* Prentice Hall, Upper Saddle River, NJ, 1992.

T. S. Laverghetta, *Practical Microwaves,* Prentice Hall, Upper Saddle River, NJ, 1996.

K. F. Sander, *Microwave Components and Systems,* Addison-Wesley, 1987.

K. F. Sander and G. A. L. Read, *Transmission and Propagation of Electromagnetic Waves,* 2nd ed. Cambridge University Press, Cambridge, UK, 1986.

W. Sinnema, *Electronic Transmission Line Technology,* 2nd ed., Prentice Hall, Upper Saddle River, NJ, 1988.

F. T. Ulaby, *Fundamentals of Applied Electromagnetics,* Prentice Hall, Upper Saddle River, NJ, 1997.

F. W. Grover, *Inductance Calculations, Working Formulas and Tables*, Van Nostrand Company, 1946.

Problems

1.1 Compute the phase velocity and wavelength in an FR4 printed circuit board whose relative dielectric constant is 4.6 and where the operating frequency is 1.92 GHz.

1.2 The current flowing in a microstrip line (assumed to be infinite and lossless) is specified to be $i(t) = 0.6\cos(9 \times 10^9 t - 500z)$ A. Find the (a) phase velocity, (b) frequency, (c) wavelength, and (d) phasor expression of the current.

1.3 A coaxial cable that is assumed lossless has a wavelength of the electric and magnetic fields of $\lambda = 20$ cm at 960 MHz. Find the relative dielectric constant of the insulation.

1.4 The electric field of a positive z-traveling wave in a medium with relative dielectric constant of $\varepsilon_r = 4$ and with frequency of 5 GHz is given by

$$E_x = E_{0x}\cos(\omega t - kz) \text{ V/m}.$$

(a) Find the magnetic field if $E_{0x} = 10^6$ V/m.
(b) Determine phase velocity and wavelength.
(c) Compute the spatial advance of the traveling wave between time intervals $t_1 = 3\mu\text{s}$ and $t_2 = 7\mu\text{s}$.

1.5 Find the frequency response of the impedance magnitude of the following series and parallel LC circuits:

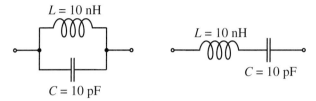

Compare your results to the situation when the ideal inductance is replaced by the same inductance and a 5 Ω resistance connected in series. Assume that these circuits operate in the VHF/UHF frequency band of 30–3000 MHz.

1.6 For the circuit shown below, derive the resonance frequency and plot the resonance frequency behavior as a function of the resistance R.

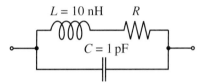

1.7 Repeat Problem 1.6 for the following circuit.

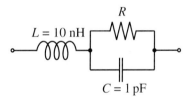

1.8 For the following circuit we chose $R \ll ((\sqrt{L/C})/2)$.

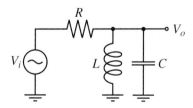

Find $|V_o/V_i|$ as a function of frequency and identify the dominant circuit portions for the low-, mid-, and high-frequency domains.

1.9 Compute (a) the inductance of a coil formed by 10 turns of AWG 26 copper wire on a 5 mm air core. The length of the coil is 5 mm. (b) For an operating frequency of 1.98 GHz and assuming a fixed DC equivalent resistance, find the quality factor.

1.10 The leads of a resistor in an RF circuit are treated as straight aluminum wires ($\sigma_{Al} = 4.0 \times 10^7$ S/m) of AWG size 14 and of total length of 5 cm. (a) Compute the DC resistance. (b) Find the AC resistance and inductance at 100 MHz, 1 GHz, and 10 GHz operating frequencies.

1.11 Compute the skin depths for copper ($\sigma_{Cu} = 64.5 \times 10^6$ S/m), aluminum ($\sigma_{Al} = 40 \times 10^6$ S/m), and gold ($\sigma_{Au} = 48.5 \times 10^6$ S/m) at 1 GHz and 10 GHz, and find the resistance of a 10 cm wire with diameter of 1 mm.

1.12 Compute and plot the per-unit-length (per meter) resistance of an AWG 36 copper wire in the frequency range from 100 kHz to 1 GHz. How does the exact theoretical result compare with the DC and skin depth based resistance approximations? Use the expression for the current density distribution in a wire, given in Appendix B

$$J_z(r) = \frac{pI}{2\pi a} \frac{J_0(pr)}{J_1(pa)}$$

where $p^2 = -j\omega\mu\sigma$, r is the current radius, and a is the radius of the wire. Hints: use the power dissipation relations and numerical integration in Matlab. Look up "besselj" and "quadl" in Matlab help.

1.13 A typical PCB substrate consists of Al_2O_3 with a relative dielectric constant of 10 and a loss tangent of 0.0004 at 10 GHz. Find the conductivity of the substrate.

1.14 For the series RLC circuit with $R = 1\,\Omega$, $L = 1$ nH, and $C = 1$ pF, compute the resonance frequency and quality factor at $\pm 10\%$ of the resonance frequency treating the circuit as a simple lumped component. Does the presence of the resistor affect the resonance frequency?

1.15 A 4.7 pF capacitor with relative dielectric constant of 4.6 and loss tangent of 0.003 is used in a circuit operated at 10 GHz. For a combined copper lead length of 1 cm and diameter of 0.5 mm, determine (a) the lead resistance and lead reactance, and (b) the conductance and the total impedance. The conductivity of copper is given as $\sigma_{Cu} = 64.5 \times 10^6\,\Omega^{-1} \cdot m^{-1}$.

1.16 One of the highest dielectric constants can be achieved with $BaTiO_3$, namely $\varepsilon_r = 1200$. Unfortunately, the loss tangent and temperature stability is very poor. For instance, at 100 MHz, $\tan\Delta = 0.03$, and at 1GHz, $\tan\Delta = 0.1$. What percentage impedance variation must be expected?

1.17 A manufacturer data sheet records the loss tangent of a capacitor to be 10^{-4} at 5 GHz. For a total plate area of $10^{-2} cm^2$, plate separation of 0.01 mm, and a relative dielectric constant of 10, find the conductance.

1.18 A two-element impedance of the generic form

$$Z = R + jX$$

has to be converted into an equivalent admittance form $Y = 1/Z$ such that

$$Y = G + jB.$$

Find the conductance G and susceptance B in terms of resistance R and reactance X.

1.19 Convert a given circuit configuration of R_p in shunt with C_p and quality factor $Q_p = \omega C_p R_p$ into an equivalent series configuration involving R_s and C_s. Express your answers as functions of Q_p, that is $R_s = f(R_p, Q_p)$ and $C_s = f(C_p, Q_p)$.

1.20 A more elaborate model of a capacitor is sometimes represented by the following circuit:

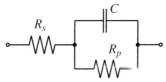

Here the loss tangent is specified as consisting of two parts involving the admittance $Y_p = 1/R_p + j\omega C$ with a parallel-circuit loss tangent $\tan\Delta_p = |\text{Re}(Y_p)/\text{Im}(Y_p)|$ and series impedance $Z_s = R_s + 1/(j\omega C)$ with a series-circuit loss tangent $\tan\Delta_s = |\text{Re}(Z_s)/\text{Im}(Z_s)|$ (it is noted that R_s is different from Example 1-4). Show that for low-loss capacitances we approximately obtain $\tan\Delta \approx \tan\Delta_s + \tan\Delta_p$, where $\tan\Delta = |\text{Re}(Z)/\text{Im}(Z)|$ and Z is the total impedance.

1.21 The maximum impedance of a parallel RLC circuit occurs at the resonance frequency $\omega_0 = 1/(\sqrt{LC})$. Using this angular resonance frequency expression and the quality factor at resonance $Q = \omega_0 CR = R\sqrt{C/L}$, find (a) the 3 dB angular corner frequencies $\omega_{1,2}$ as functions of ω_0 and Q (here, the susceptance is equal to the conductance), and (b) show that the practical formula $Q = \omega_0/(\omega_2 - \omega_1)$ used to measure the quality factor is consistent with the circuit definition given above.

1.22 Determine the LC elements of a series resonant circuit that passes all harmonic signals between the 3 dB points of 2 GHz and 5 GHz. Assume the circuit is connected between a voltage source with negligible source resistance and a load impedance of 50 Ω. Plot magnitude and phase of the impedance.

1.23 For the circuit schematic given in Problem 1.7, assume a resistance of $R = 2$ kΩ and plot the impedance response. What happens if the resistance is increased to 200 kΩ?

1.24 When recording the capacitance with measurement equipment, the user often has the choice to select a suitable circuit representation. For the series representation, the instrument attempts to predict R_s and C_s, while for the parallel representation the prediction involves R_p and C_p. Which mode should be chosen if large capacitors are to be measured? Is this mode also suitable for small values? Explain your answers.

1.25 The ability to store electric charge, expressed through the capacitance, depends on the operating temperature. This behavior can be quantified through the relation $C = C_0[1 + \alpha(T - 20°C)]$, where C_0 is the nominal capacitance and α is a temperature coefficient that can be positive or negative. If the capacitance C at $T = 20°C$ is recorded to be 4.6 pF, and increases to 4.8 pF at $T = 40°C$, what is the temperature coefficient α? Determine the capacitance at $0°C$ and $80°C$.

1.26 When measuring impedance at low frequency we connect the measurement equipment to a device using a pair of wires and assume that the reading reflects the impedance of the device under test (DUT). As we have seen in this chapter, at high frequencies we have to take into account the influence of the parasitic elements. The typical circuit representation of the measurement arrangement is as follows.

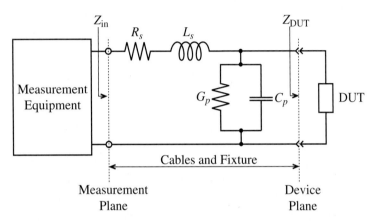

Here, the fixture and cables are replaced by an equivalent circuit of the lead impedance $(R_s + j\omega L_s)$ and stray admittance $(G_p + j\omega C_p)$. Ideally, we would like to perform the measurement at the device plane. However, due to the influence of the fixture, the measurement plane is shifted away from the DUT.

To measure accurately the impedance of the DUT, the test fixture with connecting cables has to be taken into account. The methodology adopted by most manufacturers is to compensate for these undesired fixture-related influences through an open- and short-circuit calibration. The first step is to replace the DUT by a short circuit and record the resulting impedance. Due to the influence of the fixture, the measured impedance will not be equal to zero. Next, the short circuit is replaced by an open circuit and the impedance is recorded again. These two measurements allow us to quantify the parasitic influence of the fixture.

After calibration, we can connect the DUT and measure the input impedance. The equivalent circuit in this case is as follows.

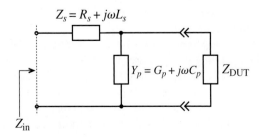

Knowing the values of the parasitic elements (Z_s and Y_p), we can now compute the true impedance of the DUT.

Explain the procedure with all necessary equations, and specify under what conditions such a calibration is possible. Next, develop the formula that allows us to find the desired DUT impedance in the absence of the fixture.

1.27 The results of a frequency sweep impedance measurement of an unknown passive device are shown in the following figure.

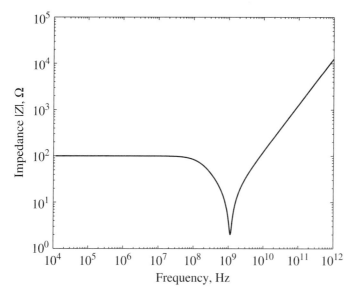

Based on the shape of the impedance response, develop a circuit that can be used as an equivalent circuit to replicate this device under test. What device can it be: resistor, inductor, or capacitor?

1.28 Measuring the impedance of a passive component at RF frequencies is quite a challenge. Conventional techniques such as bridge circuits and resonance techniques fail beyond a few MHz. A technique pursued by several instrument manufacturers is the current-voltage recording based on the following simplified schematic.

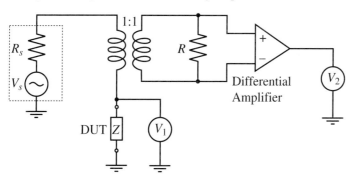

Here, the voltages are measured with vector voltmeters that allow the recording of magnitude and phase. Explain how the impedance of the component under test is determined and discuss the purpose of the transformer and differential amplifier.

1.29 An RFC is constructed by winding four turns of AWG 38 copper wire on a 2 mm diameter ceramic core ($\mu_r = 1$) of 1 mm length. Based on Example 1-5, estimate the inductance, resistance, and resonance frequency. The stray capacitance for this RFC is 0.2 pF. For the analysis, assume a frequency-independent DC equivalent resistance.

1.30 Using data and the equivalent circuit diagram developed in the previous problem, find values of the equivalent circuit parameters, if the device impedance is 1 Ω under DC conditions and 12.5 Ω at 100 GHz. Assume the resonance frequency point to be at 1.125 GHz.

1.31 A quadrupole capacitor as shown in Figure 1-18 consists of four equal-size electrodes of 25 mil × 25 mil separated 5 mil from a common ground plane through a dielectric medium of a relative dielectric constant of 11. Find the individual and total capacitance that can be achieved.

1.32 Consider the following diode circuit.

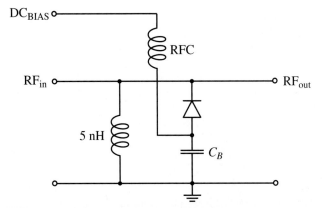

As will be shown in Chapter 6, a reverse-biased diode can be represented as a series combination of a resistor R_s and junction capacitor C, where the capacitance is bias dependent. Its value is approximately given by the expression

$$C = C_0\left(1 - \frac{V_{bias}}{V_{diff}}\right)^{-1/2}.$$

Assuming that RFC and blocking capacitor C_B have infinite values, find the biasing voltage such that the circuit exhibits a resonance at the frequency of 1 GHz. The diode is characterized as follows: $C_0 = 10$ pF, $R_s = 3\ \Omega$, and barrier voltage $V_{diff} = 0.75$ V.

2

Transmission Line Analysis

2.1 Why Transmission Line Theory? 42
2.2 Examples of Transmission Lines 45
    2.2.1 Two-Wire Lines 45
    2.2.2 Coaxial Line 46
    2.2.3 Microstrip Lines 47
2.3 Equivalent Circuit Representation 49
2.4 Theoretical Foundation 52
    2.4.1 Basic Laws 52
2.5 Circuit Parameters for a Parallel-Plate Transmission Line 57
2.6 Summary of Different Line Configurations 61
2.7 General Transmission Line Equation 62
    2.7.1 Kirchhoff Voltage and Current Law Representations 62
    2.7.2 Traveling Voltage and Current Waves 66
    2.7.3 Characteristic Impedance 67
    2.7.4 Lossless Transmission Line Model 67
2.8 Microstrip Transmission Lines 68
2.9 Terminated Lossless Transmission Line 73
    2.9.1 Voltage Reflection Coefficient 73
    2.9.2 Propagation Constant and Phase Velocity 74
    2.9.3 Standing Waves 75
2.10 Special Termination Conditions 78
    2.10.1 Input Impedance of Terminated Lossless Line 78
    2.10.2 Short-Circuit Terminated Transmission Line 79
    2.10.3 Open-Circuited Transmission Line 82
    2.10.4 Quarter-Wave Transmission Line 83
2.11 Sourced and Loaded Transmission Line 87
    2.11.1 Phasor Representation of Source 87
    2.11.2 Power Considerations for a Transmission Line 88
    2.11.3 Input Impedance Matching 91
    2.11.4 Return Loss and Insertion Loss 92
2.12 Summary 95

As discussed in Chapter 1, higher frequencies imply decreasing wavelengths. The consequence for an RF system is that voltages and currents no longer remain spatially uniform when compared to the geometric size of the discrete circuit elements; they have to be treated as propagating waves. Since Kirchhoff's voltage and current laws do not account for these spatial variations, we must significantly modify the conventional lumped circuit analysis.

The purpose of this chapter is to make the transition from lumped to distributed circuit representation, and in the process develop one of the most useful equations: the spatially dependent impedance representation of a generic RF transmission line configuration. The application of this equation to the analysis and design of high-frequency circuits is going to assume central importance in subsequent chapters. While developing the background of transmission line theory in this chapter, we have purposely attempted to minimize (albeit not eliminate) the reliance on electromagnetics. The motivated reader who would like to delve deeper into the concepts of electromagnetic wave theory is referred to a host of excellent books listed at the end of this chapter.

2.1 Why Transmission Line Theory?

Let us consider a voltage wave field representation $V(z,t) = V_0 \sin(\omega t - \beta z)$ which is traveling in the positive z-direction. This wave couples space and time in such a manner that the sinusoidal space behavior is characterized by the wavelength λ along the z-axis. Moreover, the sinusoidal temporal behavior can be quantified by the time period $T = 1/f$ along the time-axis. In mathematical terms, this leads to change in space over time denoted by the speed of evolution, in our case the constant phase velocity in the form v_p:

$$v_p = \frac{\omega}{\beta} = \lambda f = \frac{1}{\sqrt{\varepsilon\mu}} = \frac{c}{\sqrt{\varepsilon_r \mu_r}}. \tag{2.1}$$

NEXT GENERATION MICROPROCESSORS

The rapidly evolving development of processor clock speeds is predicted to exceed 5 GHz by the year 2010, and can thus pose transmission line effects even on sub-millimeter wafer scale. For instance, the wavelength at 5 GHz in a Si-based CMOS substrate ($\varepsilon_r = 11.7$) is roughly 17.6 mm. Clearly, transmission phenomena begin to play important roles when dealing with signals propagating over the interconnect fabric.

For a frequency of, let us say, $f = 1$ MHz and medium parameters of $\varepsilon_r = 10$ and $\mu_r = 1$ ($v_p = 9.49 \times 10^7$ m/s), a wavelength of $\lambda = 94.86$ m is obtained. This situation is spatially and temporally depicted in Figure 2-1.

We next focus our attention on a simple electric circuit consisting of load resistor R_L and sinusoidal voltage source V_G with internal resistance R_G connected to the load by means of 1.5 cm long copper wires. We further assume that those wires are aligned along the z-axis and their wire resistance is negligible. If the generator is set to a frequency of 1 MHz, then, as computed before, the wavelength will be 94.86 m. A 1.5 cm long wire connecting source with load will experience spatial voltage variations on such a minute scale that they are insignificant.

When the frequency is increased to 10 GHz, the situation becomes dramatically different. In this case the wavelength reduces to $\lambda = v_p/10^{10} = 0.949$ cm and thus is approximately two-thirds the

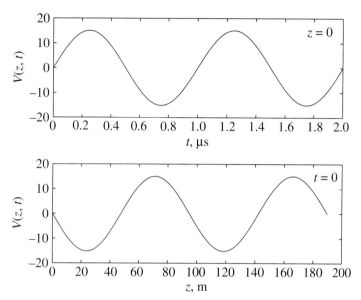

Figure 2-1 Voltage distribution as a function of time ($z = 0$) and as a function of space ($t = 0$).

length of the wire. Consequently, if voltage measurements are now conducted along the 1.5 cm wire, location becomes very important in determining the phase reference of the signal. This fact would readily be observed if an oscilloscope were to measure the voltage at the beginning (location A), at the end (location B), or somewhere along the wire, where distance AB is 1.5 cm and measured along the z-axis in Figure 2-2.

We are obviously faced with a dilemma. A simple circuit, seen in Figure 2-2, with a voltage source V_G and source resistance R_G connected to a load resistor R_L through a two-wire line of length l, whose resistance is assumed negligible, can only be analyzed with Kirchhoff's voltage law

$$\sum_{i=1}^{N} V_i = 0 \qquad (2.2)$$

when the line connecting source with load does not possess a spatial voltage variation, as is the case in low-frequency circuits. In (2.2) V_i ($i = 1, \ldots, N$) represents the voltage drops over N discrete components. When the frequency attains such high values that the spatial behavior of the voltage, and equally the current, has to be taken into account,

OSCILLOSCOPE MEASUREMENTS

A dual-channel scope can record the temporal signal response as a function of space as follows: the probe tip connected to channel 1 is measuring the signal at location A, for instance the harmonic response of a 10 GHz carrier signal. The spatial shift of this signal, indicative of wave propagation, can be captured with a second probe connected to channel 2 by moving the probe tip from location A and B. Because the time base is identical for both channels the screen of the scope will display two sinusoidal signals that are shifted with respect to one another.

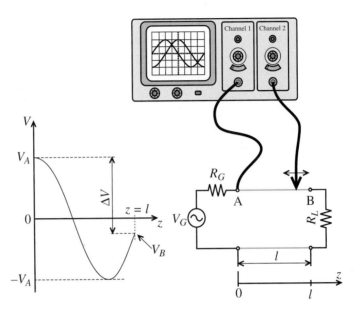

Figure 2-2 Amplitude measurements of 10 GHz voltage signal at the beginning (location A) and somewhere along a wire connecting load to source.

Kirchhoff's circuit laws cannot be directly applied. The situation can be remedied, however, if the line is subdivided into elements of small (mathematically speaking infinitesimal) length, over which voltage and current can be assumed to remain constant, as depicted in Figure 2-3.

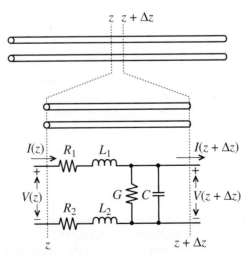

Figure 2-3 Partitioning an electric line into small elements Δz over which Kirchhoff's laws of constant voltage and current can be applied.

For each section of length Δz, we can devise an equivalent electric circuit representation. With reference to our discussions in Chapter 1, it is immediately concluded that there will be some series resistance and inductance associated with each wire. In addition, due to the relative proximity of the two wires, a capacitive effect will also be observed. Since in reality no perfect insulator does exist, a small current flow through the dielectric occurs. A more accurate analysis of all these effects will be given in Section 2.2. At this point we need to stress that equivalent elements, briefly described here, represent only a small segment of the line. To build the complete model of the entire line we would have to replicate Δz a large number of times. Therefore, the transmission line in general cannot be represented in terms of *lumped* parameters, but must be viewed as *distributed* parameters R, L, C, and G, where all circuit parameters are given in terms of unit length.

> **TRANSITION: LUMPED TO DISTRIBUTED THEORY**
>
> When the characteristic sizes of the circuit elements exceeds approximately 1/10 of the electromagnetic wavelength, Kirchhoff's lumped circuit theory should be replaced with a distributed wave theory.

The question of when a wire, or a discrete component, has to be treated as a transmission line cannot precisely be answered with a single number. The transition from lumped circuit analysis obeying Kirchhoff's laws to distributed circuit theory involving voltage and current waves depends on the wavelength in comparison with the average component size. The transition takes place gradually as the wavelength becomes increasingly comparable with the circuit elements. As a rule of thumb, *when the average size l_A of the discrete circuit component is more than a tenth of the wavelength, transmission line theory should be applied* ($l_A \geq \lambda/10$). For the example of the 1.5 cm wire, we would estimate the following frequency for the transition:

$$f = \frac{v_p}{10l} = \frac{9.49 \times 10^7 \text{ m/s}}{0.15 \text{ m}} = 633 \text{ MHz}.$$

Can the RF design engineer deal with the simple circuit in Figure 2-2 as a lumped element representation at 700 MHz? Perhaps. Can Kirchhoff's circuit theory be applied to the circuit at 1 GHz? Not without having to take into account a significant loss in precision. Additional reasons why the use of transmission line theory is needed will become apparent in later chapters.

> **DIGITAL SUBSCRIBER LINE (DSL)**
>
> There is a massive amount of legacy in twin-wire copper telephone lines that have been deployed over many years by the phone companies world-wide. To utilize these lines more efficiently for broadband Internet services, special modems have been developed to transmit high-speed digital data over the existing copper wire pairs. At speeds of 2 Mbits per second and more, Asynchronous DSL (ADSL) links allow much higher data transmission than normal phone lines of 64 kbits per second.

2.2 Examples of Transmission Lines

2.2.1 Two-Wire Lines

The two-wire transmission line discussed in Section 2.1 is one example of a system capable of transporting high-frequency electric energy from one location to another. Unfortunately, it is perhaps the most unsuitable way of transmitting high-frequency voltage and current waves. As shown schematically in Figure 2-4, the two conductors separated over a fixed distance suffer from the drawback that the electric and magnetic field lines emanating from the conductors extend to infinity and thus influence electronic equipment in the vicinity of the line.

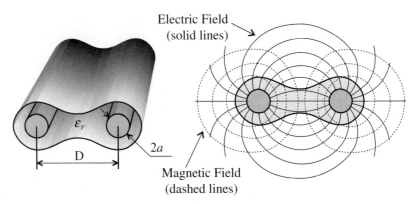

Figure 2-4 Geometry and field distribution in two-wire parallel conductor transmission line.

Further, due to the fact that the wire pair acts as a large antenna, radiation loss tends to be very high. Therefore, the two-wire line finds only limited applications in the RF domain (for instance, when connecting private TV sets to receiving antennas). However, it is commonly used in 50–60 Hz power lines and local telephone connections. Even though the frequency is low, the distance can easily extend over several kilometers, thus making the wire size comparable to the wavelength (for example, the wavelength in air for a 60 Hz wave is $\lambda = c/f = 3 \times 10^8/60 = 5000$ km). Here again, distributed circuit behavior may have to be taken into account. A common solution that helps minimize cross-talk and radiation is the twisted wire pair whereby the conductors are arranged in a helical pattern.

2.2.2 Coaxial Line

COAX CABLES

Coaxial cables were already used by Heinrich Hertz as he demonstrated standing wave effects around 1887. Today, a typical RG-58A cable is characterized by low losses of less than 1 dB/100 ft at 10 MHz. Even at frequencies of 1 GHz, the losses are relatively moderate at about 20 dB/100 ft.

A more common example of a transmission line is the coaxial cable. It is used for almost all cases of externally connected RF systems or measurement equipment at frequencies of up to 40 GHz. As shown in Figure 2-5, a typical coaxial line consists of an inner cylindrical conductor of radius a, an outer conductor of radius b, and a dielectric medium layered in between. The outer conductor completely encloses the electromagnetic fields, thus minimizing radiation loss and field interference. Several of the most commonly used dielectric materials include polystyrene ($\varepsilon_r = 2.5$, tan $\Delta = 0.0003$ at 10 GHz), polyethylene ($\varepsilon_r = 2.3$, tan $\Delta = 0.0004$ at 10 GHz), or teflon ($\varepsilon_r = 2.1$, tan $\Delta = 0.0004$ at 10 GHz).

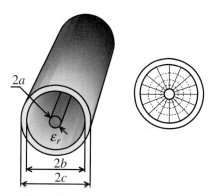

Figure 2-5 Coaxial cable transmission line.

2.2.3 Microstrip Lines

It is a common practice to use planar printed circuit boards (PCBs) as the basic medium to implement most electronic systems. When dealing with RF circuits, we need to consider the high-frequency behavior of the conducting strips etched on the PCBs, as depicted qualitatively in Figure 2-6.

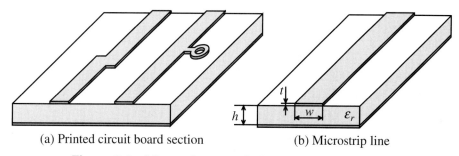

(a) Printed circuit board section   (b) Microstrip line

Figure 2-6 Microstrip transmission line representation.

The ground plane below the current carrying conductor traces helps prevent excessive field leakage and thus reduces radiation loss. The use of PCBs simplifies access to the active and passive devices on the board and reduces the cost of the manufacturing process. In addition, PCBs allow tuning of circuits by simply changing the position of the components and manually adjusting variable tuning capacitors and inductors.

One of the disadvantages of single-layer PCBs is that they have rather high radiation loss and are prone to "crosstalk" (interference) between neighboring conductor traces. As noted in Figure 2-7, the severity of field leakage depends on the relative dielectric constants, as shown qualitatively in the electric flux density displays for teflon epoxy ($\varepsilon_r = 2.55$) and alumina ($\varepsilon_r = 10.0$) dielectrics. The flux density is the electric field scaled by the dielectric material property, $D = \varepsilon_0 \varepsilon_r E$.

MICROSTRIP PROPAGATION

The total losses of microstrip lines can be substantial at higher frequencies and generally involve conduction, dielectric, and radiation losses.

A broad range of passive components are designed with microstrip lines such as filters, resonators, matching networks, and diplexers.

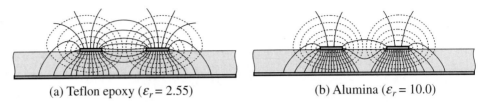

(a) Teflon epoxy ($\varepsilon_r = 2.55$) (b) Alumina ($\varepsilon_r = 10.0$)

Figure 2-7 Electric flux density field leakage as a function of dielectric constants.

> **MULTILAYER BOARDS**
>
> The radiation loss and field leakage/interference effects of microstrip lines placed on dielectric media can be minimized by adding an additional ground plane.
>
> The insertion of additional ground planes in multiboard circuit board designs allows the construction of sandwiched structures and leads to LTCC and HTCC fabrication technologies mentioned in Chapter 1.

Direct comparison of the field lines in Figure 2-7 suggests that to achieve high board density of the component layout, we should use substrates with high dielectric constants since they minimize field leakage and cross coupling.

Another way to reduce radiation losses and interference is to use multilayer techniques to achieve balanced circuit board designs where the microstrip line is "sandwiched" between two ground planes, resulting in the triple-layer configuration seen in Figure 2-8.

A microstrip configuration that is primarily used for low impedance, high-power applications is the parallel-plate line. Here, the current and voltage flow is confined to two plates separated by a dielectric medium. This configuration and the corresponding field distribution are shown in Figure 2-9.

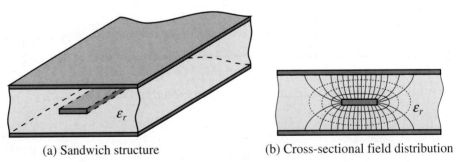

(a) Sandwich structure (b) Cross-sectional field distribution

Figure 2-8 Triple-layer transmission line configuration.

There are many more transmission line configurations used for a number of special-purpose applications. However, a detailed coverage of the pros and cons of all possible combinations would go beyond the objectives of this book.

The preceding transmission line examples all have the commonality that the electric and magnetic field components between the current-carrying conductors are transversely orientated (or polarized); that is, they form a transverse electromagnetic (TEM) field pattern similar to the one shown in Figure 1-3. As mentioned in Chapter 1, the TEM behavior has to be seen in contrast to guided modes, where the electromagnetic wave propagation is

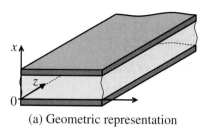
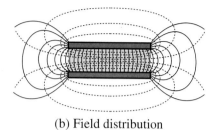

(a) Geometric representation   (b) Field distribution

Figure 2-9 Parallel-plate transmission line.

accomplished through wave reflections and refractions between conducting plates or indexed dielectric media in optical fibers. The analysis is broken down into so-called transverse magnetic (TM) and transverse electric (TE) modes. Such modes of operation are of major interest in the microwave range for satellite communication, radar, and remote sensing applications. Due to their extremely high frequency of operation, well above the RF range, waveguides and optical fiber cables require special electromagnetic treatment and are not considered further. Instead, we refer the reader to a number of references listed at the end of this chapter.

2.3 Equivalent Circuit Representation

As mentioned previously, voltages and currents are no longer spatially constant on the geometric scale of interest. As a consequence, Kirchhoff's circuit laws cannot be applied over the macroscopic line dimension. However, this problem can be circumvented when the transmission line is broken down into smaller (in the limit infinitesimally small) segments. Those segments are still large enough to contain all relevant electrical characteristics such as loss, as well as inductive and capacitive line effects. The main advantage of this reduction to a microscopic representation is the fact that a distributed parameter description can now be introduced whose analysis follows Kirchhoff's laws on a microscopic scale. Besides providing an intuitive picture, the approach also lends itself to two-port network analysis, as discussed in Chapter 4.

To develop an electrical model, let us consider once again a two-wire transmission line. As Figure 2-10 indicates, the transmission line is aligned along the z-axis and segmented into elements of length Δz.

If we focus our attention on a single section residing between z and $z + \Delta z$, we notice that each conductor (1 and 2) is described as a series connection of resistor and inductor (R_1, L_1, and R_2, L_2). In addition, the charge separation created by conductors 1 and 2 gives rise to a capacitive effect denoted by C. Recognizing that all dielectrics suffer losses (see our discussion in Section 1.4.2), we need to include a conductance G. Again, attention is drawn to the fact that all circuit parameters R, L, C, and G are given in values per unit length.

> **INDUCTIVE AND CAPACITIVE VALUES OF A COAX CABLE**
>
> A popular coax cable like the RG-58A has 0.193 inch outside and 0.116 inch dielectric core diameters. The capacitance is given as 30.8 pF/ft at a velocity factor of 66% (of the speed of light). Using the defintion of the phase velocity in the form $v_p = 1/\sqrt{LC}$ yields an inductance of 77 nH/ft.
>
> The polyethylene dielectric has an RMS maximum voltage of 1400 V.

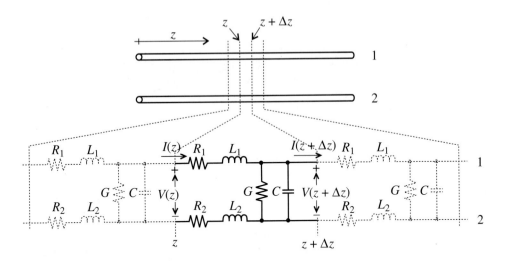

Figure 2-10 Segmentation of two-wire transmission line into Δz-long sections suitable for lumped parameter analysis.

Similarly to the two-wire transmission line, the coaxial cable in Figure 2-11 can also be recognized as a two-conductor configuration with the same lumped parameter representation.

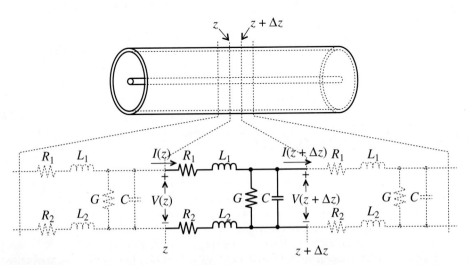

Figure 2-11 Segmentation of a coaxial cable into Δz length elements suitable for lumped parameter analysis.

Equivalent Circuit Representation

A generic form of an electric equivalent circuit is developed as shown in Figure 2-12, where the resistances and inductances of the two conductors are usually combined into single elements. This representation is not suitable for all transmission line applications. For instance, when dealing with transient wave propagation and signal integrity issues of inductive and capacitive crosstalks, it generally is required to retain the parameter representation shown in Figure 2-11 and add a third conductor (ground). For our treatment of transmission lines, we will exclusively use the model shown in Figure 2-12.

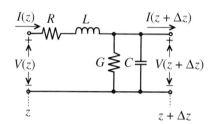

Figure 2-12 Generic electric equivalent circuit representation.

It should be recalled from the discussion in Chapter 1 that the aforementioned R, L, C, and G elements are frequency-dependent parameters, which can change significantly depending on the operating frequency and the employed transmission line type. Further, L not only incorporates the internal and external inductance of the individual wires (see Section 1.4.3), but also takes into account the mutual inductance between the wires. In general, the internal inductances are so small compared with the total inductances that they can be safely neglected. To summarize the advantages of the electric circuit representation, it is observed that this model

- provides a clear intuitive physical picture
- lends itself to a standardized two-port network representation
- permits analysis using Kirchhoff's voltage and current laws
- provides building blocks that allow the transition from microscopic to macroscopic forms.

There are also significant disadvantages worth noting:

- It is essentially a one-dimensional analysis that does not take into account field fringing in the plane orthogonal to the direction of propagation and therefore cannot predict interference with other components of the circuit.
- Material-related nonlinearities due to hysteresis effects are neglected.
- It is not suitable for direct time-domain analysis.

Despite these disadvantages, the equivalent circuit representation is a powerful mathematical model for describing the characteristic transmission line behavior. With this model in place, we can now embark on developing generalized transmission line equations.

2.4 Theoretical Foundation

2.4.1 Basic Laws

> **MAXWELL's EQUATIONS**
>
> In differential form, Maxwell's equations have the following form:
>
> 1) Faraday's law
> $$\nabla \times \mathbf{E} = -\frac{\partial(\mu \mathbf{H})}{\partial t}$$
>
> 2) Ampère's law
> $$\nabla \times \mathbf{H} = \mathbf{J} + \frac{\partial(\varepsilon \mathbf{E})}{\partial t}$$
>
> 3) Divergence of electric flux
> $$\nabla \cdot (\varepsilon \mathbf{E}) = \rho$$
>
> 4) Continuity of magnetic flux
> $$\nabla \cdot (\mu \mathbf{H}) = 0.$$
>
> Here, ρ is the charge density. These four equations can be converted into integral forms through the use of integral theorems.

The next question that we should ask ourselves is how to determine the distributed circuit parameters if we know the physical dimensions and electric properties of the transmission line. The answer is provided through the use of two central laws of electromagnetics: Faraday's law and Ampère's law.

Rooted in experimental observations, Faraday's and Ampère's laws establish two fundamental relations linking electric and magnetic field quantities. As such, both laws provide cornerstones of Maxwell's theory by stating so-called source-field relations. In other words, the time-varying electric field as a source gives rise to a rotational magnetic field. Alternatively, the time-varying magnetic field as a source results in a time-varying electric field that is proportional to the rate of change of the magnetic field. The mutual linkage between electric and magnetic fields is ultimately responsible for wave propagation and traveling voltage and current waves in RF circuits.

By stating Faraday's and Ampère's laws in integral and differential forms, we possess the necessary tools to calculate, at least in principle, the line parameters R, L, C, and G for the electric circuit elements. They are needed to characterize various transmission line systems. By going through the subsequent calculations, we will observe how abstract theoretical laws can be used as a starting point to derive practical circuit parameters for a particular type of transmission line.

Ampère's Law

This fundamental law states that moving charges, which are characterized by the current density **J**, give rise to a rotational magnetic field **H** surrounding the charge flow as expressed by the integral relation

$$\oint \mathbf{H} \cdot d\mathbf{l} = \iint \mathbf{J} \cdot d\mathbf{S} \tag{2.3}$$

where the line integral is taken along the path characterized by the differential element $d\mathbf{l}$ that defines the edge of the surface element **S** in such a manner that the surface **S** always stays on the left side. In equation (2.3) the total current density can be written as $\mathbf{J} = \sigma \mathbf{E} + \partial(\varepsilon \mathbf{E})/\partial t$. It is composed of (a) the conduction current density $\sigma \mathbf{E}$, which is induced by an electric field **E** in the conductor and is responsible for conduction losses, and (b) the displacement current density $\partial(\varepsilon \mathbf{E})/\partial t$, which is responsible for radiation and capacitive storage. Here, and in the following equations, we use bold letters to denote vector quantities such that

$$\mathbf{E}(r, t) = E_x(x, y, z, t)\hat{x} + E_y(x, y, z, t)\hat{y} + E_z(x, y, z, t)\hat{z}$$

where E_x, E_y, E_z are the vector components, and $\hat{x}, \hat{y}, \hat{z}$ are unit vectors in the x, y, z directions in a Cartesian coordinate system. Figure 2-13 illustrates the meaning of equation (2.3).

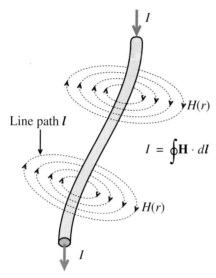

Figure 2-13 Ampère's law linking the current flow to the magnetic field.

Perhaps less intuitive than the integral relation, nonetheless perfectly identical to (2.3), is Ampère's law in differential or point form:

$$(\nabla \times \mathbf{H}) \cdot \mathbf{n} = \lim_{\Delta S \to 0} \frac{1}{\Delta S} \oint \mathbf{H} \cdot dl = \lim_{\Delta S \to 0} \frac{1}{\Delta S} \iint_{\Delta S} \mathbf{J} \cdot d\mathbf{S} = \mathbf{J} \cdot \mathbf{n} \qquad (2.4)$$

where $\nabla \times$ is the *curl* operator and $\mathbf{n}$ is a unit vector perpendicular to the surface element ΔS. When using vector components in a rectangular coordinate system, this differential operator can be represented in matrix form

$$\nabla \times = \begin{bmatrix} 0 & -\frac{\partial}{\partial z} & \frac{\partial}{\partial y} \\ \frac{\partial}{\partial z} & 0 & -\frac{\partial}{\partial x} \\ -\frac{\partial}{\partial y} & \frac{\partial}{\partial x} & 0 \end{bmatrix}. \qquad (2.5)$$

Therefore, by applying the *curl* to the vector field **H**, we obtain

$$\nabla \times \mathbf{H} = \begin{bmatrix} 0 & -\frac{\partial}{\partial z} & \frac{\partial}{\partial y} \\ \frac{\partial}{\partial z} & 0 & -\frac{\partial}{\partial x} \\ -\frac{\partial}{\partial y} & \frac{\partial}{\partial x} & 0 \end{bmatrix} \begin{bmatrix} H_x \\ H_y \\ H_z \end{bmatrix} = \begin{Bmatrix} J_x \\ J_y \\ J_z \end{Bmatrix} \qquad (2.6)$$

where H_x, H_y, H_z and J_x, J_y, J_z are x, y, z components of the magnetic field vector **H** and the current density **J**.

RF&MW→

Example 2-1. Magnetic field generated by a constant current flow in a conductor

Plot the graph of the radial magnetic field $H(r)$ inside and outside an infinitely long wire of radius $a = 5$ mm aligned along the z-axis and carrying a DC current of 5 A. The surrounding medium is assumed to be air.

Solution. This is a typical example for Ampère's law in integral form as given by (2.3). Inside the conductor the current density **J** is uniform and is equal to $\mathbf{J} = I/(\pi a^2)\hat{z}$. Therefore, the application of (2.3) yields the following result:

$$H 2\pi r = \frac{I}{\pi a^2} \pi r^2 \quad \Rightarrow \quad H = \frac{Ir}{2\pi a^2}$$

where $0 \le r \le a$. Outside the conductor the current density is equal to zero and the surface integral in (2.3) gives the total current I flowing through the conductor. Thus, the magnetic field H outside the wire is obtained as

$$H 2\pi r = I \quad \Rightarrow \quad H = \frac{I}{2\pi r}$$

where $r \ge a$. The total magnetic field inside and outside the infinitely long wire is thus

$$H(r) = \begin{cases} \frac{Ir}{2\pi a^2}, & r \le a \\ \frac{I}{2\pi r}, & r \ge a \end{cases} = \begin{cases} 31.83r \text{ kA/m}^2, & r \le 5 \text{ mm} \\ 0.796/r \text{ A}, & r \ge 5 \text{ mm}. \end{cases}$$

The graph of this radial magnetic field distribution is plotted in Figure 2-14.

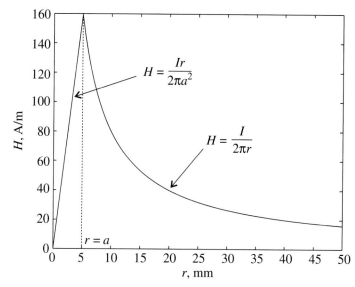

Figure 2-14 Magnetic field distribution inside and outside of an infinitely long wire of radius $a = 5$ mm carrying a current of 5 A.

We make the observation that inside the wire the magnetic field linearly increases from the center to the outer conductor periphery since more current contributes to the magnetic field.

Faraday's law

This law implies that the time rate of change of the magnetic flux density $\mathbf{B} = \mu\mathbf{H}$ ($\mu = \mu_0 \mu_r$) as a source gives rise to a rotating electric field

$$\oint \mathbf{E} \cdot d\mathbf{l} = -\frac{d}{dt}\iint \mathbf{B} \cdot d\mathbf{S}. \qquad (2.7)$$

The line integral is again taken along the edge of the surface $\mathbf{S}$ as previously described for Ampère's law. The integration of the electric field along a wire loop, as shown in Figure 2-15, yields an induced voltage $V = -\oint \mathbf{E} \cdot d\mathbf{l} = \frac{d}{dt}(\iint \mathbf{B} \cdot d\mathbf{S})$.

Similarly to Ampère's law, we can convert (2.7) into a differential, or point form:

$$\nabla \times \mathbf{E} = -\frac{\partial \mathbf{B}}{\partial t}. \tag{2.8}$$

Equation (2.8) makes it clear that we need a time-dependent magnetic flux density to obtain an electric field, which in turn creates a magnetic field according to Ampère's law.

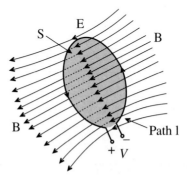

Figure 2-15 The time rate of change of the magnetic flux density induces a voltage.

Example 2-2. Induced voltage in a stationary wire loop

Find the induced voltage of a thin wire loop of radius $a = 5$ mm in air subjected to a time-varying magnetic field $H = H_0 \cos(\omega t)$, where $H_0 = 5$ A/m, and the operating frequency is $f = 100$ MHz.

Solution. The voltage induced in the loop is equal to the line integral of the electric field $\mathbf{E}$ along the loop. Employing Faraday's law (2.7) results in the following:

$$V = -\oint \mathbf{E} \cdot d\mathbf{l} = \frac{d}{dt}\iint \mathbf{B} \cdot d\mathbf{S}.$$

Since the surrounding medium is air, the relative permeability μ_r equals unity and the magnetic flux density is $\mathbf{B} = \mu_0 \mathbf{H} = \mu_0 H_0 \cos(\omega t)\hat{z}$.

Substituting **B** into the preceding integral leads to an expression for the induced voltage V in the loop:

$$V = \frac{d}{dt}\iint \mathbf{B} \cdot d\mathbf{S} = \frac{d}{dt}\mu_0 H_0 \cos(\omega t)\pi a^2$$

$$= -\pi a^2 \omega \mu_0 H_0 \sin(\omega t).$$

This can further be reduced to $V = -0.31\sin(6.28 \times 10^8 t)$ V

The result of this example is also known as the transformer form of Faraday's law whereby a time-varying field produced by a primary coil induces a voltage response in a secondary loop.

2.5 Circuit Parameters for a Parallel-Plate Transmission Line

Our goal is to compute the line parameters R, L, C, and G for a section of a transmission line seen in Figure 2-16. To avoid any notational confusion we explicitly use σ_{cond} and σ_{diel} to denote, respectively, conductivity in the conductor and conductivity in the dielectric medium.

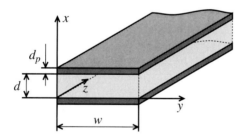

Figure 2-16 Parallel-plate transmission line geometry. The plate width w is large compared with the separation d.

We must assume that the plate width w is large compared with the plate separation d for a one-dimensional analysis to apply. Further, we assume that the skin depth δ is small compared to the thickness d_p of the plates to simplify the derivation of the parameters. Under these conditions we are able to cast the electric and magnetic fields in the conducting plates in the form

$$\mathbf{E} = \hat{z}E_z(x, z)e^{j\omega t} \quad (2.9a)$$

$$\mathbf{H} = \hat{y}H_y(x, z)e^{j\omega t}. \quad (2.9b)$$

The term $e^{j\omega t}$ represents the time dependence of the sinusoidal electric and magnetic fields, and phasors $E_z(x, z)$ and $H_y(x, z)$ encode spatial variations. We do not have any field dependence upon y, because the plates are assumed very wide, and thus the electromagnetic fields do not change appreciably along the y-axis. Application of the differential forms of Faraday's and Ampère's laws to the lower plate conductor

$$\nabla \times \mathbf{E} = -\mu \frac{\partial \mathbf{H}}{\partial t} \tag{2.10}$$

$$\nabla \times \mathbf{H} = \sigma_{cond} \mathbf{E} \tag{2.11}$$

results in two differential equations:

$$\begin{bmatrix} 0 & -\frac{\partial}{\partial z} & \frac{\partial}{\partial y} \\ \frac{\partial}{\partial z} & 0 & -\frac{\partial}{\partial x} \\ -\frac{\partial}{\partial y} & \frac{\partial}{\partial x} & 0 \end{bmatrix} \begin{Bmatrix} 0 \\ 0 \\ E_z \end{Bmatrix} = -\frac{\partial E_z}{\partial x}\hat{y} = -\mu \frac{d}{dt}\begin{Bmatrix} 0 \\ H_y \\ 0 \end{Bmatrix} = -\mu \frac{dH_y}{dt}\hat{y} = -j\omega\mu H_y \hat{y} \tag{2.12}$$

and

$$\begin{bmatrix} 0 & -\frac{\partial}{\partial z} & \frac{\partial}{\partial y} \\ \frac{\partial}{\partial z} & 0 & -\frac{\partial}{\partial x} \\ -\frac{\partial}{\partial y} & \frac{\partial}{\partial x} & 0 \end{bmatrix} \begin{Bmatrix} 0 \\ H_y \\ 0 \end{Bmatrix} = \frac{\partial H_y}{\partial x}\hat{z} = \sigma_{cond}\begin{Bmatrix} 0 \\ 0 \\ E_z \end{Bmatrix} = \sigma_{cond} E_z \hat{z}. \tag{2.13}$$

By differentiating (2.13) with respect to x and substituting (2.12), we find

$$\frac{d^2 H_y}{dx^2} = j\omega \sigma_{cond}\mu H_y = p^2 H_y \tag{2.14}$$

where $p^2 = j\omega\sigma_{cond}\mu$. The general solution for this second-order ordinary differential equation (2.14) is $H_y(x) = A e^{-px} + B e^{px}$. The coefficients A and B are integration constants. We can now perform the following manipulations:

$$p = \sqrt{j\omega\sigma_{cond}\mu} = \sqrt{j}\sqrt{\omega\sigma_{cond}\mu} = (1+j)\sqrt{(\omega\sigma_{cond}\mu)/2} = (1+j)/\delta \tag{2.15}$$

where $\delta = \sqrt{2/(\omega\sigma_{cond}\mu)}$ is recognized as the skin depth. Since p has a positive real component, constant A should be equal to zero to satisfy the condition that the magnetic field in the lower plate must decay in amplitude for negative x. A similar argument can be made for the upper plate by setting $B = 0$. Thus, for the magnetic field in the lower conducting plate we have a simple exponential solution

$$H_y = H_0 e^{px} = H_0 e^{(1+j)x/\delta} \qquad (2.16)$$

where $B = H_0$ is a yet to be determined constant factor. Since the current density can be written as

$$J_z = \sigma_{cond} E_z = \frac{\partial H_y}{\partial x} = \frac{(1+j)H_0}{\delta} e^{(1+j)x/\delta} \qquad (2.17)$$

we are now able to relate the current density J_z to the total current flow I in the lower plate

$$I = \iint_S J_z dx dy = w \int_{-d_p}^{0} J_z dx = w H_0 e^{(1+j)x/\delta} \Big|_{-d_p}^{0} = w H_0 (1 - e^{-(1+j)d_p/\delta}) \qquad (2.18)$$

where S is the cross-sectional area of the lower plate and d_p is the thickness of that plate. Since we assume that $d_p \gg \delta$, the exponential term in (2.18) drops out and $I = wH_0$. From this we conclude that $H_0 = I/w$. The electric field at the surface of the conductor ($x = 0$) can be specified as

$$E_z(0) = \frac{J_z(0)}{\sigma_{cond}} = \frac{(1+j)H_0}{\sigma_{cond}\delta} = \frac{1+j}{\sigma_{cond}\delta} \frac{I}{w}. \qquad (2.19)$$

Equation (2.19) allows us to compute the internal impedance per unit length Z_s by eliminating the current I as follows:

$$Z_s = E_z(0)/I = \frac{1}{w\sigma_{cond}\delta} + \frac{j}{w\sigma_{cond}\delta} = R_s + j\omega L_s. \qquad (2.20)$$

The resistance and internal inductance per unit length are then identified as

$$R_s = \frac{1}{w\sigma_{cond}\delta} \qquad (2.21)$$

$$L_s = \frac{1}{w\sigma_{cond}\omega\delta}. \qquad (2.22)$$

Both are dependent on the skin depth δ. It is important to point out that (2.21) and (2.22) apply for a single conductor. Since we have two conductors in our system (upper and lower plates) the total series resistance and inductance per unit length will be twice the value of R_s and L_s, respectively.

THE z-DIRECTED ELECTRIC FIELD

The introduction of a current density flowing in the z-direction and its associated longitudinal electric field is strictly speaking in violation of the boundary condition $E_z = 0$ for perfect conductors, and is also inconsistent with the TEM concept of RF transmission line propagation where the electric and magnetic fields are orthogonal to the direction of propagation. From an engineering perspective, this assumption can be defended because it is only a small longitudinal field component that is scaled by a large σ_{cond} (for copper: $5.8 \times 10^7 \ldots 6.4 \times 10^7 \text{S/m}$).

To obtain the inductive and capacitive behavior per unit length, we must employ the definitions of capacitance and inductance:

$$C = \frac{Q}{V} = \frac{\oint \mathbf{D} \cdot d\mathbf{l}}{V} = \frac{\varepsilon \int E_x dy}{\int E_x dx} = \frac{\varepsilon E_x w}{E_x d} = \frac{\varepsilon w}{d} \quad (2.23)$$

and

$$L = \frac{\oint \mathbf{B} \cdot d\mathbf{l}}{I} = \frac{\int \mu H_y dx}{I} = \frac{\mu H_y d}{H_y w} = \frac{\mu d}{w} \quad (2.24)$$

where we have used the result of (2.18) to compute the current $I = wH_y$. Both in (2.23) and (2.24) the capacitance and inductance are given per unit length.

Finally, we can express the per-unit-length conductance G in a similar way as derived in (2.23):

$$G = \frac{\oint \mathbf{J} \cdot d\mathbf{l}}{V} = \frac{\sigma_{\text{diel}} \int E_x dy}{\int E_x dx} = \frac{\sigma_{\text{diel}} E_x w}{E_x d} = \frac{\sigma_{\text{diel}} w}{d}. \quad (2.25)$$

Thus, we have succeeded in deriving all relevant parameters for the parallel-plate transmission line. From a practical point of view, at RF frequencies the magnitude of L_s is typically much smaller than L and therefore is neglected.

Example 2-3. Line parameters of a parallel-plate transmission line

For a parallel copper-plate transmission line operated at 1 GHz, the following parameters are given: $w = 6$ mm, $d = 1$ mm, $\varepsilon_r = 2.25$, and $\sigma_{\text{diel}} = 0.125$ mS/m. Find the line parameters R, L, G, and C per unit length.

Solution. The skin depth for copper with conductivity $\sigma_{\text{cond}} = 64.5 \times 10^6 \, \Omega^{-1}\text{m}^{-1}$ at the operating frequency of 1 GHz is $\delta = 1/\sqrt{\pi \sigma_{\text{cond}} \mu_0 f} = 1.98$ μm, which is assumed to be much smaller than the thickness of the conductor. Therefore, the resistance of each plate is determined by (2.21). Since we have two plates, the total resistance is $R = 2R_s = 2/(w \sigma_{\text{cond}} \delta) = 2.6 \, \Omega/\text{m}$. The series inductance due to the skin effect is $L_s = 2/(w \sigma_{\text{cond}} \omega \delta) = 0.42$ nH/m, where the factor

2 takes into account both plates. The mutual inductance between the plates is determined by (2.24) and for our problem is equal to $L = 209.4$ nH/m. As seen, the series inductance is much smaller than the mutual inductance and therefore can safely be neglected. According to (2.23), the capacitance of the line is given by $C = (\varepsilon_0 \varepsilon_r w)/d = 119.5$ pF/m. Finally, the conductance G is determined from (2.25) and equals $G = 0.75$ mS/m.

The RF resistance due to the skin depth phenomenon is, in general, much more significant than the DC resistance.

2.6 Summary of Different Line Configurations

The previous computations were carried out for the relatively simple case of a parallel-plate transmission line. Similar analyses apply when dealing with more complicated line geometries, such as coaxial cables and twisted wire pairs. Table 2-1 summarizes the three common transmission line types.

Table 2-1 Transmission line parameters for three line types

Parameter	Two-Wire Line	Coaxial Line	Parallel-Plate Line
R Ω/m	$\dfrac{1}{\pi a \sigma_{cond} \delta}$	$\dfrac{1}{2\pi \sigma_{cond} \delta}\left(\dfrac{1}{a} + \dfrac{1}{b}\right)$	$\dfrac{2}{w \sigma_{cond} \delta}$
L H/m	$\dfrac{\mu}{\pi} \cosh^{-1}\left(\dfrac{D}{2a}\right)$	$\dfrac{\mu}{2\pi} \ln\left(\dfrac{b}{a}\right)$	$\mu \dfrac{d}{w}$
G S/m	$\dfrac{\pi \sigma_{diel}}{\cosh^{-1}(D/(2a))}$	$\dfrac{2\pi \sigma_{diel}}{\ln(b/a)}$	$\sigma_{diel} \dfrac{w}{d}$
C F/m	$\dfrac{\pi \varepsilon}{\cosh^{-1}(D/(2a))}$	$\dfrac{2\pi \varepsilon}{\ln(b/a)}$	$\varepsilon \dfrac{w}{d}$

The geometric dimensions for the two-wire (D, a), coaxial (a, b), and parallel-plate (w, d) lines are depicted in Figures 2-4, 2-5, and 2-16. For more complex transmission line configurations, significant mathematical effort must be exerted, and resorting to numerical analysis procedures is often the only available solution. This is seen when dealing with microstrip transmission lines (Section 2.8).

2.7 General Transmission Line Equation

2.7.1 Kirchhoff Voltage and Current Law Representations

> **VALIDITY OF THEORETICAL PARAMETERS**
>
> Although these transmission line parameters are widely reported, they represent approximations only accurate when certain geometric aspect ratios are satisfied. For instance, if the width to height ratio of the parallel plate line gets smaller, field fringing can no longer be neglected. Furthermore, the resistance results for the two-wire line assume uniform current density, which is inaccurate for very close wire spacing.

Having developed the background of Faraday's and Ampère's laws in Section 2.4.1, we are well positioned to exploit both equations from a circuit point of view. This is identical to applying Kirchhoff's voltage and current laws (KVL and KCL, respectively) to the loop and node a shown in Figure 2-17.

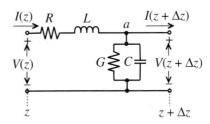

Figure 2-17 Segment of a transmission line with voltage loop and current node.

Adopting phasor notation, we can use Kirchoff's voltage law to conclude

$$(R + j\omega L)I(z)\Delta z + V(z + \Delta z) = V(z) \tag{2.26}$$

which is re-expressed as a differential equation by combining the voltage drop on either side of the differential transmission line segment into a differential quotient:

$$\lim_{\Delta z \to 0}\left(-\frac{V(z + \Delta z) - V(z)}{\Delta z}\right) = -\frac{dV(z)}{dz} = (R + j\omega L)I(z) \tag{2.27}$$

or

$$-\frac{dV(z)}{dz} = (R + j\omega L)I(z) \tag{2.28}$$

where R and L are the combined resistance and inductance of the two lines. Applying Kirchhoff's current law to node a in Figure 2-17 yields

$$I(z) - V(z + \Delta z)(G + j\omega C)\Delta z = I(z + \Delta z) \tag{2.29}$$

which can be converted into a differential equation similar to (2.27). The result is

$$\lim_{\Delta z \to 0}\frac{I(z + \Delta z) - I(z)}{\Delta z} = \frac{dI(z)}{dz} = -(G + j\omega C)V(z). \tag{2.30}$$

Equations (2.28) and (2.30) are coupled first-order differential equations. They can also be derived from a more fundamental point of view, revealing the definitions of R, G, C, and L as discussed in Example 2-4 for the previously analyzed parallel-plate transmission line example.

General Transmission Line Equation

Example 2-4. Derivation of the parallel-plate transmission line equations

Establish the transmission line equations for the parallel-plate conductors.

Solution The purpose of this example is to show how the transmission line equations (2.28) and (2.30) can be derived from the fundamental physical concepts of Faraday's and Ampère's laws.

Let us first consider Faraday's law (2.7). The surface element over which the line and surface integrations are performed is shown as a shadowed area in Figure 2-18.

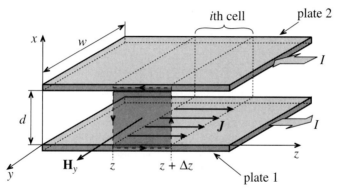

Figure 2-18 Integration surface element for Faraday's law application.

The line integral in (2.7) is taken along the edge of the shaded region with the integration direction denoted by arrows in Figure 2-18. Evaluation of this line integration yields the following contributions:

$$\oint_i \mathbf{E} \cdot d\mathbf{l} = \mathbf{E}^1 \cdot \hat{z}\Delta z + \mathbf{E}(z + \Delta z) \cdot \hat{x}d + \mathbf{E}^2 \cdot (-\hat{z})\Delta z + \mathbf{E}(z) \cdot (-\hat{x})d$$

where $E_z^1 = \mathbf{E}^1 \cdot \hat{z}$ and $E_z^2 = \mathbf{E}^2 \cdot (-\hat{z})$ are the electric fields in the lower (denoted by index 1) and upper (index 2) plates, respectively; and $E_x(z) = \mathbf{E}(z) \cdot \hat{x}$ and $E_x(z + \Delta z) = \mathbf{E}(z + \Delta z) \cdot \hat{x}$ are the electric fields in the dielectric medium between locations z and $z + \Delta z$. It is important to note that the direction of the electric field in the upper conductor is opposite to that of the electric field in the lower conductor, whereas the direction of the field in the dielectric is the same regardless of position. The minus sign in front of the unit vectors indicates that the integration is performed counterclockwise. Combining terms, we obtain

$$\oint_i \mathbf{E} \cdot d\mathbf{l} = E_z^1 \Delta z + E_z^2 \Delta z + E_x(z+\Delta z)d - E_x(z)d.$$

Since the magnetic field in the dielectric is assumed uniform, the integration over the surface in (2.7) gives

$$\iint \mu \mathbf{H} \cdot d\mathbf{S} = \mu H_y \Delta z d.$$

Substitution of these two integrals into (2.7) results in

$$E_z^1 \Delta z + E_z^2 \Delta z + E_x(z+\Delta z)d - E_x(z)d = -\frac{d}{dt}\mu H_y \Delta z d.$$

Similarly to the discussions in Section 2.5, the magnetic field in the dielectric can be expressed as $H_y = I/w$. The electric field in the conductor at high frequency is dependent on the skin effect and is $E_z^1 = E_z^2 = I/(w\sigma_{cond}\delta) + jI/(w\sigma_{cond}\delta) = E_z$. At low frequency, the skin effect does not affect the electric field behavior. The field is solely determined by the DC resistivity of the plates and current I: $E_z = I/(w\sigma_{cond}d_p)$. Since we are primarily concerned with the high-frequency performance, we must assume that the skin depth δ is much smaller than the thickness of the plates. Thus, d_p has to be replaced with δ. Combining expressions for H_y and E_z, and taking into account the relation for the potential between the plates, $V = E_x d$, we obtain

$$2\left(\frac{I}{w\sigma_{cond}\delta} + \frac{jI}{w\sigma_{cond}\delta}\right)\Delta z + V(z+\Delta z) - V(z) = -\mu \frac{d\Delta z}{w}\frac{dI}{dt}$$

$$= -j\omega\mu \frac{d\Delta z}{w}I$$

or

$$2R_s I + j\omega I(L + 2L_s) = -\frac{V(z+\Delta z) - V(z)}{\Delta z} = -\frac{\partial V}{\partial z}$$

where $R_s = 1/(w\sigma_{cond}\delta)$ is the resistance of the plates, $L_s = 1/(w\sigma_{cond}\omega\delta)$ is the high-frequency internal inductance of the plates, and $L = \mu d/w$ is the mutual inductance between both plate conductors.

For the application of Ampère's law (2.3) we use the surface element shown in Figure 2-19.

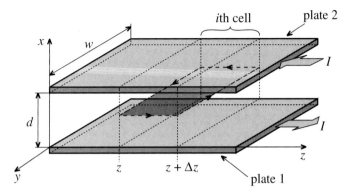

Figure 2-19 Surface element used to apply Ampère's law.

The surface integral of the current density $\mathbf{J}$ in the dielectric medium results in the following expression

$$\iint \mathbf{J} \cdot d\mathbf{S} = J_x \Delta z w = \sigma_{\text{diel}} E_x w \Delta z + \varepsilon \frac{\partial E_x}{\partial t} w \Delta z$$

where the $\sigma_{\text{diel}} E_x w \Delta z$ term represents the conduction current in the dielectric, and $\varepsilon (\partial E_x / \partial t) w \Delta z$ is the contribution of the displacement current. The line integration of the magnetic field yields

$$\oint \mathbf{H} \cdot d\mathbf{l} = -H_y(z + \Delta z) w + H_y(z) w = -I(z + \Delta z) + I(z).$$

Taking into account the relation between the electric field and the potential drop V between z and $z + \Delta z$, that is $E_x = V/d$, we combine both integrals:

$$\frac{\sigma_{\text{diel}} w}{d} V + \frac{\varepsilon w}{d} \frac{dV}{dt} = -\frac{I(z + \Delta z) - I(z)}{\Delta z}$$

or, after introducing the differential quotient,

$$-\frac{\partial I}{\partial z} = \frac{\sigma_{\text{diel}} w}{d} V + \frac{\varepsilon w}{d} \frac{dV}{dt} = \frac{\sigma_{\text{diel}} w}{d} V + \frac{\varepsilon w}{d} j\omega V$$

$$= (G + j\omega C) V.$$

Thus, we succeeded in deriving the equations for the parallel-plate transmission line. To obtain the voltage and current distribution of such a line, the following system of coupled first-order differential equations must be solved:

$$\begin{cases} -\dfrac{\partial V}{\partial z} = [2R_s + j\omega(L + 2L_s)]I \\ -\dfrac{\partial I}{\partial z} = (G + j\omega C)V. \end{cases}$$

Usually, the internal inductance due to the skin effect L_s is much smaller than the mutual inductance L and is often neglected.

This example underscores the effort and assumptions required to develop closed-form expressions for the parallel-plate transmission line. However, if w is comparable in size to d, the preceding treatment breaks down and one has to resort to numerical simulations.

2.7.2 Traveling Voltage and Current Waves

The solution of equations (2.28) and (2.30) is greatly facilitated if these first-order differential equations are decoupled. This can be accomplished by spatially differentiating both sides of (2.28) and substituting (2.30) for the space derivative of the current. The result is a standard second-order differential equation

$$\frac{d^2 V(z)}{dz^2} - \gamma^2 V(z) = 0 \tag{2.31}$$

describing the voltage behavior in phasor form. Here the factor γ is known as a complex propagation constant

$$\gamma = \alpha + j\beta = \sqrt{(R + j\omega L)(G + j\omega C)} \tag{2.32}$$

that depends on the type of transmission line. For simple line configurations, Table 2-1 provides explicit parameters. Reversing the order of decoupling by differentiating (2.30) and substituting (2.28) results in an identical differential equation describing the current:

$$\frac{d^2 I(z)}{dz^2} - \gamma^2 I(z) = 0. \tag{2.33}$$

Solutions to these decoupled equations are two exponential functions for the voltage

$$V(z) = V^+ e^{-\gamma z} + V^- e^{+\gamma z} \tag{2.34}$$

and for the current

$$I(z) = I^+ e^{-\gamma z} + I^- e^{+\gamma z}. \quad (2.35)$$

We observe that (2.34) and (2.35) are general solutions for transmission lines aligned along the z-axis. The first term represents wavefronts propagating in the $+z$-direction, whereas the second term denotes wave propagation in the $-z$-direction. This makes physical sense since the negative sign in conjunction with β, $\alpha \geq 0$ ensures diminishing amplitudes for the positive ($+z$) traveling wave. Conversely, negative traveling waves are attenuated due to the diminishing exponential term.

2.7.3 Characteristic Impedance

Equation (2.35) is related to (2.34). This can be seen if (2.34) is substituted into (2.28). Differentiating and rearranging provides us with a current expression in the following form:

$$I(z) = \frac{\gamma}{R + j\omega L}(V^+ e^{-\gamma z} - V^- e^{+\gamma z}). \quad (2.36)$$

Since voltage and current are generally related via an impedance, we can introduce the so-called **characteristic line impedance** Z_0 by defining

$$Z_0 = \frac{R + j\omega L}{\gamma} = \sqrt{\frac{R + j\omega L}{G + j\omega C}}. \quad (2.37)$$

Substituting the current expression (2.35) into the left-hand side of (2.36), we also find

$$Z_0 = \frac{V^+}{I^+} = -\frac{V^-}{I^-}. \quad (2.38)$$

The characteristic impedance allows us to express the current (2.36) in the concise form

$$I(z) = \frac{1}{Z_0}(V^+ e^{-\gamma z} - V^- e^{+\gamma z}). \quad (2.39)$$

The importance of Z_0 will become apparent in the following sections. Here, it is noteworthy to point out that Z_0 is not an impedance in the conventional circuit sense. Its definition is based on the *positive and negative traveling voltage and current waves*. As such, this definition has nothing in common with the *total voltage and current expressions* used to define a conventional circuit impedance.

2.7.4 Lossless Transmission Line Model

The characteristic line impedance defined in (2.37) is, in general, a complex quantity and therefore takes into account losses that are always present when dealing with realistic lines.

However, for short line segments, as mostly encountered in RF and MW circuits, it does not create an appreciable error to assume lossless line conditions. This implies $R = G = 0$ and the characteristic impedance (2.37) simplifies to

$$Z_0 = \sqrt{L/C}. \tag{2.40}$$

Since Z_0 is independent of frequency, current and voltage waves are only scaled by a constant factor. It is instructive to substitute values for a particular transmission line type. If we use the parallel-plate transmission line with L and C given in Table 2-1, we find the explicit form

$$Z_0 = \sqrt{\frac{\mu}{\varepsilon}\frac{d}{w}} \tag{2.41}$$

where the square root term is known as the wave impedance, which yields ($\mu = \mu_0$, $\varepsilon = \varepsilon_0$) a value of approximately 377 Ω in free space. This value is typical when dealing with radiation systems where an antenna emits electromagnetic energy into free space. However, unlike electromagnetic field radiation into open space, the transmission line introduces geometric constraints as expressed through w and d for the parallel-plate line configuration. The propagation constant is purely imaginary $\gamma = j\beta$, where $\beta = \omega\sqrt{LC}$.

2.8 Microstrip Transmission Lines

As we have seen in Figures 2-6 and 2-7, a simple treatment of the microstrip as a parallel-plate capacitor that formed the basis of computing C in Table 2-1 does not apply in the general case. If the substrate thickness h increases, or if the conductor width w decreases, fringing fields become more prominent and cannot be ignored in the mathematical model. Over the years a number of researchers have developed approximate expressions for the calculation of the characteristic impedance of the microstrip line, taking into account conductor width and thickness. As often encountered in engineering, we have to strike a balance between complexity and accuracy of our computations. The most precise expressions describing microstrip lines are derived using **conformal mapping**, but these expressions are also the most complex, requiring substantial computational efforts. For the purposes of obtaining fast and generally reliable estimates of the line parameters, simpler empirical formulas are more beneficial.

As a first approximation, we assume that the thickness t of the conductor forming the line is negligible compared to the substrate height h ($t/h < 0.005$). In this case, we can use empirical formulas that depend only on the line dimensions (w and h) and the dielectric constant ε_r. They require two separate regions of applicability depending on whether the ratio w/h is larger or less than unity. For narrow microstrips, $w/h < 1$, we obtain the line impedance

$$Z_0 = \frac{Z_f}{2\pi\sqrt{\varepsilon_{\text{eff}}}}\ln\left(8\frac{h}{w} + \frac{w}{4h}\right) \tag{2.42}$$

where $Z_f = \sqrt{\mu_0/\varepsilon_0} = 376.8\ \Omega$ is the wave impedance in free space, and ε_{eff} is the **effective dielectric constant** given by

$$\varepsilon_{\text{eff}} = \frac{\varepsilon_r + 1}{2} + \frac{\varepsilon_r - 1}{2}\left[\left(1 + 12\frac{h}{w}\right)^{-1/2} + 0.04\left(1 - \frac{w}{h}\right)^2\right]. \tag{2.43}$$

For a wide line, $w/h > 1$, we need to resort to a different characteristic line impedance expression:

$$Z_0 = \frac{Z_f}{\sqrt{\varepsilon_{\text{eff}}}\left(1.393 + \frac{w}{h} + \frac{2}{3}\ln\left(\frac{w}{h} + 1.444\right)\right)} \tag{2.44}$$

with

$$\varepsilon_{\text{eff}} = \frac{\varepsilon_r + 1}{2} + \frac{\varepsilon_r - 1}{2}\left(1 + 12\frac{h}{w}\right)^{-1/2}. \tag{2.45}$$

It is important to note that the characteristic impedances given by (2.42) and (2.44) are only approximations and do not produce continuous functions over the entire range of (w/h). In particular, we notice that at $w/h = 1$ the characteristic impedance computed according to (2.42) and (2.44) displays a small discontinuity. Since the error introduced by this discontinuity is less than 0.5%, we still can use the preceding expressions for the computation of both the characteristic line impedance and the effective dielectric constant, as shown in Figures 2-20

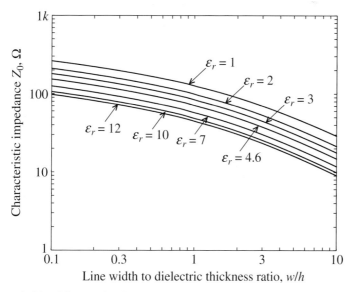

Figure 2-20 Microstrip characteristic impedance as a function of w/h.

and 2-21. In these figures, the quantities Z_0 and ε_{eff} are plotted as functions of w/h ratios and ε_r values. The parameter range of w/h and ε_r is chosen such that it spans the domain of typically encountered, practical values.

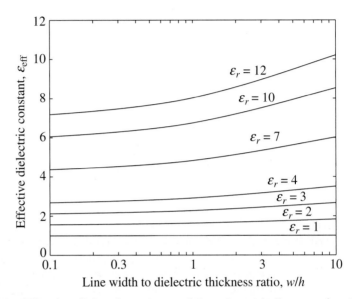

Figure 2-21 Effective dielectric constant of the microstrip line as a function of w/h for different dielectric constants.

In the preceding formulas, the effective dielectric constant is viewed as the dielectric constant of a homogeneous material that fills the entire space around the line, replacing dielectric substrate and surrounding air. With the knowledge of the effective dielectric constant we can compute the phase velocity of the microstrip as $v_p = c/\sqrt{\varepsilon_{\text{eff}}}$. This leads to an expression for the wavelength of

$$\lambda = \frac{v_p}{f} = \frac{c}{f\sqrt{\varepsilon_{\text{eff}}}} = \frac{\lambda_0}{\sqrt{\varepsilon_{\text{eff}}}},$$

where, as before, c is the speed of light, f is the operating frequency, and λ_0 is the free space wavelength.

For design purposes we would like to have a relation that allows us to compute w/h ratios based on a given characteristic impedance Z_0 and dielectric constant ε_r of the substrate. Assuming an infinitely thin line conductor, we can write (see Sobol's article in Further Reading at the end of the chapter) for $w/h \leq 2$

$$\frac{w}{h} = \frac{8e^A}{e^{2A} - 2} \quad (2.46a)$$

MICROSTRIP SUBSTRATES

In general, microstrip substrates are classified into soft and hard materials.

Soft materials such as RT Duroid 5870 and 5880 are cheap to fabricate, but suffer from poor thermal stability.

Hard materials such as quartz, sapphire, and alumina have low thermal expansion coefficents, but are difficult to fabricate.

Microstrip Transmission Lines

where the factor A is given by

$$A = 2\pi \frac{Z_0}{Z_f} \sqrt{\frac{\varepsilon_r + 1}{2}} + \frac{\varepsilon_r - 1}{\varepsilon_r + 1}\left(0.23 + \frac{0.11}{\varepsilon_r}\right).$$

For $w/h \geq 2$ we obtain:

$$\frac{w}{h} = \frac{2}{\pi}\left\{B - 1 - \ln(2B - 1) + \frac{\varepsilon_r - 1}{2\varepsilon_r}\left[\ln(B - 1) + 0.39 - \frac{0.61}{\varepsilon_r}\right]\right\} \qquad (2.46b)$$

where the factor B is given by

$$B = \frac{Z_f \pi}{2Z_0 \sqrt{\varepsilon_r}}.$$

Example 2-5. Design of a microstrip line

A particular RF circuit requires that a line impedance of 50 Ω is to be maintained. The selected PCB material is FR4 with a relative dielectric constant of 4.6 and a thickness of 40 mil. What are the width of the trace, phase velocity, and wavelength at 2 GHz?

Solution. First, we can use Figure 2-20 to determine an approximate ratio of w/h. Choosing a curve corresponding to $\varepsilon_r = 4.6$, we find that for $Z_0 = 50\ \Omega$, w/h is approximately 1.9. Therefore, in (2.46) we have to choose the case where $w/h \leq 2$. This leads to

$$A = 2\pi \frac{Z_0}{Z_f} \sqrt{\frac{\varepsilon_r + 1}{2}} + \frac{\varepsilon_r - 1}{\varepsilon_r + 1}\left(0.23 + \frac{0.11}{\varepsilon_r}\right) = 1.5583.$$

Substituting this result into (2.46a), we find

$$\frac{w}{h} = \frac{8e^A}{e^{2A} - 2} = 1.8477.$$

Then, by using (2.45), we obtain the effective dielectric constant

$$\varepsilon_{\text{eff}} = \frac{\varepsilon_r + 1}{2} + \frac{\varepsilon_r - 1}{2}\left(1 + 12\frac{h}{w}\right)^{-1/2} = 3.4575.$$

We can compute the characteristic impedance of the line (2.44) to verify our result:

$$Z_0 = \frac{Z_f}{(\varepsilon_{\text{eff}})^{1/2}\left(1.393 + \frac{w}{h} + \frac{2}{3}\ln\left(\frac{w}{h} + 1.444\right)\right)} = 50.2243 \ \Omega$$

which is very close to the target impedance of 50 Ω, and therefore indicates that our design is correct.

Using the obtained ratio for w/h, we find the trace width to be $w = 73.9$ mil. Finally, the effective dielectric constant just computed allows us to evaluate the phase velocity of the microstrip line

$$v_p = c/\sqrt{\varepsilon_{\text{eff}}} = 1.61 \times 10^8 \ \text{m/s}$$

and the effective wavelength at 2 GHz

$$\lambda = v_p/f = 80.67 \ \text{mm}.$$

Strictly speaking, this example focuses on a single trace of infinite length only. In reality, proximity to neighboring traces and bends is an issue of practical importance that is most easily accounted for in RF/MW computer aided design (CAD) programs.

For many applications, the assumption of zero strip thickness may not be valid, and corrections to the preceding equations are needed. The effect of nonzero copper strip thickness is approximated as an increase in **effective width** w_{eff} of the conductor since more fringing fields will occur. In other words, a finite thickness is modeled by simply replacing the width of the strip in (2.42)–(2.45) with an effective width computed as

$$w_{\text{eff}} = w + \frac{t}{\pi}\left(1 + \ln\frac{2x}{t}\right) \tag{2.47}$$

where t is the thickness of the conductor, and either $x = h$ if $w > h/(2\pi) > 2t$, or $x = 2\pi w$ if $h/(2\pi) > w > 2t$.

The influence of nonzero thickness on the characteristic line impedance for a standard FR4 substrate with $h = 25$ mil is illustrated in Figure 2-22. As seen in the figure, the effect is most noticeable for narrow strips, while it becomes almost negligible for cases when the width is greater than the thickness of the dielectric.

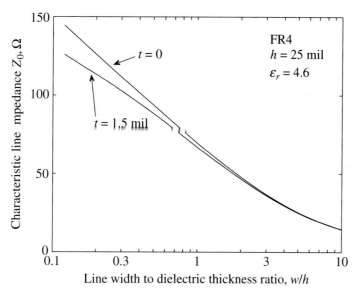

Figure 2-22 Effect of conductor thickness on the characteristic impedance of a microstrip line placed on a 25 mil thick FR4 printed circuit board.

2.9 Terminated Lossless Transmission Line

2.9.1 Voltage Reflection Coefficient

High-frequency electric circuits can be viewed as a collection of finite transmission line sections connected to various discrete active and passive devices. Therefore, let us first take a closer look at the simple configuration of a load impedance connected to a finite transmission line segment of length l, as depicted in Figure 2-23. Such a system forces us to investigate how an incident voltage wave propagating along the positive z-axis interacts with a load impedance representing a generic line termination.

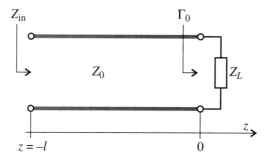

Figure 2-23 Terminated transmission line at location $z = 0$.

Without a loss of generality, the load is assumed to be located at $z = 0$ and the voltage wave is coupled into the line at $z = -l$. As we know, the voltage anywhere along the line is generically given by (2.34). The second term in (2.34) has the meaning of a reflection from the terminating load impedance for values $z < 0$. We introduce the voltage **reflection coefficient** Γ_0 as the ratio of reflected to incident voltage wave

$$\Gamma_0 = \frac{V^-}{V^+} \tag{2.48}$$

at the load location $z = 0$. As a consequence of this definition, the voltage and current waves can be re-expressed in terms of the reflection coefficient as

$$V(z) = V^+(e^{-\gamma z} + \Gamma_0 e^{+\gamma z}) \tag{2.49}$$

and

$$I(z) = \frac{V^+}{Z_0}(e^{-\gamma z} - \Gamma_0 e^{+\gamma z}). \tag{2.50}$$

If (2.49) is divided by (2.50), we find the impedance as a function of space $Z(z)$ anywhere along the z-axis $-l \le z \le 0$. For instance, at $z = -l$ the total input impedance Z_{in} is recorded, and for location $z = 0$ the impedance becomes the load impedance

$$Z(0) = Z_L = Z_0 \frac{1+\Gamma_0}{1-\Gamma_0}. \tag{2.51}$$

Equation (2.51) can be solved for the reflection coefficient Γ_0 with the result

$$\Gamma_0 = \frac{Z_L - Z_0}{Z_L + Z_0}. \tag{2.52}$$

This is a more useful representation than (2.48) since it involves known circuit quantities independent of particular voltage wave amplitude ratios.

We conclude that for an open line $(Z_L \to \infty)$ the reflection coefficient becomes 1, which means the reflected wave returns with the same polarity as the incident voltage. In contrast, for a short circuit $(Z_L = 0)$ the reflected voltage returns with inverted amplitude, resulting in $\Gamma_0 = -1$. For the case where the load impedance matches the line impedance, $Z_0 = Z_L$, no reflection occurs and $\Gamma_0 = 0$. If there is no reflection, we have the case where the incident voltage wave is completely absorbed by the load. This can be regarded as if a second transmission line with the same characteristic impedance, but infinite length, is attached at $z = 0$.

2.9.2 Propagation Constant and Phase Velocity

The definition of the complex propagation constant (2.32) assumes a very simple form for the lossless line $(R = G = 0)$. For this case we obtain

$$\gamma = \alpha + j\beta = j\omega\sqrt{LC} \tag{2.53}$$

with

$$\alpha = 0 \qquad (2.54)$$

and

$$\beta = \omega\sqrt{LC}. \qquad (2.55)$$

The terms α represents the attenuation constant and β is the phase constant. The propagation constant is now purely imaginary, resulting in

$$V(z) = V^+(e^{-j\beta z} + \Gamma_0 e^{+j\beta z}) \qquad (2.56)$$

and

$$I(z) = \frac{V^+}{Z_0}(e^{-j\beta z} - \Gamma_0 e^{+j\beta z}). \qquad (2.57)$$

Here, the characteristic impedance is again given by (2.40). Furthermore, from (2.1) it is known that the wavelength λ can be related to the frequency f via the phase velocity v_p

$$\lambda = v_p/f \qquad (2.58)$$

and the phase velocity v_p is given in terms of the line parameters L, C as

$$v_p = \frac{1}{\sqrt{LC}}. \qquad (2.59)$$

Because of (2.55), we can relate the phase constant to the phase velocity:

$$\beta = \frac{\omega}{v_p}. \qquad (2.60)$$

Substituting the appropriate line parameters from Table 2-1, we notice that for all three transmission line types the phase velocity is independent of frequency. The implication of this fact is as follows: if we assume a pulsed voltage signal propagating down a line, we can decompose the pulse into its frequency components, and each frequency component propagates with the same fixed phase velocity. Thus, the original pulse will appear at a different location without having changed in shape. This phenomenon is known as **dispersion-free** transmission. However, in reality we always have to take into account a certain degree of frequency dependence, or dispersion, of the phase velocity that causes signal distortion.

2.9.3 Standing Waves

It is instructive to insert the reflection coefficient for a short-circuited line ($\Gamma_0 = -1$) into the voltage expression (2.56) and change to a new coordinate d representation such that $z = 0$ in the old system coincides with the origin of the new coordinate system but extends in the opposite, $-z$ direction, as shown in Figure 2-24.

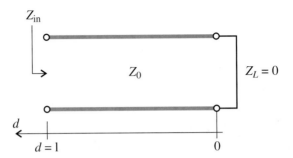

Figure 2-24 Short-circuited transmission line and new coordinate system d.

Equation (2.56) now reads

$$V(d) = V^+(e^{+j\beta d} - e^{-j\beta d}). \qquad (2.61)$$

We notice that the bracket can be replaced by $2j\sin(\beta d)$, and upon converting the phasor expression back into the time domain, we obtain

$$\begin{aligned}v(d,t) &= \text{Re}(Ve^{j\omega t}) = \text{Re}(2jV^+\sin(\beta d)e^{j\omega t}) \\ &= 2V^+\sin(\beta d)\cos(\omega t + \pi/2).\end{aligned} \qquad (2.62)$$

The sin term ensures that the voltage maintains the short-circuit condition for $d = 0$ at all time instances t, see Figure 2-25. Because time and space are now decoupled, no wave propagation, as discussed in Chapter 1, occurs. This phenomenon can physically be explained by the fact that the incident wave is 180° out of phase with the reflected wave, giving rise to fixed zero crossings of the wave at spatial locations 0, $\lambda/2$, λ, $3\lambda/2$, and so on.

Introducing the new coordinate d into (2.56), this equation becomes

$$V(d) = V^+ e^{+j\beta d}(1 + \Gamma_0 e^{-j2\beta d}) = A(d)[1 + \Gamma(d)] \qquad (2.63)$$

where we set $A(d) = V^+ e^{+j\beta d}$ and define a reflection coefficient

$$\Gamma(d) = \Gamma_0 e^{-j2\beta d} \qquad (2.64)$$

valid anywhere along the length of the line d. The far-reaching implications of equation (2.64) as part of the Smith Chart will be the subject of Chapter 3. Similarly, the current in the new spatial reference frame can be defined as

$$I(d) = \frac{V^+}{Z_0} e^{+j\beta d}(1 - \Gamma_0 e^{-j2\beta d}) = \frac{A(d)}{Z_0}[1 - \Gamma(d)]. \qquad (2.65)$$

Under the matched condition ($\Gamma_0 = 0$), the reflection coefficient $\Gamma(d)$ is zero, thus maintaining only a right-propagating wave. To quantify the degree of mismatch, it is customary to introduce the **standing wave ratio** (SWR) as the ratio of the maximum voltage (or current)

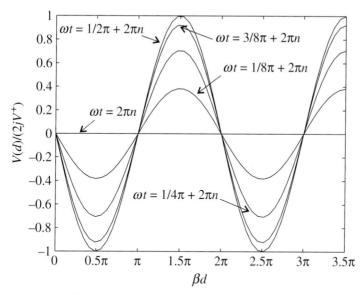

Figure 2-25 Standing wave pattern for various instances of time.

amplitude over the minimum voltage (or current) amplitude along the line, as follows:

$$\text{SWR} = \frac{|V_{max}|}{|V_{min}|} = \frac{|I_{max}|}{|I_{min}|}. \tag{2.66}$$

We note that the maximum magnitude of $\Gamma(d)$ can only be unity; we find for (2.66) the form

$$\text{SWR} = \frac{1 + |\Gamma_0|}{1 - |\Gamma_0|} \tag{2.67}$$

which has a range of $1 \leq \text{SWR} < \infty$, as seen in Figure 2-26.

In many cases, engineers use the term **voltage standing wave ratio** (VSWR) instead of SWR, which is different from the power standing wave ratio (PSWR = VSWR^2). It is concluded from the definition (2.66) and from Figure 2-26 that the ideal case of matched termination yields an SWR of 1, whereas the worst case of either open- or short-circuit termination results in SWR $\to \infty$. Strictly speaking, SWR can only be applied to lossless lines, since it is impossible to define a SWR for lossy transmission systems. This is because the magnitude of the voltage or current waves diminishes as a function of distance due to attenuation and thus invalidates (2.67), which, as a single descriptor, is independent of where along the transmission line the measurement is

VSWR SPECIFICATION FOR AMPLIFIER

VSWR specifications can be found in many data books and application notes by RF equipment manufacturers. For instance, NXP specifies an SWR of 1.4 at the input and an SWR of 1.6 at the output for the BGA2003 monolithic microwave integrated circuit (MMIC) for low-noise amplifier for W-CDMA application at 3.4 GHz.

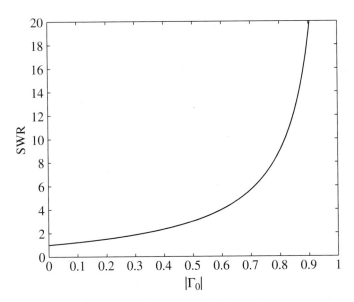

Figure 2-26 SWR as a function of load reflection coefficient $|\Gamma_0|$.

taken. Because most RF systems possess very low losses, (2.67) can be safely applied. Upon inspection of the exponent in (2.64) we see that the distance between the maximum and minimum of the real (and imaginary) part of the reflection coefficient is $2\beta d = \pi$, or $d = \lambda/4$, and the distance between two maxima is $d = \lambda/2$.

2.10 Special Termination Conditions

2.10.1 Input Impedance of Terminated Lossless Line

At a distance d away from the load, the input impedance is given by the expression

$$Z_{in}(d) = \frac{V(d)}{I(d)} = Z_0 \frac{V^+ e^{j\beta d}(1 + \Gamma_0 e^{-2j\beta d})}{V^+ e^{j\beta d}(1 - \Gamma_0 e^{-2j\beta d})} \qquad (2.68)$$

where (2.63) and (2.65) are used for the voltage and current expressions. Equation (2.68) can be converted into the form

$$Z_{in}(d) = Z_0 \frac{1 + \Gamma(d)}{1 - \Gamma(d)} \qquad (2.69)$$

and, upon using (2.52) to replace Γ_0, we obtain

Special Termination Conditions

$$Z_{in}(d) = \frac{e^{j\beta d} + \left(\frac{Z_L - Z_0}{Z_L + Z_0}\right)e^{-j\beta d}}{e^{j\beta d} - \left(\frac{Z_L - Z_0}{Z_L + Z_0}\right)e^{-j\beta d}} Z_0$$

$$= \frac{Z_L(e^{j\beta d} + e^{-j\beta d}) + Z_0(e^{j\beta d} - e^{-j\beta d})}{Z_L(e^{j\beta d} - e^{-j\beta d}) + Z_0(e^{j\beta d} + e^{-j\beta d})} Z_0 \quad (2.70)$$

$$= \frac{Z_L \cos(\beta d) + jZ_0 \sin(\beta d)}{Z_0 \cos(\beta d) + jZ_L \sin(\beta d)} Z_0$$

Division by the cosine term gives us the final form of the input impedance for the terminated transmission line:

$$Z_{in}(d) = Z_0 \frac{Z_L + jZ_0 \tan(\beta d)}{Z_0 + jZ_L \tan(\beta d)}. \quad (2.71)$$

This important result allows us to predict how the load impedance Z_L is transformed along a transmission line of characteristic impedance Z_0 and length d. It takes into account the frequency of operation through the phase constant β. Depending on the application, β can be expressed either in terms of frequency and phase velocity, $\beta = (2\pi f)/v_p$, or wavelength, $\beta = 2\pi/\lambda$.

2.10.2 Short-Circuit Terminated Transmission Line

If $Z_L = 0$ (which means the load is represented by a short circuit) expression (2.71) simplifies to

$$Z_{in}(d) = jZ_0 \tan(\beta d). \quad (2.72)$$

Equation (2.72) can also directly be derived by the division of the voltage by the current wave for the short-circuit condition ($\Gamma_0 = -1$):

$$V(d) = V^+[e^{+j\beta d} - e^{-j\beta d}] = 2jV^+ \sin(\beta d) \quad (2.73)$$

and

$$I(d) = \frac{V^+}{Z_0}[e^{+j\beta d} + e^{-j\beta d}] = \frac{2V^+}{Z_0} \cos(\beta d) \quad (2.74)$$

so that $Z_{in}(d) = V(d)/I(d) = jZ_0 \tan(\beta d)$. A plot of voltage, current, and impedance as a function of line length is shown in Figure 2-27.

It is interesting to observe the periodic transitions of the impedance as the distance from the load increases. If $d = 0$, the impedance is equal to the load impedance, which is zero. For increasing d the impedance of the line is purely imaginary and increases in magnitude. The positive sign of the reactance at this location indicates that the line exhibits inductive behavior.

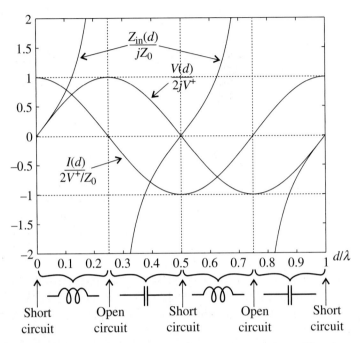

Figure 2-27 Voltage, current, and impedance as a function of line length for a short-circuit termination.

When d reaches a quarter-wave length, the impedance is equal to infinity, which represents an open-circuit condition. Further increase in distance leads to negative imaginary impedance, which is equivalent to capacitive behavior. At distance $d = \lambda/2$ the impedance becomes zero and the entire periodic process is repeated for $d > \lambda/2$.

From a practical point of view, it is difficult to conduct electric measurements at various locations along the line, or alternatively by considering a multitude of lines of different lengths. Much easier (for instance, through the use of a network analyzer) is the recording of the impedance as a function of frequency. In this case d is fixed, and the frequency is swept over a specified range, as discussed in the following example.

Example 2-6. Input impedance of a short-circuited transmission line as a function of frequency

For a short-circuited transmission line of $l = 10$ cm, compute the magnitude of the input impedance when the frequency is swept from $f = 1$ GHz to 4 GHz. Assume the line parameters are the same as the ones given in Example 2.3 (i.e., $L = 209.4$ nH/m and $C = 119.5$ pF/m).

Solution. Based on the line parameters L and C, the characteristic impedance is found to be $Z_0 = \sqrt{L/C} = 41.86 \; \Omega$. Further, the phase velocity is given by $v_p = 1/\sqrt{LC}$ and is equal to 1.99×10^8 m/s. The input impedance of the transmission line $Z_{in}(d = l)$ as a function of frequency can then be expressed in the form

$$Z_{in}(d = l) = jZ_0 \tan(\beta l) = jZ_0 \tan\left(\frac{2\pi f}{v_p} l\right) \quad (2.75)$$

The magnitude of the impedance is shown in Figure 2-28 for the frequency range of 1 GHz to 4 GHz. Again we notice the periodic short- or open-circuit behavior of this line segment. In other words, depending on the frequency, the line exhibits an open-circuit behavior (for instance at 1.5 GHz) or a short-circuit behavior (for instance at 2 GHz).

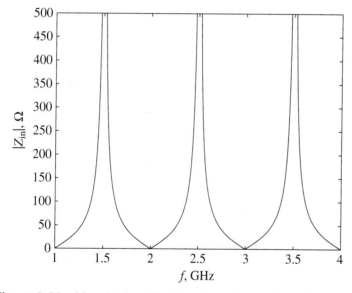

Figure 2-28 Magnitude of the input impedance for a 10 cm long, short-circuited transmission line as a function of frequency.

Practical measurements with a network analyzer permit the recording of graphs as the one seen in Figure 2-28. Had we fixed the frequency and varied the line length, we would have obtained an identical response.

2.10.3 Open-Circuited Transmission Line

If $Z_L \to \infty$, the input impedance (2.71) simplifies to the expression

$$Z_{in}(d) = -jZ_0 \frac{1}{\tan(\beta d)} \tag{2.76}$$

which can be directly derived when we divide the voltage (2.63) by the current wave (2.65) for the open circuit condition ($\Gamma_0 = +1$):

$$V(d) = V^+[e^{+j\beta d} + e^{-j\beta d}] = 2V^+\cos(\beta d) \tag{2.77}$$

and

$$I(d) = \frac{V^+}{Z_0}[e^{+j\beta d} - e^{-j\beta d}] = \frac{2jV^+}{Z_0}\sin(\beta d) \tag{2.78}$$

so that $Z_{in}(d) = V(d)/I(d) = -jZ_0\cot(\beta d)$. A plot of voltage, current, and impedance as a functions of line length is shown in Figure 2-29.

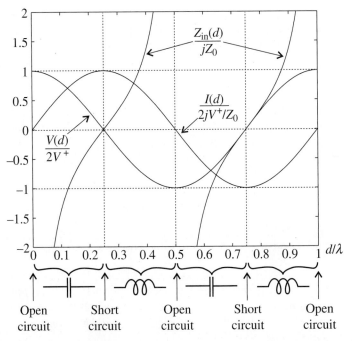

Figure 2-29 Voltage, current, and impedance as a function of line length for an open-circuit termination.

Special Termination Conditions

It is instructive to keep the length d fixed, and sweep the frequency over a specified range, as the next example illustrates.

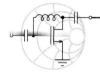

Example 2-7. Input impedance of an open-circuited transmission line as a function of frequency

For an open-circuited transmission line of $l = 10$ cm, repeat the calculations of Example 2-6.

Solution. All calculations remain the same, except that the input impedance is changed to

$$Z_{in}(d = l) = -jZ_0 \cot(\beta l) = -jZ_0 \cot\left(\frac{2\pi f}{v_p} l\right). \qquad (2.79)$$

The magnitude of the impedance is displayed in Figure 2-30 for the frequency range of 1 GHz to 4 GHz. The points where the cotangent approaches infinity correspond to values where the argument reaches 90°, 180°, 270°, and so on. In reality, small losses due to the presence of R and G tend to limit the amplitude to finite peaks. The physical reason for these peaks is due to a phase shift between voltage and current waves. Specifically, when the current wave approaches zero and the voltage is finite, the line impedance assumes a maximum. This is equivalent to the mechanical effect where, for instance, a sound wave at particular discrete frequencies (so-called eigen frequencies) forms standing waves between the walls of a confining structure.

Figures 2-28 and 2-30 teach us that impedance matching to a particular impedance value is only possible at a fixed frequency. Any frequency deviations can result in significantly different impedances.

2.10.4 Quarter-Wave Transmission Line

As evident from (2.70), if the line is matched, $Z_L = Z_0$, we see that $Z_{in}(d) = Z_0$ regardless of the line length. We can also ask ourselves the question: is it possible to make the

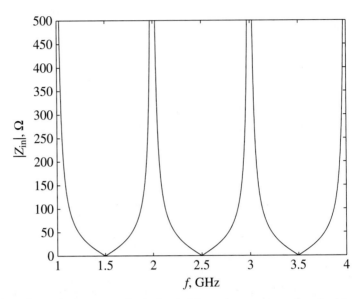

Figure 2-30 Impedance magnitude for a 10 cm long, open-circuited transmission line as a function of frequency.

input impedance of the line equal to the load impedance, $Z_{in}(d) = Z_L$? The answer is found by setting $d = \lambda/2$ (or more generally $d = m(\lambda/2)$, $m = 1, 2, \ldots$), i.e.,

$$Z_{in}(d = \lambda/2) = Z_0 \frac{Z_L + jZ_0 \tan\left(\frac{2\pi}{\lambda} \cdot \frac{\lambda}{2}\right)}{Z_0 + jZ_L \tan\left(\frac{2\pi}{\lambda} \cdot \frac{\lambda}{2}\right)} = Z_L. \qquad (2.80)$$

In other words, if the line is exactly a half wavelength long, the input impedance is equal to the load impedance, independent of the characteristic line impedance Z_0.

As a next step, let us reduce the length to $d = \lambda/4$ (or $d = \lambda/4 + m(\lambda/2)$, $m = 0, 1, 2, \ldots$). This yields

$$Z_{in}(d = \lambda/4) = Z_0 \frac{Z_L + jZ_0 \tan\left(\frac{2\pi}{\lambda} \cdot \frac{\lambda}{4}\right)}{Z_0 + jZ_L \tan\left(\frac{2\pi}{\lambda} \cdot \frac{\lambda}{4}\right)} = \frac{Z_0^2}{Z_L} \qquad (2.81)$$

The implication of (2.81) leads to the **lambda-quarter transformer**, which allows the matching of a real load impedance to a desired real input impedance by choosing a transmission line

Special Termination Conditions

segment whose characteristic impedance can be computed as the geometric mean of load and input impedances:

$$Z_0 = \sqrt{Z_L Z_{in}} \qquad (2.82)$$

This is shown in Figure 2-31, where Z_{in} and Z_L are known impedances and Z_0 is determined based on (2.82).

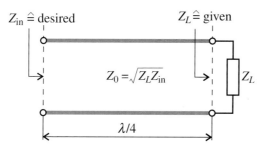

Figure 2-31 Input impedance matched to a load impedance through a $\lambda/4$ line segment.

The idea of impedance matching has important practical design implications and is investigated extensively in Chapter 8. In terms of a simple example, we place the preceding formula in context with the reflection coefficient.

Example 2-8. Impedance matching via a $\lambda/4$ transformer

A transistor has an input impedance of $Z_L = 25\ \Omega$ which is to be matched to a 50 Ω microstrip line at an operating frequency of 500 MHz (see Figure 2-32). Find the length, width, and characteristic impedance of the quarter-wave parallel-plate line transformer for which matching is achieved. The thickness of the dielectric is $d = 1$ mm and the relative dielectric constant of the material is $\varepsilon_r = 4$. Assume that the resistance R and shunt conductance G (see Table 2-1) can be neglected.

Solution. We can directly apply (2.81) by using the given impedances from the problem statement. For the line impedance we find

$$Z_{line} = \sqrt{Z_0 Z_L} = 35.355\ \Omega.$$

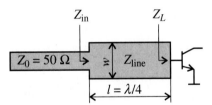

Figure 2-32 Input impedance of quarter-wave transformer.

On the other hand, the characteristic impedance of the parallel-plate line is

$$Z_{line} = \sqrt{L/C} = (d_p/w)\sqrt{\mu/\varepsilon}.$$

Thus, the width of the line is

$$w = \frac{d_p}{Z_{line}}\sqrt{\frac{\mu_0}{\varepsilon_0 \varepsilon_r}} = 5.329 \text{ mm}.$$

The length of the transmission line depends on the phase velocity, which for a parallel-plate line is

$$v_p = \frac{c}{\sqrt{\varepsilon_r}}.$$

From this, the line length l is found to be

$$l = \frac{\lambda}{4} = \frac{v_p}{4f} = \frac{c}{4f\sqrt{\varepsilon_r}} = 74.95 \text{ mm}.$$

The input impedance of the combined transmission line and the load, as shown in Figure 2-32, is

$$Z_{in} = Z_{line}\frac{Z_L + jZ_{line}\tan(\beta d)}{Z_{line} + jZ_L\tan(\beta d)} = Z_{line}\frac{1+\Gamma(d)}{1-\Gamma(d)}$$

where $d = l = \lambda/4$, and the reflection coefficient is given by

$$\Gamma(d) = \Gamma_0 e^{-2j\beta d} = \frac{Z_L - Z_{line}}{Z_L + Z_{line}}\exp\left(-j2\frac{2\pi f}{v_p}d\right).$$

A plot of the impedance magnitude is shown in Figure 2-33.

We note that Z_{in} is matched to the line impedance of 50 Ω not only at 500 MHz, but also at 1.5 GHz. Since the quarter-wave transformer is designed to achieve matching only at 500 MHz for a

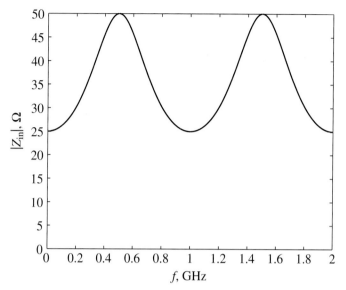

Figure 2-33 Magnitude of Z_{in} for frequency range of 0 to 2 GHz and fixed length d.

particular line length l, we cannot expect matching to occur for frequencies away from the 500 MHz point. In fact, for circuits required to operate over a wide frequency band, this approach may not be a suitable strategy.

> *The $\lambda/4$ transformer plays an important role in many applications as an easy-to-build, narrowband matching circuit.*

2.11 Sourced and Loaded Transmission Line

Thus far, our discussion has only relied on the transmission line and its termination through a load impedance. In completing our investigation, we need to attach a source to the line. This results in the added complication of not only having to deal with an impedance mismatch between transmission line and load, but having to take into considerations possible line-to-source mismatches as well.

2.11.1 Phasor Representation of Source

The generic transmission line circuit is shown in Figure 2-34, and involves a voltage source consisting of a generator voltage V_G and source impedance Z_G.

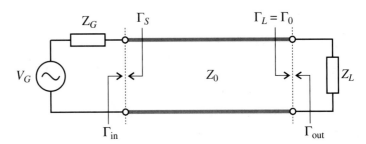

Figure 2-34 Generic transmission line circuit involving source and load terminations.

The input voltage recorded at the beginning of the transmission line can be written in the general form

$$V_{in} = V_{in}^+ + V_{in}^- = V_{in}^+(1 + \Gamma_{in}) = V_G\left(\frac{Z_{in}}{Z_{in} + Z_G}\right) \quad (2.83)$$

where the last expression follows from the voltage divider rule. The input reflection coefficient Γ_{in} is obtained by looking from the source into the transmission line of length $d = l$:

$$\Gamma_{in} = \Gamma(d = l) = \frac{Z_{in} - Z_0}{Z_{in} + Z_0} = \Gamma_0 e^{-2j\beta l}. \quad (2.84)$$

In (2.84), Γ_0 is the load reflection coefficient as defined in (2.52). In addition to the preceding reflection coefficient, the connected source introduces an additional difficulty. Since the voltage reflected from the load is traveling toward the source, we need to consider a mismatch between the transmission line and the source impedance. Accordingly, when looking from the line into the source, we can define the source reflection coefficient:

$$\Gamma_S = \frac{Z_G - Z_0}{Z_G + Z_0}. \quad (2.85)$$

The output reflection coefficient shown in Figure 2-34 is then computed similarly to (2.84), but moving in the opposite direction: $\Gamma_{out} = \Gamma_S e^{-j2\beta l}$.

2.11.2 Power Considerations for a Transmission Line

From the definition of time-averaged power

$$P_{av} = \frac{1}{2}\text{Re}(VI^*) \quad (2.86)$$

we can compute the power dissipated by the loaded transmission line. To accomplish this task, the complex input voltage $V_{in} = V_{in}^+(1 + \Gamma_{in})$ and current $I_{in} = (V_{in}^+/Z_0)(1 - \Gamma_{in})$ have to be inserted in (2.86). The result is

$$P_{in} = P_{in}^+ + P_{in}^- = \frac{1}{2}\frac{|V_{in}^+|^2}{Z_0}(1 - |\Gamma_{in}|^2). \qquad (2.87)$$

We notice here once again that, just like voltage and current, power is also treated as being composed of a positive and negative traveling wave.

Since V_{in}^+ in (2.87) is not directly accessible, it is more useful to re-express (2.87) in terms of the generator voltage V_G as follows:

$$V_{in}^+ = \frac{V_{in}}{1 + \Gamma_{in}} = \frac{V_G}{1 + \Gamma_{in}}\left(\frac{Z_{in}}{Z_{in} + Z_G}\right) \qquad (2.88)$$

where (2.83) is used. As already known from (2.69), the input impedance is rewritten

$$Z_{in} = Z_0\frac{1 + \Gamma_{in}}{1 - \Gamma_{in}}. \qquad (2.89)$$

The generator impedance follows from (2.85) as

$$Z_G = Z_0\frac{1 + \Gamma_S}{1 - \Gamma_S}. \qquad (2.90)$$

Inserting (2.89) and (2.90) into (2.88) yields, after some algebra,

$$V_{in}^+ = \frac{V_G}{2}\frac{(1 - \Gamma_S)}{(1 - \Gamma_S\Gamma_{in})}. \qquad (2.91)$$

Using (2.91) in (2.87), the final expression for the input power is therefore

$$P_{in} = \frac{1}{8}\frac{|V_G|^2}{Z_0}\frac{|1 - \Gamma_S|^2}{|1 - \Gamma_S\Gamma_{in}|^2}(1 - |\Gamma_{in}|^2). \qquad (2.92)$$

Upon using (2.84), we obtain the following expression for the input power for a lossless line:

$$P_{in} = \frac{1}{8}\frac{|V_G|^2}{Z_0}\frac{|1 - \Gamma_S|^2}{|1 - \Gamma_S\Gamma_0 e^{-2j\beta l}|^2}(1 - |\Gamma_0|^2). \qquad (2.93)$$

Since the line is lossless, the power delivered to the load will be equal to the input power. If source and load impedances both are matched to the transmission line impedance (implying $\Gamma_S = 0$ and $\Gamma_0 = 0$), then (2.93) simplifies to

$$P_{in} = \frac{1}{8}\frac{|V_G|^2}{Z_0} = \frac{1}{8}\frac{|V_G|^2}{Z_G} \qquad (2.94)$$

which represents the power produced by the source under perfectly matched conditions and which constitutes the *maximum available power provided by the source*. When the load Z_L is matched to the transmission line, but the source impedance Z_G is mismatched, then part of the power will be reflected and only a portion of the maximum available power will be transmitted into the line at location $d = l$:

$$P_{in} = \frac{1}{8}\frac{|V_G|^2}{Z_0}|1-\Gamma_S|^2. \tag{2.95}$$

For the case where both source and load impedances are mismatched, reflections will occur on both sides of the transmission line and the power that will be delivered to the load is defined by (2.93). Besides watts (W), the unit that is widely used to quantify power in RF circuit design is dBm, which is defined as follows:

$$P[\text{dBm}] = 10\log\frac{P}{1\text{ mW}}. \tag{2.96}$$

In other words, power is measured relative to 1 milliwatt.

RF&MW→

Example 2-9. Power considerations of transmission line

For the circuit shown in Figure 2-34, assume a lossless line with $Z_0 = 75\ \Omega$, $Z_G = 50\ \Omega$, and $Z_L = 40\ \Omega$. Compute the input power and power delivered to the load. Give your answer both in W and dBm. Assume the length of the line to be $\lambda/2$ with a source voltage of $V_G = 5$ V.

THE 50 Ω COAX CABLE

The 50 Ω coax cable is so popular because it represents a trade-off between minimum attenuation achieved at 77 Ω and maximum power achieved at 30 Ω (where the ratio between outer and inner conductor is 1.65 and $\varepsilon_r = 1$). Taking the geometric mean, that is $Z_0 = \sqrt{30 \cdot 77}\ \Omega$, results in approximately 50 Ω.

Solution. Since the line is lossless, the power delivered to the load is exactly the same as the input power. To find the input power, we use expression (2.93). Because the length of the line is $\lambda/2$, all exponential terms in (2.93) are equal to unity; that is, $e^{-2j\beta l} = e^{-2j(2\pi/\lambda)(\lambda/2)} = 1$ and (2.93) can be rewritten as

$$P_{in} = \frac{1}{8}\frac{|V_G|^2}{Z_0}\frac{|1-\Gamma_S|^2}{|1-\Gamma_S\Gamma_0|^2}(1-|\Gamma_0|^2)$$

where the reflection coefficient at the source end is $\Gamma_S = (Z_G - Z_0)/(Z_G + Z_0) = -0.2$ and the reflection coefficient at the load is $\Gamma_0 = (Z_L - Z_0)/(Z_L + Z_0) = -0.304$. Substitution of the obtained values into the preceding equation yields

$$P_L = P_{in} = 61.7\text{ mW}$$

or
$$P_L = P_{in} = 17.9 \text{ dBm}.$$

Most RF data sheets and application notes specify the output power in dBm. It is therefore important to gain a "feel" of the relative magnitudes of mW and dBm.

POWER RECORDING

10^{-7} mW = -70 dBm
0.1 mW = -10 dBm
1 mW = 0 dBm
10 mW = 10 dBm
1 W = 30 dBm
10 W = 40 dBm

The previous analysis is easy to extend to a lossy transmission line. Here, we find that the input power is no longer equal to the load power due to signal attenuation. However, with reference to Figure 2-34, the power absorbed by the load can be expressed similarly to (2.87) as

$$P_L = \frac{|V_L^+|^2}{2Z_0}(1 - |\Gamma_0|^2) \qquad (2.97)$$

where the voltage $|V_L^+|$ for a lossy transmission is $|V_L^+| = |V_{in}^+|e^{-\alpha l}$, with α again being the attenuation coefficient and Z_0 is computed as if the transmission line is lossless. Inserting (2.91) into (2.97) gives as the final expression

$$P_L = \frac{1}{8}\frac{|V_G|^2}{Z_0}\frac{|1-\Gamma_S|^2}{|1-\Gamma_S\Gamma_{in}|^2}e^{-2\alpha l}(1-|\Gamma_0|^2) \qquad (2.98)$$

where all parameters are defined in terms of the source voltage and the reflection coefficients, and $\Gamma_{in} = \Gamma_0 \exp(-2\gamma l)$.

2.11.3 Input Impedance Matching

Employing an electric equivalent circuit representation for the transmission line configuration shown in Figure 2-34 allows us to examine optimal conditions for the matching of the generator to the line.

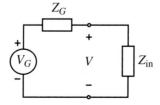

Figure 2-35 Equivalent lumped input network for a transmission line configuration.

In a lumped parameter expression, and consistent with Figure 2-35, we can express (2.93) as

$$P_{in} = \frac{1}{2}\text{Re}\left(V_{in}\frac{V_{in}^*}{Z_{in}^*}\right) = \frac{1}{2}\frac{|V_G|^2}{\text{Re}(Z_{in}^*)}\left|\frac{Z_{in}}{Z_G + Z_{in}}\right|^2. \quad (2.99)$$

If we assume the generator impedance to be of fixed complex value $Z_G = R_G + jX_G$, we can find the conditions that have to be imposed on Z_{in} to obtain maximum power transfer into the transmission line. Treating P_{in} as a function of two independent variables R_{in} and X_{in}, we find the maximum power value by taking the first derivatives of P_{in} with respect to R_{in} and X_{in} and setting the values to zero:

$$\frac{\partial P_{in}}{\partial R_{in}} = \frac{\partial P_{in}}{\partial X_{in}} = 0. \quad (2.100)$$

The two conditions that result are

$$R_G^2 - R_{in}^2 + (X_G^2 + 2X_G X_{in} + X_{in}^2) = 0 \quad (2.101a)$$

and

$$X_{in}(X_G + X_{in}) = 0. \quad (2.101b)$$

Solving (2.101b) gives $X_{in} = -X_G$ and, upon substituting this result into (2.101a), yields $R_{in} = R_G$. This derivation shows that optimal power transfer requires complex conjugate matching of the transmission line to the generator impedance:

$$Z_{in} = Z_G^*. \quad (2.102)$$

Although this is done for the case of generator-to-input impedance matching, an identical analysis can be carried out to match the output impedance to the load impedance. Again we will find that the impedances require complex conjugate matching for maximum power transfer:

$$Z_{out} = Z_L^*$$

Here, Z_{out} represents the impedance looking into the transmission line from the load side.

2.11.4 Return Loss and Insertion Loss

Practical circuit realizations always suffer a certain degree of mismatch between available source power and power delivered to the transmission line; that is, Γ_{in} in (2.89) is not zero. This mismatch is customarily defined as **return loss** (RL), which is the ratio of reflected power, $P_r = P_{in}^-$, to incident power, $P_i = P_{in}^+$, or

$$RL[dB] = -10\log\left(\frac{P_r}{P_i}\right) = -10\log|\Gamma_{in}|^2 = -20\log|\Gamma_{in}| \quad (2.103a)$$

$$RL[Np] = -\ln|\Gamma_{in}|. \quad (2.103b)$$

INSERTION AND RETURN LOSS

The following list provides an appreciation of insertion loss, return loss (both recorded in dB) and the voltage standing wave ratio (VSWR).

IL [dB]	RL [dB]	VSWR
0.01	26.4	1.10
0.50	9.64	1.98
1.00	6.87	2.66
3*	3*	5.83
5.00	1.65	10.6

*Note: 3 dB RL or IL in casual use refer to the half-power point, which is more precisely 3.0103 dB.

Sourced and Loaded Transmission Line

Here, equation (2.103a) specifies the *return loss* in decibel (dB) based on the logarithm to the base 10, whereas (2.103b) specifies RL in Nepers (Np) based on the natural logarithm. A conversion between Np and dB is accomplished by noting that

$$\text{RL[dB]} = -20\log|\Gamma_{in}| = -20(\ln|\Gamma_{in}|)/(\ln 10) = -(20\log e)\ln|\Gamma_{in}|. \quad (2.104)$$

Therefore, 1 Np = $20\log e$ = 8.686 dB. As seen from (2.104), if the line is matched $\Gamma_{in} \to 0$, then RL $\to \infty$.

Example 2-10. Return loss of transmission line section

For the circuit in Figure 2-35 a return loss of 20 dB is measured. Assuming real impedance values only, what is the source resistance R_G if the transmission line has a characteristic impedance of $Z_0 = R_{in} = 50\ \Omega$? Is the answer unique?

Solution. The reflection coefficient is found from (2.103a) as

$$|\Gamma_{in}| = 10^{-\text{RL}/20} = 10^{-20/20} = 0.1.$$

The source resistance is now computed by using (2.89):

$$R_G = R_{in}\frac{1+\Gamma_{in}}{1-\Gamma_{in}} = 50\frac{1+0.1}{1-0.1}\ \Omega = 61.1\ \Omega.$$

In the preceding calculations, we assumed that the reflection coefficient Γ_{in} is positive and therefore is equal to its absolute value. However, it can also be negative, and in that case the source resistance would be

$$R_G = R_{in}\frac{1+\Gamma_{in}}{1-\Gamma_{in}} = 50\frac{1-0.1}{1+0.1}\ \Omega = 40.9\ \Omega.$$

The return loss, which can be recorded with a network analyzer, provides immediate access to the reflection coefficient magnitude and thus the degree of impedance mismatch between the transmission line and generator.

In addition to the return loss, which involves the reflected power, it is useful to introduce the **insertion loss** (IL) defined as a ratio of transmitted power P_t to incident power P_i. In practice, insertion loss is defined in dB according to the following formula:

$$\text{IL[dB]} = -10 \log \frac{P_t}{P_i} \approx -10 \log \frac{P_i - P_r}{P_i} = -10 \log(1 - |\Gamma_{in}|^2). \quad (2.105)$$

The meaning of (2.105) in circuit design is straightforward. As the name implies, if an unmatched circuit is connected to an RF source, reflections occur that result in a loss of power delivered to the circuit. For instance, if the circuit represents an open- or short-circuit condition, the insertion loss reaches a maximum (IL $\rightarrow \infty$). Alternatively, if the circuit is matched to the source, all power is transferred to the circuit, and the insertion loss becomes a minimum (IL = 0).

Practically Speaking. Measuring the impedance of a terminated coaxial cable

The reflection coefficient and hence the impedance of any connectorized component can be measured using a network analyzer in the frequency range from 10 kHz to 100 GHz (most instruments cover only part of this range). Here, we make some practical measurements on a coaxial cable terminated by a resistor.

For this example, a 100 Ω resistor was connected using a 54 cm long cable to the network analyzer's (HP 8714ES) port 1. The measured impedance magnitude is plotted in Figure 2-36(a) as a function of frequency in

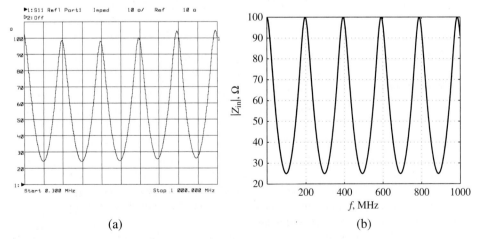

(a)   (b)

Figure 2-36 Impedance of a coaxial cable terminated by a 100 Ω resistor: (a) network analyzer measurement, (b) theoretical prediction.

the range from 300 kHz to 1 GHz. To predict (approximately) the impedance theoretically, we need to know the cable's characteristic impedance and phase velocity, specified as 50 Ω and 71% of the speed of light, respectively. The network analyzer as a source has a generator impedance that is also 50 Ω. Thus, according to equations (2.69) and (2.58), the expected input impedance is plotted in Figure 2-36(b). We observe the input impedance starts out at 100 Ω at very low frequencies and oscillates between 25 Ω and 100 Ω approximately every 200 MHz. As a result, the cable is about a half wavelength at 200 MHz. As the frequency increases, we observe some deviations in the measurements, most likely due to a greater influence of the parasitics. The network analyzer is shown in Figure 2-37 displaying the impedance magnitude of the attached load.

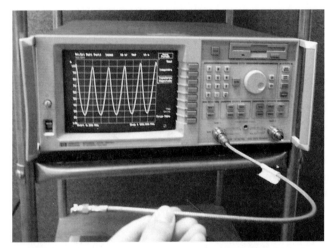

Figure 2-37 Network analyzer with the resistive 100 Ω test load attached.

2.12 Summary

In this chapter, a detailed description is given of the fundamental concepts of distributed circuit theory. The topic is motivated by the fact that when the wavelengths of the voltage and current waves shrink to roughly 10 times the size of the circuit interconnections, a transition must be made from lumped element analysis, based on Kirchhoff's current and voltage laws, to distributed theory according to wave principles. This transition from low- to high-frequency circuit analysis may not be as clear-cut as the 1/10 λ criterion implies; in fact, a considerable "gray area" does exist. Nonetheless, starting at some frequency, a transition is needed to obtain meaningful design results.

The underlying concepts of distributed theory can best be understood by developing an equivalent circuit representation (Section 2.3) of a microscopic section of the transmission line. The required circuit parameters per unit length R, L, G, C are obtained directly from Table 2-1 for three common transmission line types (Section 2.6) without going into much theoretical detail. However, for the readers who are interested in how the parameters can be found, Section 2.4 introduces the necessary tools of Faraday's and Ampère's laws, followed by Section 2.5, which derives all four circuit parameters for the parallel-plate transmission line. Under practical conditions and for complex geometries, numerical methods are used to evaluate the underlying partial differential equations defining the transmission line system.

In either case, the knowledge of the circuit parameters ultimately leads to the characteristic impedance and propagation constant of a generic transmission line system:

$$Z_0 = \sqrt{\frac{R+j\omega L}{G+j\omega C}}, \quad \gamma = \alpha + j\beta = \sqrt{(R+j\omega L)(G+j\omega C)}.$$

From this representation, the input impedance of a terminated transmission line is developed. The result is perhaps one of the most important RF equations for lossless transmisson lines:

$$Z_{in}(d) = Z_0 \frac{Z_L + jZ_0 \tan(\beta d)}{Z_0 + jZ_L \tan(\beta d)}.$$

The application of this equation for the special cases of open, short, and matched load impedances are investigated in terms of their spatial and frequency domain behaviors. Furthermore, the lambda-quarter or quarter-wave transformer is introduced as a way of matching a load impedance to a desired input impedance.

As an alternative to the input impedance equation, it is often useful to represent the line impedance in terms of the reflection coefficients at load and source end:

$$\Gamma_0 = \frac{Z_L - Z_0}{Z_L + Z_0}, \qquad \Gamma_S = \frac{Z_G - Z_0}{Z_G + Z_0}.$$

It is found that the reflection coefficient is transformed by a lossless transmission line of length d according to

$$\Gamma(d) = \Gamma_0 e^{-j2\beta d}.$$

The reflection coefficient concept allows concise expressions for power flow considerations. Similar to the input impedance we found the input power

$$P_{in} = \frac{1}{8}\frac{|V_G|^2}{Z_0}\frac{|1-\Gamma_S|^2}{|1-\Gamma_S\Gamma_0 e^{-2j\beta l}|^2}(1-|\Gamma_0|^2).$$

This equation permits the investigation of various matching or mismatching conditions at the load/source side. Chapter 2 concludes with a brief discussion of insertion loss and return loss.

Further Reading

R. G. Medhurst, "High frequency resistance and capacity of single-layer solenoids," *Wireless Engineering,* p. 35, 1947.

H. M. Greenhouse, "Design of planar rectangular microelectronic inductors," *IEEE Trans. on Parts, Hybrids, Packaging,* Vol. 10, pp. 101–109, 1974.

I. J. Bahl and D. K. Trivedi, "A designer's guide to microstrip line," *Microwaves,* pp. 90–96, 1977.

P. R. Geffe, "The design of single-layer solenoids for RF filters," *Microwave Journal,* Vol. 39, pp. 70–76, 1996.

V. F. Perna, "A guide to judging microwave capacitors," *Microwaves,* Vol. 9, pp. 40–42, 1970.

T. G. Bryant and J. A. Weiss, "Parameters of microstrip transmission lines and of coupled pairs of microstrip lines," *IEEE Trans. on Microwave Theory Techniques (MTT),* Vol. 16, pp. 1021–1027, 1968.

B. Bhat and S. K. Koul, "Unified approach to solve a class of strip and microstrip-like transmission lines," *IEEE Trans. on Microwave Theory Techniques (MTT),* Vol. 30, pp. 679–686, 1982.

R. Collin, *Foundations of Microwave Engineering,* McGraw-Hill, New York, 1966.

G. Gonzales, *Microwave Transistor Amplifiers, Analysis and Design,* 2nd ed, Prentice Hall, Upper Saddle River, New Jersey, 1997.

H. A. Haus and J. R. Melcher, *Electromagnetic Fields and Energy,* Prentice Hall, Englewood Cliffs, NJ, 1989.

M. F. Iskander, *Electromagnetic Fields and Waves,* Prentice Hall, Upper Saddle River, New Jersey, 1992.

C. T. A. Johnk, *Engineering Electromagnetic Fields and Waves,* 2nd ed., John Wiley & Sons, New York, 1989.

J. A. Kong, *Electromagnetic Wave Theory,* 2nd ed., John Wiley & Sons, New York, 1996.

S. Y. Liao, *Engineering Applications of Electromagnetic Theory,* West Publishing Company, St. Paul, MN, 1988.

P. A. Rizzi, *Microwave Engineering, Passive Circuits,* Prentice Hall, Englewood Cliffs, New Jersey, 1988.

H. Sobol, "Applications of Integrated Circuit Technology to Microwave Frequencies," *Proceedings of the IEEE,* August 1971.

D. H. Staelin, A. W. Morgenthaler, and J. A. Kong, *Electromagnetic Waves,* Prentice Hall, Upper Saddle River, New Jersey, 1994.

Problems

2.1 To estimate the effective relative permittivity ε_r of a dielectric material used in a transmission line, you decide to measure the voltage distribution along the line using a similar setup as depicted in Figure 2-2. Your measurements at 1 GHz excitation frequency have shown that the wavelength of the signal in the cable is equal to 10 cm. Using this information, compute the effective relative permittivity of the material. Discuss how this experimental setup could be used to measure the attenuation factor α.

2.2 As discussed in this chapter, a single signal trace on a printed circuit board (PCB) can be treated as a transmission line and can be modeled using an equivalent circuit, shown in Figure 2-12. Nevertheless, when the size of the PCB becomes smaller, the distance between the traces decreases and they can no longer be treated as separate transmission lines. Therefore, the interaction between them has to be accounted for. Using the configuration shown in Figure 2-7, suggest a new equivalent circuit that takes into account interaction between two signal traces.

2.3 In Example 2-1 we showed how to compute the magnetic field distribution produced by a wire carrying current I. Repeat your computations for a system consisting of two parallel wires each of radius 5 mm and carrying a current of 5 A in the same direction. Plot the field distribution of the magnetic field $H(r)$ as a function of distance r starting at the center between the two conductors and extending along the line connecting the two wires.

2.4 Consider a circular loop of radius $r = 1$ cm of thin wire (assume the radius of the wire to be equal to zero) and carrying a constant current $I = 5$ A. Compute the magnetic field along the center line of the loop as a function of distance h from the center of the loop.

2.5 Find α and β in terms of L, C, G, R, and ω in equation (2.32).

2.6 In the text we have derived the transmission line parameters (R, L, G, and C) for a parallel-plate line. Derive these parameters for a two-wire configuration; see Figure 2-4. Assume that $D \gg a$.

2.7 Repeat Problem 2.6 for a coaxial cable; see Figure 2-5.

2.8 An RG6A/U cable has a characteristic impedance of 75 Ω. The capacitance of a 0.5 m long cable is measured to be 33.6 pF. What is the cable inductance per unit length and the phase velocity if the cable is lossless?

2.9 Assuming that dielectric and conductor losses in a transmission line are small (i.e. $G \ll \omega C$ and $R \gg \omega L$), show that the propagation constant γ can be written as

$$\gamma = \alpha + j\beta = \frac{1}{2}\left(\frac{R}{Z_0} + GZ_0\right) + j\omega\sqrt{LC}$$

where $Z_0 = \sqrt{L/C}$ is the characteristic impedance of the line in the absence of loss.

2.10 Using the results from the previous problem and the transmission line parameters given in Table 2-1:
(a) show that the attenuation constant in a coaxial cable with small losses is

$$\alpha = \frac{1}{2\sigma_{cond}\delta}\sqrt{\frac{\varepsilon}{\mu}}\frac{1}{\ln(b/a)}\left(\frac{1}{a}+\frac{1}{b}\right) + \frac{\sigma_{diel}}{2}\sqrt{\frac{\mu}{\varepsilon}}$$

where σ_{diel} and σ_{cond} are the conductivities of the dielectric material and the conductors, respectively;

(b) show that the attenuation in this case is minimized for conductor radii such that $x\ln x = 1 + x$, where $x = b/a$;

(c) show that for a coaxial cable with dielectric constant $\varepsilon_r = 1$ the condition of minimum losses results in the characteristic impedance of $Z_0 = 76.7\ \Omega$.

2.11 Compute the transmission line parameters for a coaxial cable, with characteristics listed as follows:

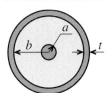

Inner conductor: copper
$a = 0.5$ mm, $\sigma_{Cu} = 64.5 \times 10^6$ S/m
Dielectric: polyethylene
$b = 1.5$ mm, $\sigma_{Poly} = 10^{-14}$ S/m, $\varepsilon_r = 2.25$
Outer conductor: copper
$t = 0.5$ mm, $\sigma_{Cu} = 64.5 \times 10^6$ S/m

2.12 An RG58A/U cable has an unknown characteristic line impedance. The measurements performed on a section of this cable produce the following results,
- capacitance of 1 meter of cable: 101 pF
- phase velocity: 66% of speed of light
- attenuation at 1 GHz: 0.705 dB/m
- dielectric layer is made out of polyethylene, $\sigma_{Poly} = 10^{-14}$ S/m.

From this information, find the following quantities:
(a) inductance L per unit length of the cable assuming it is a lossless transmission line
(b) the characteristic impedance Z_0 if the cable is loseless
(c) relative permittivity ε_r of the dielectric material
(d) resistance R per unit length of the cable at the operating frequency of 1 GHz (*hint:* use the formula for the attenuation constant derived in Problem 2.10)
(e) conductance G of the dielectric per unit length.

2.13 Compute the complex characteristic impedance of the coax cable of the previous problem. Plot the frequency behavior of the real and imaginary components of the characteristic impedance from 10 kHz to 1 GHz. Is the result what you expected to see? Explain any discrepancies.

2.14 A distortionless transmission line results if $R = 0$ and $G = 0$, which leads to $\gamma = j\omega\sqrt{LC} = \alpha + j\beta$, or $\alpha = 0$ and $\beta = \omega/v_p$ with the phase velocity independent of frequency [i.e., $v_p = 1/(\sqrt{LC})$]. A signal propagating along this transmission line will not suffer any pulse distortion or attenuation. If we allow $R \neq G \neq 0$, find the condition for which $\alpha = \sqrt{RG}$ and $\beta = \omega\sqrt{LC}$. In other words, the line is attenuative but remains distortionless.

2.15 It is desired to construct a 50 Ω microstrip line. The relative dielectric constant is 2.23 and the board height is $h = 0.787$ mm. Find the width, wavelength, and effective dielectric constant when the thickness of the copper trace is negligible. Assume an operating frequency of 1 GHz.

2.16 Given the expressions for the transmission line parameters of a two-wire line (see Table 2-1), find the formula for the characteristic impedance of a wire above a ground plane with wire radius a and ground plane to wire center distance h. Assume lossless conditions and homogeneous dielectric.

2.17 The microstrip transmission line (assumed lossless) begins to exhibit dispersion at very high frequencies, which means the phase velocity varies with frequency (the characteristic impedance also varies slightly). Speculate on why this occurs in a microstrip, but not in a coaxial cable of similar dimensions. Assume all dielectric constants are independent of frequency.

2.18 Starting with the basic definition for the standing wave ratio (SWR)

$$\text{SWR} = \frac{|V_{max}|}{|V_{min}|} = \frac{|I_{max}|}{|I_{min}|}$$

show that it can be re-expressed as

$$\text{SWR} = \frac{1 + |\Gamma_0|}{1 - |\Gamma_0|}.$$

2.19 The characteristic impedance of a coax cable is 50 Ω and assumed lossless. If the load is a short circuit, find the input impedance if the cable is 2 wavelengths, 0.75 wavelength, and 0.5 wavelength in length.

2.20 An experiment similar to the one shown in Figure 2-2 is performed with the following results: the distance between successive voltage minima is 2.1 cm; the distance of the first voltage minimum from the load impedance is 0.9 cm; the SWR is 2.5. If $Z_0 = 50$ Ω, find the load impedance.

2.21 In this chapter, we have derived the equation for the input impedance of the loaded lossless line, (2.65). Using the same approach, show that for a loaded lossy transmission line (i.e., $R \neq 0, G \neq 0$) the input impedance is

$$Z_{in}(d) = Z_0 \frac{Z_L + Z_0 \tanh(\gamma d)}{Z_0 + Z_L \tanh(\gamma d)}$$

where γ is the complex propagation constant and tanh denotes the hyperbolic tangent

$$\tanh(x) = \frac{e^x - e^{-x}}{e^x + e^{-x}}.$$

2.22 Using the result from the previous problem, compute the input impedance of a 10 cm long lossy coaxial cable connected to a $Z_L = (45 + j5)$ Ω load impedance.

The system is operated at 1 GHz, and the coaxial cable has the following parameters: $R = 123 \, \mu\Omega/m$, $L = 123 \, nH/m$, $G = 123 \, \mu S/m$, and $C = 123 \, pF/m$.

2.23 Calculate the input impedance of an RG-58 A/U coaxial cable 3λ long and loaded with a 5 Ω impedance at 1 GHz. The cable datasheet specifies $Z_0 = 50 \, \Omega$, $v_p = 0.659 \, c$, and an attenuation of 97 dB/100 m at 1 GHz. Perform calculations for both the lossless and lossy case and comment on the differences.

2.24 Show that the input impedance of a lossless transmission line repeats itself every half wavelength [i.e., $Z_{in}(l_d) = Z_{in}(l_d + m(\lambda/2))$], where l_d is an arbitrary length and m is an integer $0, 1, 2, \ldots$

2.25 A radio transmitter is capable of producing 3 W output power. The transmitter is connected to an antenna having a characteristic impedance of 75 Ω. The connection is made using a lossless coaxial cable with 50 Ω characteristic impedance. Calculate the power delivered to the antenna if the source impedance is 45 Ω and the cable length is 11λ.

2.26 For an RF circuit project, an open-circuit impedance has to be created with a 75 Ω microstrip line placed on a circuit board 31 mil thick with relative dielectric constant of 10 and operated at 1.96 GHz. The line is terminated with a short circuit on one side. To what length does the line have to be cut to measure an infinite impedance on the other side?

2.27 A short-circuited microstrip line of $Z_0 = 85 \, \Omega$ and $(3/4)\lambda$ in length is used as a lumped circuit element. What is the input impedance if the line is assumed lossless?

2.28 For the following system, compute the input power, power delivered to the load, and insertion loss. Assume that all transmission lines are lossless.

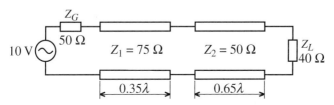

2.29 Repeat Problem 2.28 for a 50 Ω load impedance.

2.30 The complex load impedance $Z_L = (75 - j50)\Omega$ is attached to a lossless transmission line of 100 Ω characteristic impedance and 50 cm in length. The frequency is selected such that the wavelength is 30 cm. Find (a) the input impedance, (b) the impedance looking toward the load 10 cm away from the load, and (c) the voltage reflection coefficient at the load and 10 cm away from the load.

2.31 A 100 Ω microstrip line is connected to a 75 Ω line (assumed to extend to infinity). Determine Γ, SWR, percentage power reflected, return loss, percentage power transmitted, and insertion loss by looking into the 75 Ω line from the 100 Ω side.

2.32 A 50 Ω transmission line is matched to a source and feeds a load of $Z_L = 75 \, \Omega$. If the line is 3.4λ long and has an attenuation constant $\alpha = 0.5 \, dB/\lambda$, find the power that is (a) delivered by the source, (b) lost in the line, and (c) delivered to the load. The amplitude of the signal produced by the source is 10 V.

2.33 A measurement technique is proposed to determine the characteristic line impedance of a coaxial cable via the determination of open-circuit, Z_{in}^{oc}, and short-circuit, Z_{in}^{sc} input impedances with a network analyzer. It is assumed that the line impedance is real. How does one have to process these impedances to obtain Z_0?

2.34 A signal generator is used to feed two loads, as shown in the following figure.

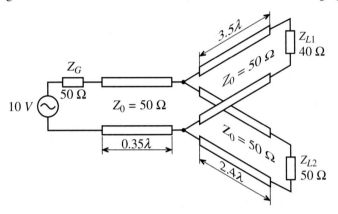

Find the both the power produced by the source and the power delivered to each load.

2.35 A lossless 50 Ω microstrip line is terminated into a load with an admittance of 0.05 mS. (a) What additional impedance has to placed in parallel with the load to assure an input impedance of 50 Ω? (b) If the input voltage of a matched generator is 10 V, find the voltage, current, and power absorbed by the combined load.

2.36 Show that return loss and insertion loss can be expressed in terms of the voltage standing wave ratio SWR as

$$RL = 20 \log \frac{SWR + 1}{SWR - 1} \text{ and } IL = 20 \log \frac{SWR + 1}{2\sqrt{SWR}}.$$

2.37 A lossy transmission line possesses a complex Z_0 with a small imaginary part. In most cases, this imaginary part has little imact on calculations, and can be ignored. However, if the transmission line is used to replace a lumped component (L or C), the effect on the quality factor is substantial. Let a $\lambda/8$, short circuit terminated RG-58 coaxial cable represent a lumped inductance at 1 GHz. Compute its quality factor using (a) real $Z_0 = 50$ Ω, and (b) complex Z_0, assuming cable loss is purely resistive (dielectric is lossless). The cable parameters can be found in problem 2.23.

3 The Smith Chart

3.1 From Reflection Coefficient to Load Impedance 104
    3.1.1 Reflection Coefficient in Phasor Form 104
    3.1.2 Normalized Impedance Equation 106
    3.1.3 Parametric Reflection Coefficient Equation 108
    3.1.4 Graphical Representation 110
3.2 Impedance Transformation 112
    3.2.1 Impedance Transformation for General Load 112
    3.2.2 Standing Wave Ratio 113
    3.2.3 Special Transformation Conditions 116
    3.2.4 Computer Simulations 120
3.3 Admittance Transformation 123
    3.3.1 Parametric Admittance Equation 123
    3.3.2 Additional Graphical Displays 126
3.4 Parallel and Series Connections 127
    3.4.1 Parallel Connection of R and L Elements 131
    3.4.2 Parallel Connection of R and C Elements 132
    3.4.3 Series Connection of R and L Elements 133
    3.4.4 Series Connection of R and C Elements 133
    3.4.5 Example of a T-Network 131
3.5 Summary 135

A transmission line changes its impedance depending on material properties and geometric dimensions. Typical practical realizations include microstrip line, coaxial cable, and parallel-plate line. In addition, both the length and operating frequency of the transmission line significantly influence the input impedance. In the previous chapter we derived the fundamental equation describing the input impedance of a terminated transmission line. We found that this equation incorporates the characteristic line impedance, load impedance, and, through the argument of the tangent function, line length and operating frequency. As we saw in Section 2.9, the input impedance can equivalently be evaluated by using the spatially dependent reflection coefficient. To facilitate the evaluation of the reflection coefficient, P. H. Smith developed a graphical procedure based on conformal mapping principles. This approach permits an easy and intuitive display of the reflection coefficient as well as the line impedance in one single graph. Although this graphical procedure, nowadays known as the Smith Chart, was developed in the 1930s prior to the computer age, it has retained its popularity and today can be found in every data book describing passive and active RF/MW components and systems. Almost all computer-aided design programs utilize the Smith Chart for the analysis of circuit impedances, design of matching networks, and computations of noise figures, gain, and stability circles. Even instruments such as the ubiquitous network analyzer have the option to represent certain measurements in a Smith Chart format.

This chapter reviews the steps necessary to convert the input impedance in its standard complex plane into a suitable complex reflection coefficient representation via a specific conformal transformation originally proposed by Smith. The graphical display of the reflection coefficient in this new complex plane can then be utilized directly to find the input impedance of the transmission line. Moreover, the Smith Chart facilitates evaluation of more complicated circuit configurations, which will be employed in subsequent chapters to build filters and matching networks for active devices.

The following sections present a step-by-step derivation of the Smith Chart followed by several examples of how to use this graphical design tool in computing the impedance of passive circuits.

3.1 From Reflection Coefficient to Load Impedance

In Section 2.9 the reflection coefficient is defined as the ratio of reflected voltage wave to incident voltage wave at a certain fixed spatial location along the transmission line. Of particular interest is the reflection coefficient at the load location $d = 0$. From a physical point of view, this coefficient Γ_0 describes the mismatch in impedance between the characteristic line impedance Z_0 and the load impedance Z_L as expressed by (2.52). In moving away from the load in the positive d-direction toward the beginning of the transmission line, we have to multiply Γ_0 by the exponential factor $\exp(-j2\beta d)$, as seen in (2.64), to obtain $\Gamma(d)$. This transformation from Γ_0 to $\Gamma(d)$ constitutes one of the key ingredients in the **Smith Chart** as a graphical design tool.

3.1.1 Reflection Coefficient in Phasor Form

The representation of the reflection coefficient Γ_0 can be cast in the following complex notation.

$$\Gamma_0 = \frac{Z_L - Z_0}{Z_L + Z_0} = \Gamma_{0r} + j\Gamma_{0i} = |\Gamma_0|e^{j\theta_L} \qquad (3.1)$$

where $\theta_L = \text{atan2}(\Gamma_{0i}, \Gamma_{0r})$. We recall that pure short- and open-circuit conditions in (3.1) correspond to Γ_0 values of -1 and $+1$, located on the real axis in the complex Γ-plane.

From Reflection Coefficient to Load Impedance

Example 3-1. Reflection coefficient representations

A transmission line with a characteristic line impedance of $Z_0 = 50\ \Omega$ is terminated into the following load impedances:

(a) $Z_L = 0$ (short circuit)
(b) $Z_L \rightarrow \infty$ (open circuit)
(c) $Z_L = 50\ \Omega$
(d) $Z_L = (16.67 - j16.67)\ \Omega$
(e) $Z_L = (50 + j150)\ \Omega$.

Find the individual reflection coefficients Γ_0 and display them in the complex Γ-plane.

Solution. Based on (3.1) we compute the following numbers for the reflection coefficients:

(a) $\Gamma_0 = -1$ (short circuit)
(b) $\Gamma_0 = 1$ (open circuit)
(c) $\Gamma_0 = 0$ (matched circuit)
(d) $\Gamma_0 = 0.54 \angle 221°$
(e) $\Gamma_0 = 0.83 \angle 34°$.

The values are displayed in polar form in Figure 3-1.

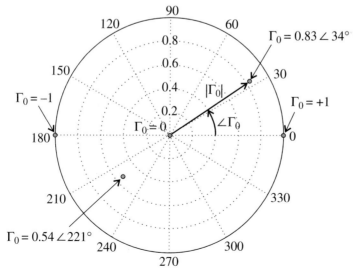

Figure 3-1 Complex Γ-plane and various locations of Γ_0.

> *The reflection coefficient is represented in phasor form as done when dealing with the conventional voltages and currents in standard circuit theory.*

3.1.2 Normalized Impedance Equation

Let us return to our general input impedance expression (2.69), into which we substitute the reflection coefficient

$$\Gamma(d) = |\Gamma_0|e^{j\theta_L}e^{-j2\beta d} = \Gamma_r + j\Gamma_i. \tag{3.2}$$

This results in

$$Z_{in}(d) = Z_0 \frac{1 + \Gamma_r + j\Gamma_i}{1 - \Gamma_r - j\Gamma_i}. \tag{3.3}$$

In order to generalize the subsequent derivations, we normalize (3.3) with respect to the characteristic line impedance as follows

$$Z_{in}(d)/Z_0 = z_{in} = r + jx = \frac{1 + \Gamma(d)}{1 - \Gamma(d)} = \frac{1 + \Gamma_r + j\Gamma_i}{1 - \Gamma_r - j\Gamma_i}. \tag{3.4}$$

The preceding equation represents a mapping from one complex plane, the z_{in}-plane, to a second complex plane, the Γ-plane. Multiplying numerator and denominator of (3.4) by the complex conjugate of the denominator allows us to isolate the real and imaginary parts of z_{in} in terms of the reflection coefficient. This means

$$z_{in} = r + jx = \frac{1 - \Gamma_r^2 - \Gamma_i^2 + 2j\Gamma_i}{(1 - \Gamma_r)^2 + \Gamma_i^2} \tag{3.5}$$

can be separated into

$$r = \frac{1 - \Gamma_r^2 - \Gamma_i^2}{(1 - \Gamma_r)^2 + \Gamma_i^2} \tag{3.6}$$

and

$$x = \frac{2\Gamma_i}{(1 - \Gamma_r)^2 + \Gamma_i^2}. \tag{3.7}$$

Equations (3.6) and (3.7) are explicit transformation rules for finding z_{in} if the reflection coefficient is specified in terms of Γ_r and Γ_i. Therefore, the mapping from the complex Γ-plane into the z_{in}-plane is straightforward, as the following example demonstrates.

Example 3-2. Input impedance of a terminated transmission line

A load impedance $Z_L = (30 + j60)\ \Omega$ is connected to a 50 Ω transmission line of 2 cm length and operated at 2 GHz. Use the reflection coefficient concept and find the input impedance Z_{in} under the assumption that the phase velocity is 50% of the speed of light.

Solution. We first determine the load reflection coefficient

$$\Gamma_0 = \frac{Z_L - Z_0}{Z_L + Z_0} = \frac{30 + j60 - 50}{30 + j60 + 50} = 0.2 + j0.6$$

$$= \sqrt{.40}\,e^{j71.56°}. \qquad (3.8)$$

Next we compute $\Gamma(d = 2\,\text{cm})$ based on the fact that

$$\beta = \frac{2\pi}{\lambda} = \frac{2\pi f}{v_p} = \frac{2\pi f}{0.5c} = 83.77\ \text{m}^{-1}.$$

This results in $2\beta d = 192.0°$, and yields for the reflection coefficient

$$\Gamma = \Gamma_0 e^{-j2\beta d} = \Gamma_r + j\Gamma_i = -0.32 - j0.55$$

$$= \sqrt{.40}\,e^{-j120.4°}.$$

Having thus determined the reflection coefficient, we can now directly find the corresponding input impedance:

$$Z_{in} = Z_0 \frac{1 + \Gamma}{1 - \Gamma} = R + jX = (14.7 - j26.7)\Omega.$$

We note that the reflection coefficient phasor form Γ_0 at the load is multiplied by a rotator that incorporates twice the electric length βd. This mathematical statement conveys the idea that voltage/current waves have to travel to the load and return back to the source to define the input impedance.

Example 3.2 could have been solved just as efficiently by using the impedance equation (2.65) developed in Section 2.9.

3.1.3 Parametric Reflection Coefficient Equation

The goal of our investigation is to pursue a different approach toward computing the input impedance. This approach involves the inversion of (3.6) and (3.7). In other words, we ask ourselves how a point in the z_{in}-domain, expressed through its normalized real r and imaginary x components, is mapped into the complex Γ-plane, where it then can be expressed in terms of the real Γ_r and imaginary Γ_i components of the reflection coefficient. Since Γ appears in the numerator and denominator, we have to suspect that straight lines in the impedance plane z_{in} may not be mapped into straight lines in the Γ-plane. All we can say at this point is that the matching of the load impedance to the transmission line impedance $Z_{in} = Z_0$, or $z_{in} = 1$, results in a zero reflection coefficient (i.e., $\Gamma_r = \Gamma_i = 0$) located in the center of the Γ-plane.

The inversion of (3.6) is accomplished by going through the following basic algebraic operations:

$$r[(1-\Gamma_r)^2 + \Gamma_i^2] = 1 - \Gamma_r^2 - \Gamma_i^2 \tag{3.9a}$$

$$\Gamma_r^2(r+1) - 2r\Gamma_r + \Gamma_i^2(r+1) = 1 - r \tag{3.9b}$$

$$\Gamma_r^2 - \frac{2r}{r+1}\Gamma_r + \Gamma_i^2 = \frac{1-r}{r+1}. \tag{3.9c}$$

At this point the trick consists of completing the square (see also Appendix C)

$$\left(\Gamma_r - \frac{r}{r+1}\right)^2 - \frac{r^2}{(r+1)^2} + \Gamma_i^2 = \frac{1-r}{r+1}. \tag{3.9d}$$

This finally can be cast in the form

$$\left(\Gamma_r - \frac{r}{r+1}\right)^2 + \Gamma_i^2 = \left(\frac{1}{r+1}\right)^2. \tag{3.10}$$

In a similar way, we proceed to invert (3.7). The result for the normalized reactance is

$$(\Gamma_r - 1)^2 + \left(\Gamma_i - \frac{1}{x}\right)^2 = \left(\frac{1}{x}\right)^2. \tag{3.11}$$

Both (3.10) and (3.11) are parametric equations of circles in the complex Γ-plane that can be written in the generic form $(\Gamma_r - a)^2 + (\Gamma_i - b)^2 = c^2$. Here, a and b denote the center of the circle, and c the radius in the complex Γ-plane.

Figure 3-2 depicts the parametric circle equations of (3.10) for various resistances. For example, if the normalized resistance r is zero, the circle is centered at the origin and possesses a radius of 1, since (3.10) reduces to $\Gamma_r^2 + \Gamma_i^2 = 1$. For $r = 1$ we find $(\Gamma_r - 1/2)^2 + \Gamma_i^2 = (1/2)^2$, which represents a circle of radius 1/2 shifted in the positive Γ_r direction by 1/2 units. We conclude that as r increases, the radii of the circles are

From Reflection Coefficient to Load Impedance

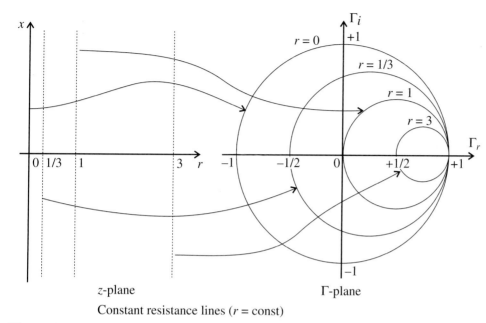

Constant resistance lines (r = const)

Figure 3-2 Parametric representation of the normalized resistance r in the complex Γ-plane.

continually reduced and the centers shifted further to the right toward the point 1 on the real axis. In the limit for $r \to \infty$ we see that the shift converges to the point $r/(r+1) \to 1$ and the circle radius approaches $1/(r+1) \to 0$.

It is important to realize that this mapping transforms fixed values of r only and does not involve x. Thus, for a fixed r an infinite range of reactance values x, as indicated by the straight lines in the z-plane, maps onto the same resistance circle. The mapping involving r alone is therefore not a unique point-to-point correspondence.

A different graphical display results for the circle equation (3.11), which involves the normalized reactance. Here, the centers of the circles reside all along a line perpendicular to the real axis and intersecting at the $\Gamma_r = 1$ point. For instance, for $x = \infty$ we note that $(\Gamma_r - 1)^2 + \Gamma_i^2 = 0$, which is a circle of zero radius, or a point located at $\Gamma_r = 1$ and $\Gamma_i = 0$. For $x = 1$ we see that the circle equation becomes $(\Gamma_r - 1)^2 + (\Gamma_i - 1)^2 = 1$. As $x \to 0$ the radii and shifts along the positive imaginary axis approach infinity. Interestingly, the shifts can also be along the negative imaginary axis. Here, for $x = -1$ we notice that the circle equation becomes $(\Gamma_r - 1)^2 + (\Gamma_i + 1)^2 = 1$ with the center located at $\Gamma_r = 1$ and $\Gamma_i = -1$. We observe that negative x-values refer to capacitive impedances residing in the lower half of the Γ-plane. Figure 3-3 shows the parametric form of the normalized imaginary impedance. For better readability, the circles are displayed inside the unit circle only.

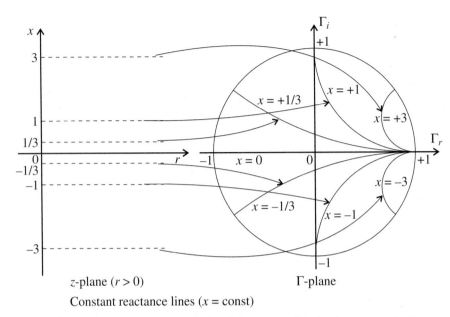

Figure 3-3 Parametric representation of the normalized reactance x in the complex Γ-plane.

In contrast to Figure 3-2, we notice that fixed x-values are mapped into circles in the Γ-plane for arbitrary resistance values $0 \leq r < \infty$, as indicated by the straight lines in the impedance plane.

The transformations (3.10) and (3.11) taken individually do not constitute unique mappings from the normalized impedance into the reflection coefficient plane. In other words, impedance points mapped into the Γ-plane by either (3.10) or (3.11) cannot uniquely be inverted back into the original impedance points. However, since the transformations complement each other, a unique mapping can be constructed by combining both transformations, as discussed in the next section.

3.1.4 Graphical Representation

Combining the parametric representations for normalized resistance and reactance circles (i.e., Figures 3-2 and 3-3) for $|\Gamma| \leq 1$ results in the Smith Chart as illustrated in Figure 3-4. An important observation of the Smith Chart is that there is a *one-to-one mapping* between the normalized impedance plane and the reflection coefficient plane. We also notice that the normalized resistance circles r have a range $0 \leq r < \infty$ and the normalized reactance circles x can represent either negative (i.e., capacitive) or positive (i.e., inductive) values in the range $-\infty < x < +\infty$.

It should be pointed out that the reflection coefficient does not have to satisfy $|\Gamma| \leq 1$. Negative resistances, encountered for instance in oscillators, lead to the case $|\Gamma| > 1$ and consequently map to points residing outside the unit circle. Graphical displays where the reflection coefficient is greater than 1 are known as **compressed Smith Charts**. These charts, however, play a rather limited role in RF/MW engineering designs and are therefore not further pursued in this text. The interested reader may consult specialized literature (see the Hewlett-Packard application note listed at the end of this chapter).

In Figure 3-4 we must note that the angle of rotation $2\beta d$ introduced by the length of the transmission line is measured from the phasor location of $\Gamma_0 = |\Gamma_0| e^{j\theta_L}$ in clockwise (mathematically negative) direction due to the negative exponent $(-2j\beta d)$ in the reflection coefficient expression (3.2). For the computation of the input impedance of a terminated transmission line, the motion is thus always *away from the load impedance* or *toward the generator*. This rotation is indicated by an arrow on the periphery of the chart. We further observe that a complete revolution around the unit circle requires

$$2\beta d = 2\frac{2\pi}{\lambda}d = 2\pi$$

where $d = \lambda/2$ or $\beta d = 180°$. The quantity βd is sometimes referred to as the **electrical length** of the line.

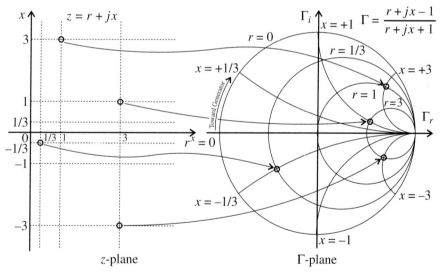

Figure 3-4 Smith Chart representation formed by combining r and x circles for $|\Gamma| \leq 1$.

3.2 Impedance Transformation

3.2.1 Impedance Transformation for General Load

The determination of the impedance response of a high-frequency circuit is often a critical issue for the RF design engineer. Without detailed knowledge of the impedance behavior, RF/MW system performance cannot adequately be predicted. In this section we will elaborate on how the impedance can be determined easily and efficiently with the aid of the previously introduced Smith Chart.

A typical Smith Chart computation involving a load impedance Z_L connected to a transmission line of characteristic line impedance Z_0 and length d proceeds according to the following six steps:

1. Normalize the load impedance Z_L with respect to the line impedance Z_0 to determine z_L.
2. Locate z_L in the Smith Chart.
3. Identify the corresponding load reflection coefficient Γ_0 in the Smith Chart both in terms of its magnitude and phase.
4. Rotate Γ_0 by twice its electrical length βd to obtain $\Gamma_{in}(d)$.
5. Record the normalized input impedance z_{in} at this spatial location d.
6. Convert z_{in} into the actual impedance Z_{in}.

Example 3-3 goes through these steps, which are the standard procedure to arrive at the graphical impedance solution.

Example 3-3. Transmission line input impedance determination with the Smith Chart

Solve Example 3-2 by following the six-step Smith Chart computation given in the preceding list.

Solution. We commence with the load impedance $Z_L = (30 + j60)\ \Omega$ and proceed according to the previously outlined steps:

1. The normalized load impedance is

$$z_L = (30 + j60)/50 = 0.6 + j1.2.$$

2. This point can be identified in the Smith Chart as the intersection of the circle of constant resistance $r = 0.6$ with the circle of constant reactance $x = 1.2$, as seen in Figure 3-5.

3. The straight line connecting the origin to point z_L determines the load reflection coefficient Γ_0. The associated angle is recorded with respect to the positive real axis.
4. Keeping in mind that the outside circle on the Smith Chart corresponds to the unity reflection coefficient ($|\Gamma_0| = 1$), we can find its magnitude as the length of the vector connecting the origin to z_L. Rotating this vector by twice the electrical length of the line (i.e., $2 \times \beta d = 2 \times 96° = 192°$) yields the input reflection coefficient Γ_{in}.
5. This point uniquely identifies the associated normalized input impedance $z_{in} = 0.3 - j0.53$.
6. The preceding normalized impedance can be converted back into actual input impedance values by multiplying it by $Z_0 = 50 \, \Omega$, resulting in the final solution $Z_{in} = (15 - j26.5) \, \Omega$.

We recall that the exact value of the input impedance obtained in Example 3-2 is $(14.7 - j26.7) \, \Omega$. The small discrepancy is understandable because of the approximate processing of the graphical data in the Smith Chart. The entire sequence of steps leading to the determination of the input impedance of the line connected to the load is shown in Figure 3-5.

These steps appear at first cumbersome and prone to error if carried out by hand. However, using mathematical spreadsheets and relying on computer-based instrumentation, the calculations are routinely done in seconds and with a high degree of accuracy.

SMITH CHART DISPLAY FORMAT FOR MEASUREMENT

The network analyzer (see also Practically Speaking at the end of this Chapter), one of the most important RF measurement devices, features prominently a Smith Chart display. This enables a design engineer to directly determine the circuit's inductive or capacitive behavior and its degree of matching.

3.2.2 Standing Wave Ratio

From the basic definition of the SWR in Section 2.8.3, it follows that for an arbitrary distance d along the transmission line, the standing wave ratio is written

$$\text{SWR} = \frac{1 + |\Gamma(d)|}{1 - |\Gamma(d)|} \quad (3.12)$$

where $\Gamma(d) = \Gamma_0 \exp(-j2\beta d)$. Equation (3.12) can be inverted to give

$$|\Gamma(d)| = \frac{\text{SWR} - 1}{\text{SWR} + 1}. \quad (3.13)$$

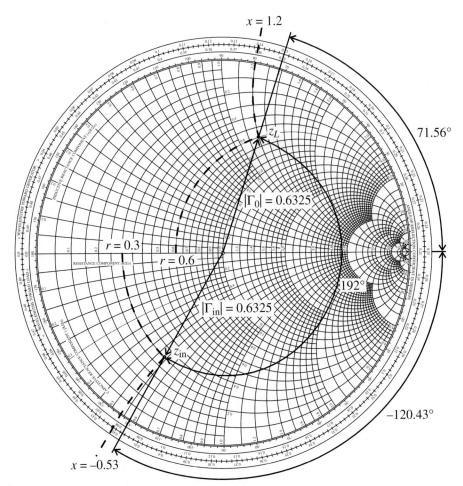

Figure 3-5 Usage of the Smith Chart to determine the input impedance for Example 3-3.

This form of the reflection coefficient permits the representation of the SWR as circles in the Smith Chart with the matched condition $\Gamma(d) = 0$ (or SWR = 1) being the origin.

It is interesting to note that equation (3.12) is very similar in appearance to the expression for determining the impedance from a given reflection coefficient:

$$Z(d) = Z_0 \frac{1 + \Gamma(d)}{1 - \Gamma(d)}. \qquad (3.14)$$

This similarity, together with the fact that for $|\Gamma(d)| \leq 1$ the SWR is greater or equal to unity, suggests that the actual numerical value for the SWR can be found from the Smith

Impedance Transformation

Chart. The r-value of the intersection of the circle of radius $|\Gamma(d)|$ with the right-hand side of the real axis indicates the SWR.

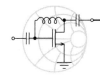

Example 3-4. Reflection coefficient, voltage standing wave ratio, and return loss

—RF&MW→

Four different load impedances, (a) $Z_L = 50\ \Omega$, (b) $Z_L = 48.5\ \Omega$, (c) $Z_L = (75 + j25)\ \Omega$, and (d) $Z_L = (10 - j5)\ \Omega$, are sequentially connected to a $50\ \Omega$ transmission line. Find the reflection coefficients and the SWR circles, and determine the return loss in dB.

Solution. The normalized load impedances and corresponding reflection coefficients, return loss, and SWR values are computed as follows:

(a) $z_L = 1$, $\Gamma = (z_L - 1)/(z_L + 1) = 0$, RL = ∞ dB, SWR = 1
(b) $z_L = 0.97$, $\Gamma = (z_L - 1)/(z_L + 1) = -0.015$, RL = 36.3 dB, SWR = 1.03
(c) $z_L = 1.5 + j0.5$, $\Gamma = (z_L - 1)/(z_L + 1) = 0.23 + j0.15$, RL = 11.1 dB, SWR = 1.77
(d) $z_L = 0.2 - j0.1$,
$\Gamma = (z_L - 1)/(z_L + 1) = -0.66 - j0.14$, RL = 3.5 dB, SWR = 5.05.

To determine the approximate values of the SWR, we can exploit the similarity with the input impedance, as discussed previously. To this end, we first plot the normalized impedance values in the Smith Chart (see Figure 3-6). Then we draw circles with centers at the origin and radii whose lengths reach the respective impedance points defined in the previous step. From these circles we see that the load refection coefficient for zero load reactance $(x_L = 0)$ is

$$\Gamma_0 = \frac{z_L - 1}{z_L + 1} = \frac{r_L - 1}{r_L + 1} = \Gamma_r.$$

The SWR can be defined in term of the real load reflection coefficient along the real Γ-axis:

$$\text{SWR} = \frac{1 + |\Gamma_0|}{1 - |\Gamma_0|} = \frac{1 + \Gamma_r}{1 - \Gamma_r}.$$

> **VSWR SPECIFICATION FOR AMPLIFIER**
>
> In Chapter 2 the VSWR of the input and output ports of a monolithic microwave integrated circuit (MMIC) was given as 1.4 and 1.6, respectively. Figure 3-6 makes clear how this directly translates into normalized impedances. For instance, if a normalization of $50\ \Omega$ is chosen, then an output impedance of $(75-j15)\ \Omega$ would meet the VSWR requirement of 1.6.

This requires $|\Gamma_0| = \Gamma_r \geq 0$. In other words, for $\Gamma_r \geq 0$ we have to enforce $r_L \geq 1$, meaning that only the intersections of the circles with the right-hand-side of the real axis define the SWR.

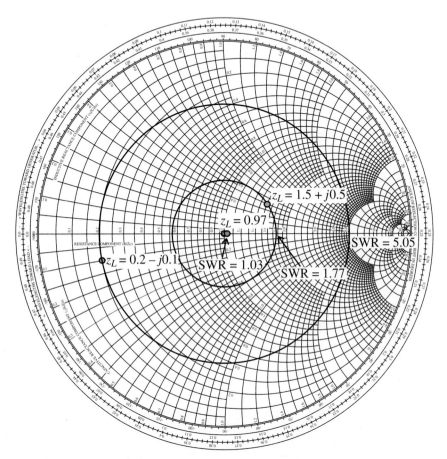

Figure 3-6 SWR circles for various reflection coefficients.

As a graphical design tool, the Smith Chart allows immediate observation of the degree of mismatch between line and load impedances by plotting the SWR circle.

3.2.3 Special Transformation Conditions

The angle by which the point of the normalized transmission line impedance revolves around the Smith Chart is controlled by the length of the line, or alternatively the operating frequency. Consequently, both inductive (upper plane) and capacitive (lower plane)

Impedance Transformation

impedances can be generated based on the line length and the termination conditions at a given frequency. These lumped circuit parameter representations, realized through distributed circuit analysis techniques, are of significant practical importance.

The cases of open- and short-circuit line termination are of particular interest in generating inductive and capacitive behavior and are examined in more detail next.

Open-Circuit Transformations

To obtain a pure inductive or capacitive impedance behavior, we need to operate along the $r = 0$ circle. The starting point is the right-hand location ($\Gamma_0 = 1$) with rotation toward the generator in the clockwise direction.

A capacitive impedance $-jX_C$ is obtained through the condition

$$\frac{1}{j\omega C}\frac{1}{Z_0} \equiv z_{\text{in}} = -j\cot(\beta d_1) \tag{3.15}$$

as direct comparison with (2.70) shows. The line length d_1 is found to be

$$d_1 = \frac{1}{\beta}\left[\cot^{-1}\left(\frac{1}{\omega C Z_0}\right) + n\pi\right] \tag{3.16}$$

where $n\pi$ ($n = 0, 1, 2, \ldots$) is required due to the periodicity of the cotangent function. Alternatively, an inductive impedance jX_L can be realized via the condition

$$j\omega L \frac{1}{Z_0} \equiv z_{\text{in}} = -j\cot(\beta d_2). \tag{3.17}$$

The line length d_2 is now found to be

$$d_2 = \frac{1}{\beta}\left[\pi - \cot^{-1}\left(\frac{\omega L}{Z_0}\right) + n\pi\right]. \tag{3.18}$$

Both conditions are schematically depicted in Figure 3-7. How to choose a particular open-circuit line length to exhibit capacitive or inductive behavior is discussed in the following example.

Example 3-5. Representation of passive circuit elements through transmission line sections

For an open-circuited 50 Ω transmission line operated at 3 GHz and with a phase velocity of 77% of the speed of light, find the line lengths to create a 2 pF capacitor and a 5.3 nH inductor. Perform your computations both by relying on (3.16) and (3.18) and by using the Smith Chart.

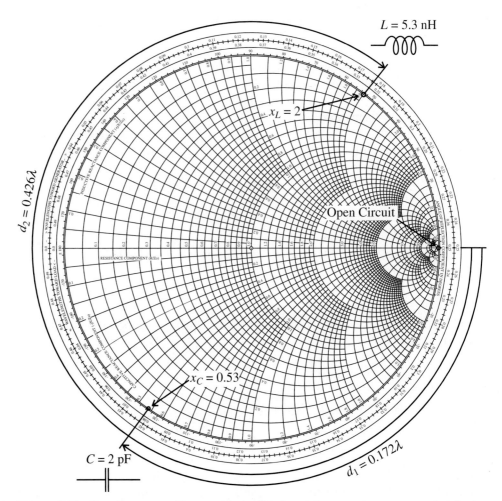

Figure 3-7 Creating capacitive and inductive impedances via an open-circuited transmission line at 3 GHz.

Solution. For the given value of phase velocity, the propagation constant is

$$\beta = 2\pi f/v_p = 2\pi f/(0.77c) = 81.6 \text{ m}^{-1}.$$

Substituting this value into (3.16) and (3.18), we conclude that for the representation of a 2 pF capacitor we need an open-circuited line or stub with line length $d_1 = 13.27 + n38.5$ mm. For the realization of a 5.3 nH inductor, a $d_2 = 32.81 + n38.5$ mm stub is required.

The alternative method for computing the lengths of the required stubs is through the use of the Smith Chart (see Figure 3-7). At 3 GHz, the reactance of a 2 pF capacitor is $X_C = 1/(\omega C) = 26.5\ \Omega$. The corresponding normalized impedance in this case is $z_C = -jX_C = -j0.53$. From the Smith Chart we can deduce that the required transmission line length has to be approximately 0.172 of one wavelength. We note that for the given phase velocity, the wavelength is $\lambda = v_p/f = 77$ mm. This results in a line length of $d_1 = 13.24$ mm which is very close to the previously computed value of 13.27 mm. Similarly, for the inductance we obtain $z_L = j2$. The line length in this case is 0.426 of one wavelength, which is equal to 32.8 mm.

Circuits are often designed with lumped elements before converting them into transmission line segments, similarly to the procedure described in this example.

Short-Circuit Transformations

Here, the transformation rules follow similar procedures as outlined previously, except that the starting point in the Smith Chart is now the $\Gamma_0 = -1$ point on the real axis, as indicated in Figure 3-8.

A capacitive impedance $-jX_C$ follows from the condition

$$\frac{1}{j\omega C}\frac{1}{Z_0} \equiv z_{\text{in}} = j\tan(\beta d_1) \qquad (3.19)$$

where use is made of (2.66). The line length d_1 is

$$d_1 = \frac{1}{\beta}\left[\pi - \tan^{-1}\left(\frac{1}{\omega C Z_0}\right) + n\pi\right]. \qquad (3.20)$$

Alternatively, an inductive impedance jX_L can be realized via the condition

$$j\omega L \frac{1}{Z_0} \equiv z_{\text{in}} = j\tan(\beta d_2). \qquad (3.21)$$

The line length d_2 is now

$$d_2 = \frac{1}{\beta}\left[\tan^{-1}\left(\frac{\omega L}{Z_0}\right) + n\pi\right]. \qquad (3.22)$$

At high frequencies, it is difficult to maintain perfect open-circuit conditions because of changing temperatures, humidity, and other parameters of the medium surrounding the open transmission line. For this reason, short-circuit conditions are more preferable in

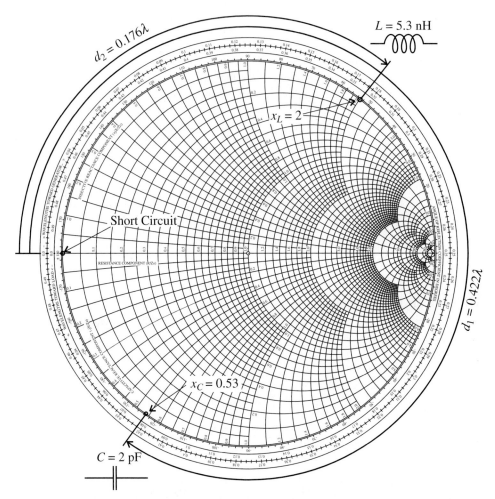

Figure 3-8 Creating capacitive and inductive impedances via a short-circuited transmission line at 3 GHz.

practical applications. However, even a short-circuit termination becomes problematic at very high frequencies. When through-hole connections in printed circuit boards are involved, they result in additional parasitic inductances. Moreover, a design engineer may not have a choice if the circuit layout area is to be minimized by requiring the selection of the shortest line segments. For instance, the realization of a capacitor always yields the shortest length for an open-circuited line.

3.2.4 Computer Simulations

There are many **computer aided design** (CAD) programs available to facilitate the RF/MW circuit design and simulation processes. These programs can perform a multitude of tasks,

Impedance Transformation

varying from simple impedance calculations to complex circuit optimizations and circuit board layouts. One commercial software package that is used throughout this textbook is called Advanced Design System **(ADS)** by Agilent Technology, Westlake Village, CA, USA, which is capable of performing linear as well as nonlinear analyses and optimizations.

It is not the purpose of this textbook to review and discuss the various CAD programs presently in industrial and academic use. However, to reproduce the subsequent simulation results, Appendix I provides a brief introduction to the basic features of Matlab, which was chosen as a tool to carry out most simulations presented in this book.

The main reason for using standard Matlab (without toolboxes) is its widespread use as a mathematical spreadsheet, which permits easy programming and direct graphical display. This eliminates the need to rely on complex and expensive programs accessible to only a few readers. The benefit of a Matlab routine will immediately become apparent when Smith Chart computations have to be performed repetitively for a range of operating frequencies or line lengths as the following discussion underscores.

In this section we revisit Example 3-2, which computed the input reflection coefficient and input impedance of a generic transmission line connected to a load. We now extend this example beyond a single operating frequency and fixed line length. Our goal is to examine the effect of a frequency sweep in the range from 0.1 GHz to 3 GHz and a change in line length from 0.1 cm to 3 cm. The example Matlab routine, which performs the analysis of the transmission line length changing from 0.1 cm to 3 cm at a fixed operating frequency 2 GHz, is as follows:

```
smith_chart;              % plot smith chart
Set_Z0(50);               % set characteristic impedance to 50 Ohm
s_Load(30+j*60);          % set load impedance to 30+j60 Ohm
vp=0.5*3e8;               % compute phase velocity
f=2e9;                    % set frequency to 2 GHz
d=0.0:0.001:0.03;         % set the line length to a range from 0 to
                          % 3 cm in 1 mm increments
beta=2*pi*f/vp;           % compute propagation constant
Gamma=(ZL-Z0)/(ZL+Z0);    % compute load reflection coefficient
rd=abs(Gamma);            % magnitude of the reflection coefficient
alpha=angle(Gamma)-2*beta*d;   % phase of the reflection
                               % coefficient
plot(rd*cos(alpha),rd*sin(alpha));  % plot the graph
```

In the first line of the Matlab code (see file fig3_9.m on the website) we generate the Smith Chart with the necessary resistance and reactance circles. The next lines define the characteristic line impedance $Z_0 = 50 \; \Omega$, load impedance $Z_L = (30 + j60) \; \Omega$, operating frequency $f = 2$ GHz, and phase velocity $v_p = 0.5 \times 3 \times 10^8$ m/s. The line d=0.0:0.001:0.03 creates an array d representing the transmission line length, which is varied from 0 mm to 3 cm in 1 mm increments. After all parameters have been identified, the magnitude and phase of the input reflection coefficients have to be computed. This is accomplished by determining the propagation constant $\beta = 2\pi f/v_p$, load reflection coefficient $\Gamma_0 = (Z_L - Z_0)/(Z_L + Z_0)$ and its magnitude $|\Gamma_0|$, and the total angle of rotation $\alpha = \angle(\Gamma_0) - 2\beta d$. Finally, the display of the impedance as part of the Smith Chart is done through the plot command, which requires both real and imaginary phasor arguments $|\Gamma_0|\cos(\alpha)$ and $|\Gamma_0|\sin(\alpha)$. The final result is shown in Figure 3-9.

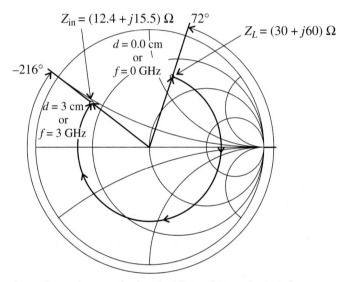

Figure 3-9 Input impedance of a loaded line of 2 cm length for a sweep in operating frequency from 0.0 to 3 GHz. If the frequency is fixed at 2 GHz and the line length is varied from 0.0 to 3 cm, the same impedance curve is obtained.

For the case where the length of the line is fixed at 2 cm and the frequency is swept from values ranging from 0.0 to 3 GHz, the only necessary modification to the above input file is to set d=0.02, followed by specifying the frequency range in increments of 10 MHz (i.e., f=0.0:1e7:3e9). We should note that in both cases, the electrical length βd of the line changes from $0°$ to $144°$. Therefore, the impedance graphs produced for both cases are identical.

At the end of the rotation, either by fixing the frequency and varying the length or vice versa, the input impedance is found to be $Z_{in} = (12.4 + j15.5)\,\Omega$. It is reassuring that for a fixed frequency $f = 2$ GHz and a line length range $d = 0 \ldots 2$ cm, we ultimately arrive at the same input impedance of $Z_{in} = (14.7 - j26.7)\,\Omega$ as obtained in Example 3-2.

3.3 Admittance Transformation

3.3.1 Parametric Admittance Equation

From the representation of the normalized input impedance (3.4), it is possible to obtain a normalized admittance equation by simple inversion:

$$y_{in} = \frac{Y_{in}}{Y_0} = \frac{1}{z_{in}} = \frac{1-\Gamma(d)}{1+\Gamma(d)} \quad (3.23)$$

where $Y_0 = 1/Z_0$. To represent (3.23) graphically in the Smith Chart, we have several options. A very intuitive way of displaying admittances in the conventional Smith Chart or **Z-Smith Chart** is to recognize that (3.23) can be found from the standard representation (3.4) via

$$\frac{1-\Gamma(d)}{1+\Gamma(d)} = \frac{1+e^{-j\pi}\Gamma(d)}{1-e^{-j\pi}\Gamma(d)}. \quad (3.24)$$

In other words, we take the normalized input impedance representation and multiply the reflection coefficient by $-1 = e^{-j\pi}$, which is equivalent to a 180° rotation in the complex Γ-plane.

Example 3-6. Use of the Smith Chart for converting impedance to admittance

Convert the normalized input impedance $z_{in} = 1+j1 = \sqrt{2}e^{j(\pi/4)}$ into normalized admittance and display it in the Smith Chart.

Solution. The admittance can be found by direct inversion, that is

$$y_{in} = \frac{1}{\sqrt{2}}e^{-j(\pi/4)} = \frac{1}{2} - j\frac{1}{2}.$$

In the Smith Chart we simply rotate the reflection coefficient corresponding to z_{in} by 180° to obtain the impedance. Its *numerical* value is equal to y_{in}, as shown in Figure 3-10. To denormalize y_{in}, we multiply by the inverse of the impedance normalization factor. Thus,

$$Y_{in} = \frac{1}{Z_0}y_{in} = Y_0 y_{in}.$$

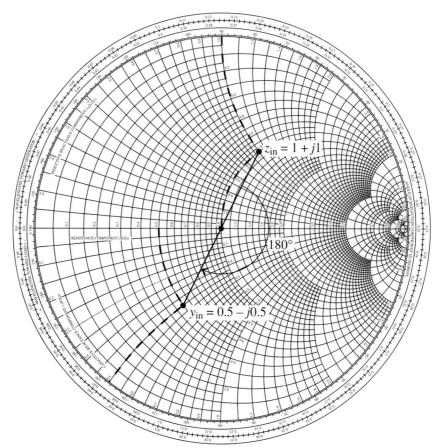

Figure 3-10 Conversion from impedance to admittance by 180° rotation.

> *Rotations by 180 degrees to convert from the impedance to the admittance representation require only a reflection about the origin in the Γ-plane.*

In addition to the preceding operation, there is a widely used additional possibility. Instead of rotating the reflection coefficient by $180°$ in the Z-Smith Chart, we can rotate the Smith Chart itself. The chart obtained by this transformation is called the **admittance Smith Chart** or the **Y-Smith Chart.** The correspondences are such that normalized resistances become normalized conductances and normalized reactances become normalized susceptances. That is,

$$r = \frac{R}{Z_0} \Rightarrow g = \frac{G}{Y_0} = Z_0 G$$

and

$$x = \frac{X}{Z_0} \Rightarrow b = \frac{B}{Y_0} = Z_0 B.$$

This reinterpretation is depicted in Figure 3-11 for a particular normalized impedance point $z = 0.6 + j1.2$.

As seen in Figure 3-11, the transformation preserves (a) the direction in which the angle of the reflection coefficient is measured, and (b) the direction of rotation (either toward or away from the generator). Attention has to be paid to the proper identification of the extreme points: a short-circuit condition $z_L = 0$ in the Z-Smith Chart is $y_L = \infty$ in the Y-Smith Chart, and conversely an open-circuit $z_L = \infty$ in the Z-Smith Chart is $y_L = 0$ in the Y-Smith Chart. Furthermore, negative values of susceptance are now plotted in the upper half of the chart, corresponding to inductive behavior, and positive values in the bottom half, corresponding to capacitive behavior. The real component of the admittance increases from right to left.

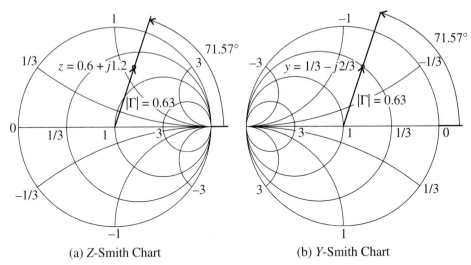

(a) Z-Smith Chart (b) Y-Smith Chart

Figure 3-11 Reinterpretation of the Z-Smith Chart as a Y-Smith Chart.

To complete our discussion of the Y-Smith Chart, we should mention an additional, often employed definition of the admittance chart. Here, the admittance is represented in exactly the same manner as the impedance chart without a 180° rotation. In this case, the reflection coefficient's phase angle is measured from the opposite end of the chart (see the book by Gonzalez listed in Further Reading at the end of this chapter).

3.3.2 Additional Graphical Displays

In many practical design applications, it is necessary to switch frequently from impedance to admittance representations and vice versa. To deal with those situations, a combined, or so-called **ZY-Smith Chart**, can be obtained by overlaying the Z- and Y-Smith Charts, as shown in Figure 3-12.

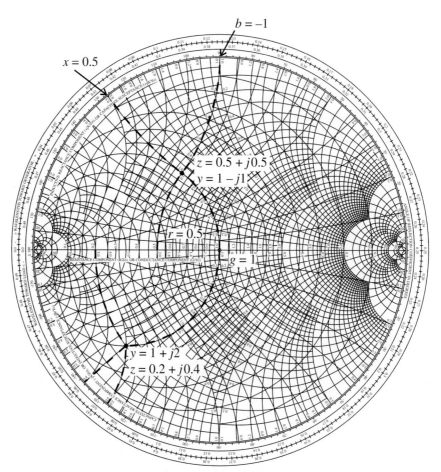

Figure 3-12 The ZY-Smith Chart superimposes the Z- and Y-Smith Charts in one graphical display.

This combined ZY-Smith Chart allows direct conversion between impedances and admittances. In other words, a point in this combined chart has two interpretations depending on whether the Z-Chart or Y-Chart display is chosen.

Example 3-7. Use of the combined ZY-Smith Chart

Identify (a) the normalized impedance value $z = 0.5 + j0.5$, and (b) the normalized admittance value $y = 1 + j2$ in the combined ZY-Smith Chart and find the corresponding values of normalized admittance and impedance.

Solution. Let us first consider the normalized impedance value $z = 0.5 + j0.5$. In the combined ZY-Smith Chart we locate the impedance by using circles of constant resistance $r = 0.5$ and constant reactance $x = 0.5$, as shown in Figure 3-12. The intersection of these two circles determines the specified impedance value $z = 0.5 + j0.5$. To find the corresponding admittance value, we simply move along the circles of constant conductance g and susceptance b. The intersection gives us $g = 1$ and $b = -1$ (i.e., the admittance for part (a) of this example is $y = 1 - j1$). The solution for the normalized admittance $y = 1 + j2$ is obtained in identical fashion and is also illustrated in Figure 3-12.

The ZY-Smith Chart requires a fair amount of practice due to its "busy" appearance and the fact that inductors and capacitors are counted either in positive or negative units depending on whether an impedance or admittance representation is needed.

3.4 Parallel and Series Connections

In the following sections, several basic circuit element configurations are analyzed, and their impedance responses are displayed in the Smith Chart as functions of frequency. The aim is to develop insight into how the impedance/admittance behaves over a range of frequencies for different combinations of lumped circuit parameters. A practical understanding of these circuit responses is needed later in the design of matching networks (see Chapter 8) and in the development of equivalent circuit models.

3.4.1 Parallel Connection of *R* and *L* Elements

With reference to Figure 3-13, we recognize that $g = Z_0/R$ and $b_L = -Z_0/(\omega L)$. We can locate the normalized admittance value in the upper Y-Smith Chart plane for a fixed normalized conductance g at a certain angular frequency ω_L:

$$y_{in}(\omega_L) = g - j\frac{Z_0}{\omega_L L}. \tag{3.25}$$

As the angular frequency is increased to the upper limit ω_U, we trace out a curve along the constant conductance circle g. Figure 3-13 schematically shows the frequency-dependent admittance behavior for various constant conductance values $g = 0.3, 0.5, 0.7$, and 1, and for frequencies ranging from 500 MHz to 4 GHz. For a fixed inductance value of $L = 10$ nH and a characteristic line impedance $Z_0 = 50\ \Omega$, the susceptance always starts at -1.59 (500 MHz) and ends at -0.20 (4 GHz).

In Figure 3-13 and the following three additional cases, the transmission line characteristic impedance is represented as a lumped impedance of $Z_0 = 50\ \Omega$. This is permissible, since our interest is focused on the impedance and admittance behavior of different load configurations. For these cases the characteristic line impedance serves only as a normalization factor.

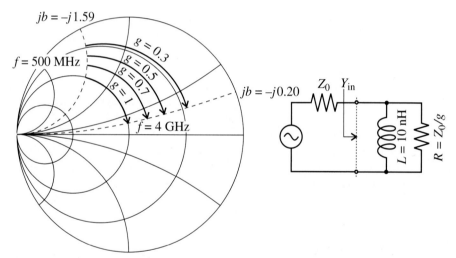

Figure 3-13 Admittance response of parallel RL circuit for $\omega_L \leq \omega \leq \omega_U$ at constant conductances $g = 0.3, 0.5, 0.7$, and 1.

3.4.2 Parallel Connection of R and C Elements

Here, we operate in the lower Y-Chart plane because susceptance $b_C = Z_0 \omega C$ remains positive. To locate the normalized admittance value for a fixed normalized conductance g and angular frequency ω_L, we have

$$y_{in}(\omega_L) = g + jZ_0\omega_L C. \tag{3.26}$$

Figure 3-14 depicts the frequency-dependent admittance behavior as a function of various constant conductance values $g = 0.3, 0.5, 0.7$, and 1. The normalized susceptance for $C = 1$ pF and characteristic line impedance $Z_0 = 50\ \Omega$ always starts at 0.16 (500 MHz) and ends at 1.26 (4 GHz).

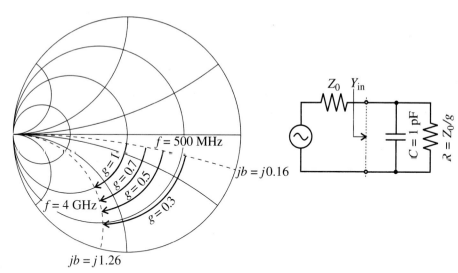

Figure 3-14 Admittance response of parallel RC circuit for $\omega_L \leq \omega \leq \omega_U$ at constant conductances g = 0.3, 0.5, 0.7, and 1.

3.4.3 Series Connection of R and L Elements

When dealing with series connections, we can conveniently choose the Z-Smith Chart for the impedance display. Identifying the normalized reactive component as $x_L = \omega L/Z_0$, it is straightforward to locate the normalized impedance value for a particular, fixed normalized resistance r at a given angular frequency ω_L:

$$z_{\text{in}}(\omega_L) = r + j\omega_L L/Z_0. \tag{3.27}$$

In Figure 3-15, the frequency-dependent impedance behavior is shown as a function of various constant resistance values r = 0.3, 0.5, 0.7, and 1. For the same inductance of 10 nH and characteristic line impedance of 50 Ω as used in Figure 3-13, we now select the reactance circles x = 0.63 (500 MHz) and x = 5.03 (4 GHz). Because the reactance is positive and since we use the Z-Smith Chart, all impedances have to reside in the upper halfplane.

3.4.4 Series Connection of R and C Elements

We again choose the Z-Smith Chart for the impedance display. The normalized reactive component is $x_C = -1/(\omega C Z_0)$, indicating that all curves will reside in the lower half of the Smith Chart. The normalized impedance value for a particular, fixed normalized resistance r at an angular frequency ω_L reads

$$z_{\text{in}}(\omega_L) = r - j\frac{1}{\omega_L C Z_0}. \tag{3.28}$$

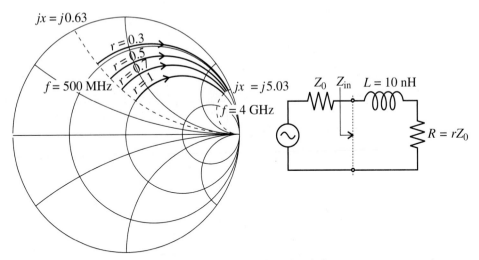

Figure 3-15 Impedance response of series *RL* circuit for $\omega_L \leq \omega \leq \omega_U$ and constant resistances r = 0.3, 0.5, 0.7, and 1.

Figure 3-16 displays the frequency-dependent impedance behavior as a function of various constant resistance values r = 0.3, 0.5, 0.7, and 1. The capacitance of 1 pF in series with the variable resistance connected to a characteristic line impedance of 50 Ω now uses the reactance circles $x = -6.03$ (500 MHz) and $x = -0.8$ (4 GHz), which intersect with the four resistance circles, uniquely determining upper and lower impedance values.

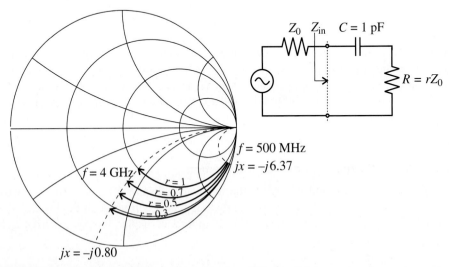

Figure 3-16 Impedance response of series *RC* circuit for $\omega_L \leq \omega \leq \omega_U$ at constant resistances r = 0.3, 0.5, 0.7, and 1.

3.4.5 Example of a T-Network

In the previous examples, only pure series or shunt configurations have been analyzed. In reality, however, one often encounters combinations of both. To show how easily the ZY-Smith Chart allows transitions between series and shunt connections, let us investigate by way of an example the behavior of a **T-type network** connected to the input of a bipolar transistor. The input port of the transistor is modeled as a parallel RC network as depicted in Figure 3-17. As we will see in Chapter 6, R_L approximates the base-emitter resistance and C_L is the base-emitter junction capacitance. The numerical parameter values are listed in Figure 3-17.

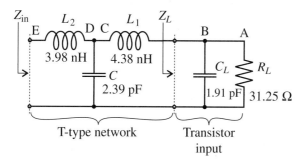

Figure 3-17 T-network connected to the base-emitter input impedance of a bipolar transistor.

To use the Smith Chart for the computation of the input impedance of this more complicated network, we first analyze this circuit at 2 GHz and then show the entire response of the circuit for a frequency range from 500 MHz to 4 GHz by employing the commercial MMICAD software simulation package.

To obtain the load impedance, or the input impedance of the transistor, we use the Y-Smith Chart to identify the conductance point corresponding to the load resistor $R_L = 31.25\ \Omega$. Assuming a 50 Ω characteristic line impedance, we determine the normalized admittance for this case to be $g_A = 1.6$, which corresponds to point A in Figure 3-18.

The next step is to connect the capacitance $C_L = 1.91$ pF in shunt with the resistor R_L. At the angular frequency of $\omega_L = 2\pi 2 \times 10^9$ s^{-1}, the susceptance of this capacitor becomes $B_{C_L} = \omega_L C_L = 24$ mS, which corresponds to a rotation of the original point A into the new location B. The amount of rotation is determined by the normalized susceptance of the capacitor $b_{C_L} = B_{C_L} Z_0 = 1.2$ and is carried out along the circle of constant conductance in the Y-Smith Chart (see Figure 3-18).

Re-evaluating point B in the Z-Smith Chart, we obtain the normalized impedance of the parallel combination of resistor R_L and capacitor C_L to be $z_B = 0.4 - j0.3$. The series connection of the inductance L_1 results in the new location C. This point is obtained through a rotation from $x_B = -0.3$ by an amount $x_{L_1} = \omega_L L_1/Z_0 = 1.1$ to $x_C = 0.8$ along the circle of constant resistance $r = 0.4$ in the Z-Smith Chart as discussed in Section 3.4.3.

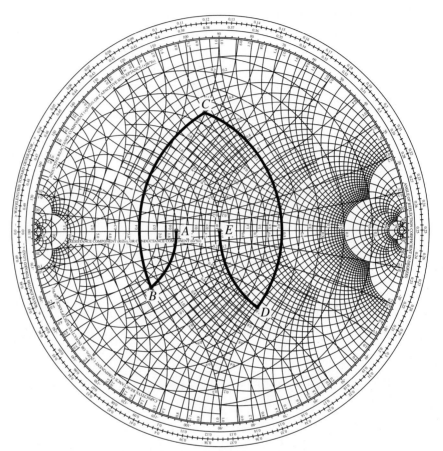

Figure 3-18 Computation of the normalized input impedance of the T-network shown in Figure 3-17 for a center frequency $f = 2$ GHz.

Converting point C into a Y-Smith Chart value results in $y_C = 0.5 - j1.0$. The shunt-connected capacitance requires the addition of a normalized susceptance $b_C = \omega C Z_0 = 1.5$, which results in the admittance value of $y_D = 0.5 + j0.5$ or point D in the Y-Smith Chart. Finally, converting point D into the impedance value $z_D = 1 - j1$ in the Z-Smith Chart allows us to add the normalized reactance $x_{L_2} = \omega_L L_2/Z_0 = 1$ along the constant $r = 1$ circle. Therefore, we reach $z_{in} = 1$ or point E in Figure 3-18. This value happens to match the 50 Ω characteristic transmission line impedance at the given frequency 2 GHz. In other words, $Z_{in} = Z_0 = 50$ Ω.

When the frequency changes, we need to go through the same steps but will arrive at a different input impedance point z_{in}. It would be tedious to go through the preceding computations for a range of frequencies. This is most efficiently done by a computer.

Parallel and Series Connections

Relying on a CAD program, we are able to produce a graphical display of the input impedance in the Z-Smith Chart over the entire frequency range in preselected increments of 10 MHz, as shown in Figure 3-19. This figure can also be generated in Matlab (see file fig3_18.m).

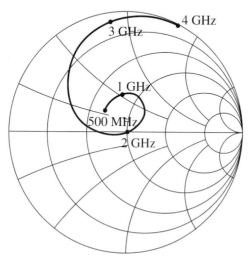

Figure 3-19 CAD simulation of the normalized input impedance Z_{in} for the network depicted in Figure 3-17 over the frequency range 500 MHz–4 GHz.

We notice that the impedance trace ranging from 0.5 to 4 GHz is in agreement with our previous calculations at 2 GHz. Also, as the frequency approaches 4 GHz, the capacitor $C = 2.39$ pF behaves increasingly like a short circuit in series with a single inductor L_2. For this reason, the normalized resistance r approaches zero and the reactance grows to large positive values.

Practically Speaking. The Smith Chart display

The frequency dependent impedance behavior of a physical RF device can be displayed easily in the Smith Chart as the following example demonstrates.

In this example, the impedance behavior of an RF coil for magnetic resonance imaging (MRI) is investigated using a network analyzer. As mentioned in the previous chapter, the network analyzer can measure

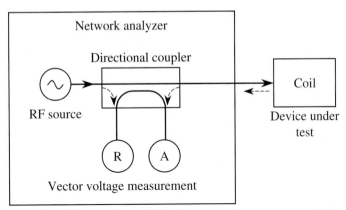

Figure 3-20 Reflection coefficient measurement chain internal to the network analyzer.

THE IMPORTANCE OF RF TECHNOLOGY IN MRI

In an effort to improve resolution and signal to noise ratio, MRI systems are equipped with stronger main magnets, having reached 3 Tesla for human full body systems, and 11.7 Tesla for animal research. Since the main magnetic field determines the frequency of the precessional nuclei of the biological load through the so-called Lamor equation, a 3 Tesla magnet results in a 128 MHz precessional frequency, a 4.7 Tesla magnet correspond to 200 MHz, and an 11.7 Tesla magnet reaches 500 MHz. To achieve a re-orientation of the nuclei, followed by a measurement of the voltage induced by the relaxing nuclei once the RF transmit pulse is turned off, requires specialized RF coil technology. Figure 3-21 shows such a coil built with microstrip lines mounted on the inside of a dielectric former and an outside shield.

the reflection coefficient and hence the impedance of an RF device. Figure 3-20 illustrates the internal procedure used by the network analyzer to record the reflection coefficient of the load (the MRI coil). The internal RF source produces continuous RF power at a specified frequency. This signal passes through a directional coupler, which diverts a portion of its power to the reference measurement port R. Whatever RF signal is reflected from the device under test again travels through the directional coupler, which diverts some of its power into the measurement port A. The remaining reflected power is absorbed by the matched RF source. Roughly speaking, the ratio A/R scaled by a calibration parameter indicates the reflection coefficient. The internal calibration in the network analyzer is more involved, allowing to compensate for various parasitics and imperfections.

Figure 3-21 displays the MRI coil (courtesy of InsightMRI, Worcester, MA) attached via a coaxial cable to the network analyzer port 1. This type of coil is used to generate an intense rotating magnetic field to excite the hydrogen nuclei in the biological sample inserted into the coil. Later in the MRI sequence, this coil receives weak RF signals emitted by the biological sample. The image is then reconstructed from these signals. Electrically, the RF coil is a resonator constructed with inductive and capacitive components. The inductive strips generate the magnetic field, while the capacitors ensure resonance at the desired frequency and matching to the RF source.

The network analyzer is set up to measure the reflection coefficient in the frequency range from 175 to 225 MHz. The coil itself is tuned to 200 MHz. Figure 3-22 depicts the measurement results

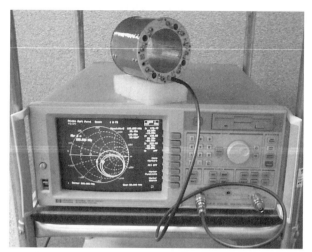

Figure 3-21 Network analyzer with the MRI RF coil attached.

produced by the network analyzer. On the left is the reflection coefficient magnitude (in dB) response, showing three sharp dips at 178.5, 200, and 222 MHz. These dips are different resonant modes of the coil, of which there are a total of five (two are at higher frequencies). However, the only mode of interest the coil has been tuned and matched for is at 200 MHz, indicated by a very low reflection coefficient. On the right in Figure 3-22 is the reflection coefficient plotted on the Smith Chart. Here, the resonant modes appear as circles. Since frequency is not obvious in the Smith Chart, three markers have been positioned to identify the three modes. We observe that at resonance, the reflection coefficient passes close to the origin, especially for mode 2 where the impedance of $(50.26 + j0.266)\ \Omega$ represents an almost perfect match to the 50 Ω line. However, for the majority of frequencies, the coil acts as a mostly capacitive impedance in the lower half of the Smith Chart.

3.5 Summary

This chapter has derived the *Smith Chart* as the most widely used RF *graphical design tool* to display the reflection coefficient and impedance behavior of a transmission line as a function of either line length or frequency. Our approach originated from the representation of the normalized input impedance of a terminated transmission line in the form

$$z_{in} = r + jx = \frac{1 + \Gamma(d)}{1 - \Gamma(d)} = \frac{1 + \Gamma_r + j\Gamma_i}{1 - \Gamma_r - j\Gamma_i}$$

which can be inverted in terms of the reflection coefficient to yield two circle equations (3.10) and (3.11), which take on the following expressions for the normalized resistance r:

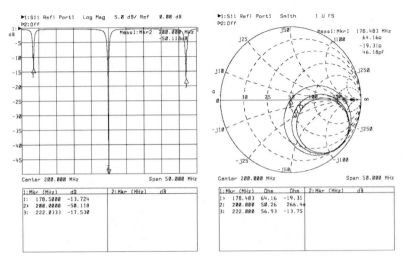

Figure 3-22 Network analyzer measurement results for the MRI RF coil.

$$\left(\Gamma_r - \frac{r}{r+1}\right)^2 + \Gamma_i^2 = \left(\frac{1}{r+1}\right)^2$$

and for the normalized reactance x

$$(\Gamma_r - 1)^2 + \left(\Gamma_i - \frac{1}{x}\right)^2 = \left(\frac{1}{x}\right)^2.$$

Superimposing the circles described by both equations over the complex reflection coefficient plane within the unit circle yields the Smith Chart. The key feature to remember is that one full rotation is equal to *half a wavelength* because of the exponent $2\beta d$ in the reflection coefficient expression (3.2). In addition to observing the impedance behavior, we can also quantify in the Smith Chart the degree of mismatch expressed by the *standing wave ratio* (SWR) equation (3.12), or

$$\text{SWR}(d) = \frac{1 + |\Gamma(d)|}{1 - |\Gamma(d)|}$$

which can be directly obtained from the chart.

To facilitate computer-based manipulations in the Smith Chart, a wide range of commercial programs can be utilized. For the relatively uncomplicated circuits analyzed in this Chapter, one can also create a custom-tailored Smith Chart and perform simple computations by relying on mathematical spreadsheets such as *Mathematica, Matlab,* or *MathCad*. To demonstrate the procedure, a number of Matlab modules have been developed, and the use of these so-called *m-files* as part of a basic Smith Chart computation is demonstrated in Section 3.2.4.

A transition to the admittance, or *Y-Smith Chart*, can be made via (3.23):

$$y_{in} = \frac{Y_{in}}{Y_0} = \frac{1}{z_{in}} = \frac{1 - \Gamma(d)}{1 + \Gamma(d)}$$

and it is found that the only difference to (3.4) is a sign reversal in front of the reflection coefficient. Consequently, rotating the reflection coefficient in the Z-Smith Chart by 180° allows us to convert z_{in} to y_{in}. In practice, this rotation can be avoided by turning the chart itself creating the *Y*-Smith Chart. Superimposing the rotated chart over the original *Z*-Smith Chart provides a combined *ZY-Smith Chart* display. The benefit of such a display is the easy transition from shunt to series connection in circuit designs. This ease is demonstrated by a *T-network* configuration connected to the input port of a bipolar transistor consisting of a parallel *RC* network. Investigating the impedance behavior as a function of frequency, however, is most easily accomplished through the use of CAD programs.

Further Reading

B. C. Wadell, "Smith Charts are easy. I. - III.," *Instrumentation and Measurement Magazine, IEEE,* Vol. 2, Issues 1, 2, 3, March, June, and September, 1999.

J. W. Verzino, "Computer Programs for Smith-Chart Solutions," *IEEE Trans. on Microwave Theory and Techniques,* Vol. 17, Issue 8, pp. 649–650, 1969.

H. J. Delgada and M. H. Thursby, "Derivation of the Smith Chart equations for use with MathCAD," *IEEE Trans. on Antennas and Propagation,* pp. 99–101, 1999.

M. Vai and S. Prasad, "Computer-aided Microwave Circuit Analysis by a Computerized Numerical Smith Chart," *IEEE Microwave and Guided Wave Letters,* Vol. 2, Issue 7, pp. 294–296, 1992.

G. Gonzalez, *Microwave Transistor Amplifiers: Analysis and Design,* 2nd ed., Prentice Hall, Upper Saddle River, NJ, 1997.

K. C. Gupta, R. Garg, and I. J. Bohl, *Microstrip Lines and Slotlines,* Artech House, Dedham, MA, 1979.

J. Helszajn, *Passive and Active Microwave Circuits,* John Wiley, New York, 1978.

Hewlett-Packard Application Note 154, "S-Parameter Design," 1972.

H. Howe, *Stripline Circuit Design,* Artech House, Dedham, MA, 1974.

S. Y. Liao, *Microwave Devices and Circuits,* Prentice Hall, Englewood Cliffs, NJ, 1980.

MMICAD for Windows, Reference Manual, Optotek, Ltd., 1997.

D. M. Pozar, *Microwave Engineering,* 2nd edition, John Wiley, New York, 1998.

P. A. Rizzi, *Microwave Engineering, Passive Circuits,* Prentice Hall, Englewood Cliffs, NJ, 1988.

P. H. Smith, *Electronic Applications of the Smith Chart,* Noble Publishing, 1995.

P. H. Smith, "Transmission-Line Calculator," *Electronics,* Vol. 12, pp. 29–31, 1939.

P. H. Smith, "An Improved Transmission-Line Calculator," *Electronics,* Vol. 17, p. 130, 1944.

Problems

3.1 Consider a load $Z_L = (80 + j40)$ Ω connected to a lossy transmission line with characteristic line impedance of

$$Z_0 = \sqrt{\frac{0.1 + j200}{0.05 - j0.003}}.$$

Determine the reflection coefficient and the standing wave ratio (SWR) at the load.

3.2 A coaxial cable of characteristic line impedance $Z_0 = 75$ Ω is terminated by a load impedance of $Z_L = (40 + j35)$ Ω. Find the input impedance of the line for each of the following pairs of frequency f and cable length d assuming that the phase velocity is 77% of the speed of light:
(a) $f = 1$ GHz and $d = 50$ cm
(b) $f = 5$ GHz and $d = 25$ cm
(c) $f = 9$ GHz and $d = 5$ cm.

3.3 The attenuation coefficient of a transmission line can be determined by shorting the load side and recording the VSWR at the beginning of the line. We recall that the reflection coefficient for a lossy line takes on the form $\Gamma(d) = \Gamma_0 \exp(-2\gamma l) = \Gamma_0 \exp(-2\alpha l)\exp(-2j\beta l)$. If the line is 100 m in length and the VSWR is 3, find the attenuation coefficient α in Np/m, and dB/m.

3.4 A load impedance of $Z_L = (150 - j50)$ Ω is connected to a 5 cm long transmission line with characteristic line impedance of $Z_0 = 75$ Ω. For a wavelength of 6 cm, compute
(a) the input impedance
(b) the operating frequency, if the phase velocity is 77% the speed of light
(c) the SWR.

3.5 Identify the following normalized impedances and admittances in the Smith Chart:
(a) $z = 0.1 + j0.7$
(b) $y = 0.3 + j0.5$
(c) $z = 0.2 + j0.1$
(d) $y = 0.1 + j0.2$.
Also find the corresponding reflection coefficients and SWRs.

3.6 An unknown load impedance is connected to a 0.3λ long, 50 Ω lossless transmission line. The SWR and phase of the reflection coefficient measured at the input of the line are 2.0 and −20°, respectively. Using the Smith Chart, determine the input and load impedances.

3.7 In Section 3.1.3 the circle equation (3.10) for the normalized resistance r is derived from (3.6). Start with (3.7), that is,

$$x = \frac{2\Gamma_i}{(1-\Gamma_r)^2 + \Gamma_i^2}$$

and show that the circle equation

$$(\Gamma_r - 1)^2 + \left(\Gamma_i - \frac{1}{x}\right)^2 = \left(\frac{1}{x}\right)^2$$

can be derived.

3.8 Starting with the equation for normalized admittance

$$y = g + jb = \frac{1-\Gamma}{1+\Gamma}$$

prove that the circle equations for the Y-Smith Chart are given by the following two formulas:
(a) for the constant conductance circle

$$\left(\Gamma_r + \frac{g}{1+g}\right)^2 + \Gamma_i^2 = \left(\frac{1}{1+g}\right)^2$$

(b) for the constant susceptance circle

$$(\Gamma_r + 1)^2 + (\Gamma_i + 1/b)^2 = (1/b)^2.$$

3.9 A lossless transmission line ($Z_0 = 50\ \Omega$) is 10 cm long ($f = 800$ MHz, $v_p = 0.77c$). If the input impedance is $Z_{in} = j60\ \Omega$
(a) find Z_L (using the Smith Chart)
(b) what length of a short-circuit terminated transmission line would be needed to replace Z_L?

3.10 A transmission line of characteristic impedance $Z_0 = 50\ \Omega$ and length $d = 0.15\lambda$ is terminated into a load impedance of $Z_L = (25 - j30)\ \Omega$. Find Γ_0, $Z_{in}(d)$, and the SWR by using the Z-Smith Chart.

3.11 A short-circuited 50 Ω transmission line section is operated at 1 GHz and possesses a phase velocity of 75% of the speed of light. Use both the analytical and the Smith Chart approach to determine the shortest lengths required to obtain (a) a 5.6 pF capacitor and (b) a 4.7 nH inductor.

3.12 A load of 100 Ω is connected to a 75 Ω lossless transmission line of 0.5λ. Determine (a) the reflection coefficients at the load and at the input of the line, (b) the VSWR, and (c) the input impedance.

3.13 Determine the shortest length of a 75 Ω open-circuited transmission line that equivalently represents a capacitor of 4.7 pF at 3 GHz. Assume the phase velocity is 66% of the speed of light.

3.14 A circuit is operated at 1.9 GHz and a lossless section of a 50 Ω transmission line is short-circuited to construct a reactance of 25 Ω. (a) If the phase velocity is 3/4 of the speed of light, what is the shortest possible length of the line to realize this impedance? (b) If an equivalent capacitive load of 25 Ω is desired, determine the shortest possible length based on the same phase velocity.

3.15 A microstrip line with 50 Ω characteristic impedance is terminated into a load impedance consisting of a 200 Ω resistor in shunt with a 5 pF capacitor. The line is 10 cm in length and the phase velocity is 50% the speed of light. (a) Find the input impedance in the Smith Chart at 500 MHz, 1 GHz, and 2 GHz, and (b) use the Matlab routine (see Section 3.2.4) and plot the frequency response from 100 MHz to 3 GHz in the Smith Chart.

3.16 For an FM broadcasting station operated at 100 MHz, the amplifier output impedance of 250 Ω has to be matched to a 75 Ω dipole antenna.
   (a) Determine the length and characteristic impedance of a quarter-wave transformer with $v_p = 0.7c$.
   (b) Find the spacing D for a two-wire lossless transmission line with AWG 26 wire size and a polystyrene dielectric ($\varepsilon_r = 2.55$).

3.17 Consider the case of matching a 73 Ω load to a 50 Ω line by means of a $\lambda/4$ transformer. Assume the matching is achieved for a center frequency of $f_C = 2$ GHz. Plot the SWR for the frequency range $1/3 \leq f/f_C \leq 3$.

3.18 A line with $Z_0 = 75$ Ω is terminated by a load consisting of a series connection of $R = 30$ Ω, $L = 10$ nH, and $C = 2.5$ pF. Find the values of the SWR and minimum line lengths at which a match of the input impedance to Z_0 can be achieved using only one additional series component at the input. Consider the following frequencies: (a) 100 MHz, (b) 500 MHz, and (c) 2 GHz.

3.19 A 50 Ω lossless coaxial cable ($\varepsilon_r = 2.8$) is connected to a 75 Ω antenna operated at 2 GHz. If the cable length is 25 cm, find the input impedance by using the analytical equation (2.71) and the Z-Smith Chart.

3.20 A **bal**anced to **un**balanced (balun) transformation is often needed to connect a dipole antenna (balanced) to a coaxial cable (unbalanced). The following figure depicts the basic concept.

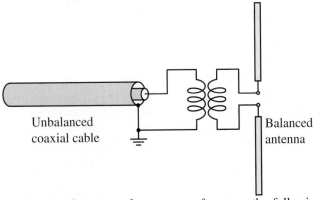

As an alternative to using a transformer, one often uses the following antenna connection.

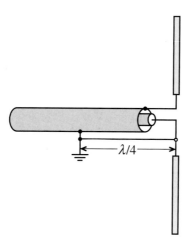

(a) Explain why one leg of the dipole antenna is connected a distance $\lambda/4$ away from the end of the coax cable.

(b) For an FM broadcast band antenna in the frequency range from 88 to 108 MHz, find the average length where the connection has to be made ($\varepsilon_r = 2.25$).

3.21 Using the ZY-Smith Chart, find the input impedance of the following network at 2 GHz.

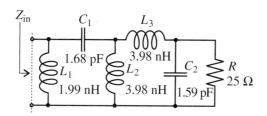

What is the input impedance of this network at 1 GHz?

3.22 A $Z_0 = 50\ \Omega$ transmission line is 0.5λ in length and terminated into a load of $Z_L = (50 - j30)\ \Omega$. At 0.35λ away from the load, a resistor of $R = 25\ \Omega$ is connected in a shunt configuration (see figure below). Find the input impedance with the help of the ZY-Smith Chart.

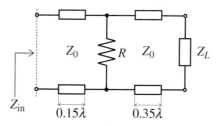

3.23 A 50 Ω transmission line of 3/4 wavelength in length is connected to two transmission line sections each of 75 Ω in impedance and length of 0.86 and 0.5 wavelength, respectively, as illustrated in the following figure.

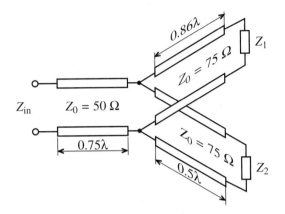

The termination for line 1 is $Z_1 = (30 + j40)$ Ω, and $Z_2 = (75 - j80)$ Ω for line 2. Employ the Smith Chart to find the input impedance.

3.24 Repeat the previous problem if all characteristic line impedances are $Z_0 = 50$ Ω and all transmission line sections are $\lambda/4$ in length.

3.25 A dipole antenna of impedance $Z_L = (75 + j20)$ Ω is connected to a 50 Ω lossless transmission line, whose length is $\lambda/3$. A voltage source $V_G = 25$V is attached to the transmission line via an unknown resistance R_G. It is determined that an average power of 3W is delivered to the load under load-side matching ($Z_L^{match} = 50$ Ω). Find the generator resistance R_G, and determine the power delivered to the antenna if the generator impedance is matched to the line via a quarter-wave transformer.

3.26 Determine the values of the inductance L and the capacitance C such that they result in a 50 Ω input impedance at 3 GHz for the following network.

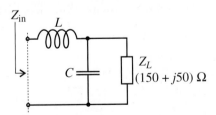

3.27 An open-circuit terminated transmission line (50 Ω) is operated at 500 MHz ($v_p = 0.7c$). Use the ZY Smith Chart and find the impedance Z_{in} if the line is 65 cm in length. Find the shortest distance for which the admittance is $Y_{in} = -j0.05$ S.

Problems

3.28 A transmission line with $Z_{0a} = 75\ \Omega$ is terminated with $Z_L = 25\ \Omega$. As line length increases, the input reflection coefficient Γ_{in} traces out a circle centered on the real (Γ_{in}) axis in the $Z_0 = 50\ \Omega$ Smith Chart. Find the center and radius of this circle in the Γ_{in} plane (50 Ω Smith chart). Hint: remember the $\lambda/4$ transformer.

3.29 We wish to transform a load impedance of $Z_L = (120 - 75j)\ \Omega$ to $Z_{in} = 50\ \Omega$ using only a length of lossless transmission line. Determine (a) Z_0 of the transmission line and (b) the minimum length of the transmission line in terms of wavelengths. Hints: the transformation is possible if the magnitude of the reflection coefficient (or alternatively SWR) is the same for both Z_L and Z_{in}. This problem can be solved graphically on the Smith Chart using the concept from Problem 3.28.

3.30 A load with $Z_L = (80 + 100j)\ \Omega$ is attached to a $\lambda/2$ microstrip transmission line with $Z_0 = 50\ \Omega$. Using the Smith Chart, find the appropriate value and location (distance from the load) for a shunt capacitor such that the impedance at the end of the transmission line becomes $Z_{in} = 50\ \Omega$. The frequency is 2 GHz and the phase velocity is 60% of the speed of light. Can this technique transform any load with finite SWR to 50 Ω?

3.31 Repeat Problem 3.30 except the line length is $\lambda/4$, the load impedance is $Z_L = 10\ \Omega$ and the target input impedance is $Z_{in} = 75\ \Omega$.

3.32 Find the minimum line length l_1 and the minimum length of the short-circuited stub l_2 in terms of wavelength λ, such that the input impedance of the circuit is equal to 50 Ω.

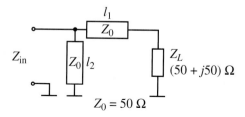

3.33 Find the input impedance in terms of magnitude and phase of the following network at the operating frequency of 950 MHz.

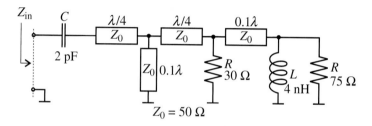

3.34 Repeat your computation and solve Problem 3.33 for a 1.5 GHz operating frequency. Comment on the differences in your results.

3.35 A specific transmission line configuration is as follows:

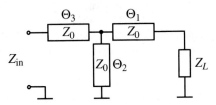

The characteristic line impedance for all three elements is $Z_0 = 50 \, \Omega$. The load impedance has a value of $Z_L = (20 + j40) \, \Omega$, and the electrical lengths of the corresponding line segments are $\Theta_1 = 164.3°$, $\Theta_2 = 57.7°$, and $\Theta_3 = 25.5°$.

(a) Find the input impedance.
(b) Find the input impedance if the transmission line segment Θ_2 is open circuit terminated.

(This problem and Problem 3.32 become very important in Chapter 8, when we discuss the concept of matching a particular load impedance to a desired input impedance.)

4

 Single- and Multiport Networks

4.1 Basic Definitions 146
4.2 Interconnecting Networks 154
    4.2.1 Series Connection of Networks. 154
    4.2.2 Parallel Connection of Networks 156
    4.2.3 Cascading Networks 157
    4.2.4 Summary of *ABCD* Network Representations 162
4.3 Network Properties and Applications. 163
    4.3.1 Interrelations between Parameter Sets 163
    4.3.2 Analysis of Microwave Amplifier. 166
4.4 Scattering Parameters 169
    4.4.1 Definition of Scattering Parameters. 169
    4.4.2 Meaning of *S*-Parameters 176
    4.4.3 Chain Scattering Matrix. 175
    4.4.4 Conversion between *Z*- and *S*-Parameters 181
    4.4.5 Signal Flowgraph Modeling 178
    4.4.6 Generalization of *S*-Parameters. 188
    4.4.7 Practical Measurements of *S*-Parameters 192
4.5 Summary 195

Ever since single- and multiple port networks were first introduced into the electrical engineering profession through Guillemin and Feldkeller, they have quickly become indispensable tools in restructuring and simplifying complicated circuits as well as in providing fundamental insight into the performance of active and passive electronic devices. Moreover, the importance of network modeling has extended far beyond electrical engineering and has influenced such diverse fields as vibrational analysis in structural and mechanical engineering as well as biomedicine. For example, today's piezoelectric medical transducer elements and their electrical-mechanical conversion mechanisms are most easily modeled as a three-port network.

> The ability to reduce most passive and active circuit devices, irrespective of their complicated and often nonlinear behavior, to simple input-output relations has many advantages. Chief among them is the experimental determination of input and output port parameters without the need to know the internal structure of the system. The "black box" methodology has tremendous appeal to engineers whose concern is mostly focused on the overall circuit performance rather than the analysis of individual components. This approach is especially important in RF and MW circuits, where complete theoretical field solutions to Maxwell's equations are either too difficult to derive or provide more information than is normally needed to develop functional, practical designs involving systems such as filters, resonators, and amplifiers.
>
> In the following sections, our objective is to establish the basic network input-output parameter relations, including impedance, admittance, hybrid, and *ABCD*-parameters for linear circuits. We then develop conversions between these sets. Rules of connecting networks are presented to show how more complicated circuits can be constructed by series and parallel connections as well as cascading of individual network blocks. Finally, the scattering parameters are presented as an important practical way of characterizing RF/MW circuits and devices through the use of power wave relations.

SINGLE-, DUAL- AND MULTIPORT NETWORKS
Examples of single port networks include voltage and current sources, whereas classical dual port networks involve transformers and transistor models. Common multiport RF networks are power dividers and circulators (three-port devices) and quadrature hybrids (four-port devices).

4.1 Basic Definitions

Before embarking on a discussion of electrical networks, we have to identify some general definitions pertaining to directions and polarity of voltages and currents. For our purposes we use the convention shown in Figure 4-1. Regardless of whether we deal with a single-port or an *N*-port network, the port-indexed current is assumed to flow into the respective port and the associated voltage is recorded as indicated.

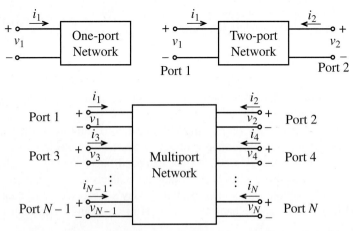

Figure 4-1 Basic voltage and current definitions for single- and multiport network.

Basic Definitions

In establishing the various parameter conventions, we begin with the voltage-current relations through double-indexed impedance coefficients Z_{nm}, where indices n and m range between 1 and N. The voltage at each port $n = 1 \ldots N$ is given by

$$v_1 = Z_{11}i_1 + Z_{12}i_2 + \ldots + Z_{1N}i_N \ . \tag{4.1a}$$

for port 1,

$$v_2 = Z_{21}i_1 + Z_{22}i_2 + \ldots + Z_{2N}i_N \tag{4.1b}$$

for port 2, and

$$v_N = Z_{N1}i_1 + Z_{N2}i_2 + \ldots + Z_{NN}i_N \tag{4.1c}$$

for port N. We see that each port n is affected by its own impedance Z_{nn} as well as by a linear superposition of all other ports. In a more concise notation, (4.1) can be converted into an **impedance** or **Z**-matrix form:

$$\begin{Bmatrix} v_1 \\ v_2 \\ \vdots \\ v_N \end{Bmatrix} = \begin{bmatrix} Z_{11} & Z_{12} & \cdots & Z_{1N} \\ Z_{21} & Z_{22} & \cdots & Z_{2N} \\ \vdots & \vdots & \ddots & \vdots \\ Z_{N1} & Z_{N2} & \cdots & Z_{NN} \end{bmatrix} \begin{Bmatrix} i_1 \\ i_2 \\ \vdots \\ i_N \end{Bmatrix} \tag{4.2}$$

or, in matrix notation,

$$\{V\} = [Z]\{I\} \tag{4.3}$$

where $\{V\}$ and $\{I\}$ are vectors of voltages $v_1, v_2, \ldots, v_N$ and currents $i_1, i_2, \ldots, i_N$, respectively, and $[Z]$ is the impedance matrix.

Each impedance element in (4.2) can be determined via the following protocol:

$$Z_{nm} = \left. \frac{v_n}{i_m} \right|_{i_k = 0 \ (\text{for } k \neq m)} \tag{4.4}$$

which means that the voltage v_n is recorded at port n, while port m is driven by current i_m and the rest of the ports are maintained under open circuit conditions (i.e. $i_k = 0$ where $k \neq m$).

Instead of voltages as the dependent variable, we can specify currents such that

$$\begin{Bmatrix} i_1 \\ i_2 \\ \vdots \\ i_N \end{Bmatrix} = \begin{bmatrix} Y_{11} & Y_{12} & \cdots & Y_{1N} \\ Y_{21} & Y_{22} & \cdots & Y_{2N} \\ \vdots & \vdots & \ddots & \vdots \\ Y_{N1} & Y_{N2} & \cdots & Y_{NN} \end{bmatrix} \begin{Bmatrix} v_1 \\ v_2 \\ \vdots \\ v_N \end{Bmatrix} \tag{4.5}$$

> **RECIPROCAL NETWORKS**
>
> If a two port network is linear, does not exhibit magnetic or electric hysteresis effects, and has no internal resonances, we refer to such a system as reciprocal, i.e. input and output ports can be interchanged. A familiar example of a reciprocal dual-port network is an antenna system which can be treated as both a transmitter and a receiver.

or
$$\{I\} = [Y]\{V\} \tag{4.6}$$

where, similarly to (4.4), we define the individual elements of the **admittance** or **Y-matrix** as

$$Y_{nm} = \left.\frac{i_n}{v_m}\right|_{v_k = 0 \text{ (for } k \ne m)} \tag{4.7}$$

Comparing (4.2) and (4.5), it is apparent that impedance and admittance matrices are inverses of each other:

$$[Z] = [Y]^{-1}. \tag{4.8}$$

Example 4-1. Matrix representation of a Pi-network

For the Pi-network (the name of the network comes from the resemblance with the Greek capital letter Π) shown in Figure 4-2 with generic impedances Z_A, Z_B, and Z_C, find the impedance and admittance matrices.

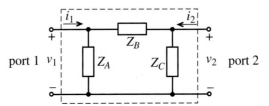

Figure 4-2 Pi-network as a two-port network.

Solution. The impedance matrix elements are found by using (4.4) and the appropriate open- and short-circuit termination conditions.

To find Z_{11} we must compute the ratio of the voltage drop v_1 across port 1 to the current i_1 flowing into this port when the current into port 2 equals zero. The requirement $i_2 = 0$ is equivalent to an open-circuit condition. Thus, the impedance Z_{11} is equal to the parallel combination of impedances Z_A and $Z_B + Z_C$.

$$Z_{11} = \left.\frac{v_1}{i_1}\right|_{i_2 = 0} = Z_A \| (Z_B + Z_C) = \frac{Z_A(Z_B + Z_C)}{Z_A + Z_B + Z_C}.$$

Basic Definitions

The value for Z_{12} can be found as the ratio of the voltage drop v_1 measured across port 1 to the current i_2. In this case, we must ensure that the current i_1 remains zero (i.e., we must treat port 1 as open). Voltage v_1 is equal to the voltage drop across impedance Z_A and can be obtained using the voltage divider rule:

$$v_1 = \frac{Z_A}{Z_A + Z_B} v_{AB}$$

where v_{AB} is a voltage drop across impedances Z_A and Z_B connected in series and computed as $v_{AB} = i_2 [Z_C \| (Z_A + Z_B)]$. Thus,

$$Z_{12} = \left.\frac{v_1}{i_2}\right|_{i_1 = 0} = \frac{Z_A}{Z_A + Z_B}[Z_C \| (Z_A + Z_B)] = \frac{Z_A Z_C}{Z_A + Z_B + Z_C}.$$

Similarly, we can obtain the remaining two coefficients of the impedance matrix:

$$Z_{21} = \left.\frac{v_2}{i_1}\right|_{i_2 = 0} = \frac{Z_C}{Z_B + Z_C}[Z_A \| (Z_B + Z_C)] = \frac{Z_A Z_C}{Z_A + Z_B + Z_C}$$

$$Z_{22} = \left.\frac{v_2}{i_2}\right|_{i_1 = 0} = Z_C \| (Z_A + Z_B) = \frac{Z_C(Z_A + Z_B)}{Z_A + Z_B + Z_C}.$$

Thus, the impedance matrix for the generic Pi-network is written in the form

$$[\mathbf{Z}] = \frac{1}{Z_A + Z_B + Z_C} \begin{bmatrix} Z_A(Z_B + Z_C) & Z_A Z_C \\ Z_A Z_C & Z_C(Z_A + Z_B) \end{bmatrix}.$$

The coefficients for the admittance matrix can be derived using (4.7). To find the value for Y_{11} we must find the ratio of the current flow into port 1 to the voltage drop across this port when the second port is shorted (i.e., $v_2 = 0$):

$$Y_{11} = \left.\frac{i_1}{v_1}\right|_{v_2 = 0} = \frac{1}{Z_A} + \frac{1}{Z_B}.$$

The value for the coefficient Y_{12} of the admittance matrix can be obtained by shorting port 1 (i.e., forcing $v_1 = 0$) and measuring the ratio of the current i_1 to the voltage drop across port 2. We note that, when a positive voltage is applied to port 2, the current i_1 will flow away from port 1, resulting in a negative current:

$$Y_{12} = \left.\frac{i_1}{v_2}\right|_{v_1 = 0} = -\frac{1}{Z_B}.$$

The rest of the admittance matrix is derived in the same manner, leading to the following final form

$$[\mathbf{Y}] = \begin{bmatrix} \dfrac{1}{Z_A} + \dfrac{1}{Z_B} & -\dfrac{1}{Z_B} \\ -\dfrac{1}{Z_B} & \dfrac{1}{Z_B} + \dfrac{1}{Z_C} \end{bmatrix} = \begin{bmatrix} Y_A + Y_B & -Y_B \\ -Y_B & Y_B + Y_C \end{bmatrix}$$

where $Y_A = Z_A^{-1}$, $Y_B = Z_B^{-1}$, and $Y_C = Z_C^{-1}$.

Direct evaluation shows that the obtained impedance and admittance matrices are indeed inversely related, which supports the validity of (4.8).

Practical determination of the matrix coefficients can be accomplished by enforcing open- and short-circuit conditions. However, as the frequency reaches RF limits, parasitic terminal effects can no longer be ignored and a different measurement approach becomes necessary.

Example 4-1 indicates that both impedance and admittance matrices are symmetric. This is generally true for linear, passive networks. *Passive* in this context implies not containing any current or voltage sources. We can state the symmetry as

$$Z_{nm} = Z_{mn} \qquad (4.9)$$

which also applies for admittances because of (4.8). In fact, it can be proved that any reciprocal (that is, nonactive, linear isotropic materials) N-port network is described by symmetric Z- and Y-matrices.

Besides impedance and admittance network descriptions, there are two additional useful parameter sets depending on how the voltage and currents are arranged. Restricting our discussion to two-port networks and with reference to Figure 4-1, we define the **chain** or **ABCD-matrix** as

$$\begin{Bmatrix} v_1 \\ i_1 \end{Bmatrix} = \begin{bmatrix} A & B \\ C & D \end{bmatrix} \begin{Bmatrix} v_2 \\ -i_2 \end{Bmatrix} \qquad (4.10)$$

and the **hybrid** or **h-matrix** as

$$\begin{Bmatrix} v_1 \\ i_2 \end{Bmatrix} = \begin{bmatrix} h_{11} & h_{12} \\ h_{21} & h_{22} \end{bmatrix} \begin{Bmatrix} i_1 \\ v_2 \end{Bmatrix}. \qquad (4.11)$$

Basic Definitions

The determination of the individual matrix coefficients is identical to the method introduced for the impedance and admittance matrices. For instance, to find h_{12} in (4.11), we set i_1 to zero and compute the ratio of v_1 over v_2; that is,

$$h_{12} = \left.\frac{v_1}{v_2}\right|_{i_1 = 0}.$$

We note that in the hybrid representation, parameters h_{21} and h_{12} define the forward current and reverse voltage gain, respectively. The remaining two parameters determine the input impedance (h_{11}) and output admittance (h_{22}) of the network. These properties of the hybrid representation explain why it is most often used for low-frequency transistor models. The following example shows the derivation of the hybrid matrix representation for a **bipolar-junction transistor** (BJT) for low-frequency operation.

Example 4-2. Low-Frequency hybrid network description of a BJT

Describe the common-emitter BJT transistor in terms of its hybrid network parameters for the low-frequency, small-signal transistor model shown in Figure 4-3.

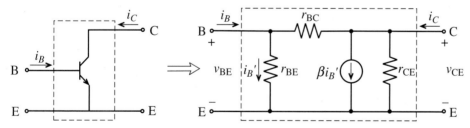

Figure 4-3 Common-emitter low-frequency, small-signal transistor model.

Solution. In the transistor model shown in Figure 4-3, r_{BE}, r_{BC}, and r_{CE} represent base-emitter, base-collector, and collector-emitter internal resistances of the transistor. The current through the current-controlled current source is dependent on the current i_B' flowing through the base-emitter resistance.

To evaluate the h_{11} parameter of the hybrid matrix according to (4.11) we must short-circuit the collector and emitter terminals, thus setting $v_2 = v_{CE} = 0$, and compute the ratio of the base-emitter voltage to the base current. Using the notation established in Figure 4-3, we notice that h_{11} is equal to the parallel combination of r_{BE} and r_{BC}:

$$h_{11} = \left.\frac{v_{BE}}{i_B}\right|_{v_{CE}=0} = \frac{r_{BC}r_{BE}}{r_{BE}+r_{BC}} \quad \text{(input impedance)}.$$

Following a similar procedure, the relations for the remaining three parameters of the hybrid representation are established as follows:

$$h_{12} = \left.\frac{v_{BE}}{v_{CE}}\right|_{i_B=0} = \frac{r_{BE}}{r_{BE}+r_{BC}} \quad \text{(voltage feedback ratio)}$$

$$h_{21} = \left.\frac{i_C}{i_B}\right|_{v_{CE}=0} = \frac{\beta r_{BC} - r_{BE}}{r_{BE}+r_{BC}} \quad \text{(small-signal current gain)}$$

$$h_{22} = \left.\frac{i_C}{v_{CE}}\right|_{i_B=0} = \frac{1}{r_{CE}} + \frac{1+\beta}{r_{BE}+r_{BC}} \quad \text{(output admittance)}.$$

In the majority of all practical transistor designs, the current amplification coefficient β is usually much greater than unity, and the collector-base resistance is much larger than the base-emitter resistance. Keeping these relations in mind, we can simplify the expressions derived for the **h**-matrix representation of the transistor:

$$h_{11} = \left.\frac{v_{BE}}{i_B}\right|_{v_{CE}=0} = r_{BE} \quad \text{(input impedance)}$$

$$h_{12} = \left.\frac{v_{BE}}{v_{CE}}\right|_{i_B=0} = \frac{r_{BE}}{r_{BC}} \quad \text{(voltage feedback ratio)}$$

$$h_{21} = \left.\frac{i_C}{i_B}\right|_{v_{CE}=0} = \beta \quad \text{(small-signal current gain)}$$

$$h_{22} = \left.\frac{i_C}{v_{CE}}\right|_{i_B=0} = \frac{1}{r_{CE}} + \frac{\beta}{r_{BC}} \quad \text{(output admittance)}.$$

The hybrid network description is a very popular way to characterize the BJT, and its h-parameter coefficients are widely reported in data sheets.

Due to the presence of the current source in Example 4-2, the **h**-matrix is no longer reciprocal ($h_{12} \neq -h_{21}$). In low-frequency electronic circuit design, the coefficients of the hybrid matrix representation are often listed as h_{ie} for h_{11}, h_{re} for h_{12}, h_{fe} for h_{21}, and h_{oe} for h_{22}.

Up to this point we considered the problem of deriving the matrix representation based on a known topology and element values of the circuit. However, in practical design tasks it is often required to go the opposite way: obtain the equivalent circuit model from measurements. This so-called **inverse modeling** becomes extremely important when a device manufacturer has to characterize its components to customers. The system's performance must be generically specified since customer applications can vary widely. In this case, the use of the equivalent circuit representation enables an engineer to predict with reasonable accuracy the response of the device or circuit under changing operating conditions. In the following example we will derive the values of the internal resistances of the BJT from known **h**-matrix parameters.

PARAMETER EXTRACTION
The quest of finding a circuit model representation that fits measured device performance data is also known as parameter extraction. It is still an actively pursued research endeavor since the relation of external measurements to a number of circuit components is inherently non-unique and often requires major simplifications such as linearization around the device's operating point and limiting the frequency range. Furthermore, to simplify the computations, the underlying circuit models are frequently reduced in their complexity.

RF&MW→

Example 4-3. Determination of internal resistances and current gain of a BJT based on *h*-parameter measurements

Use the equivalent circuit representation of the BJT shown in Figure 4-3 and employ the following measured hybrid parameters: $h_{ie} = 5 \text{ k}\Omega$, $h_{re} = 2 \times 10^{-4}$, $h_{fe} = 250$, $h_{oe} = 20 \text{ μS}$ (these parameters correspond to the 2n3904 transistor manufactured by Motorola). Find the internal resistances r_{BE}, r_{BC}, and r_{CE}, and the current gain β.

Solution. As derived in Example 4-2, the values of the **h**-matrix for the equivalent circuit shown in Figure 4-3 are given by the following four equations:

$$h_{ie} = \frac{r_{BC} r_{BE}}{r_{BE} + r_{BC}} \quad \text{(input impedance)} \quad (4.12)$$

$$h_{re} = \frac{r_{BE}}{r_{BE} + r_{BC}} \quad \text{(voltage feedback ratio)} \quad (4.13)$$

$$h_{fe} = \frac{\beta r_{BC} - r_{BE}}{r_{BE} + r_{BC}} \quad \text{(small-signal current gain)} \quad (4.14)$$

$$h_{oe} = \frac{1}{r_{CE}} + \frac{1+\beta}{r_{BE} + r_{BC}} \quad \text{(output admittance)}. \quad (4.15)$$

If we divide (4.12) by (4.13), we determine that the base-collector resistance is equal to the ratio of h_{ie} over h_{re}. Accordingly, for values given in the problem formulation, we obtain: $r_{BC} = h_{ie}/h_{re} = 25$ MΩ. Substituting this value into either equation (4.12) or (4.13), we can then find $r_{BE} = h_{ie}/(1-h_{re}) = 5$ kΩ. Knowing r_{BC} and r_{BE}, (4.14) allows us to compute the current gain coefficient $\beta = (h_{re} + h_{fe})/(1-h_{re}) = 250.05$. Finally, the collector-emitter resistance can be evaluated from (4.15) as

$$r_{CE} = \frac{h_{ie}}{h_{oe}h_{ie} - h_{re}h_{fe} - h_{re}} = 100.4 \text{ k}\Omega.$$

We note from the obtained values that r_{BE} is indeed much smaller than r_{BC}.

This example provides a first idea of how the measured h-parameters can be used as a basis to characterize the BJT circuit model. The concept of "inverting" the measurements to determine circuit model parameters will be further analyzed in Chapter 7.

4.2 Interconnecting Networks

4.2.1 Series Connection of Networks

A series connection consisting of two two-port networks is shown in Figure 4-4. The individual networks are shown in impedance matrix representation.

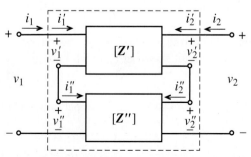

Figure 4-4 Series connection of two two-port networks.

Interconnecting Networks

In this case, the individual voltages are additive while the currents remain the same. This results in

$$\begin{Bmatrix} v_1 \\ v_2 \end{Bmatrix} = \begin{Bmatrix} v_1' + v_1'' \\ v_2' + v_2'' \end{Bmatrix} = [Z] \begin{Bmatrix} i_1 \\ i_2 \end{Bmatrix} \tag{4.16}$$

where the new composite network [Z] takes the form

$$[Z] = [Z'] + [Z''] = \begin{bmatrix} Z_{11}' + Z_{11}'' & Z_{12}' + Z_{12}'' \\ Z_{21}' + Z_{21}'' & Z_{22}' + Z_{22}'' \end{bmatrix}. \tag{4.17}$$

Caution has to be exercised in not indiscriminately connecting individual networks, as common mode signals may be created. The two-port network representation only deals with differential signals at the ports. This situation is exemplified in Figure 4-5 (a). The problem can be avoided by including a transformer, as seen in Figure 4-5 (b). The transformer in this case decouples input and output ports of the second network. However, this approach will only work for AC signals since the transformer acts as a high-pass filter and rejects all DC contributions.

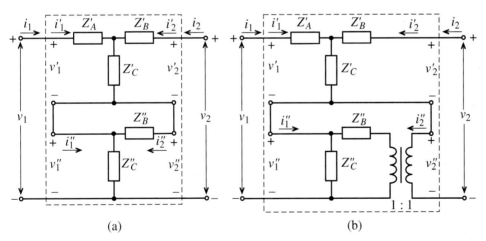

Figure 4-5 (a) Short circuit in series connection. (b) Transformer to avoid short circuit.

When two networks are connected with the output interchanged, as shown in Figure 4-6, the most suitable representation is the hybrid form. In the network connection in Figure 4-6, the voltages on the input ports and currents on the output ports are additive (i.e., $v_1 = v_1' + v_1''$ and $i_2 = i_2' + i_2''$), while the voltages on the output ports

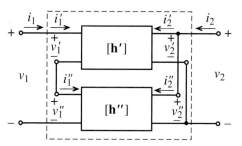

Figure 4-6 Connection of two-port networks suitable for hybrid representation.

and currents on input ports are the same (i.e., $v_2 = v_2' = v_2''$ and $i_1 = i_1' = i_1''$). From this observation we can conclude that the resulting **h**-matrix for the overall system is equal to the sum of the **h**-matrices of the individual networks:

$$\left\{\begin{array}{c} v_1 \\ i_2 \end{array}\right\} = \left\{\begin{array}{c} v_1' + v_1'' \\ i_2' + i_2'' \end{array}\right\} = \begin{bmatrix} h_{11}' + h_{11}'' & h_{12}' + h_{12}'' \\ h_{21}' + h_{21}'' & h_{22}' + h_{22}'' \end{bmatrix} \left\{\begin{array}{c} i_1 \\ v_2 \end{array}\right\}. \tag{4.18}$$

An example of this type of connection is a pair of transformers shown in Figure 4-7.

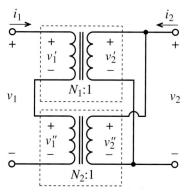

Figure 4-7 Series connection of two hybrid networks.

The circuit details of this transformer pair configuration are discussed in the problem section at the end of this chapter. Because transformers possess differential inputs and outputs, they do not suffer from common mode problems when adding **h**-matrices according to (4.18).

4.2.2 Parallel Connection of Networks

A parallel connection of two dual-port networks is shown in Figure 4-8 for the admittance matrices **Y'** and **Y''**, where, unlike in (4.16), the currents are now additive

$$\left\{\begin{array}{c} i_1 \\ i_2 \end{array}\right\} = \left\{\begin{array}{c} i_1' + i_1'' \\ i_2' + i_2'' \end{array}\right\} = [\mathbf{Y}] \left\{\begin{array}{c} v_1 \\ v_2 \end{array}\right\} \tag{4.19}$$

Interconnecting Networks

and the new admittance matrix is defined as the sum of the individual admittances

$$[\mathbf{Y}] = [\mathbf{Y}'] + [\mathbf{Y}''] = \begin{bmatrix} Y_{11}' + Y_{11}'' & Y_{12}' + Y_{12}'' \\ Y_{21}' + Y_{21}'' & Y_{22}' + Y_{22}'' \end{bmatrix} \qquad (4.20)$$

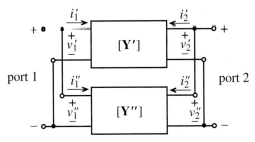

Figure 4-8 Parallel connection of two two-port networks.

4.2.3 Cascading Networks

The **ABCD**-parameter description is most suitable when cascading networks, as depicted in Figure 4-9 for the example of a two-transistor configuration. In this case, the current on the output of the first network is equal in value, but opposite in sign, to the input current of the second network (i.e., $i_2' = -i_1''$). The voltage drop v_2' across the output port of the first

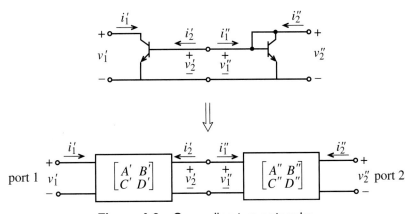

Figure 4-9 Cascading two networks.

network is equal to the voltage drop v_1'' across the input port of the second network. Thus, we can write the following relations:

$$\begin{Bmatrix} v_1 \\ i_1 \end{Bmatrix} = \begin{Bmatrix} v_1' \\ i_1' \end{Bmatrix} = \begin{bmatrix} A' & B' \\ C' & D' \end{bmatrix} \begin{Bmatrix} v_2' \\ -i_2' \end{Bmatrix} = \begin{bmatrix} A' & B' \\ C' & D' \end{bmatrix} \begin{Bmatrix} v_1'' \\ i_1'' \end{Bmatrix}$$

$$= \begin{bmatrix} A' & B' \\ C' & D' \end{bmatrix} \begin{bmatrix} A'' & B'' \\ C'' & D'' \end{bmatrix} \begin{Bmatrix} v_2'' \\ -i_2'' \end{Bmatrix}.$$

(4.21)

The overall system **ABCD**-matrix is equal to the product of the **ABCD**-matrices of the individual networks.

4.2.4 Summary of *ABCD* Network Representations

As we will see in subsequent chapters, microwave circuits can usually be represented as the result of cascading simpler networks. It is therefore important to develop **ABCD**-matrix representations for simple two-port networks that can be used as building blocks of more complex configurations. In this section, we derive the **ABCD**-parameters for a transmission line, a series impedance, and a passive T-network. Other useful circuits, such as shunt impedance, passive pi-network, and transformer, are left as exercises at the end of this chapter (see Problems 4.10, 4.12, 4.13). The results of all the computations are summarized in Table 4-1 at the end of this section.

Example 4-4. *ABCD* network representation of an impedance element

Compute the **ABCD**-matrix representation for the following network:

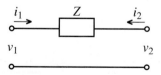

Solution. Guided by the definition (4.10), to determine parameter A we have to compute the ratio of the voltage drop across port 1 to the voltage drop across port 2 when the current into this port is equal to zero (i.e., port 2 is disconnected). In this case, it is apparent that for the

Interconnecting Networks

circuit under consideration, the voltages on both ports are equal to their ratio, which is equal to unity

$$A = \left.\frac{v_1}{v_2}\right|_{i_2 = 0} = 1.$$

To obtain the value for B, we need to find the ratio of the voltage drop across port 1 to the current flowing from port 2 when the terminals of port 2 are shorted. From the circuit topology, this ratio is equal to the impedance Z:

$$B = \left.\frac{v_1}{-i_2}\right|_{v_2 = 0} = Z.$$

The remaining two parameters are found according to (4.10) of the **ABCD** representation and are

$$C = \left.\frac{i_1}{v_2}\right|_{i_2 = 0} = 0 \text{ and } D = \left.\frac{i_1}{-i_2}\right|_{v_2 = 0} = 1.$$

*The **ABCD**-matrix coefficients are determined in a similar manner as the previously discussed **Z**-, **Y**-, and **h**-matrix coefficients. Accurate measurements of the coefficients depend on the ability to enforce open- and short-circuit terminal conditions.*

In the following example, the *ABCD*-parameters of the passive T-network are determined. In the derivation of the parameters we will rely on the knowledge of *ABCD*-parameters for series and shunt connections of the impedance.

Example 4-5. ABCD-matrix computation of a T-network

Compute the **ABCD**-matrix representation for the following T-network:

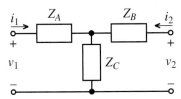

Solution. This problem can be solved using two different approaches. The first approach involves directly applying the definition of the **ABCD**-matrix coefficients and computing them as done in the previous example. Another approach is to utilize the knowledge of the *ABCD*-parameters for shunt and series connections of a single impedance. If we choose this method, we first have to break the initial circuit into subcircuits as follows:

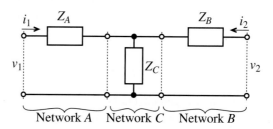

As discussed previously, the **ABCD**-matrix representation of the entire circuit is equal to the product of the **ABCD**-matrices of the individual subcircuits. Using the results from Example 4-4 and Problem 4.10, we can write

$$[\text{ABCD}] = \begin{bmatrix} 1 & Z_A \\ 0 & 1 \end{bmatrix} \begin{bmatrix} 1 & 0 \\ Z_C^{-1} & 1 \end{bmatrix} \begin{bmatrix} 1 & Z_B \\ 0 & 1 \end{bmatrix} = \begin{bmatrix} 1 + \dfrac{Z_A}{Z_C} & Z_A + Z_B + \dfrac{Z_A Z_B}{Z_C} \\ \dfrac{1}{Z_C} & 1 + \dfrac{Z_B}{Z_C} \end{bmatrix}.$$

*Here, we see the advantage of using the **ABCD**-matrix representation is that a more complex network can be constructed by cascading simpler building blocks.*

As a final example, let us consider the computation of the *ABCD*-parameters for a transmission line.

Example 4-6. ABCD-matrix coefficient computation of a transmission line section

Compute the **ABCD**-matrix representation of the following transmission line with characteristic impedance Z_0, propagation constant β, and length l.

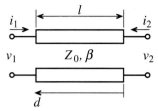

Solution. Similar to Example 4-4, we have to apply open- and short-circuit conditions at port 2. For a transmission line these conditions are equivalent to the analysis of open- and short-circuited stub lines. Such lines are simply the open/short-circuited transmission line representations discussed in Sections 2.10.3 and 2.10.2. In these sections we found that for the open-circuited stub the voltage and current are given by the following expressions [see (2.77) and (2.78)]:

$$V(d) = 2V^+ \cos(\beta d) \text{ and } I(d) = \frac{2jV^+}{Z_0}\sin(\beta d)$$

where distance d is measured from the open port (i.e., in our case from port 2).

For a short-circuited stub of length l, voltages and currents are determined by (2.73) and (2.74):

$$V(d) = 2jV^+ \sin(\beta d) \text{ and } I(d) = \frac{2V^+}{Z_0}\cos(\beta d)$$

where distance d is again measured from port 2 to port 1. In addition to these relations, it is important to recall that the current is defined as flowing *toward the load*. Therefore, the current is equal to i_1 at port 1 and equal to $-i_2$ at port 2.

Having determined the relations for voltages and currents, it is now possible to establish the equations for the *ABCD*-parameters of the transmission line. Parameter A is defined as the ratio of the voltages at ports 1 and 2 when port 2 is open (i.e., we have to use the formulas for the open-circuited stub):

$$A = \left.\frac{v_1}{v_2}\right|_{i_2=0} = \frac{2V^+\cos(\beta l)}{2V^+} = \cos(\beta l)$$

where we employ the fact that $d = 0$ at port 2 and $d = l$ at port 1.

Parameter B is defined as the ratio of the voltage drop across port 1 to the current flowing from port 2 (i.e., toward the load) when port 2 is shorted. For this case we have to use the formulas for voltage and current defined for a short-circuited stub. This yields

$$B = \left.\frac{v_1}{-i_2}\right|_{v_2 = 0} = \frac{2jV^+\sin(\beta l)}{2V^+/Z_0} = jZ_0\sin(\beta l).$$

The remaining two coefficients are obtained in a similar manner:

$$C = \left.\frac{i_1}{v_2}\right|_{i_2 = 0} = \frac{\frac{2jV^+}{Z_0}\sin(\beta l)}{2V^+} = jY_0\sin(\beta l)$$

$$D = \left.\frac{i_1}{-i_2}\right|_{v_2 = 0} = \frac{\frac{2V^+}{Z_0}\cos(\beta l)}{\frac{2V^+}{Z_0}} = \cos(\beta l).$$

Thus, a transmission line with characteristic impedance Z_0, propagation constant β, and length l has the following matrix representation:

$$\begin{bmatrix} A & B \\ C & D \end{bmatrix} = \begin{bmatrix} \cos(\beta l) & jZ_0\sin(\beta l) \\ jY_0\sin(\beta l) & \cos(\beta l) \end{bmatrix}.$$

The ABCD transmission line representation has the expected periodic parameter behavior similar to the line input impedance formula derived in Chapter 2.

GENERALIZING FOR N-PORT NETWORKS
While the various two-port network systems involve simple two-by-two matrix manipulations, it should be clear that this can be extended to three and higher order networks, resulting in three-by-three and higher order matrix manipulations.

In Table 4-1 six of the most common circuit configurations are summarized in terms of their *ABCD* two-port network representations. From these six basic models, more complicated circuits are readily constructed by suitably combining these elementary networks.

Table 4-1 ABCD-parameters of several two-port circuits

Circuit	ABCD Parameters	
Series impedance Z	$A = 1$	$B = Z$
	$C = 0$	$D = 1$
Shunt admittance Y	$A = 1$	$B = 0$
	$C = Y$	$D = 1$
T-network (Z_A, Z_B, Z_C)	$A = 1 + \dfrac{Z_A}{Z_C}$	$B = Z_A + Z_B + \dfrac{Z_A Z_B}{Z_C}$
	$C = \dfrac{1}{Z_C}$	$D = 1 + \dfrac{Z_B}{Z_C}$
Π-network (Y_A, Y_B, Y_C)	$A = 1 + \dfrac{Y_B}{Y_C}$	$B = \dfrac{1}{Y_C}$
	$C = Y_A + Y_B + \dfrac{Y_A Y_B}{Y_C}$	$D = 1 + \dfrac{Y_A}{Y_C}$
Transmission line Z_0, β, length l	$A = \cos(\beta l)$	$B = j Z_0 \sin(\beta l)$
	$C = \dfrac{j \sin(\beta l)}{Z_0}$	$D = \cos(\beta l)$
Transformer $N:1$	$A = N$	$B = 0$
	$C = 0$	$D = \dfrac{1}{N}$

4.3 Network Properties and Applications

4.3.1 Interrelations between Parameter Sets

Depending on the particular circuit configuration, we may be forced to convert between different parameter sets to arrive at a particular input/output description. For instance, the low-frequency transistor parameters are often recorded in **h**-matrix form. However, when cascading the transistor with additional networks, a more useful **ABCD**-matrix form may be appropriate. Thus, converting the **h**-matrix into an **ABCD**-matrix form and vice versa can greatly simplify the analysis.

To show how the conversion between the individual parameter sets is accomplished, let us find an **ABCD**-matrix representation of a given **h**-matrix. From the definition (4.11), we can express parameter A as follows:

$$A = \left.\frac{v_1}{v_2}\right|_{i_2 = 0} = \frac{h_{11}i_1 + h_{12}v_2}{v_2}. \tag{4.22}$$

In this expression, we are able to re-express the current i_1 in (4.11) in terms of the voltage v_2 because $i_2 = 0$. The result is

$$A = \left.\frac{v_1}{v_2}\right|_{i_2 = 0} = \frac{h_{11}\left(-\frac{h_{22}}{h_{21}}v_2\right) + h_{12}v_2}{v_2} = \frac{1}{h_{21}}(-h_{22}h_{11} + h_{12}h_{21}) = -\frac{\Delta h}{h_{21}} \tag{4.23}$$

where $\Delta h = h_{11}h_{22} - h_{12}h_{21}$ denotes the determinant of the **h**-matrix. Similarly, for the remaining coefficients we compute

$$B = \left.-\frac{v_1}{i_2}\right|_{v_2 = 0} = -\frac{h_{11}i_1}{i_2} = -\frac{h_{11}\left(\frac{i_2}{h_{21}}\right)}{i_2} = -\frac{h_{11}}{h_{21}} \tag{4.24}$$

$$C = \left.\frac{i_1}{v_2}\right|_{i_2 = 0} = \frac{\frac{-h_{22}}{h_{21}}v_2}{v_2} = -\frac{h_{22}}{h_{21}} \tag{4.25}$$

$$D = \left.-\frac{i_1}{i_2}\right|_{v_2 = 0} = -\frac{\frac{i_2}{h_{21}}}{i_2} = -\frac{1}{h_{21}}. \tag{4.26}$$

This concludes the conversion from h-parameters to **ABCD** form. A similar procedure could have been performed to convert *ABCD*-parameters to **h**-matrix form.

As an additional case, let us investigate the conversion from *ABCD*-parameters to the **Z**-representation. Starting with (4.2) and using (4.10), we can develop the following relations:

$$Z_{11} = \left.\frac{v_1}{i_1}\right|_{i_2 = 0} = \frac{Av_2}{Cv_2} = \frac{A}{C} \tag{4.27}$$

Network Properties and Applications

$$Z_{12} = \left.\frac{v_1}{i_2}\right|_{i_1=0} = \frac{Av_2 - Bi_2}{\frac{C}{D}v_2} = \frac{Av_2 - \frac{BC}{D}v_2}{\frac{C}{D}v_2} = \frac{AD - BC}{C} = \frac{\Delta ABCD}{C} \quad (4.28)$$

$$Z_{21} = \left.\frac{v_2}{i_1}\right|_{i_2=0} = \frac{v_1/A}{Cv_2} = \frac{A(v_2/A)}{Cv_2} = \frac{1}{C} \quad (4.29)$$

$$Z_{22} = \left.\frac{v_2}{i_2}\right|_{i_1=0} = \frac{v_2}{Cv_2/D} = \frac{D}{C} \quad (4.30)$$

where $\Delta ABCD = AD - BC$ is the determinant of the **ABCD**-matrix.

By relying on the respective defining voltage and current relations, it is relatively straightforward to work out all parameter conversions. For convenience, Table 4-2 summarizes the formulas for the previously defined four network parameter sets (see also Appendix H for a complete list of all conversion formulas).

Table 4-2 Conversion between different network representations

	[Z]	[Y]	[h]	[ABCD]
[Z]	$\begin{bmatrix} Z_{11} & Z_{12} \\ Z_{21} & Z_{22} \end{bmatrix}$	$\begin{bmatrix} \frac{Z_{22}}{\Delta Z} & -\frac{Z_{12}}{\Delta Z} \\ -\frac{Z_{21}}{\Delta Z} & \frac{Z_{11}}{\Delta Z} \end{bmatrix}$	$\begin{bmatrix} \frac{\Delta Z}{Z_{22}} & \frac{Z_{12}}{Z_{22}} \\ -\frac{Z_{21}}{Z_{22}} & \frac{1}{Z_{22}} \end{bmatrix}$	$\begin{bmatrix} \frac{Z_{11}}{Z_{21}} & \frac{\Delta Z}{Z_{21}} \\ \frac{1}{Z_{21}} & \frac{Z_{22}}{Z_{21}} \end{bmatrix}$
[Y]	$\begin{bmatrix} \frac{Y_{22}}{\Delta Y} & -\frac{Y_{12}}{\Delta Y} \\ -\frac{Y_{21}}{\Delta Y} & \frac{Y_{11}}{\Delta Y} \end{bmatrix}$	$\begin{bmatrix} Y_{11} & Y_{12} \\ Y_{21} & Y_{22} \end{bmatrix}$	$\begin{bmatrix} \frac{1}{Y_{11}} & -\frac{Y_{12}}{Y_{11}} \\ \frac{Y_{21}}{Y_{11}} & \frac{\Delta Y}{Y_{11}} \end{bmatrix}$	$\begin{bmatrix} -\frac{Y_{22}}{Y_{21}} & -\frac{1}{Y_{21}} \\ -\frac{\Delta Y}{Y_{21}} & -\frac{Y_{11}}{Y_{21}} \end{bmatrix}$
[h]	$\begin{bmatrix} \frac{\Delta h}{h_{22}} & \frac{h_{12}}{h_{22}} \\ -\frac{h_{21}}{h_{22}} & \frac{1}{h_{22}} \end{bmatrix}$	$\begin{bmatrix} \frac{1}{h_{11}} & -\frac{h_{12}}{h_{11}} \\ \frac{h_{21}}{h_{11}} & \frac{\Delta h}{h_{11}} \end{bmatrix}$	$\begin{bmatrix} h_{11} & h_{12} \\ h_{21} & h_{22} \end{bmatrix}$	$\begin{bmatrix} -\frac{\Delta h}{h_{21}} & -\frac{h_{11}}{h_{21}} \\ -\frac{h_{22}}{h_{21}} & -\frac{1}{h_{21}} \end{bmatrix}$
[ABCD]	$\begin{bmatrix} \frac{A}{C} & \frac{\Delta ABCD}{C} \\ \frac{1}{C} & \frac{D}{C} \end{bmatrix}$	$\begin{bmatrix} \frac{D}{B} & -\frac{\Delta ABCD}{B} \\ -\frac{1}{B} & \frac{A}{B} \end{bmatrix}$	$\begin{bmatrix} \frac{B}{D} & \frac{\Delta ABCD}{D} \\ -\frac{1}{D} & \frac{C}{D} \end{bmatrix}$	$\begin{bmatrix} A & B \\ C & D \end{bmatrix}$

4.3.2 Analysis of Microwave Amplifier

In this section we consider, by way of an example, the use of the conversion between different network representations to analyze a relatively complicated circuit. The basis of the analysis is the circuit diagram of a particular microwave amplifier shown in Figure 4-10.

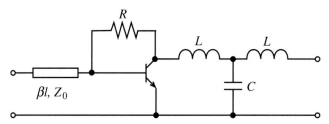

Figure 4-10 Microwave amplifier circuit diagram.

The first step is to break down the circuit into smaller, simpler subnetworks. This can be accomplished in several ways, one of which is shown in Figure 4-11.

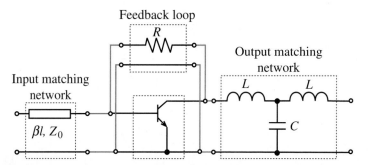

Figure 4-11 Subnetwork representation of the microwave amplifier.

As shown in this figure, the amplifier is divided into a set of four subcircuits. The input matching network consists of a transmission line (for convenience only the upper trace is shown) and is cascaded with a parallel combination of the transistor and a feedback loop. This circuit is then cascaded with an output matching network.

For the transistor we will use a high-frequency hybrid Pi-network model (see also Chapter 7), which is shown in Figure 4-12.

The derivation of the h parameters is left as a problem (Problem 4.14 at the end of this chapter). Here, we only list the resulting **h**-matrix for the transistor:

$$h_{11} = h_{ie} = \frac{r_{BE}}{1 + j\omega(C_{BE} + C_{BC})r_{BE}} \qquad (4.31a)$$

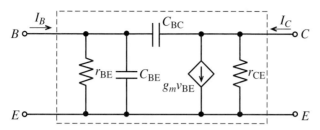

Figure 4-12 High-frequency hybrid transistor model.

$$h_{12} = h_{re} = \frac{j\omega C_{BC} r_{BE}}{1 + j\omega(C_{BE} + C_{BC})r_{BE}} \qquad (4.31b)$$

$$h_{21} = h_{fe} = \frac{r_{BE}(g_m - j\omega C_{BC})}{1 + j\omega(C_{BE} + C_{BC})r_{BE}} \qquad (4.31c)$$

$$h_{22} = h_{oe} = \frac{1}{r_{CE}} + \frac{j\omega C_{BC}(1 + g_m r_{BE} + j\omega C_{BE} r_{BE})}{1 + j\omega(C_{BE} + C_{BC})r_{BE}}. \qquad (4.31d)$$

To compute the matrix for the parallel combination of the transistor and the feedback loop resistor, we have to convert the **h**-matrix into a **Y**-matrix called $[\mathbf{Y}]_{tr}$ in order to apply the summation rule (4.20). To accomplish this, we can use formulas from Table 4-2 and add the result to the **Y**-matrix of the feedback resistor. The admittance matrix for the feedback resistor can be derived either directly using the definition of the **Y**-matrix or by converting the *ABCD*-parameters derived in Example 4-4 into the **Y**-form. The result of these computations is

$$\begin{bmatrix} Y_{11} & Y_{12} \\ Y_{21} & Y_{22} \end{bmatrix}_R = \begin{bmatrix} R^{-1} & -R^{-1} \\ -R^{-1} & R^{-1} \end{bmatrix}. \qquad (4.32)$$

After the summation, we obtain the admittance matrix for the parallel combination of the transistor and the feedback resistor $[\mathbf{Y}]_{tr+R}$.

The same result could have been obtained if we had noticed that the feedback resistor is connected in parallel with the capacitor C_{BC} of the transistor. Thus, to obtain the **h** matrix of the parallel combination of the feedback resistor and the transistor, we simply need to replace C_{BC} of the transistor with $C_{BC} + 1/(j\omega R)$.

The final step in the analysis is to multiply the **ABCD**-matrices for the input matching network (label: IMN), the transistor with feedback resistor (label: tr + R), and the output matching network (label: OMN)

$$\begin{bmatrix} A & B \\ C & D \end{bmatrix}_{amp} = \begin{bmatrix} A & B \\ C & D \end{bmatrix}_{IMN} \begin{bmatrix} A & B \\ C & D \end{bmatrix}_{tr+R} \begin{bmatrix} A & B \\ C & D \end{bmatrix}_{OMN} \qquad (4.33)$$

where the **ABCD**-matrices for the matching networks are found using the results from Table 4-1:

$$\begin{bmatrix} A & B \\ C & D \end{bmatrix}_{\text{IMN}} = \begin{bmatrix} \cos\beta l & jZ_0\sin\beta l \\ \dfrac{j\sin\beta l}{Z_0} & \cos\beta l \end{bmatrix} \tag{4.34}$$

$$\begin{bmatrix} A & B \\ C & D \end{bmatrix}_{\text{OMN}} = \begin{bmatrix} 1-\omega^2 LC & 2j\omega L - j\omega^3 L^2 C \\ j\omega C & 1-\omega^2 LC \end{bmatrix}. \tag{4.35}$$

Due to rather lengthy expressions, we are not presenting the final result for the *ABCD*-parameters of the entire amplifier. Instead, we urge the interested readers to perform these computations by relying on a mathematical spreadsheet program of their choice (MathCad, Matlab, Mathematica, etc.). One of the results of these computations is shown in Figure 4-13, where the small-signal current gain of the amplifier with short-circuited output (inverse of the *D* coefficient) is plotted versus frequency for different values of the feedback resistor.

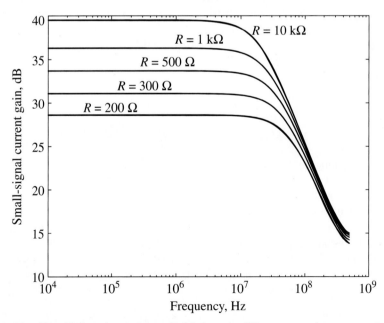

Figure 4-13 Small-signal current gain of the amplifier versus frequency for different values of the feedback resistor.

The computations are based on the circuit in Figure 4-11 with $L = 1$ nH, $C = 10$ pF, transmission line length of $l = 5$ cm, phase velocity equal to 65% of the speed of light, and $Z_o = 50\ \Omega$. The transistor is described by the following set of values: $r_{BE} = 520\ \Omega$, $r_{CE} = 80\ \text{k}\Omega$, $C_{BE} = 10$ pF, $C_{BC} = 1$ pF, and $g_m = 0.192$ S.

4.4 Scattering Parameters

In almost all databooks and technical literature regarding RF systems, the **scattering** or *S*-**parameter representation** plays a central role. This importance is derived from the fact that practical system characterizations can no longer be accomplished through simple open- or short-circuit measurements, as it is customarily done in low-frequency applications, and as discussed at the beginning of this chapter. We should recall what happens when we attempt to create a short circuit with a wire: the wire itself possesses an inductance that can be of substantial magnitude at high frequency. Also, the open circuit leads to capacitive loading at the terminal. In either case, the open/short-circuit conditions needed to determine *Z*-, *Y*-, *h*-, and *ABCD*-parameters can no longer be guaranteed. Moreover, when dealing with wave propagation phenomena, it is not desirable to introduce a reflection coefficient whose magnitude is unity. For instance, the terminal discontinuity will cause undesirable voltage and/or current wave reflections, possibly leading to oscillations that can result in the destruction of the device. With the *S*-parameters, the RF engineer has a tool to characterize the two-port network description of practically all RF devices without requiring unachievable terminal conditions, or causing harm to the **device under test** (DUT). The *S*-parameters denote the fraction of incident power reflected at a port and transmitted to other ports. The addition of phase information allows the complete description of any linear circuit.

4.4.1 Definition of Scattering Parameters

Simply put, *S*-parameters are power wave descriptors that permit us to define the input-output relations of a network in terms of incident and reflected power waves. With reference to Figure 4-14, we define an incident *normalized* power wave a_n and a reflected *normalized* power wave b_n as follows:

$$a_n = \frac{1}{2\sqrt{Z_0}}(V_n + Z_0 I_n) \qquad (4.36\text{a})$$

$$b_n = \frac{1}{2\sqrt{Z_0}}(V_n - Z_0 I_n) \qquad (4.36\text{b})$$

where the index *n* refers either to port number 1 or 2. The impedance Z_0 is the characteristic impedance of the connecting lines on the input and output side of the network. Under more general conditions, the line impedance on the input side can differ from the line impedance on the output side. Even transitions between transmission lines and waveguides

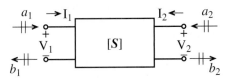

Figure 4-14 Convention used to define S-parameters for a two-port network.

can be described by S-parameters. However, for our initial discussion, we will keep things simple and assume that both impedances are the same.

Inverting (4.36) leads to the following voltage and current expressions:

$$V_n = \sqrt{Z_0}(a_n + b_n) \tag{4.37a}$$

$$I_n = \frac{1}{\sqrt{Z_0}}(a_n - b_n) \tag{4.37b}$$

The physical meaning of (4.36) becomes clear when we recall the equations for power:

$$P_n = \frac{1}{2}\text{Re}\{V_n I_n^*\} = \frac{1}{2}\left(|a_n|^2 - |b_n|^2\right). \tag{4.38}$$

Isolating forward and backward traveling wave components in (4.37), we immediately see

$$a_n = \frac{V_n^+}{\sqrt{Z_0}} = \sqrt{Z_0} I_n^+ \tag{4.39a}$$

$$b_n = \frac{V_n^-}{\sqrt{Z_0}} = -\sqrt{Z_0} I_n^- \tag{4.39b}$$

which is consistent with the definitions (4.37) since

$$V_n = V_n^+ + V_n^- = Z_0 I_n^+ - Z_0 I_n^-. \tag{4.40}$$

Based on the directional convention shown in Figure 4-14, we are now in a position to define the S-parameters:

$$\begin{Bmatrix} b_1 \\ b_2 \end{Bmatrix} = \begin{bmatrix} S_{11} & S_{12} \\ S_{21} & S_{22} \end{bmatrix} \begin{Bmatrix} a_1 \\ a_2 \end{Bmatrix} \tag{4.41}$$

where the terms are

$$S_{11} = \left.\frac{b_1}{a_1}\right|_{a_2 = 0} \equiv \frac{\text{reflected power wave at port 1}}{\text{incident power wave at port 1}} \tag{4.42a}$$

$$S_{21} = \left.\frac{b_2}{a_1}\right|_{a_2=0} \equiv \frac{\text{transmitted power wave at port 2}}{\text{incident power wave at port 1}} \quad (4.42b)$$

$$S_{22} = \left.\frac{b_2}{a_2}\right|_{a_1=0} \equiv \frac{\text{reflected power wave at port 2}}{\text{incident power wave at port 2}} \quad (4.42c)$$

$$S_{12} = \left.\frac{b_1}{a_2}\right|_{a_1=0} \equiv \frac{\text{transmitted power wave at port 1}}{\text{incident power wave at port 2}}. \quad (4.42d)$$

We observe that the conditions $a_2 = 0$ and $a_1 = 0$ imply that no power waves are returned to the network at either port 2 or port 1. These conditions can only be ensured when the connecting transmission lines are terminated into their characteristic impedances.

Since the S-parameters are closely related to power relations, we can express the normalized input and output waves in terms of time-averaged power. With reference to Section 2.1.2, we note that the average power at port 1 is given by

$$P_1 = \frac{1}{2}\frac{|V_1^+|^2}{Z_0}(1 - |\Gamma_{in}|^2) = \frac{1}{2}\frac{|V_1^+|^2}{Z_0}(1 - |S_{11}|^2) \quad (4.43)$$

where the reflection coefficient at the input side is expressed in terms of S_{11} under matched output according to the following argument:

$$\Gamma_{in} = \frac{V_1^-}{V_1^+} = \left.\frac{b_1}{a_1}\right|_{a_2=0} = S_{11}. \quad (4.44)$$

This also allows us to redefine the VSWR at port 1 in terms of S_{11} as

$$\text{VSWR} = \frac{1 + |S_{11}|}{1 - |S_{11}|}. \quad (4.45)$$

Furthermore, based on (4.39a) we can identify the incident power in (4.43) and express it in terms of a_1:

$$\frac{1}{2}\frac{|V_1^+|^2}{Z_0} = P_{inc} = \frac{|a_1|^2}{2} \quad (4.46)$$

which is the maximum available power from the generator. Using (4.46) and (4.44) in (4.43) gives us the total power at port 1 (under matched output condition) expressed as a combination of incident and reflected power:

$$P_1 = P_{inc} + P_{ref} = \frac{1}{2}(|a_1|^2 - |b_1|^2) = \frac{|a_1|^2}{2}(1 - |\Gamma_{in}|^2). \quad (4.47)$$

If the reflection coefficient S_{11} is zero, all available power from the source is delivered to port 1 of the network. An identical analysis at port 2 yields

$$P_2 = \frac{1}{2}(|a_2|^2 - |b_2|^2) = \frac{|a_2|^2}{2}(1 - |\Gamma_{out}|^2). \tag{4.48}$$

4.4.2 Meaning of S-Parameters

As already mentioned in the previous section, the S-parameters can only be determined under conditions of perfect matching on the input or output side. For instance, in order to record S_{11} and S_{21} we have to ensure that on the output side, the line impedance Z_0 is matched for $a_2 = 0$ to be enforced, as shown in Figure 4-15.

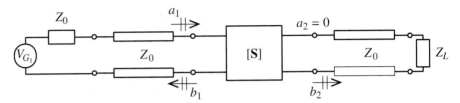

Figure 4-15 Measurement of S_{11} and S_{21} by matching the line impedance Z_0 at port 2 through a corresponding load impedance $Z_L = Z_0$.

This configuration allows us to compute S_{11} by finding the input reflection coefficient:

$$S_{11} = \Gamma_{in} = \frac{Z_{in} - Z_0}{Z_{in} + Z_0}. \tag{4.49}$$

In addition, taking the logarithm of the magnitude of S_{11} gives us the return loss in dB

$$RL = -20\log|S_{11}|. \tag{4.50}$$

Moreover, with port 2 properly terminated, we find

$$S_{21} = \frac{b_2}{a_1}\bigg|_{a_2 = 0} = \frac{V_2^-/\sqrt{Z_0}}{(V_1 + Z_0 I_1)/(2\sqrt{Z_0})}\bigg|_{I_2^+ = 0,\, V_2^+ = 0}. \tag{4.51}$$

Since $a_2 = 0$, we can set to zero the positive traveling voltage and current waves at port 2. Replacing V_1 by the generator voltage V_{G1} minus the voltage drop over the source impedance Z_0, $V_{G1} - Z_0 I_1$ gives

$$S_{21} = \frac{2V_2^-}{V_{G1}} = \frac{2V_2}{V_{G1}}. \tag{4.52}$$

Here, we observe that the voltage recorded at port 2 is directly related to the generator voltage and thus specifies the **forward voltage gain** of the network. To find the **forward power gain**, we square (4.52) to obtain

$$G_0 = |S_{21}|^2 = \left|\frac{V_2}{V_{G1}/2}\right|^2. \qquad (4.53)$$

The term G_0 is often expressen in dB. For lossy networks we usually use the insertion loss (inverse of G_0):

$$\text{IL[dB]} = -10\log|S_{21}|^2 = -20\log|S_{21}|. \qquad (4.54)$$

If we reverse the measurement procedure and attach a generator voltage V_{G2} to port 2 and properly terminate port 1, as shown in Figure 4-16, we can determine the remaining two S-parameters S_{22} and S_{12}.

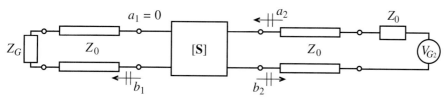

Figure 4-16 Measurement of S_{22} and S_{12} by matching the line impedance Z_0 at port 1 through a corresponding input impedance $Z_G = Z_0$.

To compute S_{22} we need to find the output reflection coefficient Γ_{out} in a similar way as for S_{11}:

$$S_{22} = \Gamma_{\text{out}} = \frac{Z_{\text{out}} - Z_0}{Z_{\text{out}} + Z_0} \qquad (4.55)$$

and for S_{12}

$$S_{12} = \left.\frac{b_1}{a_2}\right|_{a_1=0} = \left.\frac{V_1^-/\sqrt{Z_0}}{(V_2 + Z_0 I_2)/(2\sqrt{Z_0})}\right|_{I_1^+ = 0,\, V_1^+ = 0}. \qquad (4.56)$$

The term S_{12} can further be manipulated through the substitution of V_2 by $V_{G2} - Z_0 I_2$, leading to the form

$$S_{12} = \frac{2V_1^-}{V_{G2}} = \frac{2V_1}{V_{G2}} \qquad (4.57)$$

known as the **reverse voltage gain** and whose square $|S_{12}|^2$ is identified as **reverse power gain**. While S_{11} and S_{22} can be directly computed as part of the impedance definitions, S_{12} and S_{21} require the replacement of the defining voltages by the appropriate network parameters. In the following example, the S-parameters are computed for a simple three-element network.

—RF&MW→

Example 4-7. Determination of T-network attenuator elements

Find the S-parameters and the resistive elements for the 3 dB attenuator network shown in Figure 4-17(a) assuming that the network is placed into a transmission line section with a characteristic line impedance of $Z_0 = 50\ \Omega$.

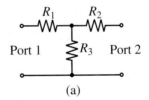

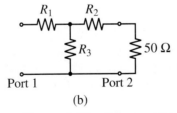
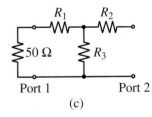

Figure 4-17 *S*-parameter computation for a T-network.
(a) circuit diagram; (b) circuit for S_{11} and S_{21} measurements;
(c) circuit for S_{12} and S_{22} measurements.

Solution. An attenuator should be matched to the line impedance and must therefore meet the requirement $S_{11} = S_{22} = 0$. As a result, based on Figure 4-17(b) and consistent with (4.49), we set

$$Z_{in} = R_1 + \frac{R_3(R_2 + 50\ \Omega)}{(R_3 + R_2 + 50\ \Omega)} = 50\ \Omega.$$

Because of symmetry, it is immediately clear that $R_1 = R_2$. We now investigate the voltage $V_2 = V_2^-$ at port 2 in terms of $V_1 = V_1^+$. According to the circuit configuration shown in Figure 4-17(c), the following expression is obtained

$$V_2 = \left(\frac{\dfrac{R_3(R_1 + 50\ \Omega)}{R_3 + R_1 + 50\ \Omega}}{\dfrac{R_3(R_1 + 50\ \Omega)}{R_3 + R_1 + 50\ \Omega} + R_1} \right) \left(\frac{50\ \Omega}{50\ \Omega + R_1} \right) V_1.$$

For a 3 dB attenuation, we require

$$S_{21} = \frac{2V_2}{V_{G1}} = \frac{V_2}{V_1} = \frac{1}{\sqrt{2}} = 0.707 = S_{12}.$$

Setting the ratio of V_2/V_1 to 0.707 in the preceding equation allows us, in combination with the input impedance expression, to determine R_1 and R_3. After simplification it is seen that

$$R_1 = R_2 = \frac{\sqrt{2}-1}{\sqrt{2}+1} Z_0 = 8.58\ \Omega$$

and $R_3 = 2\sqrt{2} \cdot Z_0 = 141.4\ \Omega$.

The choice of the resistor network ensures that at the input and output ports an impedance of 50 Ω is maintained. This implies that this network can be inserted into a 50 Ω transmission line section without causing reflections, only introducing the desired insertion loss.

The definitions for the S-parameters require appropriate termination. For instance, if S_{11} is desired, the transmission line connected to port 2 has to be terminated into its characteristic line impedance. This does not necessarily mean that the output impedance Z_{out} of the network has to be matched to the line impedance Z_0. Rather, the line impedance must be matched to ensure that no wave is reflected from the load, as implied by $a_2 = 0$. If this is not the case, we will see in Section 4.4.5 how S_{11} is modified.

4.4.3 Chain Scattering Matrix

To extend the concept of the S-parameter representation to cascaded networks, it is more efficient to rewrite the power wave expressions arranged in terms of input and output ports. This results in the **chain scattering matrix** notation. That is,

$$\begin{Bmatrix} a_1 \\ b_1 \end{Bmatrix} = \begin{bmatrix} T_{11} & T_{12} \\ T_{21} & T_{22} \end{bmatrix} \begin{Bmatrix} b_2 \\ a_2 \end{Bmatrix} \qquad (4.58)$$

It is seen that cascading dual-port networks becomes a simple multiplication. This is apparent in Figure 4-18, where network A (given by matrix $[\mathbf{T}]_A$) is connected to network B (given by matrix $[\mathbf{T}]_B$).

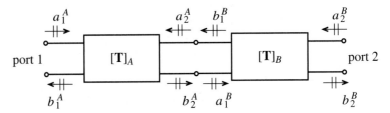

Figure 4-18 Cascading of two networks A and B.

If network A is described by the relation

$$\begin{Bmatrix} a_1^A \\ b_1^A \end{Bmatrix} = \begin{bmatrix} T_{11}^A & T_{12}^A \\ T_{21}^A & T_{22}^A \end{bmatrix} \begin{Bmatrix} b_2^A \\ a_2^A \end{Bmatrix} \tag{4.59a}$$

and network B by

$$\begin{Bmatrix} a_1^B \\ b_1^B \end{Bmatrix} = \begin{bmatrix} T_{11}^B & T_{12}^B \\ T_{21}^B & T_{22}^B \end{bmatrix} \begin{Bmatrix} b_2^B \\ a_2^B \end{Bmatrix} \tag{4.59b}$$

we notice, based on the parameter convention shown in Figure 4-18, that

$$\begin{Bmatrix} b_2^A \\ a_2^A \end{Bmatrix} = \begin{Bmatrix} a_1^B \\ b_1^B \end{Bmatrix}. \tag{4.60}$$

Thus, for the combined system, we conclude

$$\begin{Bmatrix} a_1^A \\ b_1^A \end{Bmatrix} = \begin{bmatrix} T_{11}^A & T_{12}^A \\ T_{21}^A & T_{22}^A \end{bmatrix} \begin{bmatrix} T_{11}^B & T_{12}^B \\ T_{21}^B & T_{22}^B \end{bmatrix} \begin{Bmatrix} b_2^B \\ a_2^B \end{Bmatrix} \tag{4.61}$$

which is the desired matrix multiplication. Therefore, the chain scattering matrix plays a similar role as the **ABCD**-matrix discussed earlier.

The conversion from the **S**-matrix to the chain matrix notation follows identical steps as outlined in Section 4.3.1. For instance, to compute T_{11} we see that

$$T_{11} = \left.\frac{a_1}{b_2}\right|_{a_2=0} = \frac{a_1}{S_{21}a_1} = \frac{1}{S_{21}}. \tag{4.62}$$

Similarly,

$$T_{12} = -\frac{S_{22}}{S_{21}} \tag{4.63}$$

$$T_{21} = \frac{S_{11}}{S_{21}} \tag{4.64}$$

$$T_{22} = \frac{-(S_{11}S_{22} - S_{12}S_{21})}{S_{21}} = \frac{-\Delta S}{S_{21}}. \tag{4.65}$$

Conversely, when the chain scattering parameters are given and we need to convert to S-parameters, we find the following relations:

$$S_{11} = \frac{b_1}{a_1}\bigg|_{a_2 = 0} = \frac{T_{21}b_2}{T_{11}b_2} = \frac{T_{21}}{T_{11}} \tag{4.66}$$

$$S_{12} = \frac{T_{11}T_{22} - T_{21}T_{12}}{T_{11}} = \frac{\Delta T}{T_{11}} \tag{4.67}$$

$$S_{21} = \frac{1}{T_{11}} \tag{4.68}$$

$$S_{22} = -\frac{T_{12}}{T_{11}}. \tag{4.69}$$

Alternatively, a matrix manipulation as discussed in the next section could have been carried out with the same result.

4.4.4 Conversion between Z- and S-Parameters

We have already seen how certain S-parameters can be defined in terms of input and output impedances of a network [i.e., equations (4.49) and (4.55)]. In this section, we go through a formal conversion between the Z- and S-parameter sets. Once this interrelation is established, we are able to formulate conversion links between all six network parameter sets (S, Z, Y, ABCD, h, T).

To find the conversion between the previously defined S-parameters and the Z-parameters, let us begin with the defining S-parameter relation in matrix notation [i.e., (4.41)]

$$\{\mathbf{b}\} = [S]\{\mathbf{a}\}. \tag{4.70}$$

Multiplying by $\sqrt{Z_0}$ gives

$$\sqrt{Z_0}\{\mathbf{b}\} = \{\mathbf{V}^-\} = \sqrt{Z_0}[S]\{\mathbf{a}\} = [S]\{\mathbf{V}^+\}. \tag{4.71}$$

Adding $\{\mathbf{V}^+\} = \sqrt{Z_0}\{\mathbf{a}\}$ to both sides results in

$$\{\mathbf{V}\} = [S]\{\mathbf{V}^+\} + \{\mathbf{V}^+\} = ([S] + [E])\{\mathbf{V}^+\} \tag{4.72}$$

where [E] is the identity matrix. To compare this form with the impedance expression $\{V\} = [Z]\{I\}$, we have to express $\{V^+\}$ in terms of $\{I\}$. This is accomplished by first subtracting $[S]\{V^+\}$ from both sides of $\{V^+\} = \sqrt{Z_0}\{a\}$; that is,

$$\{V^+\} - [S]\{V^+\} = \sqrt{Z_0}(\{a\} - \{b\}) = Z_0\{I\}. \tag{4.73}$$

Now, by isolating $\{V^+\}$, it is seen that

$$\{V^+\} = Z_0([E] - [S])^{-1}\{I\}. \tag{4.74}$$

Substituting (4.74) into (4.72) yields the desired result of

$$\{V\} = ([S] + [E])\{V^+\} = Z_0([S] + [E])([E] - [S])^{-1}\{I\} \tag{4.75}$$

or

$$[Z] = Z_0([S] + [E])([E] - [S])^{-1}. \tag{4.76}$$

Explicit evaluation yields

$$\begin{bmatrix} Z_{11} & Z_{12} \\ Z_{21} & Z_{22} \end{bmatrix} = Z_0 \begin{bmatrix} 1 + S_{11} & S_{12} \\ S_{21} & 1 + S_{22} \end{bmatrix} \begin{bmatrix} 1 - S_{11} & -S_{12} \\ -S_{21} & 1 - S_{22} \end{bmatrix}^{-1}$$

$$= \frac{Z_0 \begin{bmatrix} 1 + S_{11} & S_{12} \\ S_{21} & 1 + S_{22} \end{bmatrix}}{(1 - S_{11})(1 - S_{22}) - S_{21}S_{12}} \begin{bmatrix} 1 - S_{22} & S_{12} \\ S_{21} & 1 - S_{11} \end{bmatrix}. \tag{4.77}$$

Identifying individual terms is now easily carried out. A complete summary of all network coefficient sets is given in Appendix C.

4.4.5 Signal Flowgraph Modeling

The analysis of RF networks and their overall interconnection is greatly facilitated through signal flowgraphs as commonly used in system and control theory. As originally introduced to seismology and remote sensing, wave propagation can be associated with directed paths and associated nodes connecting these paths. In this section we will briefly summarize key principles needed for signal flow network analysis.

The main concepts required to construct a flowgraph are as follows:

1. **Nodes** that are deployed to identify power waves such as a_1, b_1, a_2, b_2 when dealing with S-parameters
2. **Branches** are needed to indicate power wave propagation through the network
3. Addition and subtraction of power wave values in accordance with the directions of the branches.

We will now discuss these three items in detail. To this end let us consider a section of a transmission line that is terminated in a load impedance Z_L, as seen in Figure 4-19.

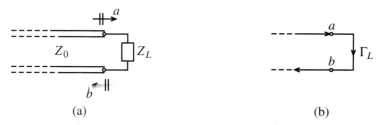

Figure 4-19 Terminated transmission line segment with incident and reflected power wave description. (a) Conventional form, and (b) Signal flow form.

Even though we could use voltage values as node identifiers, it is the S-parameter representation that finds widespread use. In Figure 4-19(b) the nodes a and b are connected through the load reflection coefficient Γ_L. This makes sense since the reflection coefficient is the ratio b/a, so that it simply states that node b is found as a result of multiplying node a by Γ_L. This is depicted in generic form in Figure 4-20.

(a) Source node a, which launches wave. (b) Sink node b, which receives wave.

(c) Branch connecting source and sink.

Figure 4-20 Generic source node (a), receiver node (b), and the associated branch connection (c).

In terms of notation, we can encode the situation shown in Figure 4-20 as

$$b = \Gamma a. \tag{4.78}$$

A more complicated situation arises when we need to make the transmission line circuit shown in Figure 4-19 more realistic by including a source term, as seen in Figure 4-21. Unlike Figure 4-19, the nodes a and b are preceded by two additional nodes that we shall denote a' and b'. The ratio b'/a' defines the source reflection coefficient Γ_S as already discussed in Section 2.11. Here, we also see that b' is given by multiplying a' with the source reflection coefficient. By relying on the concept of summation, we define b' as the sum of b_S and $a'\Gamma_S$. Thus, the source b_S is

$$b_S = b' - a'\Gamma_S. \tag{4.79}$$

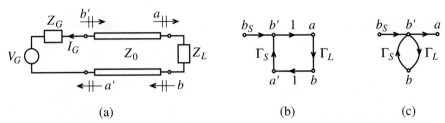

Figure 4-21 Terminated transmission line with source. (a) conventional form, (b) signal flow form, and (c) simplified signal flow form.

An explicit expression for b_S is obtained by noting that

$$V_S = V_G + I_G Z_G \tag{4.80}$$

based on an outflowing current convention (see Figure 4-21). This can be converted into the form

$$V_S^+ + V_S^- = V_G + Z_G \left(\frac{V_S^+}{Z_0} - \frac{V_S^-}{Z_0} \right). \tag{4.81}$$

Rearranging terms and dividing by $\sqrt{Z_0}$ gives

$$\frac{\sqrt{Z_0}}{Z_G + Z_0} V_G = \frac{V_S^-}{\sqrt{Z_0}} - \Gamma_S \frac{V_S^+}{\sqrt{Z_0}}. \tag{4.82}$$

When comparing (4.82) with (4.79), we immediately see that

$$b_S = \frac{\sqrt{Z_0}}{Z_G + Z_0} V_G. \tag{4.83}$$

An important conclusion can be drawn when expressing a' in (4.79) by $\Gamma_L b'$ so that we obtain

$$b' = b_S + \Gamma_L \Gamma_S b' = \frac{b_S}{1 - \Gamma_L \Gamma_S}. \tag{4.84}$$

This is a known as a self- or feedback loop (see Figure 4-22), which allows us to represent the nodes b_S and b' by a single branch whose value is given by (4.84).

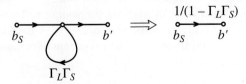

Figure 4-22 A self-loop that collapses to a single branch.

Scattering Parameters

All signal flowgraph principles can therefore be reduced to six building blocks, as summarized in Table 4-3. By way of an example, let us analyze a more complicated RF circuit consisting of a sourced and terminated dual-port network.

Table 4-3 Signal flowgraph building blocks.

Description	Graphical Representation
Nodal Assignment	
Branch	
Series Connection	
Parallel Connection	
Splitting of Branches	
Self-Loop	

Example 4-8. Flowgraph analysis of a dual-port network

For the network shown in Figure 4-23, find the ratios of b_1/a_1 and a_1/b_S. Assume unity for the multiplication factor of the transmission line segments.

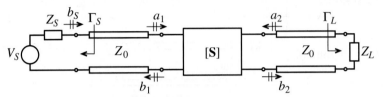

(a) Circuit representation

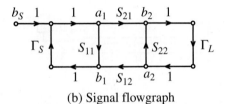

(b) Signal flowgraph

Figure 4-23 Sourced and terminated two-port network.

Solution. The process of setting up the individual ratios is explained best by going through a step-by-step simplification for the ratio a_1/b_S employing the rules summarized in Table 4-3. Figure 4-24 depicts the five steps.

Step 1 Splitting of the rightmost loop between b_2 and a_2, leading to the self-loop $S_{22}\Gamma_L$.

Step 2 Decomposition of the self-loop between branches a_1 and b_2, resulting in the multiplication factor $S_{21}/(1 - S_{22}\Gamma_L)$, which can be combined with Γ_L and S_{12}.

Step 3 Series and parallel connections between a_1 and b_1, leading to the input reflection coefficient

$$\Gamma_{in} = \frac{b_1}{a_1} = S_{11} + \frac{S_{12}S_{21}}{1 - S_{22}\Gamma_L}\Gamma_L.$$

Step 4 Splitting the loop into a self-loop, resulting in the multiplication factor

$$\left(S_{11} + \frac{S_{12}S_{21}}{1 - S_{22}\Gamma_L}\Gamma_L\right)\Gamma_S.$$

Scattering Parameters

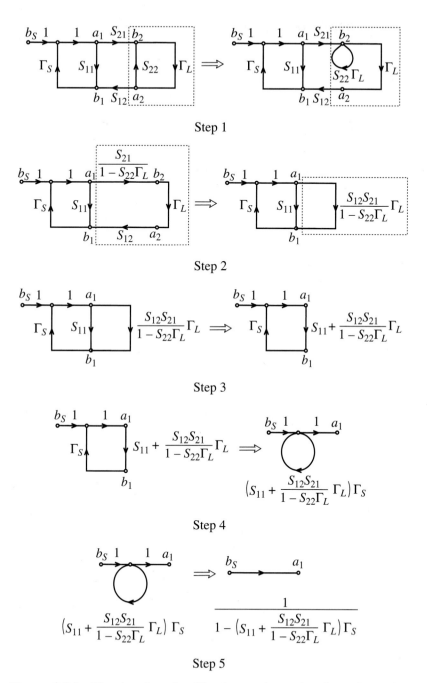

Figure 4-24 Step-by-step simplification to determine the ratio a_1/b_S.

Step 5 Decomposition of the self-loop at a_1, leading to the expression

$$a_1 = \frac{1}{1 - \left(S_{11} + \frac{S_{12}S_{21}}{1 - S_{22}\Gamma_L}\Gamma_L\right)\Gamma_S} b_S.$$

Rearranging and simplification leads to the final form:

$$\frac{a_1}{b_S} = \frac{1 - S_{22}\Gamma_L}{1 - (S_{11}\Gamma_S + S_{22}\Gamma_L + S_{12}S_{21}\Gamma_S) + S_{11}S_{22}\Gamma_S\Gamma_L}.$$

The preceding derivation follows a pattern similar to finding the transfer function of a control system or a signal processor. Even complicated circuits can be reduced efficiently and quickly to establish the nodal dependcies.

The preceding example points out what will happen if the matching condition for recording the S-parameters is not satisfied. As we know, if we compute S_{11} we need to ensure that $a_2 = 0$. However, if $a_2 \neq 0$, as is the case in the preceding example, we see that S_{11} is modified by the additional term $S_{12}S_{21}\Gamma_L/(1 - S_{22}\Gamma_L)$.

4.4.6 Generalization of S-Parameters

In our discussion thus far it was assumed that the characteristic line impedance at both ports has the same value Z_0. However, this does not have to be the case. Indeed, if we assume that port 1 is connected to line impedance Z_{01} and port 2 to impedance Z_{02}, we have to represent the voltage and current waves at the respective port ($n = 1, 2$) as

$$V_n = V_n^+ + V_n^- = \sqrt{Z_{0n}}(a_n + b_n) \tag{4.85}$$

and

$$I_n = \frac{V_n^+}{Z_{0n}} - \frac{V_n^-}{Z_{0n}} = \frac{a_n}{\sqrt{Z_{0n}}} - \frac{b_n}{\sqrt{Z_{0n}}} \tag{4.86}$$

where we immediately observe

$$a_n = \frac{V_n^+}{\sqrt{Z_{0n}}}, \; b_n = \frac{V_n^-}{\sqrt{Z_{0n}}}. \tag{4.87}$$

Scattering Parameters

These equations allow the definition of the S-parameters as follows:

$$S_{ij} = \left.\frac{b_i}{a_j}\right|_{a_n = 0 (n \neq j)} = \left.\frac{V_i^-/\sqrt{Z_{0i}}}{V_j^+/\sqrt{Z_{0j}}}\right|_{V_n^+ = 0 (n \neq j)} \quad (4.88)$$

When compared to the previous S-parameter definitions, we notice that scaling by the appropriate line impedances has to be taken into account. It should also be apparent that although the focus of our derivations was a two-port network, the preceding formulas can be extended to an N-port network where $n = 1, \ldots, N$.

A second consideration is related to the fact that practical measurements involve the determination of the network S-parameters through transmission lines of finite length. In this case we need to investigate a system where the measurement planes are shifted away from the actual network, as depicted in Figure 4-25.

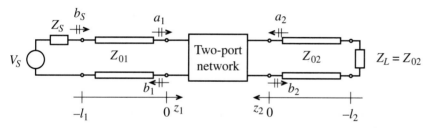

Figure 4-25 Two-port network with finite-length transmission line segments.

An incident voltage wave launched from the power supply will have to travel a distance l_1 in order to reach port 1. Consistent with the notation introduced in Section 2.9, we note that at port 1 the incident voltage is given as

$$V_{in}^+(z_1 = 0) = V_1^+ \quad (4.89)$$

and, at the generator side, as

$$V_{in}^+(z_1 = -l_1) = V_1^+ e^{-j\beta_1(-l_1)}. \quad (4.90)$$

The reflected voltage wave at port 1 can be cast in the form

$$V_{in}^-(z_1 = 0) = V_1^- \quad (4.91)$$

and

$$V_{in}^-(z_1 = -l_1) = V_1^- e^{j\beta_1(-l_1)} \quad (4.92)$$

where, as usual, β_1 stands for the phase constant of line 1. In an identical fashion, the voltage behavior at port 2 can be formulated by simply substituting V_{in} with V_{out} and V_1 with V_2 as well as β_1 with β_2. The preceding equations can be combined in matrix form

$$\begin{Bmatrix} V_{in}^+(-l_1) \\ V_{out}^+(-l_2) \end{Bmatrix} = \begin{bmatrix} e^{j\beta_1 l_1} & 0 \\ 0 & e^{j\beta_2 l_2} \end{bmatrix} \begin{Bmatrix} V_1^+ \\ V_2^+ \end{Bmatrix} \qquad (4.93)$$

which links the impinging waves at the network ports to the corresponding voltages shifted by the electric lengths of the attached transmission line segments. For the reflected voltage waves we obtain the matrix form

$$\begin{Bmatrix} V_{in}^-(-l_1) \\ V_{out}^-(-l_2) \end{Bmatrix} = \begin{bmatrix} e^{-j\beta_1 l_1} & 0 \\ 0 & e^{-j\beta_2 l_2} \end{bmatrix} \begin{Bmatrix} V_1^- \\ V_2^- \end{Bmatrix}. \qquad (4.94)$$

As the discussion in Section 4.4.1 taught us, the S-parameters are linked to the coefficients a_n and b_n, which in turn can be expressed through voltages (if we assume $Z_{01} = Z_{02}$).

$$\begin{Bmatrix} V_1^- \\ V_2^- \end{Bmatrix} = \begin{bmatrix} S_{11} & S_{12} \\ S_{21} & S_{22} \end{bmatrix} \begin{Bmatrix} V_1^+ \\ V_2^+ \end{Bmatrix}. \qquad (4.95)$$

It is apparent that if transmission line segments are added, we have to replace the above voltages by the previously derived expressions, leading to the form

$$\begin{Bmatrix} V_{in}^-(-l_1) \\ V_{out}^-(-l_2) \end{Bmatrix} = \begin{bmatrix} e^{-j\beta_1 l_1} & 0 \\ 0 & e^{-j\beta_2 l_2} \end{bmatrix} \begin{bmatrix} S_{11} & S_{12} \\ S_{21} & S_{22} \end{bmatrix} \begin{bmatrix} e^{-j\beta_1 l_1} & 0 \\ 0 & e^{-j\beta_2 l_2} \end{bmatrix} \begin{Bmatrix} V_{in}^+(-l_1) \\ V_{out}^+(-l_2) \end{Bmatrix}. \qquad (4.96)$$

This final result reveals that the S-parameters for the shifted network are composed of three matrices. In terms of the coefficients, we see that

$$[S]^{SHIFT} = \begin{bmatrix} S_{11} e^{-j2\beta_1 l_1} & S_{12} e^{-j(\beta_1 l_1 + \beta_2 l_2)} \\ S_{21} e^{-j(\beta_1 l_1 + \beta_2 l_2)} & S_{22} e^{-j2\beta_2 l_2} \end{bmatrix}. \qquad (4.97)$$

The physical meaning of this form is easy to understand. The first matrix coefficient reveals that we have to take into account $2\beta_1 l_1$ or twice the travel time for the incident voltage to reach port 1 and, upon reflection, return. Similarly, for port 2 we see that the phase shift is $2\beta_2 l_2$. Moreover, the cross terms, which are closely related to the forward and reverse gains, require the additive phase shifts associated with transmission line 1 ($\beta_1 l_1$) and transmission line 2 ($\beta_2 l_2$), since the overall input/output configuration now consists of both line segments.

Scattering Parameters

Example 4-9. Input impedance computation of a transmission line based on the use of a signal flowgraph

A lossless transmission line system with characteristic line impedance Z_0 and length l is terminated into a load impedance Z_L and attached to a source voltage V_G and source impedance Z_G, as shown in Figure 4-26. (a) Draw the signal flowgraph and (b) derive the input impedance formula at port 1 from the signal flowgraph representation.

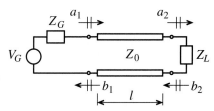

Figure 4-26 Transmission line attached to a voltage source and terminated by a load impedance.

Solution. (a) Consistent with our previously established signal flowgraph notation, we can readily convert Figure 4-26 into the form seen in Figure 4-27.

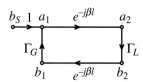

Figure 4-27 Signal flowgraph diagram for transmission line system in Figure 4-26.

(b) The input reflection coefficient at port 1 is given by

$$b_1 = \Gamma_L e^{-j2\beta l} a_1$$

which is exactly in the form given in Section 3.1, with $\Gamma_L = \Gamma_0$ and $l = d$. Thus

$$\Gamma_{in}(l) = \Gamma_L e^{-j2\beta l} = \frac{Z_{in} - Z_0}{Z_{in} + Z_0}.$$

Solving for Z_{in} yields the final result

$$Z_{in} = Z_0 \frac{1 + \Gamma_L e^{-j2\beta l}}{1 - \Gamma_L e^{-j2\beta l}}.$$

This example shows how the input impedance of a transmission line can be found quickly and elegantly by using signal flowgraph concepts.

4.4.7 Practical Measurements of S-Parameters

Measurement of the *S*-parameters of a two-port network requires reflection and transmission evaluations of traveling waves at both ports. One of the most popular methods is to use a vector network analyzer. The vector network analyzer is an instrument that can measure voltages in terms of magnitude and phase. Usually network analyzers have one output port, which provides the RF signal either from an internal source or an external signal generator, and three measurement channels, which are denoted as *R, A,* and *B* (see Figure 4-28).

The RF source is typically set to sweep over a specified frequency range. The measurement channel *R*, the reference port, is employed for measuring the incident wave. Channels *A* and *B* usually measure the reflected and transmitted waves. In general, the measurement channels *A* and *B* can be configured to record any two parameters with a single measurement setup. An example of the test arrangement that allows us to measure S_{11} and S_{21} is shown in Figure 4-28.

In this case, the value of S_{11} can be obtained by evaluating the ratio A/R, and S_{21} through computing B/R. To measure S_{12} and S_{22} we have to reverse the DUT. In Figure 4-28, the dual directional coupler allows the separation of the incident and reflected waves at the input port of the DUT. The bias tees are employed to provide necessary biasing conditions, such as a quiescent point for the DUT. Since the most common use of network analyzers is the characterization of two-port devices, bias tees, directional couplers, and necessary electronic switches as well as the RF sweep signal generator are all integral parts of most modern analyzers.

As we can see, a practical test arrangement is more complicated when compared with the simple ideal system described in Sections 4.4.4 and 4.4.6, where we assume that the DUT is connected to perfectly matched transmission lines of equal (Section 4.4.4) or unequal (Section 4.4.6) characteristic impedance. In a realistic measurement system, we cannot guarantee either matching conditions or ideality of the components. In fact, we have to consider all effects of the external components connected to the input and output ports of the DUT. Furthermore, the primary reference plane for measurements of complex voltages, which are then converted into *S*-parameters, is usually somewhere inside the networks analyzer. As a result, it is necessary to take into account not only attenuation and phase shifts due to the external components, but also portions of the internal structure of the network analyzer itself.

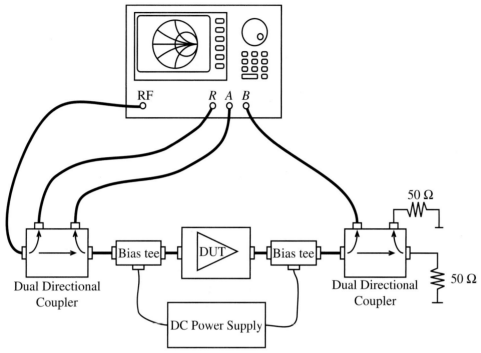

Figure 4-28 Measurement system for S_{11} and S_{21} parameters using a network analyzer.

In general, the measurement test arrangement can be reduced to the cascade of three networks depicted in Figure 4-29. Here, the signals R, A, B correspond to the reference port and channels A and B of the network analyzer. RF_{in} is the output line from the signal source. The branch denoted E_X represents possible leakage between the output of the signal source and channel B.

The network analyzer treats everything between the measurement reference planes as a single device. Therefore, our task is reduced to finding a way to calibrate the network analyzer such that it becomes possible to eliminate the effect of all undesired influences or parasitics. The main goal of a **calibration procedure** is to characterize the error boxes prior to measuring the DUT. This information can then be used by an internal computer to evaluate the error-free S-parameters of the actual DUT.

Assuming that the error box A network is reciprocal, we can state $E_{12} = E_{21}$. Therefore, we have to find six parameters ($E_{11}, E_{12}, E_{22}, E_X, E_R$, and E_T) to characterize the error boxes.

The simplest calibration method involves three or more known loads (open, short, and matched). The problem with this approach is that such standards are usually imperfect and are likely to introduce additional errors into the measurement procedures. These errors become especially significant at higher frequencies. To avoid the dependency on

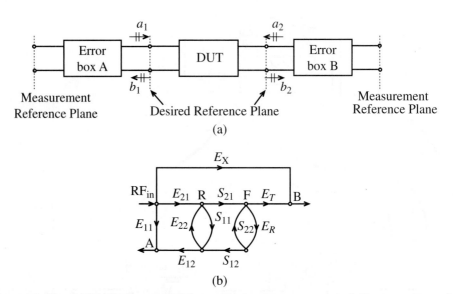

Figure 4-29 (a) Block diagram of the setup for measurement of S-parameters of a two-port network; (b) signal flowgraph of the measurement test setup.

the accuracy of calibration standards, several methods have been developed (see Eul and Schiek and Engen and Hoer, listed in the Further Reading section at the end of this chapter). In this section we will only consider the so-called **through-reflect-line** (TRL) technique (see Engen and Hoer).

The TRL calibration scheme does not rely on known standard loads. Instead, it is based on the use of three types of connections, which are shown in Figure 4-30. The Through connection is made by directly connecting ports 1 and 2 of the DUT. Next, the Reflect connection uses a load with high reflectivity. The reflection coefficient does not have to be known because it will be determined during the calibration process. The only requirement is that the load possesses the same reflection coefficient for both input and output ports. The Line connection is made by connecting ports 1 and 2 via a transmission line matched to the impedance of the error boxes. Usually, this impedance is close to 50 Ω. Before we continue with the actual analysis of each particular connection type, let us first consider the system as a general two-port network.

From Figure 4-29(b) it is seen that the signal at node B is a linear combination of the input RF signal (setting RF_{in} to unity) and the signal at node F:

$$B = E_X + E_T F. \qquad (4.98)$$

Applying the self-loop rule, we can write that signal at node F as

$$F = \frac{S_{21}}{1 - E_R S_{22}} R. \qquad (4.99)$$

Scattering Parameters

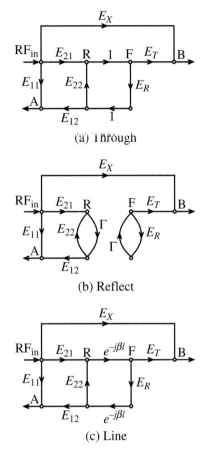

Figure 4-30 Signal flow graphs of TRL method: (a) Through, (b) Reflect, (c) Line configurations.

To compute the signal at port R, the same method as discussed in Example 4-8 can be used. In this example we first replaced the loop with the signal F through a self-loop and then performed the same transformation for the signal R. The result of these computations is

$$R = \frac{E_{21}}{1 - E_{22}\left(S_{11} + \dfrac{S_{12}S_{21}E_R}{1 - E_R S_{22}}\right)}. \tag{4.100}$$

Substituting (4.100) into (4.99) followed by the substitution of (4.99) into (4.98), we obtain an expression for signal B:

$$B = E_X + E_T \frac{S_{21}}{1 - E_R S_{22}} \frac{E_{21}}{1 - E_{22}\left(S_{11} + \dfrac{S_{12}S_{21}E_R}{1 - E_R S_{22}}\right)}. \tag{4.101}$$

Finally, the value for the signal at node A is obtained by using the summation rule:

$$A = E_{11} + \frac{E_{12}E_{21}}{1 - E_{22}\left(S_{11} + \frac{S_{12}S_{21}E_R}{1 - E_R S_{22}}\right)}\left(S_{11} + S_{12}E_R\frac{S_{21}}{1 - E_R S_{22}}\right). \quad (4.102)$$

If the measurement system does not introduce any errors, then $E_{12} = E_{21} = E_T = 1$ and $E_{11} = E_{22} = E_R = E_X = 0$. Substituting these values into (4.100), (4.101), and (4.102), we find that $R = 1$, $A = S_{11}$, and $B = S_{21}$, which shows the validity of the formulas.

Now we are ready to investigate the TRL connections in more detail. To avoid confusion, let us denote the measured signals R, A, and B for Through by subscript T, for Reflect by R, and for Line by L.

For the Through connection, we know that $S_{11} = S_{22} = 0$ and $S_{12} = S_{21} = 1$. Setting $E_{12} = E_{21}$ yields

$$R_T = \frac{E_{12}}{1 - E_{22}E_R} \quad (4.103a)$$

$$A_T = E_{11} + \frac{E_{12}^2}{1 - E_{22}E_R}E_R \quad (4.103b)$$

$$B_T = E_X + E_T\frac{E_{12}}{1 - E_{22}E_R}. \quad (4.103c)$$

For the Reflect connection, we have $S_{11} = S_{22} = \Gamma$ and $S_{12} = S_{21} = 0$. This results in the equations

$$R_R = \frac{E_{12}}{1 - E_{22}\Gamma} \quad (4.104a)$$

$$A_R = E_{11} + \frac{E_{12}^2\Gamma}{1 - E_{22}\Gamma} \quad (4.104b)$$

$$B_R = E_X. \quad (4.104c)$$

Finally, for the Line connection we see that $S_{11} = S_{22} = 0$ and $S_{12} = S_{21} = e^{-\gamma l}$, where l is the transmission line length and γ is the complex propagation constant ($\gamma = \alpha + j\beta$) that takes into account attenuation effects. The result is

$$R_L = \frac{E_{12}}{1 - E_{22}E_R e^{-2\gamma l}} \quad (4.105a)$$

$$A_L = E_{11} + \frac{E_{12}^2 E_R e^{-2\gamma l}}{1 - E_{22}E_R e^{-2\gamma l}} \quad (4.105b)$$

Scattering Parameters

$$B_L = E_X + E_T e^{-\gamma l} \frac{E_{12}}{1 - E_{22} E_R e^{-2\gamma l}}. \quad (4.105c)$$

Equations (4.103a)–(4.105c) allow us to solve for the unknown coefficients of the error boxes E_{11}, E_{12}, E_{22}, E_X, E_R, E_T, the reflection coefficient Γ, and the transmission line parameter $e^{-\gamma l}$. Knowing the error coefficients we are then in a position to process the measured data in order to obtain an error-free S-parameter set of the DUT.

Practically Speaking. Resistive attenuator

In this section, we measure the S-parameters of a printed circuit board implementation of a resistive attenuator. However, unlike the T-network discussed in Example 4-7, we have implemented a Pi-network.

The desired S-parameter values for an attenuator are $S_{11} = S_{22} = 0$ (matched input and output ports) and $S_{21} = S_{12} = a$. Ideally, the frequency-independent voltage gain a is less than unity. If this attenuator is inserted into a transmission line, we expect an insertion loss (IL = $-20 \log a$) with no reflections occurring at the ports. Unlike the previously discussed 3 dB T-network attenuator with 50 Ω ports, this Pi-network attenuator is sufficiently general to allow termination into arbitrary and different characteristic impedances at the ports. The schematic of this attenuator is presented in Figure 4-31.

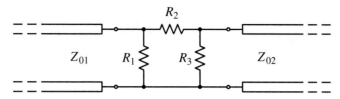

Figure 4-31 Resistive Pi-network attenuator with arbitrary characteristic impedances at the ports.

The mathematical expressions for the three resistors R_1, R_2, R_3 with port characteristic impedances Z_{01} and Z_{02} are:

$$R_1 = \frac{Z_{01}(1 - a^2)}{a^2 - 2a\sqrt{Z_{01}/Z_{02}} + 1}$$

$$R_2 = \frac{(1-a^2)\sqrt{Z_{01}Z_{02}}}{2a}$$

$$R_3 = \frac{Z_{02}(1-a^2)}{a^2 - 2a\sqrt{Z_{02}/Z_{01}} + 1}.$$

Furthermore, if we define $r = \max(Z_{01}, Z_{02})/\min(Z_{01}, Z_{02})$, we can show that the voltage gain is constrained by $a \leq \sqrt{r} - \sqrt{r-1}$ (see problem at the end of this chapter). Thus, if two different characteristic impedances are chosen, a certain minimum insertion loss is required. For example, with $Z_{01} = 50\ \Omega$ and $Z_{02} = 75\ \Omega$ we are restricted to a value of $a \leq 0.518$, or IL ≥ 5.72 dB.

In this example, however, we consider a basic 3 dB attenuator with 50 Ω characteristic impedance at both ports. According to the above expressions, we require $R_1 = R_3 = 291.4\ \Omega$ and $R_2 = 17.68\ \Omega$ to satisfy $a = 1/\sqrt{2}$. This corresponds to IL = 3.01 dB. Figure 4-32 shows the PCB implementation (board material is FR4) of this attenuator. Each of the three resistances is implemented as a parallel combination of two SMD (1206 footprint) resistors to reduce the parasitic inductance. Carbon (thick) film resistors with actual values of 576 Ω and 35.7 Ω (1% tolerance and 1/4 W power rating) are used in this design. The bottom side of the PCB contains a continuous ground plane with vias providing the necessary ground connections to the resistors. Two SMA edge connectors are used to connect the attenuator to a coaxial cable. The circuit board measures 1.1×0.75 inch.

Figure 4-32 Pi-network of a 3 dB attenuator with SMA connectors.

Figure 4-33 presents S_{11} and S_{21} measurements (recorded in dB for direct determination of RL and IL) of this attenuator recorded with a network analyzer over the frequency range from 300 kHz to 3 GHz. We observe that at low frequencies, the system behaves almost like an ideal attenuator with IL ≈ 3.06 dB and RL > 40 dB. At higher frequencies, the performance deteriorates with RL crossing 20 dB at 1.5 GHz to reach 12.5 dB at 3 GHz, and IL increases steadily to 3.8 dB at 3 GHz. We can qualitatively describe the attenuator performance as very good up to 500 MHz (RL > 30 dB, 3 dB < IL < 3.1 dB), good up to 1.5 GHz (RL > 20 dB, 3 dB < IL < 3.3 dB) and fair up to 3 GHz (RL > 12 dB, 3 dB < IL < 3.9 dB).

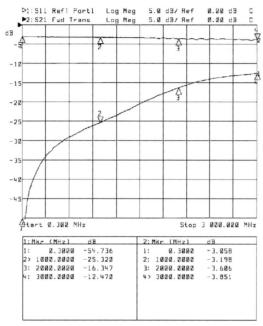

Figure 4-33 S_{11} and S_{21} recording of the attenuator.

This attenuator was not extensively optimized for high-frequency operation. A RF circuit simulator can be used to further refine the circuit by simulating the PCB layout and minimizing the parasitics that cause performance deterioration.

4.5 Summary

Networks play an integral part in analyzing basic low-frequency circuits as well as RF/MW circuits. For instance, the admittance or **Y**-matrix for an N-port network can be written in generic form as

$$\begin{Bmatrix} i_1 \\ i_2 \\ \vdots \\ i_N \end{Bmatrix} = \begin{bmatrix} Y_{11} & Y_{12} & \cdots & Y_{1N} \\ Y_{21} & Y_{22} & \cdots & Y_{2N} \\ \vdots & \vdots & \ddots & \vdots \\ Y_{N1} & Y_{N2} & \cdots & Y_{NN} \end{bmatrix} \begin{Bmatrix} v_1 \\ v_2 \\ \vdots \\ v_N \end{Bmatrix}$$

where currents and voltages become the defining external port conditions. The evaluation of the matrix coefficients is accomplished through the appropriate terminal conditions:

$$Y_{nm} = \left. \frac{i_n}{v_m} \right|_{v_k = 0 \text{ (for } k \neq m\text{)}}.$$

The concepts of **Z**-, **Y**-, **h**-, and **ABCD**-matrix representations of networks can be directly extended to high-frequency circuits. Unfortunately, we encounter practical difficulties in applying the required open- and short-circuit network conditions needed when defining the respective parameter sets. It is for this reason that the scattering parameters that use normalized forward and backward propagating power waves are introduced:

$$a_n = \frac{V_n^+}{\sqrt{Z_0}} = \sqrt{Z_0} I_n^+$$

$$b_n = \frac{V_n^-}{\sqrt{Z_0}} = -\sqrt{Z_0} I_n^-.$$

For a two-port network this results in the matrix form

$$\begin{Bmatrix} b_1 \\ b_2 \end{Bmatrix} = \begin{bmatrix} S_{11} & S_{12} \\ S_{21} & S_{22} \end{bmatrix} \begin{Bmatrix} a_1 \\ a_2 \end{Bmatrix}.$$

Unlike open- or short-circuit network conditions, impedance line matching at the respective port is now required to establish the **S**-matrix set. The *S*-parameters can be directly related to the reflection coefficients at the input and output of the two-port network (S_{11}, S_{22}). Furthermore, forward and reverse power gains are readily identified ($|S_{21}|^2$, $|S_{12}|^2$).

The *S*-parameters are also very useful descriptors when dealing with signal flowgraphs. A signal flow diagram is a circuit representation involving nodes and paths. A flowgraph for the sourced and terminated transmission line follows:

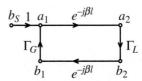

With signal flowgraphs even complicated systems can be examined in terms of specific input output relations in a similar manner as done in control system theory.

Chapter 4 finishes with a brief discussion of the practical recording of the *S*-parameters for a two-port network (DUT) through the use of a vector network analyzer. To compensate for various error sources associated with the measurement arrangement, the so-called TRL method is presented. Here, the Through, Reflect, and Line calibrations are shown to account for the various errors and therefore permit the recording of the actual S-parameters needed to characterize the DUT.

Further Reading

A. Cote, "Matrix Analysis of Oscillators and Transistor Applications," *IRE Trans. on Circuit Theory*, Vol. 5, Issue 3, pp. 181–188, 1958.

S. Walker, "A Low Phase Shift Attenuator," *IEEE Trans. Microwave Theory and Techniques,* Vol. MTT-42, pp. 182–18, 1994.

G. Weiss, "Network Theorems for Transistor Circuits," *IEEE Trans. on Reliability,* Vol. 43, Issue 1, pp. 36–41, 1994.

G. F. Engen and C. A. Hoer, "Thru-Reflect-Line: An Improved Technique for Calibrating the Dual Six-Port Automatic Network Analyzer," *IEEE Trans. Microwave Theory and Techniques,* Vol. MTT-27, pp. 987–998, 1979.

H. J. Eul and B. Schiek, "Thru-Match-Reflect: One Result of a Rigorous Theory for De-embedding and Network Analyzer Calibration," *Proceedings of the 18th European Microwave Conference*, Stockholm, Sweden, 1988.

R. Marks and D. Williams, "Characteristic Impedance Measurement Determination using Propagation Measurements," *IEEE Trans. Microwave and Guided Wave Letters*, pp. 141–143, 1991.

S-Parameter Design, Hewlett-Packard Application Note 154, 1972.

G. Antonini, A.C. Scogna, A. Orlandi, "S-Parameter Characterization of Through, Blind, and Buried Via Holes," *IEEE Trans. on Mobile Computing*, Vol. 2, Issue, 2, pp. 174–184, 2003.

J. Biernacki, D. Czarkowski, "High Frequency Transformer Modeling," *IEEE Int. Symposium on Circuits and Systems*, Vol. 3, pp. 6–9, 2001.

D. V. Morgan and M. J. Howes, eds., *Microwave Solid State Devices and Applications,* P. Peregrinus Ltd., New York, 1980.

P. A. Rizzi, *Microwave Engineering-Passive Circuits,* Prentice Hall, Upper Saddle River, NJ, 1988.

D. Roddy, *Microwave Technology,* Prentice Hall, Upper Saddle River, NJ.

C. Bowick, *RF Circuit Design,* Howard Sams & Co., Indianapolis, IN, 1982.

R. S. Elliot, *An Introduction to Guided Waves and Microwave Circuits,* Prentice Hall, Upper Saddle River, NJ, 1997.

Problems

4.1 From the defining equations (4.3) and (4.6) for the impedance and admittance matrices, show that $[\mathbf{Z}] = [\mathbf{Y}]^{-1}$.

4.2 For the following generic T-network, find the impedance and admittance matrices.

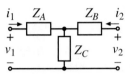

4.3 The common-emitter transistor model shown below is to be converted into its Z network form.

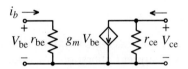

Accomplish this task by first finding the h network parameters, followed by converting them to the Z-parameters based on our conversion table.

4.4 Show that for a bipolar-junction transistor in a common-base configuration under small-signal low-frequency conditions (whose equivalent circuit is shown below), a hybrid parameter matrix can be established as follows:

$$[\mathbf{h}] = \begin{bmatrix} \dfrac{r_{ce} r_{be}}{r_{be} + (1+\beta) r_{ce}} & \dfrac{r_{be}}{r_{be} + (1+\beta) r_{ce}} \\ -\dfrac{r_{be} + \beta r_{ce}}{r_{be} + (1+\beta) r_{ce}} & \dfrac{1}{r_{bc}} + \dfrac{1}{r_{be} + (1+\beta) r_{ce}} \end{bmatrix}$$

where the individual transistor parameters are denoted in the figure.

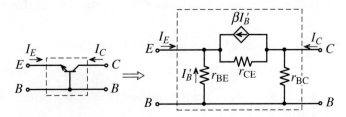

4.5 Using the results from Problem 4.4, compute the equivalent circuit parameters for a BJT in common-base configuration if the **h**-matrix is given as

$$[\mathbf{h}] = \begin{bmatrix} 16.6 & 0.262 \times 10^{-3} \\ -0.99668 & 66.5 \times 10^{-9} \end{bmatrix}$$

4.6 Employ the conversion table for the different parameter representations of the two-port network and find the **h**-matrix representation for the circuit shown in Figure 4-7 under the assumption that the transformers are ideal.

4.7 Using the definition of the *ABCD* network representation, find the *Y*-parameter description.

4.8 The low-frequency *h*-parameters for the *npn* transistor BC108C in common collector configuration are given as follows: $h_{11} = 10\text{k}\,\Omega$, $h_{12} = 1$, $h_{21} = -500$, $h_{22} = 80\,\mu S$. Determine the corresponding Y-parameters.

4.9 From the results of Problem 4.4 and Example 4.2, establish the conversion equations between the **h**-matrix parameters for the common-base and common-emitter transistor configurations.

4.10 Unlike the series connection discussed in Example 4-4, derive the *ABCD*-parameters for a two-port network where the impedance *Z* is connected in shunt.

4.11 For the double-T network in the figure below, develop a *Y* network representation.

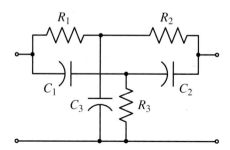

4.12 Find the *ABCD*-parameters for a generic three-element Pi-network, as depicted in Figure 4-2.

4.13 Consider the following ideal transformer with a turn ratio between the primary and secondary windings of n:1. (a) Establish the *ABCD*-parameter representation for the transformer, and (b) Find the input impedance if a load resistance is attached to the secondary winding.

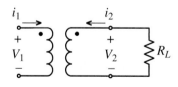

4.14 Prove that the **h**-matrix parameters for a high-frequency hybrid transistor model shown in Figure 4-12 are given by (4.31).

4.15 In this chapter we have mentioned several **h**-matrix representations of the bipolar-junction transistor for different frequency conditions. In all cases we have neglected the influence of the parasitic components associated with the casing of the transistor. A modification to the equivalent circuit of the transistor that takes into account these parasitics is shown below:

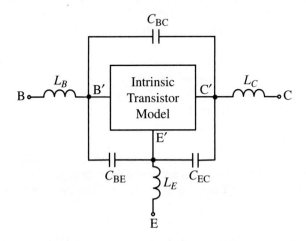

Assuming that the intrinsic transistor model is given by a generic **h**-matrix, derive the modified **h**-matrix model that accounts for the casing.

4.16 Compute the input impedance of a lossless transmission line that is terminated into a load impedance by using the **ABCD**-matrix representations for the transmission line and the load impdance.

4.17 Find the forward voltage gain of the circuit discussed in Problem 4-8, if the input and output are terminated into 50 Ω.

4.18 Your task is to implemented a lambda-quarter transformer at 500 MHz relying on a 50 Ω transmission line. You realize that a cable with a phase velocity of 70% the speed of light would require a cable length of L = 10.5 cm - an impractical approach for your purposes. Therefore, you convert the distributed system into a lumped Pi-network as shown below:

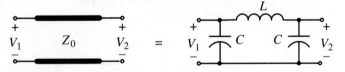

Show (a) the equivalence of both circuits, and (b) determine the values of L and C for the above circuit conditions.

4.19 Given that the input of an amplifier has a VSWR of 2 and the output is given by VSWR = 3, find the magnitudes of the input and output reflection coefficients. What does your result mean in terms of S_{11} and S_{22}?

4.20 Using the same approach as described in Section 4.4.4, show that the S-parameters of the network are computed from the known Y-parameters using

$$[S] = ([Y] + Y_0[E])^{-1}(Y_0[E] - [Y])$$

and the corresponding inverse relation is

$$[Y] = Y_0([E] - [S])([S] + [E])^{-1}$$

where $Y_0 = 1/Z_0$ is the characteristic line admittance.

4.21 The ideal transformer of Problem 4.13 can also be represented in S-parameter form. Show that the S-matrix is given by

$$[S] = \frac{1}{1+N^2}\begin{bmatrix} N^2-1 & 2N \\ 2N & 1-N^2 \end{bmatrix}$$

where $N = N_1/N_2$.

4.22 For the following two passive circuits, prove that the S-parameters are given as

$$[S] = \begin{bmatrix} \Gamma_1 & 1-\Gamma_1 \\ 1-\Gamma_1 & \Gamma_1 \end{bmatrix} \text{ and } [S] = \begin{bmatrix} \Gamma_2 & 1+\Gamma_2 \\ 1+\Gamma_2 & \Gamma_2 \end{bmatrix}$$

respectively, where $\Gamma_1 = (1 + 2Z_0/Z_1)^{-1}$ and $\Gamma_2 = -(1 + 2Y_0/Y_1)^{-1}$.

4.23 For the following T-network inserted into a transmission line with characteristic impedance of $Z_0 = 50\ \Omega$, the three resistances are $R_1 = R_2 = 8.56\ \Omega$, and $R_3 = 141.8\ \Omega$. Find the S-parameters of this configuration and plot the insertion loss as a function of inductance L for the frequency of $f = 2$ GHz and L changing from 0 to 100 nH.

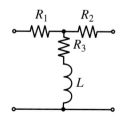

4.24 In practice, the resistors in the T-network of the previous problem are not frequency independent. At RF frequencies, parasitic effects have to be taken into account. Compute the S-parameters at 2 GHz when all resistors have a 0.5 nH parasitic series inductance. Assume L is fixed at 10 nH.

4.25 A BJT is operated in a 50 Ω circuit at 1.5 GHz. For the bias conditions of 4 mA collector current and collector-emitter voltage of 10 V, the manufacturer provides the S-parameters in magnitude and angle as follows:
$S_{11} = 0.6 \angle -127°$; $S_{21} = 3.88 \angle 87°$; $S_{12} = 0.039 \angle 28°$;
$S_{22} = 0.76 \angle -35°$.
Find (a) the Z-parameter and (b) the h-parameter representation.

4.26 As a generalization of Problem 4.20, it is possible to work with different characteristic line impedances attached to an n-port network, as shown in the figure below. It is your task to show that the following two representations are valid:
$$[S] = [Z_0]^{-1/2}([E]-[Z_0][Y])([E]-[Z_0][Y])^{-1}[Z_0]^{1/2}$$
$$[Y] = [Z_0]^{-1/2}([E]+[S])^{-1}([E]-[S])[Z_0]^{-1/2}$$
where $[Z_0]$ is a diagonal matrix involving the characteristic line impedances at the n ports ($Z_{01}, Z_{02},... Z_{0n}$) and [E] is the identity matrix.

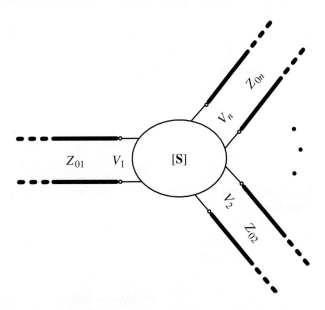

4.27 Derive the S-parameter matrix of a transmission line T-junction.

4.28 An RF device, whose S-parameters are known, is driven on the input side by a voltage source that is connected via a matched generator resistor, and on the output is terminated by a matched (i.e., to Z_0) load resistor as shown in the schematic below.

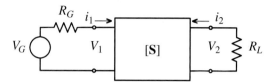

If we assume that the maximum power available from the source with the attached generator resistance is given by P_{max} and the input and output powers are, respectively, P_1 and P_2, prove that

(a) $P_2 = |S_{21}|^2 P_{max}$

(b) $P_1 = (1 - |S_{11}|^2) P_{max}$.

4.29 The Practically Speaking section introduced a resistive Pi-network attenuator with arbitrary port characteristic impedances, which is subject to the voltage gain constraint $a \leq \sqrt{r} - \sqrt{r-1}$ where $r = \max(Z_{01}, Z_{02})/\min(Z_{01}, Z_{02})$. Determine the origin of this constraint and verify its correctness.

4.30 The 3 dB attenuator model in the Practically Speaking section was built using 1/4 W resistors connected in parallel pairs. Determine the maximum allowable port 1 incident power when (a) a matched load is connected to port 2, and (b) when port 2 is left open.

4.31 The T-network attenuator, introduced in Example 4-7, can support arbitrary and different port characteristic impedances, similar to the Pi-network attenuator in the Practically Speaking section. Find the analytical expressions for the three T-network resistances R_1, R_2 and R_3 given the port characteristic impedances Z_{01} and Z_{01} and the S-parameters $S_{11} = S_{22} = 0$, $S_{21} = S_{12} = a$.

4.32 A Wilkinson power divider (discussed in Appendix G) is a three-port passive device designed to split the port 1 incident power equally between ports 2 and 3, or combine power from the ports 2 and 3 into port 1. The ideal S-matrix is given as

$$[S] = \frac{-1}{\sqrt{2}} \begin{bmatrix} 0 & j & j \\ j & 0 & 0 \\ j & 0 & 0 \end{bmatrix}.$$

Find the port 1 return loss and input impedance if port 2 is terminated with 75 Ω and port 3 with 25 Ω. Assume $Z_0 = 50\ \Omega$.

4.33 Consider the following attenuator configuration.

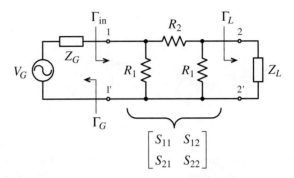

Under ideal conditions ($Z_G = Z_{in} = Z_L = Z_o$) we can show that the attenuation in dB is given by $\alpha = -20\log_{10}|S_{21}|$. Use signal flow analysis principles and show that the attenuation under nonideal conditions ($Z_G \neq Z_{in} \neq Z_L$) is given by $\alpha = -20\log_{10}|T|$, where

$$T = \frac{S_{21}}{1 - \Gamma_G S_{11} - \Gamma_L S_{22} - \Gamma_G \Gamma_L S_{21}^2 + \Gamma_G \Gamma_L S_{11} S_{22}}.$$

5

An Overview of RF Filter Design

5.1 Basic Resonator and Filter Configurations 206
    5.1.1 Filter Types and Parameters 206
    5.1.2 Low-Pass Filter 210
    5.1.3 High-Pass Filter 213
    5.1.4 Bandpass and Bandstop Filters 214
    5.1.5 Insertion Loss 221
5.2 Special Filter Realizations 224
    5.2.1 Butterworth-Type Filters 225
    5.2.2 Chebyshev-Type Filters 228
    5.2.3 Denormalization of Standard Low-Pass Design 236
5.3 Filter Implementation 245
    5.3.1 Unit Elements 247
    5.3.2 Kuroda's Identities 247
    5.3.3 Examples of Microstrip Filter Design 249
5.4 Coupled Filter 257
    5.4.1 Odd and Even Mode Excitation 257
    5.4.2 Bandpass Filter Section 260
    5.4.3 Cascading Bandpass Filter Elements 262
    5.4.4 Design Example 264
5.5 Summary 268

Capitalizing on our knowledge of networks developed in Chapter 4, we are ready to extend and apply these concepts to design RF filters. It is of particular interest in any analog circuit design to manipulate high-frequency signals in such a way as to enhance or attenuate certain frequency ranges or bands. This chapter examines the filtering of analog signals. As we know from elementary circuit courses, there are four types of filters: *low-pass*, *high-pass*, *bandpass*, and *bandstop*. The low-pass filter allows low-frequency signals to be transmitted from the input to the output port with little attenuation. However, as the frequency exceeds a certain *cutoff* point, the attenuation increases significantly with the result of delivering an amplitude-reduced signal to the output port. The opposite behavior is true for a high-pass filter, where the low-frequency signal components are highly attenuated, or reduced in amplitude, while beyond a cutoff frequency point the signal passes the filter with little attenuation. Bandpass and bandstop filters restrict the passband between specific lower and upper frequency points where the attenuation is either low (bandpass) or high (bandstop) compared to the remaining frequency band.

In this chapter, we first review several fundamental concepts and definitions pertaining to filters and resonators. Specifically, the key concept of loaded and unloaded quality factors will be examined in some detail. Then, we introduce the basic, multisection low-pass filter configuration for which tabulated coefficients have been developed both for the so-called maximally flat binomial, or Butterworth filter, and the equiripple, or Chebyshev filter. The intent of Chapter 5 is not to introduce the reader to the entire filter theory, particularly how to derive these coefficients, but rather how to utilize the information to design specific filter types. We will see that the normalized low-pass filter serves as the basic building block from which all four filter types can be derived.

Once we know the procedures of converting a standard low-pass filter design in Butterworth or Chebyshev configuration into a particular filter type that meets our requirements, we then need to investigate ways of implementing the filter through distributed elements. This step is critical, since at frequencies above approximately 500 MHz lumped elements, such as inductors and capacitors, are unsuitable. Relying on the Richards transformation, which converts lumped into distributed elements, and Kuroda's identities, we are given powerful tools to develop a wide range of practically realizable filter configurations.

> **FILTER CLASSIFICATION**
>
> Depending on the frequency of operation and the components used, a distinction is made between passive and active filters. Among passive filters, a further separation is made into lumped systems, primarily *LC* components and crystal resonators, and distributed systems, primarily coaxial and microstrip transmission lines, but also waveguide or helical resonators. Active filters rely on cascaded operational amplifiers or other active devices.

5.1 Basic Resonator and Filter Configurations

5.1.1 Filter Types and Parameters

It is convenient to begin our discussion by introducing the ideal behavior of the four basic filter types: low-pass, high-pass, bandpass, and bandstop. Figure 5-1 summarizes their insertion loss (IL) versus normalized angular frequency behavior.

We have chosen the parameter $\Omega = \omega/\omega_c$ as a normalized frequency with respect to the angular frequency ω_c, which denotes **cutoff frequency** for low-pass and high-pass filters and **center frequency** for bandpass and bandstop filters. As we will see, this normalization will greatly simplify our task of developing standard filter approaches. Actual attenuation profiles are shown in Figure 5-2 for the so-called **binomial (Butterworth)**, **Chebyshev**, and **elliptic (Cauer)** low-pass filters.

The binomial filter exhibits a monotonic attenuation profile that is generally easy to implement. Unfortunately, to achieve a steep attenuation transition from pass- to stopband, a large number of components is needed. A better, steeper slope can be implemented if one permits a certain degree of variations, or ripples, in the passband attenuation profile. If these ripples maintain equal amplitude, either in the stopband or passband, we speak of a **Chebyshev filter** since the design relies on the so-called Chebyshev polynomials. For both the

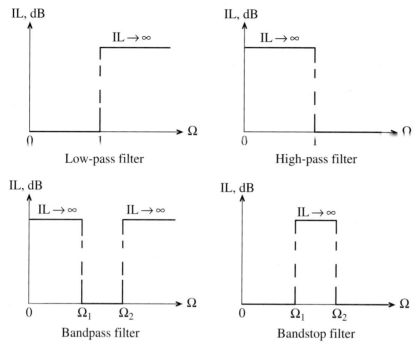

Figure 5-1 Four basic filter types.

binomial and the Chebyshev filter we observe that the attenuation approaches infinity as $\Omega \to \infty$. This is in contrast to the elliptic filters, which allow the steepest transitions from passband to stopband at the expense of ripples in both bands. Because of the mathematical complexity in designing elliptic filters, we will not investigate them any further (for more information see the book by Rizzi, listed in Further Reading at the end of this chapter).

In analyzing the various trade-offs when dealing with filters, the following parameters play key roles:

SKIRT SELECTIVITY

The transition from pass- to stopband or stop- to passband is typically defined as the slope of the loss versus frequency behavior, or skirt, at the 3 dB point. Generally, the higher the component count the better the skirt selectivity.

- **Insertion loss.** Ideally, a perfect filter inserted into the RF circuit path would introduce no power loss in the passband. In other words, it would have zero insertion loss. In reality, however, we have to expect a certain amount of power loss associated with the filter. The insertion loss quantifies how much below the 0 dB line the power amplitude response drops. In mathematical terms, assuming the filter is connected to a source with $Z_S = Z_0$ and a load with $Z_L = Z_0$,

$$\text{IL [dB]} = 10 \log \frac{P_A}{P_L} = -20 \log |S_{21}| \qquad (5.1)$$

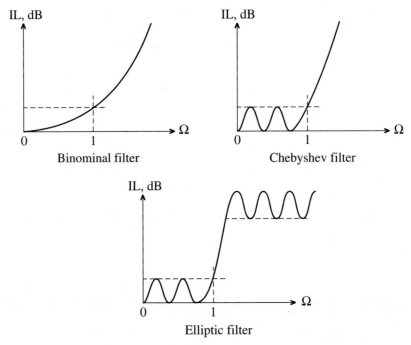

Figure 5-2 Actual attenuation profile for three types of low-pass filters.

where P_L is the power delivered to the load, P_A is the available power from the source, and S_{21} is one of the filter's S-parameters, characterising signal transmission from port 1 (source) to port 2 (load).
- **Ripple.** The flatness of the signal in the passband can be quantified by specifying the ripple or difference between maximum and minimum amplitude response in either dB or Nepers. As already mentioned, and as will be discussed further, the Chebyshev filter design allows us to precisely control the magnitude of the ripple.
- **Bandwidth.** For a bandpass filter, bandwidth defines the difference between upper and lower frequencies, typically recorded at the points of 3 dB attenuation above the passband:

$$\text{BW}^{3\text{dB}} = f_U^{3\text{dB}} - f_L^{3\text{dB}}. \tag{5.2}$$

- **Shape factor.** This factor describes the sharpness of the filter response by taking the ratio between the 60 dB and the 3 dB bandwidths:

$$\text{SF} = \frac{\text{BW}^{60\text{dB}}}{\text{BW}^{3\text{dB}}} = \frac{f_U^{60\text{dB}} - f_L^{60\text{dB}}}{f_U^{3\text{dB}} - f_L^{3\text{dB}}}. \tag{5.3}$$

Basic Resonator and Filter Configurations

- **Rejection.** For an ideal filter we would obtain infinite attenuation level for the undesirable signal frequencies. However, in reality we expect an upper bound due to the use of a finite number of filter components. Practical designs often specify 60 dB as the rejection rate since it can readily be combined with the shape factor (5.3).

The preceding filter parameters are best illustrated by way of a generic bandpass attenuation profile, as summarized in Figure 5-3. The magnitude of the filter's attenuation behavior is plotted with respect to the normalized frequency Ω. As a result, the center frequency f_c is normalized to $\Omega = 1$. The 3 dB lower and upper cutoff frequencies are symmetric with respect to this center frequency. Beyond these 3 dB points, we observe the attenuation response rapidly increasing and reaching the 60 dB rejection points at which the stopband begins.

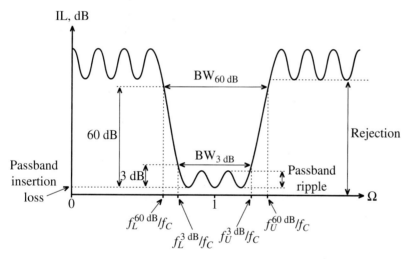

Figure 5-3 Generic attenuation profile for a bandpass filter.

There is one additional parameter describing the selectivity of the filter. This parameter is known as the **quality factor Q**, which generally defines the ratio of the average stored energy to the energy loss per cycle at the resonant frequency:

$$Q = 2\pi \left. \frac{\text{average stored energy}}{\text{energy loss per cycle}} \right|_{\omega = \omega_c} = \omega \left. \frac{\text{average stored energy}}{\text{power loss}} \right|_{\omega = \omega_c}$$

$$= \omega \left. \frac{W_{\text{stored}}}{P_{\text{loss}}} \right|_{\omega = \omega_c} \quad (5.4)$$

where the power loss P_{loss} is equal to the energy loss per unit time. In applying this definition, care must be taken to distinguish between an unloaded and loaded filter. What is meant here is best seen by viewing the filter as a two-port network connected to a source at the input side and a load at the output, as shown in Figure 5-4.

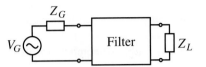

Figure 5-4 Filter as a two-port network connected to an RF source and load.

> **QUALTITY FACTORS FOR CELL PHONE FRONT END**
>
> It is interesting to note how the quality factor is affected by the frequency. If we assume a 1900 MHz receiver is required to bandpass filter a 30 kHz channel from the center of the band, we would need, under ideal conditions, a Q of more than 60,000, a number that is generally difficult to achieve, even for surface acoustic wave (SAW) filters.

It is customary to consider the power loss as consisting of the power loss associated with the external load and the filter itself. The resulting quality factor is named **loaded** Q, or Q_{LD}. Interestingly, if we take the inverse of the loaded Q, we see that

$$\frac{1}{Q_{LD}} = \frac{1}{\omega}\left(\frac{\text{power loss in filter}}{\text{average stored energy}}\right)\bigg|_{\omega = \omega_c}$$

$$+ \frac{1}{\omega}\left(\frac{\text{power loss in load}}{\text{average stored energy}}\right)\bigg|_{\omega = \omega_c} \quad (5.5)$$

since the total power loss is composed of the power losses due to the presence of the filter and the load. This can be written in the concise form

$$\frac{1}{Q_{LD}} = \frac{1}{Q_F} + \frac{1}{Q_E} \quad (5.6)$$

where Q_F and Q_E are the **filter** Q and the **external** Q. The precise meaning of (5.6) will be analyzed in Section 5.1.4. As we will also see in this section, (5.6) can be cast in the form

$$Q_{LD} = \frac{f_c}{f_U^{3dB} - f_L^{3dB}} \equiv \frac{f_c}{BW^{3dB}} \quad (5.7)$$

where f_c is the center or resonance frequency of the filter. In the following sections, a summary is given of the salient features of the three most common filters. Emphasis is placed on the network description as previously developed in Chapter 4.

5.1.2 Low-Pass Filter

As one of the simplest examples, we start our investigation by analyzing a first-order low-pass filter connected to a generator and a load, as depicted in Figure 5-5. In this standard filter circuit, the generator and the load impedances are equal to Z_0, for which we will use the typical value of 50 Ω. Here, Z_0 refers to the transmission line characteristic impedance in case we decide to connect transmission lines between the generator and filter, or between the filter and load. It also allows us to define the S-parameters of the filter block, treating it as a two-port network with port 1 on the generator side and port 2 on the load side.

Basic Resonator and Filter Configurations

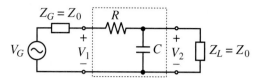

Figure 5-5 Low-pass filter connected between a source and load.

To characterize the filter response, we need to determine $S_{21}(\omega)$, which is the S-parameter representing signal transfer from the source to the load. We recall from Chapter 4 that for the source and load configuration of Figure 5-5,

$$S_{21} = \frac{2V_2}{V_G}. \tag{5.8}$$

For our simple circuit, this can best be evaluated by cascading four *ABCD* networks (labeled 1 through 4) as suggested in Figure 5-6.

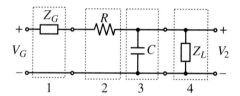

Figure 5-6 Cascading four *ABCD* networks.

The overall *ABCD* network is therefore

$$\begin{bmatrix} A & B \\ C & D \end{bmatrix} = \begin{bmatrix} 1 & Z_G \\ 0 & 1 \end{bmatrix} \begin{bmatrix} 1 & R \\ 0 & 1 \end{bmatrix} \begin{bmatrix} 1 & 0 \\ j\omega C & 1 \end{bmatrix} \begin{bmatrix} 1 & 0 \\ \frac{1}{Z_L} & 1 \end{bmatrix} = \begin{bmatrix} 1+(R+Z_0)\left(j\omega C + \frac{1}{Z_0}\right) & R+Z_0 \\ j\omega C + \frac{1}{Z_0} & 1 \end{bmatrix} \tag{5.9}$$

where we use the fact that $Z_G = Z_L = Z_0$. Since A is already the ratio V_G/V_2,

$$S_{21}(\omega) = \frac{2}{A} = \frac{2}{1+(R+Z_0)\left(j\omega C + \frac{1}{Z_0}\right)}. \tag{5.10}$$

Equation (5.10) can be examined for the limiting cases where the frequency is either zero or approaches infinity. For $\omega \to 0$ we obtain

$$S_{21} = \frac{2}{1+(R+Z_0)/Z_0} = \frac{2Z_0}{2Z_0+R} \tag{5.11a}$$

and for $\omega \to \infty$ it is seen that

$$S_{21} = 0. \tag{5.11b}$$

In the first case we notice that the parasitic resistance R causes some signal attenuation ($S_{21} < 1$) at DC, while for the second case the filter exhibits the expected low-pass behavior of zero output signal at high frequencies.

> **GROUP DELAY AND GROUP VELOCITY**
>
> The **group delay** is the time it takes for information to traverse the network. It is how long it takes for the envelope of a narrow-band pulse to propagate through the network. A related concept for transmission lines is the **group velocity**, which is the velocity of the envelope of a pulse:
>
> $$v_g = \left(\frac{\partial \beta}{\partial \omega}\right)^{-1}$$
>
> The group velocity is different from the phase velocity, the velocity of wave fronts, in dispersive transmission lines (where phase velocity varies with frequency). Microstrips become slightly dispersive at high frequencies. Waveguides are highly dispersive.

In this analysis, $S_{21}(\omega)$ is used as the RF/microwave equivalent of $H(\omega)$, known from system theory as the **transfer function**. Besides specifying the transfer function, it is more common to compute the attenuation factor in Neper (Np) such that

$$\alpha(\omega) \text{ [Np]} = -\ln|S_{21}(\omega)| = -\frac{1}{2}\ln|S_{21}(\omega)|^2 \quad (5.12a)$$

or in dB as

$$\alpha(\omega) \text{ [dB]} = -20 \log|S_{21}(\omega)| = -10 \log|S_{21}(\omega)|^2 \quad (5.12b)$$

which is equivalent to the insertion loss (IL). The corresponding phase is

$$\phi(\omega) = \operatorname{atan2}(\operatorname{Im}(S_{21}(\omega)), \operatorname{Re}(S_{21}(\omega))). \quad (5.12c)$$

Directly related to phase is the so-called **group delay** t_g, which is defined as the frequency derivative of the phase

$$t_g = -\frac{d\phi(\omega)}{d\omega}. \quad (5.12d)$$

It is often desirable to design a filter with nearly linear phase (i.e., $\phi = -A\omega$, with A being an arbitrary constant factor). The group delay is then simply a constant $t_g = A$.

A typical filter response for $C = 10$ pF and various values of R is shown in Figure 5-7.

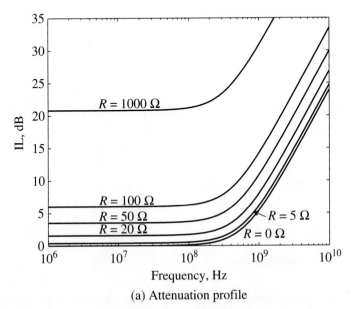

(a) Attenuation profile

Figure 5-7 First-order low-pass filter response as a function of various parasitic resistance values.

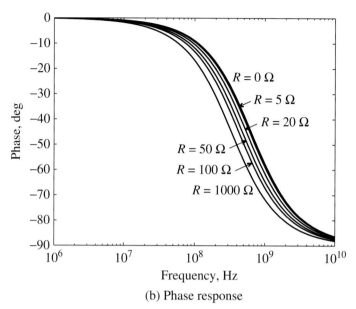

(b) Phase response

Figure 5-7 First-order low-pass filter response as a function of various parasitic resistance values. (Continued)

In this filter design, the resistance R is parasitic and only contributes to unwanted insertion loss in the passband. An ideal RF/microwave filter has purely reactive components ($R = 0$). It is important to note that typical filters require certain real (resistive) source and load impedances. Otherwise, the filter response would be distorted.

5.1.3 High-Pass Filter

Replacing the capacitor with an inductor in Figure 5-5 permits the construction of a first-order high-pass filter, as depicted in Figure 5-8. The analysis follows the same steps as outlined in (5.9), except that the capacitive reactance is replaced by an inductive reactance. The result is

$$\begin{bmatrix} A & B \\ C & D \end{bmatrix} = \begin{bmatrix} 1 & Z_G \\ 0 & 1 \end{bmatrix} \begin{bmatrix} 1 & R \\ 0 & 1 \end{bmatrix} \begin{bmatrix} 1 & 0 \\ \frac{1}{j\omega L} & 1 \end{bmatrix} \begin{bmatrix} 1 & 0 \\ \frac{1}{Z_L} & 1 \end{bmatrix} = \begin{bmatrix} 1 + (R + Z_0)\left(\frac{1}{j\omega L} + \frac{1}{Z_0}\right) & R + Z_0 \\ \frac{1}{j\omega L} + \frac{1}{Z_0} & 1 \end{bmatrix}. \quad (5.13)$$

This gives us directly the result

$$S_{21}(\omega) = \frac{2}{A} = \frac{2}{1 + (R + Z_0)\left(\frac{1}{j\omega L} + \frac{1}{Z_0}\right)}. \quad (5.14)$$

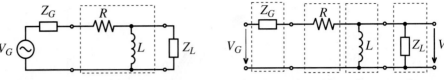

(a) High-pass filter with load resistance   (b) Network and input/output voltages

Figure 5-8 First-order high-pass filter.

As $\omega \to 0$, it is seen that

$$S_{21} = 0 \qquad (5.15a)$$

and for $\omega \to \infty$ we conclude

$$S_{21} = \frac{2}{1 + (R + Z_0)/Z_0} = \frac{2Z_0}{2Z_0 + R}. \qquad (5.15b)$$

This reveals that the inductive influence can be neglected and the insertion loss is purely due to the parasitic resistance R. The filter response for an arbitrarily chosen inductance $L = 100$ nH and various values of R is plotted in Figure 5-9.

5.1.4 Bandpass and Bandstop Filters

A bandpass filter can be constructed through an *RLC* series circuit or through a shunt connection of an *RLC* parallel circuit. The generic series circuit diagram, including generator and load impedances, is displayed in Figure 5-10. For this case, the network representation in *ABCD* notation takes on the form

$$\begin{bmatrix} A & B \\ C & D \end{bmatrix} = \begin{bmatrix} 1 & Z_G \\ 0 & 1 \end{bmatrix} \begin{bmatrix} 1 & Z \\ 0 & 1 \end{bmatrix} \begin{bmatrix} 1 & 0 \\ 1/Z_L & 1 \end{bmatrix} = \begin{bmatrix} 2 + \dfrac{Z}{Z_0} & Z_0 + Z \\ \dfrac{1}{Z_0} & 1 \end{bmatrix} \qquad (5.16)$$

where we used $Z_G = Z_L = Z_0$. The impedance Z is specified from conventional circuit analysis as

$$Z = R + j\left(\omega L - \frac{1}{\omega C}\right). \qquad (5.17)$$

Next, the transfer function $S_{21}(\omega) = 2V_2/V_G$ is found to be

$$S_{21}(\omega) = \frac{2}{A} = \frac{2Z_0}{2Z_0 + R + j[\omega L - 1/(\omega C)]}. \qquad (5.18)$$

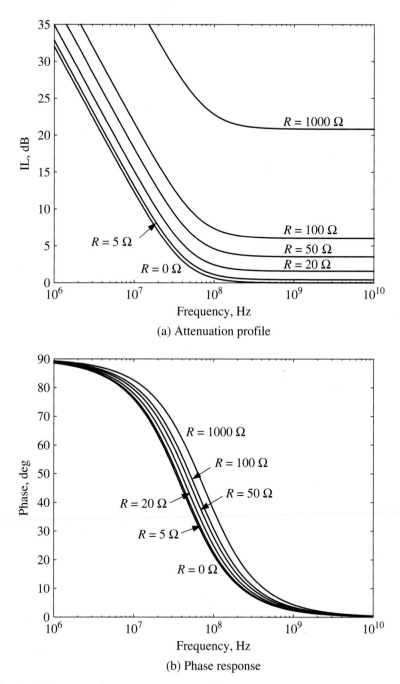

Figure 5-9 High-pass filter response as a function of various parasitic resistance values.

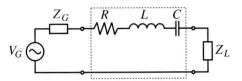

Figure 5-10 Bandpass filter implemented in series configuration.

Explicit plots of the transfer function and the attenuation profile are discussed in the following example.

Example 5-1. Bandpass filter response

For a particular bandpass filter with $Z_L = Z_G = Z_0 = 50\ \Omega$, the following discrete components are selected: $R = 20\ \Omega$, $L = 5$ nH, and $C = 2$ pF. Find the resonance frequency, and plot the frequency response of the phase of the transfer function and the associated attenuation profile in dB.

Solution. To solve this problem, we use the definition of the transfer function for the bandpass filter presented in (5.18). The attenuation profile of the filter expressed in dB is computed as IL $= -20\ \log|S_{21}(\omega)|$. Both the attenuation and phase profiles of the filter are shown in Figure 5-11. From the graph we can estimate the resonance frequency f_0 of the filter to be approximately 1.5 GHz. The exact numerical value is $f_0 = 1/(2\pi\sqrt{LC}) = 1.59$ GHz.

As expected, our bandpass filter assumes minimum attenuation at the resonance point. Unfortunately the transitions from stopband to passband are very gradual.

NARROWBAND VERSUS WIDEBAND FILTER RESPONSE

Bandpass or bandstop filters are generally terms narrowband if the bandwidth falls below 10% of the center frequency. Wideband filters typically involve bandwidths of 40% and more of the center frequency.

A bandstop filter is realized if the series circuit is replaced by a parallel circuit, as shown in Figure 5-12. We only have to replace Z by $1/Y$ in (5.16), which leads to

Basic Resonator and Filter Configurations

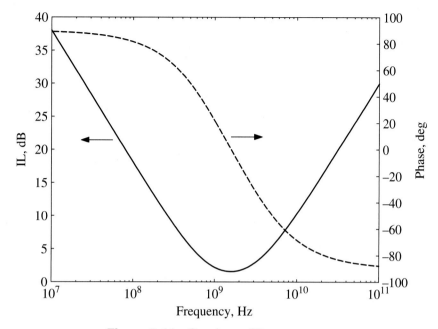

Figure 5-11 Bandpass filter response.

$$S_{21}(\omega) = \frac{2Z_0}{2Z_0 + 1/Y} \tag{5.19}$$

where the admittance is

$$Y = G + j\left(\omega C - \frac{1}{\omega L}\right) \tag{5.20}$$

and upon insertion into (5.19), the filter transfer function yields

$$S_{21}(\omega) = \frac{2Z_0[G + j(\omega C - 1/(\omega L))]}{2Z_0[G + j(\omega C - 1/(\omega L))] + 1}. \tag{5.21}$$

A typical transfer function response of magnitude and phase for the values listed in Example 5-1 is depicted in Figure 5-12.

Working with second order energy storage systems or LC-based networks, we can use the quality factor as introduced in Section 5.1.1 to specify the bandwidth of the 3 dB passband or stopband of a filter:

$$\text{BW} = \frac{f_0}{Q} \tag{5.22}$$

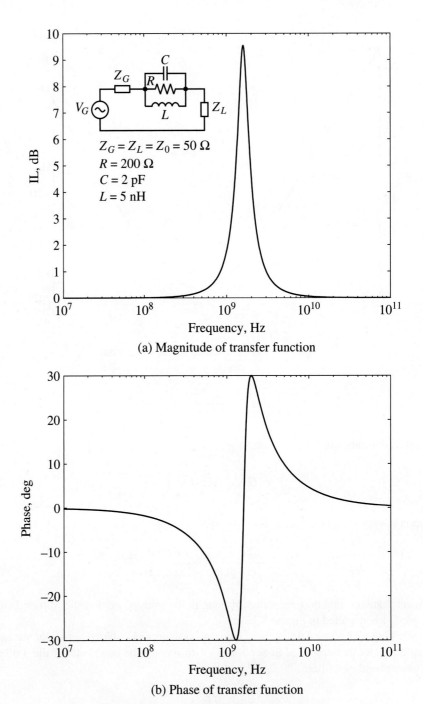

(a) Magnitude of transfer function

(b) Phase of transfer function

Figure 5-12 Bandstop filter response.

Basic Resonator and Filter Configurations

where f_0 is the resonance frequency. This quality factor is the inverse of the **dissipation factor** d. Table 5-1 summarizes all relevant definitions for the series and parallel resonance circuits. The quality factor provides important insight into the losses in a particular resonator circuit configuration. The circuits shown in Table 5-1 depict unloaded filters (i.e., filters in the absence of any external load connections).

Table 5-1 Series and parallel resonators

Parameter	Series (R-L-C)	Parallel (C, G, L)
Impedance or Admittance	$Z = R + j\omega L + \dfrac{1}{j\omega C}$	$Y = G + j\omega C + \dfrac{1}{j\omega L}$
Resonance Frequency	$\omega_0 = \dfrac{1}{\sqrt{LC}}$	$\omega_0 = \dfrac{1}{\sqrt{LC}}$
Dissipation Factor	$d = \dfrac{R}{\omega_0 L} = R\omega_0 C$	$d = \dfrac{G}{\omega_0 C} = G\omega_0 L$
Quality Factor	$Q = \dfrac{\omega_0 L}{R} = \dfrac{1}{R\omega_0 C}$	$Q = \dfrac{\omega_0 C}{G} = \dfrac{1}{G\omega_0 L}$
Bandwidth	$BW = \dfrac{f_0}{Q} = \dfrac{1}{2\pi}\dfrac{R}{L}$	$BW = \dfrac{f_0}{Q} = \dfrac{1}{2\pi}\dfrac{G}{C}$

When dealing with the loaded situation, we are confronted with the additional complication of generator and load impedances attached to the resonator. With reference to Figure 5-10, let us take a more detailed look at how the three different quality factors arise. To this end, our aim is to analyze the series resonance (or bandpass filter) connected to the generator resistance R_G and load R_L. Without loss of generality, we can combine both resistances into the configuration shown in Figure 5-13.

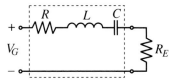

Figure 5-13 Circuit used for the definitions of loaded and unloaded quality factors.

In Figure 5-13, $R_E = R_G + R_L$ and the voltage V_G is understood as a Thévenin equivalent source. The losses can now be partitioned as originating from an external resistance R_E, an internal resistance R, or both. Therefore, we differentiate three cases:

External quality factor $(R_E \neq 0, R = 0)$

$$Q_E = \frac{\omega_0 L}{R_E} = \frac{1}{R_E \omega_0 C} \tag{5.23}$$

Internal or filter quality factor $(R_E = 0, R \neq 0)$

$$Q_F = \frac{\omega_0 L}{R} = \frac{1}{R \omega_0 C} \tag{5.24}$$

Loaded quality factor $(R_E \neq 0, R \neq 0)$

$$Q_{LD} = \frac{\omega_0 L}{R + R_E} = \frac{1}{(R + R_E)\omega_0 C}. \tag{5.25}$$

Identical expressions are derived for a parallel resonator circuit if we replace R and R_E with G and G_E. It is customary to introduce the normalized frequency deviation from the resonance point

$$\varepsilon = \frac{\omega}{\omega_0} - \frac{\omega_0}{\omega} \tag{5.26}$$

and expand it as follows:

$$\varepsilon = \frac{f_0 + f - f_0}{f_0} - \frac{f_0}{f_0 + f - f_0} = \left(1 + \frac{\Delta f}{f_0}\right) - \left(1 + \frac{\Delta f}{f_0}\right)^{-1} \approx 2\frac{\Delta f}{f_0} \tag{5.27}$$

where $\Delta f = f_0 - f$. We define the differential change in quality factor as

$$\Delta Q_{LD} = Q_{LD}\varepsilon \approx 2\frac{\Delta f}{f_0} Q_{LD}. \tag{5.28}$$

If (5.28) is solved for Q_{LD}, and using $X = \omega L - 1/(\omega C)$, we obtain

$$Q_{LD} = \frac{\Delta Q_{LD}}{\varepsilon} = \frac{f_0}{2(R_E + R)} \left.\frac{dX}{df}\right|_{f=f_0} \tag{5.29a}$$

for the series circuit configuration.

Alternatively, for the parallel circuit with $B = \omega C - 1/(\omega L)$ we have

$$Q_{LD} = \frac{\Delta Q_{LD}}{\varepsilon} = \frac{f_0}{2(G_E + G)} \left.\frac{dB}{df}\right|_{f=f_0}. \tag{5.29b}$$

Basic Resonator and Filter Configurations

The equations (5.29a) and (5.29b) show that, generically, the loaded quality factor for complex impedances (or admittances) can be computed as

$$Q_{LD} = \frac{\Delta Q_{LD}}{\varepsilon} = \frac{f_0}{2\operatorname{Re}(Z)} \left.\frac{d\operatorname{Im}(Z)}{df}\right|_{f=f_0} \quad (5.30)$$

or

$$Q_{LD} = \frac{\Delta Q_{LD}}{\varepsilon} = \frac{f_0}{2\operatorname{Re}(Y)} \left.\frac{d\operatorname{Im}(Y)}{df}\right|_{f=f_0} \quad (5.31)$$

where $\operatorname{Re}(Z)$, $\operatorname{Im}(Z)$, $\operatorname{Re}(Y)$, and $\operatorname{Im}(Y)$ are real and imaginary parts of the total impedance or admittance of the resonance circuit.

5.1.5 Insertion Loss

The previously developed quality factor expressions are very useful in RF circuit design, since the Q of a filter can more easily be measured (for instance, with a network analyzer) than the actual impedance or admittance. It is therefore helpful to re-express the impedance or admittance values of bandpass or bandstop filters in terms of the various Q-factors. For example, the impedance of the series resonance circuit can be rewritten as

$$Z = R + j\left(\omega L - \frac{1}{\omega C}\right) = (R_E + R)\left[\frac{R}{R_E + R} + j\left(\frac{\omega L}{R_E + R} - \frac{1}{\omega C(R_E + R)}\right)\right] \quad (5.32)$$

which leads to

$$Z = (R_E + R)\left[\frac{Q_{LD}}{Q_F} + jQ_{LD}\varepsilon\right]. \quad (5.33)$$

Following the same steps as described for a series resonator, a very similar expression can be derived for the admittance Y of a parallel resonator:

$$Y = (G_E + G)\left[\frac{Q_{LD}}{Q_F} + jQ_{LD}\varepsilon\right]. \quad (5.34)$$

We now turn our attention to the following situation: a transmission line system with characteristic line impedance Z_0 is matched at the load and generator sides ($Z_L = Z_G = Z_0$) as seen in Figure 5-14(a).

In Figure 5-14(a), the power delivered to the load P_L is the total available power from the source P_{in}:

$$P_L = P_{in} = |V_G|^2/(8Z_0). \quad (5.35)$$

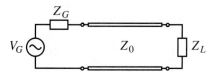

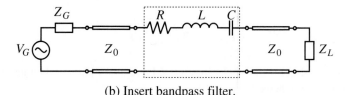

(b) Insert bandpass filter.

Figure 5-14 Insertion loss considerations.

If the filter is inserted as shown in Figure 5-14(b), the power delivered to the load becomes

$$P_L = \frac{1}{2}\left|\frac{V_G}{2Z_0 + Z}\right|^2 Z_0 = \frac{|V_G|^2/(8Z_0)}{\frac{1}{4Z_0^2}\left|2Z_0 + (2Z_0 + R)\left[\frac{Q_{LD}}{Q_F} + j\varepsilon Q_{LD}\right]\right|^2} \quad (5.36)$$

which, after some algebra and the use of (5.6), yields

$$P_L = P_{in}\frac{1}{(1 + \varepsilon^2 Q_{LD}^2)Q_E^2/Q_{LD}^2}. \quad (5.37)$$

The insertion loss in dB due to the presence of the filter is then computed as

$$\text{IL [dB]} = 10\log\left(\frac{1 + \varepsilon^2 Q_{LD}^2}{Q_{LD}^2/Q_E^2}\right) = 10\log(1 + \varepsilon^2 Q_{LD}^2) - 10\log(1 - Q_{LD}/Q_F)^2. \quad (5.38)$$

At resonance, $\varepsilon = 0$, the first term drops out and the second term quantifies the associated resonator losses. However, if the filter is off resonance, then the first term quantifies the sensitivity. If we consider the frequency at which the power delivered to the load is half, or -3 dB, of the power at the resonance frequency, we can immediately state that $1 + \varepsilon^2 Q_{LD}^2 = 2$, and, taking into account relation (5.27), it is concluded

$$\text{BW}_{3\text{dB}} = 2\Delta f = \varepsilon f_0 = f_0/Q_{LD}. \quad (5.39)$$

Recalling Section 2.11, we notice that for a lossless filter, (5.38) can be related to the input reflection coefficient:

$$1 - |\Gamma_{in}|^2 = 1 - \left|\frac{Z_{in} - Z_G}{Z_{in} + Z_G}\right|^2 = \frac{1}{1 + \varepsilon^2 Q_{LD}^2} = \frac{1}{\text{LF}} \quad (5.40)$$

where LF is known as the **loss factor**. This loss factor plays a central role when developing the desired filter attenuation profiles.

Example 5-2. Calculation of various quality factors for a filter

For the filter configuration shown in Figure 5-14(b), the following parameters are given: $Z_0 = 50\ \Omega$, $Z_G = Z_L = Z_0$, $R = 10\ \Omega$, $L = 50$ nH, $C = 0.47$ pF, and the generator voltage is $V_G = 5$ V. Find: the loaded, unloaded (filter), and external quality factors; power generated by the source; power absorbed by the load at resonance; and plot the insertion loss in the range of $\pm 20\%$ of the resonance frequency.

Solution. The first step in the solution of this problem is to find the resonance frequency of the filter:

$$f_0 = \frac{1}{2\pi\sqrt{LC}} = 1.038 \text{ GHz}.$$

Knowing this value, we are now capable of computing the various quality factors of the filter:

External quality factor $\quad Q_E = \dfrac{\omega_0 L}{2Z_0} = 3.26$

Internal or filter quality factor $\quad Q_F = \dfrac{\omega_0 L}{R} = 32.62$

Loaded quality factor $\quad Q_{LD} = \dfrac{\omega_0 L}{R + 2Z_0} = 2.97$.

To determine the input power, or maximum available power from the source, we use (5.35):

$$P_{in} = |V_G|^2/(8Z_0) = 62.5 \text{ mW}.$$

Due to nonzero internal resistance of the filter ($R = 10\ \Omega$), the signal will suffer some attenuation even at the resonance frequency and the power delivered to the load will be less than the available power:

$$P_L = P_{in} \frac{1}{(1 + \varepsilon^2 Q_{LD}^2) Q_E^2/Q_{LD}^2}\bigg|_{f=f_0} = P_{in} \frac{1}{Q_E^2/Q_{LD}^2}$$

$$= 51.7 \text{ mW}.$$

Finally, substituting the loaded and external quality factors into (5.38), we proceed to find the insertion loss of the filter in the range of ±20% of f_0, as shown in Figure 5-15. As we see from the graph, the 3 dB bandwidth of this filter is approximately equal to 350 MHz, which agrees with the result obtained using our formula derived earlier in this section (i.e., $BW_{3dB} = f_0/Q_{LD} = 350.07$ MHz).

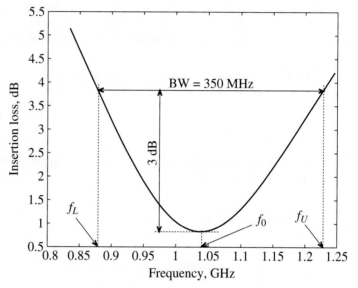

Figure 5-15 Insertion loss versus frequency.

This example shows that the loaded quality factor is lower than both the external and internal filter quality factors.

5.2 Special Filter Realizations

The analytical synthesis of special filter types such as low-pass, high-pass, bandpass, and bandstop filters is generally complicated. In our brief introductory treatment, we are going to concentrate on two filter types: the maximally flat **Butterworth** and the equiripple **Chebyshev** filter realizations. Both filter types are analyzed first in a normalized low-pass configuration, before the low-pass behavior is frequency scaled to implement the remaining filter types through frequency transformations.

5.2.1 Butterworth-Type Filters

This filter type is also known as a maximally flat filter since no ripple is permitted in its attenuation profile. For the low-pass filter, and in the absence of any filter internal losses, the insertion loss is determined through the loss factor,

$$\text{IL [dB]} = -10 \log(1 - |\Gamma_{\text{in}}|^2) = 10 \log(\text{LF}) = 10 \log(1 + a^2 \Omega^{2N}) \quad (5.41)$$

where Ω is again the normalized frequency as introduced in Section 5.1.1 and where N denotes the order of the filter. It is customary to select the constant $a = 1$ so that at $\Omega = \omega/\omega_c = 1$ the insertion loss becomes IL = $10 \log(2)$, which is the 3 dB point at the cutoff frequency. In Figure 5-16, the insertion loss for several values of N is plotted.

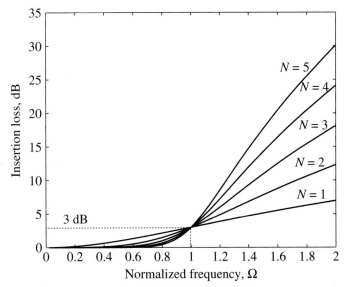

Figure 5-16 Butterworth low-pass filter design.

Two possible realizations of the generic normalized low-pass filter are shown in Figure 5-17, where we set $R_G = 1$. The element values in the circuits in Figure 5-17 are numbered from g_0 at the generator side to g_{N+1} at the load location. The elements in the circuit alternate between series inductance and shunt capacitance. The corresponding elements g are defined as follows:

$$g_0 = \begin{cases} \text{internal generator resistance for circuit in Figure 5-17(a)} \\ \text{internal generator conductance for circuit in Figure 5-17(b)} \end{cases}$$

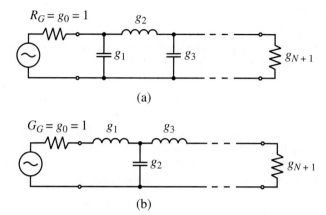

Figure 5-17 Two equivalent realizations of the generic multisection low-pass filter with normalized elements.

$$g_m = \left\{ \begin{array}{l} \text{inducance for series inductor} \\ \text{capacitance for shunt capacitor} \\ (m = 1, \ldots, N) \end{array} \right\}$$

$$g_{N+1} = \left\{ \begin{array}{l} \text{load resistance if the last element is a shunt capacitor} \\ \text{load conductance if the last element is a series inductor.} \end{array} \right.$$

The values for the g's are tabulated and can be found in the literature (see Pozar and Rizzi, listed in Further Reading). For N up to 10, Table 5-2 summarizes the respective g values for the maximally flat low-pass filter based on $g_0 = 1$ and cutoff frequency $\omega_c = 1$.

The corresponding attenuation versus frequency behavior for various filter orders N is seen in Figure 5-18. We note that $\Omega = 1$ is the 3 dB cutoff frequency point. The attenuation curves in Figure 5-18 are very useful in determining the required order of the filter. For instance, if a maximally flat low-pass filter is to be designed with attenuation of at least 60 dB at $\Omega = 2$, we see that an order of $N = 10$ is required.

The filter responses in Figure 5-18 exhibit a steep increase in attenuation after cutoff. We notice that for $\Omega \gg 1$ or $\omega \gg \omega_c$ the loss factor increases as Ω^{2N}, which is a rate of $20N$ dB per decade. However, nothing is said about the phase response of such a filter. In many wireless communication applications, a **linear phase** behavior may be a more critical issue than a rapid attenuation or amplitude transition. Unfortunately, linear phase and rapid amplitude change are opposing requirements. If linear phase is desired, we demand a functional behavior similar to (5.41)

$$\phi(\Omega) = -A_1\Omega(1 + A_2\Omega^{2N}) \qquad (5.42)$$

Table 5-2 Coefficients for maximally flat low-pass filter ($N = 1$ to 10)

N	g_1	g_2	g_3	g_4	g_5	g_6	g_7	g_8	g_9	g_{10}	g_{11}
1	2.0000	1.0000									
2	1.4142	1.4142	1.0000								
3	1.0000	2.0000	1.0000	1.0000							
4	0.7654	1.8478	1.8478	0.7654	1.0000						
5	0.6180	1.6180	2.0000	1.6180	0.6180	1.0000					
6	0.5176	1.4142	1.9318	1.9318	1.4142	0.5176	1.0000				
7	0.4450	1.2470	1.8019	2.0000	1.8019	1.2470	0.4450	1.0000			
8	0.3902	1.1111	1.6629	1.9615	1.9615	1.6629	1.1111	0.3902	1.0000		
9	0.3473	1.0000	1.5321	1.8794	2.0000	1.8794	1.5321	1.0000	0.3473	1.0000	
10	0.3129	0.9080	1.4142	1.7820	1.9754	1.9754	1.7820	1.4142	0.9080	0.3129	1.0000

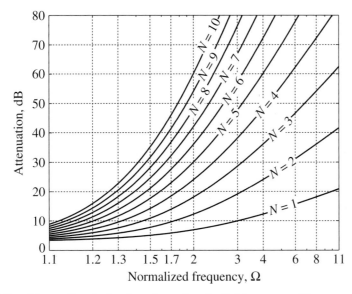

Figure 5-18 Attenuation behavior of maximally flat low-pass filter versus normalized frequency.

> **CONSTANT GROUP DELAY**
>
> This desirable property implies that all spectral components of a multi-frequency signal will pass through the filter with the same time delay; pulsed signals will not disperse.

with A_1 and A_2 being arbitrary constants. The associated group delay t_g is

$$t_g = -\frac{d\phi(\Omega)}{d\Omega} = A_1[1 + A_2(2N+1)\Omega^{2N}]. \quad (5.43)$$

In Table 5-3, the first 10 coefficients for a linear phase response with group delay $t_g = 1$ are listed.

Since steep filter transition and linear phase are generally competing requirements, it has to be expected that the shape factor is reduced. The question of how a linear phase design based on Table 5-3 compares with a standard design of Table 5-2 is discussed in Example 5-3 for the case $N = 3$.

Table 5-3 Coefficients for linear phase low-pass filter (N = 1 to 10)

N	g_1	g_2	g_3	g_4	g_5	g_6	g_7	g_8	g_9	g_{10}	g_{11}
1	2.0000	1.0000									
2	1.5774	0.4226	1.0000								
3	1.2550	0.5528	0.1922	1.0000							
4	1.0598	0.5116	0.3181	0.1104	1.0000						
5	0.9303	0.4577	0.3312	0.2090	0.0718	1.0000					
6	0.8377	0.4116	0.3158	0.2364	0.1480	0.0505	1.0000				
7	0.7677	0.3744	0.2944	0.2378	0.1778	0.1104	0.0375	1.0000			
8	0.7125	0.3446	0.2735	0.2297	0.1867	0.1387	0.0855	0.0289	1.0000		
9	0.6678	0.3203	0.2547	0.2184	0.1859	0.1506	0.1111	0.0682	0.0230	1.0000	
10	0.6305	0.3002	0.2384	0.2066	0.1808	0.1539	0.1240	0.0911	0.0557	0.0187	1.0000

5.2.2 Chebyshev-Type Filters

The design of an equiripple filter type is based on an insertion loss whose functional behavior is described by the **Chebyshev polynomials** $T_N(\Omega)$ in the following form:

$$\text{IL [dB]} = 10 \log(\text{LF}) = 10 \log(1 + a^2 T_N^2(\Omega)) \quad (5.44)$$

where

$$T_N(\Omega) = \cos(N[\cos^{-1}(\Omega)]), \text{ for } |\Omega| \leq 1$$

$$T_N(\Omega) = \cosh(N[\cosh^{-1}(\Omega)]), \text{ for } |\Omega| \geq 1.$$

To appreciate the behavior of the Chebyshev polynomials in the normalized frequency range $-1 < \Omega < 1$, we list the first five terms:

$$T_0 = 1, T_1 = \Omega, T_2 = -1 + 2\Omega^2, T_3 = -3\Omega + 4\Omega^3, T_4 = 1 - 8\Omega^2 + 8\Omega^4.$$

The functional behavior of the first two terms is a constant and a linear function, and the subsequent three terms are quadratic, cubic, and fourth-order functions, as seen in Figure 5-19.

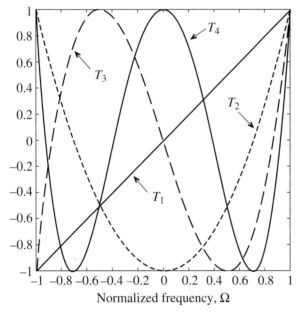

Figure 5-19 Chebyshev polynomials $T_1(\Omega)$ through $T_4(\Omega)$ in the normalized frequency range $-1 \leq \Omega \leq 1$.

It can be observed that all polynomials oscillate within a ±1 interval, a fact that is exploited in the equiripple design. The magnitude of the transfer function $|H(\Omega)|$ is obtained from the Chebyshev polynomial as follows:

$$|H(\Omega)| = \sqrt{H(\Omega)H(\Omega)^*} = \frac{1}{\sqrt{1 + a^2 T_N^2(\Omega)}} \quad (5.45)$$

where $T_N(\Omega)$ is the Chebyshev polynomial of order N and a is a constant factor that allows us to control the height of the passband ripples. For instance, if we choose $a = 1$, then at $\Omega = 1$ we have

$$|H(0)| = \frac{1}{\sqrt{2}} = 0.707$$

which is the 3 dB level that applies uniformly throughout the passband (equiripple). We do not intend to go any further into the general theory of Chebyshev filter design, but rather refer the reader to classical textbooks that covers this topic comprehensively (see, for instance, Matthaei et al. and Zverev). In Figure 5-20, the loss factor and insertion loss are plotted for a Chebyshev filter with coefficient $a = 1$, which again results in a 3 dB attenuation response at the cutoff frequency ($\Omega = 1$).

As mentioned, the magnitude of the ripple can be controlled by suitably choosing the factor a. Since Chebyshev polynomials oscillate in the range from -1 to $+1$ for $-1 \leq \Omega \leq 1$, the squared value of these polynomials will change from 0 to $+1$ in the same frequency range. Therefore, in the frequency range of $-1 \leq \Omega \leq 1$ the minimum attenuation introduced by the filter is 0 dB, and the maximum attenuation, or equivalently the magnitude of the ripples, is IL [dB] $= 10 \log(1 + a^2)$. Thus, if the desired magnitude of the ripples is denoted as RPL [dB], then a should be chosen as

$$a = \sqrt{10^{(\text{RPL [dB]})/10} - 1}.$$

For instance, to obtain a ripple level of 0.5 dB, we have to select $a = (10^{0.5/10} - 1)^{1/2} = 0.3493$. The associated attenuation profiles for the first 10 orders are shown in Figure 5-21 for 3 dB ripple, and in Figure 5-22 for 0.5 dB ripple.

Upon comparing Figure 5-21 with 5-22, it is apparent that higher ripple in the passband has the advantage of steeper transition to the stopband. For instance, a fifth-order 3 dB ripple Chebyshev filter design at $\Omega = 1.2$ has an attenuation of 20 dB, whereas the same order 0.5 dB ripple filter reaches only 12 dB at the same frequency point. The trend remains the same for higher frequencies and different orders. As a case in point, at $\Omega = 5$ the fourth-order 0.5 dB filter has an attenuation of 65 dB compared with the 3 dB design, which has an attenuation of approximately 73 dB.

With reference to the prototype filter circuit, Figure 5-17, the corresponding coefficients are listed in Table 5-4.

Unlike the previously discussed Butterworth filter, the Chebyshev filter approach provides us with a steeper passband/stopband transition. For higher normalized frequencies $\Omega \gg 1$, the Chebyshev polynomials $T_N(\Omega)$ can be approximated as $(1/2)(2\Omega)^N$. This means that the filter has an improvement in attentuation of roughly $(2^{2N})/4$ over the Butterworth design.

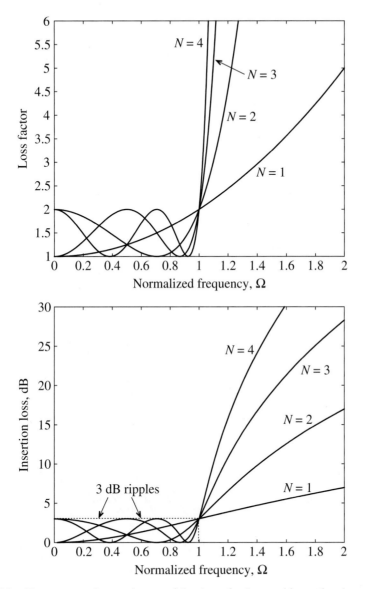

Figure 5-20 Frequency dependence of the loss factor and insertion loss of the Chebyshev low-pass filter.

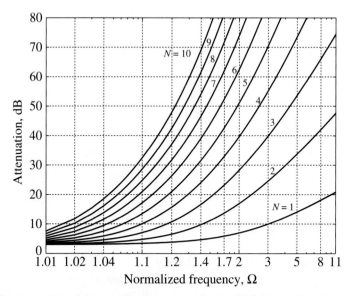

Figure 5-21 Attenuation response for 3 dB Chebyshev design.

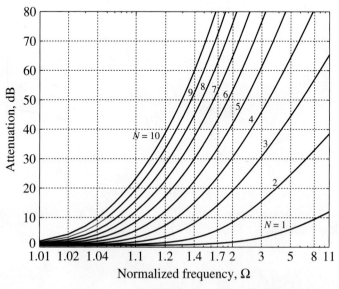

Figure 5-22 Attenuation response for 0.5 dB Chebyshev design.

Table 5-4 (a)   Chebyshev filter coefficients; 3 dB filter design ($N = 1$ to 10)

N	g_1	g_2	g_3	g_4	g_5	g_6	g_7	g_8	g_9	g_{10}	g_{11}
1	1.9953	1.0000									
2	3.1013	0.5339	5.8095								
3	3.3487	0.7117	3.3487	1.0000							
4	3.4389	0.7483	4.3471	0.5920	5.8095						
5	3.4817	0.7618	4.5381	0.7618	3.4817	1.0000					
6	3.5045	0.7685	4.6061	0.7929	4.4641	0.6033	5.8095				
7	3.5182	0.7723	4.6386	0.8039	4.6386	0.7723	3.5182	1.0000			
8	3.5277	0.7745	4.6575	0.8089	4.6990	0.8018	4.4990	0.6073	5.8095		
9	3.5340	0.7760	4.6692	0.8118	4.7272	0.8118	4.6692	0.7760	3.5340	1.0000	
10	3.5384	0.7771	4.6768	0.8136	4.7425	0.8164	4.7260	0.8051	4.5142	0.6091	5.8095

Table 5-4 (b)   Chebyshev filter coefficients; 0.5 dB filter design ($N = 1$ to 10)

N	g_1	g_2	g_3	g_4	g_5	g_6	g_7	g_8	g_9	g_{10}	g_{11}
1	0.6986	1.0000									
2	1.4029	0.7071	1.9841								
3	1.5963	1.0967	1.5963	1.0000							
4	1.6703	1.1926	2.3661	0.8419	1.9841						
5	1.7058	1.2296	2.5408	1.2296	1.7058	1.0000					
6	1.7254	1.2479	2.6064	1.3137	2.4758	0.8696	1.9841				
7	1.7372	1.2583	2.6381	1.3444	2.6381	1.2583	1.7372	1.0000			
8	1.7451	1.2647	2.6564	1.3590	2.6964	1.3389	2.5093	0.8796	1.9841		
9	1.7504	1.2690	2.6678	1.3673	2.7939	1.3673	2.6678	1.2690	1.7504	1.0000	
10	1.7543	1.2721	2.6754	1.3725	2.7392	1.3806	2.7231	1.3485	2.5239	0.8842	1.9841

—RF&MW→

Example 5-3. Comparison between Butterworth, linear phase Butterworth, and Chebyshev filters

Compare the attenuation versus frequency behavior of the third-order low-pass filter for (a) standard 3 dB Butterworth, (b) linear phase Butterworth, and (c) 3 dB Chebyshev design.

Solution. If we choose the first element of the filter to be an inductor connected in series with the source, then the circuit topology of the third-order filter is given by

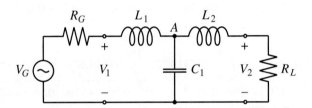

where the inductances and the capacitance are obtained from Tables 5-2, 5-3, and 5-4. Specifically,

- Standard Butterworth: $L_1 = L_2 = 1$ H, $C_1 = 2$ F
- Linear phase Butterworth: $L_1 = 1.255$ H, $C_1 = 0.5528$ F, $L_2 = 0.1922$ H
- 3 dB Chebyshev filter: $L_1 = L_2 = 3.3487$ H, $C_1 = 0.7117$ F
- generator and load: $R_G = R_L = 1\ \Omega$.

As we can see from the preceding circuit diagram, under DC conditions the inductances become short circuits and the capacitor acts like an open circuit. The voltage across the load is equal to one-half the voltage at the source due to the voltage divider formed by the load and source impedances (i.e. $V_2 = 0.5 V_G$). When the frequency is not equal to zero the voltage across the load can be obtained by applying the voltage divider rule twice; first, to obtain the voltage at node A,

$$V_A = \frac{Z_{C_1} \| (Z_{L_2} + R_L)}{Z_{C_1} \| (Z_{L_2} + R_L) + Z_{L_1} + R_G} V_G$$

and, second, to obtain the voltage across the load with reference to V_A,

$$V_2 = \frac{R_L}{R_L + Z_{L_2}} V_A.$$

If we find the ratio of the circuit gain at AC to the gain under DC conditions, it is possible to compute the attenuation that is introduced by the filter:

$$\alpha = 2 \frac{R_L}{R_L + Z_{L_2}} \frac{Z_{C_1} \| (Z_{L_2} + R_L)}{Z_{C_1} \| (Z_{L_2} + R_L) + Z_{L_1} + R_G}.$$

The graph of the attenuation coefficient expressed in dB for the three filter realizations is shown in Figure 5-23. As expected, the Chebyshev filter has the steepest slope of the attenuation profile, while the linear phase filter exhibits the lowest rolloff with frequency. Therefore, if a sharp transition from passband to stopband is required, and ripples can be tolerated, the most appropriate choice would be a Chebyshev filter implementation. We also note that the attenuation of the Chebyshev filter at the cutoff frequency is equal to the ripple size in the passband.

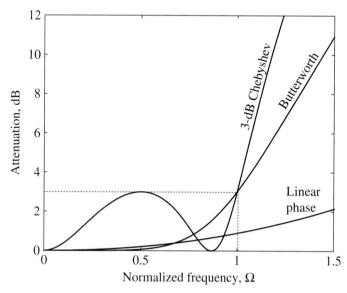

Figure 5-23 Comparison of the frequency response of the Butterworth, linear phase, and 3-dB Chebyshev third-order filters.

> *Even though the linear phase Butterworth filter suffers from a shallow transition, it is the linear phase that makes it particularly attractive for modulation and mixer circuits.*

5.2.3 Denormalization of Standard Low-Pass Design

To arrive at realizable filters, we have to denormalize the aforementioned coefficients to meet realistic frequency and impedance requirements. In addition, the standard low-pass filter prototype should be converted into high-pass or bandpass/bandstop filter types, depending on the application. Those objectives can be achieved by considering two distinct steps:

- **Frequency transformation** to convert from nomalized frequency Ω to actual frequency ω. This step implies the scaling of the standard inductances and capacitances.
- **Impedance transformation** to convert standard generator and load resistances g_0 and g_{N+1} to actual resistances R_L and R_G.

We begin by examining the frequency transformation and its implications in terms of the various filter types. To eliminate confusing notation, we drop the index denoting individual components (i.e., $L_n(n = 1, \ldots, N) \to L$ and $C_n(n = 1, \ldots, N) \to C$). This makes sense, since the transformation rules to be developed will be applicable to all components equally.

Frequency Transformation

A standard fourth-order low-pass Chebyshev filter with 3 dB ripples in the passband response is shown in Figure 5-24, where we have included negative frequencies to display more clearly the symmetry of the attenuation profile in the frequency domain. Furthermore, by appropriately scaling and shifting, we notice that all four filter types, Figures 5-25, 5-26, 5-28, and 5-29, can be generated. This is now examined in detail.

For the *low-pass filter*, we see that a simple multiplication by the angular cutoff frequency ω_c accomplishes the desired scaling (see Figure 5-25):

$$\omega = \Omega \omega_c. \tag{5.46}$$

For the scaling we picked an arbitrary cutoff frequency of 1 GHz. In the corresponding insertion loss and loss factor expressions, Ω is simply replaced by $\Omega \omega_c$. For the inductive and capacitive elements, we have to compare normalized with actual reactances:

$$jX_L = j\Omega L = j(\omega/\omega_c)L = j\omega \tilde{L} \tag{5.47a}$$

$$jX_C = \frac{1}{j\Omega C} = \frac{1}{j(\omega/\omega_c)C} = \frac{1}{j\omega \tilde{C}}. \tag{5.47b}$$

Special Filter Realizations

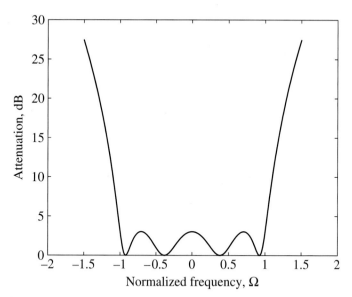

Figure 5-24 Fourth-order low-pass Chebyshev filter with 3 dB ripples in the passband.

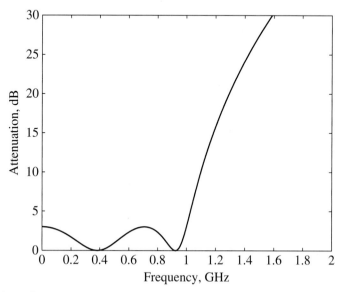

Figure 5-25 Conversion of standard low-pass filter prototype into low-pass realization. Cutoff frequency is $f_c = 1$ GHz.

This reveals that the actual inductance and capacitance $\tilde{L}$ and $\tilde{C}$ are computed from the normalized L and C as

$$\tilde{L} = L/\omega_c \tag{5.48a}$$

$$\tilde{C} = C/\omega_c. \tag{5.48b}$$

For the *high-pass filter*, the parabolically shaped frequency response has to be mapped into a hyperbolic frequency domain behavior. This can be accomplished through the transformation

$$\omega = \frac{-\omega_c}{\Omega}. \tag{5.49}$$

The correctness of this transformation becomes apparent when the normalized cutoff frequency $\Omega = \pm 1$ is substituted in (5.49). This assigns the actual cutoff frequency $\omega = \mp \omega_c$ to the high-pass filter, consistent with Figure 5-26.

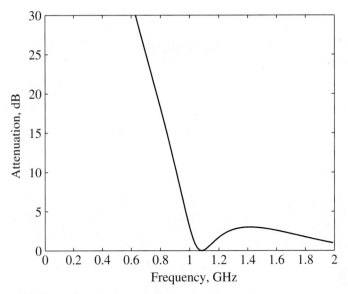

Figure 5-26 Conversion of standard low-pass filter prototype into high-pass realization. Cutoff frequency is $f_c = 1$ GHz.

Care has to be exercised in denormalizing the circuit parameters. We note

$$jX_L = j\Omega L = -j\frac{\omega_c}{\omega}L = \frac{1}{j\omega \tilde{C}} \tag{5.50a}$$

$$jX_c = \frac{1}{j\Omega C} = -\frac{\omega}{j\omega_c C} = j\omega \tilde{L}. \tag{5.50b}$$

Thus, it follows that

$$\tilde{C} = \frac{1}{\omega_c L} \tag{5.51a}$$

$$\tilde{L} = \frac{1}{\omega_c C}. \tag{5.51b}$$

This should make intuitive sense, since it is known from fundamental circuit theory that a first-order high-pass filter can be obtained from a low-pass filter by replacing the inductors with capacitors, or vice versa. Equations (5.51) are the logical extension to higher-order filters.

The *bandpass filter* requires a more sophisticated transformation. In addition to scaling, we also have to shift the standard low-pass filter response. The mapping from the normalized frequency Ω to the actual frequency ω is best explained by considering Figure 5-27.

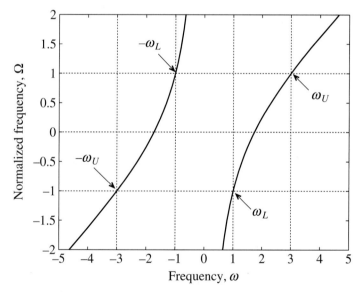

Figure 5-27 Mapping from standard frequency Ω into actual frequency ω. Lower cutoff frequency is $\omega_L = 1$ and upper cutoff frequency is $\omega_U = 3$.

The functional relation that achieves scaling and shifting is

$$\Omega = \frac{1}{\omega_U/\omega_c - \omega_L/\omega_c}\left(\frac{\omega}{\omega_c} - \frac{1}{\omega/\omega_c}\right) = \frac{\omega_c}{\omega_U - \omega_L}\left(\frac{\omega}{\omega_c} - \frac{\omega_c}{\omega}\right) \tag{5.52}$$

where the upper and lower frequencies ω_U, ω_L define the bandwidth expressed in rad/s (BW = $\omega_U - \omega_L$) of the passband located at $\omega_c = \omega_0$. In other words, the cutoff frequency

ω_c now defines the center frequency ω_0, as mentioned earlier. Using ω_0 and (5.26), it is possible to rewrite (5.46) as

$$\Omega = \frac{\omega_0}{\omega_U - \omega_L} \varepsilon. \qquad (5.53)$$

The upper and lower frequencies are the inverse of each other:

$$\frac{\omega_U}{\omega_0} = \frac{\omega_0}{\omega_L} \qquad (5.54)$$

a fact that can be employed to specify the center frequency as the geometric mean of the upper and lower frequencies, $\omega_0 = \sqrt{\omega_U \omega_L}$. The mapping of this transformation is verified if we first consider $\Omega = 1$. Equation (5.52) is +1 and –1 for $\omega = \omega_U$ and $\omega = \omega_L$. For $\Omega = 0$, we obtain $\omega = \pm\omega_0$. The frequency transformations are therefore as follows:

$$0 \leq \Omega \leq 1 \rightarrow \omega_0 \leq \omega \leq \omega_U$$

$$-1 \leq \Omega \leq 0 \rightarrow \omega_L \leq \omega \leq \omega_0.$$

The result of this transformation applied to the low-pass filter prototype is shown in Figure 5-28.

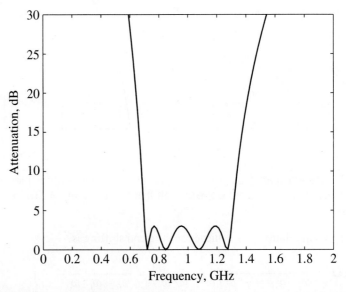

Figure 5-28 Conversion of standard low-pass filter prototype into bandpass realization with lower cutoff frequency $f_L = 0.7$ GHz, upper cutoff frequency $f_U = 1.3$ GHz, and center frequency of $f_0 = 0.95$ GHz.

Special Filter Realizations

The circuit parameters are next transformed according to the assignment

$$jX_L = j\Omega L = j\left(\frac{\omega_0}{\omega_U - \omega_L}\varepsilon\right)L = j\omega\tilde{L} + \frac{1}{j\omega\tilde{C}} \qquad (5.55)$$

which yields for the series inductor $\tilde{L}$ in (5.55), the denormalized series inductor

$$\tilde{L} = \frac{L}{\omega_U - \omega_L} \qquad (5.56a)$$

and the denormalized series capacitance

$$\tilde{C} = \frac{\omega_U - \omega_L}{\omega_0^2 L}. \qquad (5.56b)$$

The shunt capacitor is transformed based on the equation

$$jB_C = j\Omega C = j\left(\frac{\omega_0}{\omega_U - \omega_L}\varepsilon\right)C = j\omega\tilde{C} + \frac{1}{j\omega\tilde{L}} \qquad (5.57)$$

with the two shunt elements

$$\tilde{L} = \frac{\omega_U - \omega_L}{\omega_0^2 C} \qquad (5.58a)$$

$$\tilde{C} = \frac{C}{\omega_U - \omega_L}. \qquad (5.58b)$$

Referring to Figure 5-17, we see that a normalized inductor is transformed into a series inductor and capacitor with values given by (5.56). On the other hand, the normalized capacitor is transformed into shunt inductor and capacitor, whose values are stated by (5.58).

The *bandstop filter* transformation rules are not explicitly derived, since they can be developed through an inverse transform of (5.53), or by using the previously derived highpass filter and applying (5.55). In either case, we find for the series inductor the parallel combination of

$$\tilde{L} = \frac{(\omega_U - \omega_L)L}{\omega_0^2} \qquad (5.59a)$$

$$\tilde{C} = \frac{1}{(\omega_U - \omega_L)L} \qquad (5.59b)$$

and for the shunt capacitor, the series combination of

$$\tilde{L} = \frac{1}{(\omega_U - \omega_L)C} \qquad (5.60a)$$

$$\tilde{C} = \frac{(\omega_U - \omega_L)C}{\omega_0^2}. \qquad (5.60b)$$

The resulting frequency response for the bandstop filter is shown in Figure 5-29. Table 5-5 summarizes the conversion from the standard low-pass filter to the four filter realizations.

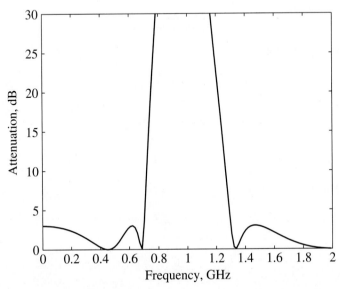

Figure 5-29 Conversion of standard low-pass filter prototype into bandstop realization with center frequency of $f_0 = 0.95$ GHz. Lower cut-off frequency is $f_L = 0.7$ GHz and upper cutoff frequency is $f_U = 1.3$ GHz.

Impedance transformation

In the original filter prototype shown in Figure 5-17, we have unit source and load resistances except for the even-numbered Chebyshev filter coefficients listed in Table 5-4. If, however, either the generator resistance g_0 or the load resistance R_L is required to be unequal to unity, we need to scale the entire impedance expression. This is accomplished by scaling all filter coefficients by the actual resistance R_G. That is,

$$\tilde{R}_G = 1 R_G \tag{5.61a}$$

$$\tilde{L} = L R_G \tag{5.61b}$$

$$\tilde{C} = \frac{C}{R_G} \tag{5.61c}$$

$$\tilde{R}_L = R_L R_G \tag{5.61d}$$

Special Filter Realizations   243

Table 5-5 Transformation between normalized low-pass filter and actual bandpass and bandstop filter (BW = $\omega_U - \omega_L$, $\omega_0 = \sqrt{\omega_U \omega_L}$)

Low-pass prototype	Low-pass	High-pass	Bandpass	Bandstop
$L = g_k$	$\dfrac{L}{\omega_c}$	$\dfrac{1}{\omega_c L}$	$\dfrac{L}{BW}$; $\dfrac{BW}{\omega_0^2 L}$	$\dfrac{1}{(BW)L}$; $\dfrac{(BW)L}{\omega_0^2}$
$C = g_k$	$\dfrac{C}{\omega_c}$	$\dfrac{1}{\omega_c C}$	$\dfrac{C}{BW}$; $\dfrac{BW}{\omega_0^2 C}$	$\dfrac{1}{(BW)C}$; $\dfrac{(BW)C}{\omega_0^2}$

where the tilde expressions are again the resulting actual parameters and L, C, and R_L are the values of the original prototype. In Example 5-4, we demonstrate the design of a Chebyshev bandpass filter based on the low-pass prototype.

─────────────────────────────────── RF & MW ➔

Example 5-4. Chebysehev bandpass filter design

An $N = 3$ Chebyshev bandpass filter is to be designed with a 3 dB passband ripple for a communication link. The center frequency is at 2.4 GHz and the filter has to meet a bandwidth requirement of 20%. The filter has to be inserted into a 50 Ω characteristic line impedance. Find the inductive and capacitive elements and plot the attenuation response in the frequency range 1 to 4 GHz.

Solution. From Table 5-4(a) we find that the coefficients for a standard low-pass $N = 3$ Chebyshev filter with 3 dB ripples in the passband are $g_0 = g_4 = 1$, $g_1 = g_3 = 3.3487$, and $g_2 = 0.7117$. In this filter prototype we assumed that both generator and load impedances are equal to unity. In our problem, however, we have to match the filter to 50 Ω line impedances. Thus we must apply scaling as described by (5.61). The resulting circuit is shown in the following figure:

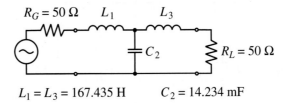

$L_1 = L_3 = 167.435$ H $\qquad C_2 = 14.234$ mF

This is still a low-pass filter with cutoff frequency of $\omega_c = 1$ or $f_c = 1/(2\pi) = 0.159$ Hz. We can next apply the frequency transformation to change the low-pass filter into a bandpass filter:

$$\omega_U = 1.1(2\pi 2.4 \times 10^9) \text{ rad/s} = 16.59 \times 10^9 \text{ rad/s}$$

$$\omega_L = 0.9(2\pi 2.4 \times 10^9) \text{ rad/s} = 13.57 \times 10^9 \text{ rad/s}$$

and

$$\omega_0 = \sqrt{\omega_L \omega_U} = 15 \times 10^9 \text{ rad/s}.$$

The actual inductive and capacitive values are defined in (5.56) and (5.58):

$$\tilde{L}_1 = \tilde{L}_3 = \frac{L_1}{\omega_U - \omega_L} = 55.5 \text{ nH}$$

$$\tilde{C}_1 = \tilde{C}_3 = \frac{\omega_U - \omega_L}{\omega_0^2 L_1} = 0.08 \text{ pF}$$

$$\tilde{L}_2 = \frac{\omega_U - \omega_L}{\omega_0^2 C_2} = 0.94 \text{ nH}$$

$$\tilde{C}_2 = \frac{C_2}{\omega_U - \omega_L} = 4.7 \text{ pF}.$$

The final circuit is shown in Figure 5-30 together with the resulting graph of the attenuation response.

Filter Implementation

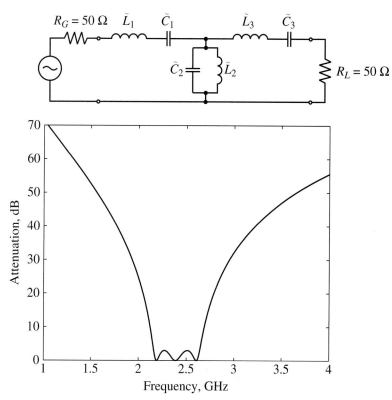

Figure 5-30 Attenuation response of a three-element 3 dB ripple bandpass Chebyshev filter centered at 2.4 GHz. The lower cutoff frequency is $f_L = 2.16$ GHz and the upper cutoff frequency is $f_U = 2.64$ GHz.

> The filter design becomes almost a cookbook approach if we start from the standard low-pass filter and subsequently apply the appropriate frequency transformation and component scaling. However, we often obtain unrealistic component values.

5.3 Filter Implementation

Filter designs beyond 1 GHz are difficult to realize with discrete components because the wavelength becomes comparable with the physical filter element dimensions, resulting in various losses that severely degrade the circuit performance. Thus, to arrive at practical filters, the lumped component filters discussed in Section 5.2 must be converted into distributed element realizations. In this section, some of the necessary tools are introduced—namely, the Richards transformation, the concept of the unit element, and Kuroda's identities.

> **FREQUENCY RANGE OF FILTER APPLICATIONS**
>
> In general, the required bandwidth and frequency range make various filter implementations more or less suitable for particular system requirements. A very rough classification for passive components can be given as follows:
>
> **Lumped LC-type filters**
> BW: 1% - 20%
> Range: 100 Hz - 2 GHz
>
> **Crystal and SAW filters**
> BW: 0.01% - 5%
> Range: 100 Hz - 0.1 GHz
>
> **Helical and Coaxial filters**
> BW: 0.1% - 15%
> Range: 10 MHz - 5 GHz
>
> **Waveguide filters (or resonators)**
> BW: 0.1% - 10%
> Range: 5 GHz - 100 GHz

To accomplish the conversion between lumped and distributed circuit designs, Richards proposed a special transformation that allows open- and short-circuited transmission line segments to emulate the inductive and capacitive behavior of the discrete components. We recall that the input impedance Z_{in} of a short-circuited transmission line ($Z_L = 0$) of characteristic line impedance Z_0 is purely reactive:

$$Z_{in} = jZ_0 \tan(\beta l) = jZ_0 \tan\theta. \quad (5.62)$$

Here, the electric length θ can be rewritten in such a way as to make the frequency behavior explicit. If we choose the line length to be $\lambda_0/8$ at a particular reference frequency $f_0 = v_p/\lambda_0$, the electric length becomes

$$\theta = \beta\frac{\lambda_0}{8} = \frac{2\pi f v_p}{v_p 8 f_0} = \frac{\pi f}{4 f_0} = \frac{\pi}{4}\Omega. \quad (5.63)$$

By substituting (5.63) into (5.62), a direct link between the frequency-dependent inductive behavior of the transmission line and the lumped element representation can be established:

$$jX_L = j\omega L \equiv jZ_0 \tan\left(\frac{\pi f}{4 f_0}\right) = jZ_0 \tan\left(\frac{\pi}{4}\Omega\right) = SZ_0 \quad (5.64)$$

where $S = j\tan(\pi\Omega/4)$ is the actual Richards transform. The capacitive lumped element effect can be replicated through the open-circuited transmission line section

$$jB_C = j\omega C \equiv jY_0 \tan\left(\frac{\pi}{4}\Omega\right) = SY_0. \quad (5.65)$$

Thus, the Richards transformation allows us to replace lumped inductors with short-circuited stubs of characteristic impedance $Z_0 = L$ and capacitors with open-circuited stubs of characteristic impedance $Z_0 = 1/C$.

It is interesting to note that the choice of $\lambda_0/8$ as line length is somewhat arbitrary. Indeed, several authors use different values as the basic length. However, $\lambda_0/8$ is more convenient since it results in smaller physical circuits and the cutoff frequency point in the standard low-pass filter response is preserved (i.e., $S = j1$ for $f = f_0 = f_c$). In Section 5.5.3, we will encounter a bandstop filter that requires a $\lambda_0/4$ line length to meet the expected attenuation profile.

The Richards transformation maps the lumped element frequency response in the range of $0 \leq f < \infty$ into the range $0 \leq f \leq 2f_0$ due to the periodic behavior of the tangent function and the fact that all lines are $\lambda_0/8$ in length, a property that is known as **commensurate line**

Filter Implementation

length. Because of this periodic property, the frequency response of such a filter cannot be regarded as broadband.

5.3.1 Unit Elements

When converting lumped elements into transmission line sections, there is a need to separate the transmission line elements spatially to achieve practically realizable configurations. This is accomplished by inserting so-called **unit elements** (UEs). The unit element has an electric length of $\theta = \frac{\pi}{4}(f/f_0)$ and a characteristic impedance Z_{UE}. The two-port network expression in chain parameter representation is immediately apparent from our discussion in Chapter 4. We recall that the transmission line representation is

$$[UE] = \begin{bmatrix} A_{UE} & B_{UE} \\ C_{UE} & D_{UE} \end{bmatrix} = \begin{bmatrix} \cos\theta & jZ_{UE}\sin\theta \\ \dfrac{j\sin\theta}{Z_{UE}} & \cos\theta \end{bmatrix} = \dfrac{1}{\sqrt{1-S^2}} \begin{bmatrix} 1 & Z_{UE}S \\ \dfrac{S}{Z_{UE}} & 1 \end{bmatrix} \quad (5.66)$$

where the definition of S is given by (5.64). The use of the unit elements is discussed best by way of a few examples, as presented in Section 5.3.4.

5.3.2 Kuroda's Identities

In addition to the unit element, it often becomes necessary to convert a difficult-to-implement design to a more suitable filter realization. For instance, a series inductance implemented by a short-circuited transmission line segment is more complicated to realize than a shunt stub line. To facilitate the conversion between the various transmission line realizations, Kuroda has developed four identities which are summarized in Table 5-6.

We should note that in Table 5-6 all inductances and capacitances are represented by their equivalent Richards transformations. As an example, we will prove one of the identities and defer proof of the remaining identities to the problems at the end of this chapter.

Example 5-5. Prove the fourth of Kuroda's identities from Table 5-6

Solution. It is convenient to employ chain parameter representation of the shunt-connected inductor (see Table 4-1 for the corresponding

Table 5-6 Kuroda's identities

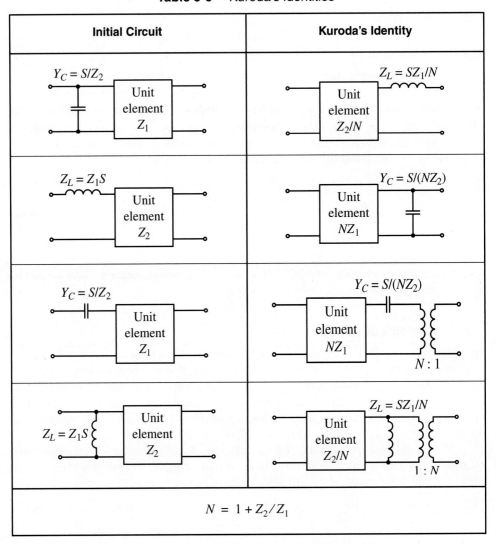

ABCD-matrix) and the unit element as given in (5.66) to write the left-hand side as follows:

$$\begin{bmatrix} A & B \\ C & D \end{bmatrix}_L = \begin{bmatrix} 1 & 0 \\ \frac{1}{SZ_1} & 1 \end{bmatrix} \frac{1}{\sqrt{1-S^2}} \begin{bmatrix} 1 & Z_2 S \\ \frac{S}{Z_2} & 1 \end{bmatrix}$$

Filter Implementation

$$= \frac{1}{\sqrt{1-S^2}} \begin{bmatrix} 1 & Z_2 S \\ \dfrac{1}{SZ_1} + \dfrac{S}{Z_2} & 1 + \dfrac{Z_2}{Z_1} \end{bmatrix}.$$

Similarly, we can write the **ABCD**-matrix for Kuroda's fourth identity, or the right-hand side:

$$\begin{bmatrix} A & B \\ C & D \end{bmatrix}_R = \frac{1}{\sqrt{1-S^2}} \begin{bmatrix} 1 & \dfrac{Z_2 S}{N} \\ \dfrac{SN}{Z_2} & 1 \end{bmatrix}_{UE} \begin{bmatrix} 1 & 0 \\ \dfrac{N}{SZ_1} & 1 \end{bmatrix}_{ind} \begin{bmatrix} 1/N & 0 \\ 0 & N \end{bmatrix}_{trans}$$

where subscripts UE, ind, and trans indicate chain parameter matrices for unit element, inductor, and transformer, respectively. After carrying out the multiplication between the matrices, we obtain the following **ABCD**-matrix describing Kuroda's identity:

$$\begin{bmatrix} A & B \\ C & D \end{bmatrix}_R = \frac{1}{\sqrt{1-S^2}} \begin{bmatrix} \dfrac{1}{N}\left(1 + \dfrac{Z_2}{Z_1}\right) & Z_2 S \\ \dfrac{S}{Z_2} + \dfrac{1}{SZ_1} & N \end{bmatrix}$$

which is identical to the left-hand side, if we set $N = 1 + Z_2/Z_1$. The remaining three Kuroda identities can be proved in a similar fashion.

We see again the utility of the ABCD network representation, which allows us to directly multiply the individual network matrices.

5.3.3 Examples of Microstrip Filter Design

In the following two examples, we will concentrate on the design of a low-pass and a bandstop filter. The bandstop design will be conducted based on the aforementioned Richards transformation followed by use of Kuroda's identities. Specifically, the bandstop design requires some attention in converting from lumped to distributed elements.

The practical filter realization proceeds in four steps:

1. Select the normalized filter order and parameters to meet the design criteria.
2. Replace the inductances and capacitances by equivalent $\lambda_0/8$ transmission lines.

3. Convert series stub lines to shunt stubs through Kuroda's identities.
4. Denormalize and select equivalent microstrip lines (length, width, and dielectric constant).

Specifically, step 4 requires knowledge of the appropriate geometric dimensions of the respective microstrip lines, a subject that is discussed in detail in Chapter 2. According to these four steps, let us now discuss the two examples. The first design task involves a low-pass filter which is formulated as follows:

Project I

Design a low-pass filter whose input and output are matched to 50 Ω, and that meets the following specifications: cutoff frequency of 3 GHz; equiripple of 0.5 dB; and rejection of at least 25 dB at approximately 1.5 times the cutoff frequency. Assume a dielectric material that results in a phase velocity of 60% of the speed of light.

In solving this problem, we proceed according to the previously outlined four steps.

Step 1 From Figure 5-22, it is seen that the filter has to be of order $N = 5$, with coefficients

$$g_1 = 1.7058 = g_5, g_2 = 1.2296 = g_4, g_3 = 2.5408, g_6 = 1.0.$$

The normalized low-pass filter is given in Figure 5-31.

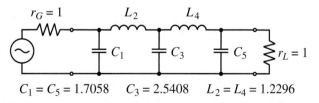

$C_1 = C_5 = 1.7058 \quad C_3 = 2.5408 \quad L_2 = L_4 = 1.2296$

Figure 5-31 Normalized low-pass filter of order $N = 5$.

> **BOARD MATERIALS**
> Common substrate materials for high temperature stability, low loss, and minimal variations in dielectric constants are quartz, sapphire, and alumina. In particular, Duroid is frequently used for very high frequency applications. Popular epoxy glass substrates such as the widely used FR4 (fire retardant) are low cost, but also low performance board materials with not very stable dielectric constants and relatively high dielectric loss tangent of 0.03 at 1 GHz.

Step 2 The inductances and capacitances in Figure 5-31 are replaced by open- and short-circuited series and shunt stubs as shown in Figure 5-32. This is a direct consequence of applying Richards transformation (5.64) and (5.65). The characteristic line impedances and admittances are

$$Y_1 = Y_5 = g_1, Y_3 = g_3, Z_2 = Z_4 = g_4.$$

Step 3 To make the filter realizable, unit elements are introduced with the intent to apply the first and second Kuroda's identities (see Table 5-6) to convert all series stubs into shunt stubs. Since we have a fifth-order filter, we must deploy a total of four unit elements to convert all series-connected short-circuited stubs into shunt-connected open-circuited stubs. To clarify this process, we divide this step into several substeps.

Filter Implementation

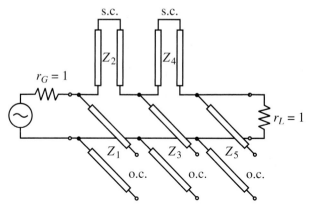

Figure 5-32 Replacing inductors and capacitors by series and shunt stubs (o.c. = open-circuited line, s.c. = short-circuited line).

First, we introduce two unit elements on the input and output ends of the filter, as shown in Figure 5-33.

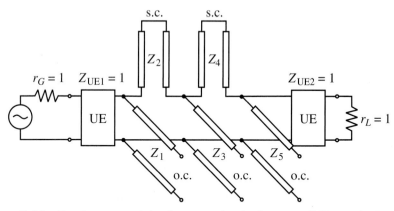

Figure 5-33 Deployment of the first set of unit elements (UE = unit element).

The introduction of unit elements does not affect the filter performance since they are matched to source and load impedances. The result of applying Kuroda's identities to the first and last shunt stubs is shown in Figure 5-34.

This version of the circuit is still nonrealizable because we have four series stubs. To convert them to shunt connections, we have to deploy two more unit elements, as shown in Figure 5-35.

Again, the introduction of unit elements does not affect the performance of the filter since they are matched to the source and load impedances. Applying Kuroda's identities to

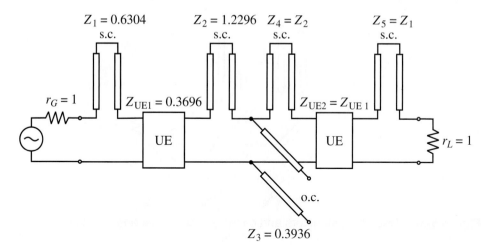

Figure 5-34 Converting shunt stubs to series stubs.

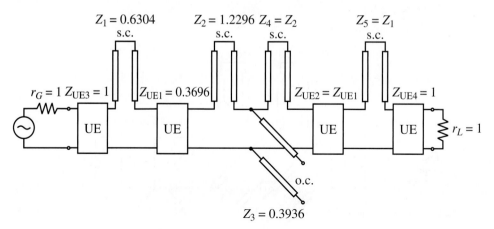

Figure 5-35 Deployment of the second set of unit elements to the fifth-order filter.

the circuit shown in Figure 5-35, we finally arrive at the realizable filter design, depicted in Figure 5-36.

Step 4 Denormalization involves scaling the unit elements to the 50 Ω input and output impedances and computing the length of the lines based on (5.63). Using $v_p = 0.6c = 1.8 \times 10^8$ m/s, the length is found to be $l = (\lambda_0/8) = v_p/(8f_0) = 7.5$ mm. The final design implemented in microstrip lines is shown in Figure 5-37(a). Figure 5-37(b) plots the attenuation profile in the frequency range 0 to 3.5 GHz. We notice that the passband ripple does not exceed 0.5 dB up to the cutoff frequency of 3 GHz.

Filter Implementation

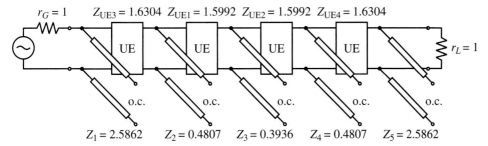

Figure 5-36 Realizable filter circuit obtained by converting series and shunt stubs using Kuroda's identities.

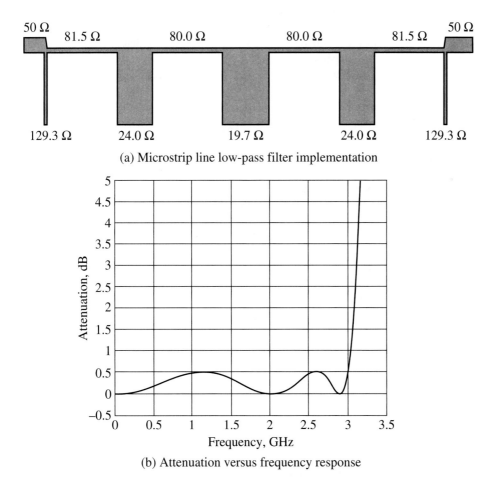

Figure 5-37 Final microstrip line low-pass filter.

The second design project involves a more complicated bandstop filter, which requires the transformation of the standard low-pass prototype with a unity cutoff frequency into a design with specified center frequency and lower and upper 3 dB frequency points.

Project II
Design a maximally flat third-order **bandstop filter** whose input and output are matched to 50 Ω. The filter should meet the following design specifications: center frequency of 4 GHz and 3 dB bandwidth of 50%. Again, we assume a dielectric material that results in a phase velocity of 60% of the speed of light.

This design requires careful analysis when converting from lumped to distributed elements. Specifically, when dealing with bandstop designs, we require either maximum or minimum impedance at the center frequency f_0 depending on whether series or shunt connections are involved. With our previous definition of Richards transformation based on $\lambda_0/8$ line segments, we encounter the difficulty that at $f = f_0$, (5.64) yields a tangent value of 1, and not a maximum. However, if a line length of $\lambda_0/4$ is used, then the tangent will go to infinity as required for a bandstop design. Another aspect that we have to take into account is the fact that we desire the $\Omega = 1$ cutoff frequency of the low-pass prototype filter to be transformed into lower and upper cutoff frequencies of the bandstop filter. This is done by introducing a so-called **bandwidth factor** bf:

$$bf = \cot\left(\frac{\pi}{2}\frac{\omega_L}{\omega_0}\right) = \cot\left[\frac{\pi}{2}\left(1 - \frac{sbw}{2}\right)\right] \tag{5.67}$$

where $sbw = (\omega_U - \omega_L)/\omega_0$ is the **stopband width** and $\omega_0 = (\omega_U + \omega_L)/2$ is the center frequency. Multiplying the Richards transformation for $\lambda_0/4$ line lengths by bf at the lower or upper frequency points reveals that the magnitude of the product is equal to unity. For instance, for the lower frequency point ω_L, it follows that

$$(bf)S|_{\omega = \omega_L} = \cot\left(\frac{\pi}{2}\frac{\omega_L}{\omega_0}\right)\tan\left(\frac{\pi}{2}\frac{\omega_L}{\omega_0}\right) = 1.$$

This corresponds to the $\Omega = 1$ cutoff frequency in the normalized low-pass filter response. Similarly, for the upper cutoff frequency ω_U, we have

$$(bf)S|_{\omega = \omega_U} = \cot\left(\frac{\pi}{2}\frac{\omega_L}{\omega_0}\right)\tan\left(\frac{\pi}{2}\frac{\omega_U}{\omega_0}\right) = \cot\left(\frac{\pi}{2}\frac{\omega_L}{\omega_0}\right)\tan\left[\frac{\pi}{2}\left(\frac{2\omega_0 - \omega_L}{\omega_0}\right)\right] = -1$$

which corresponds to the $\Omega = -1$ cutoff frequency in the normalized low-pass filter. With these preliminary remarks, we are now ready to proceed according to the four steps.

Step 1 From Table 5-2, the coefficients for a maximally flat normalized low-pass filter prototype of third order are

$$g_1 = 1.0 = g_3, g_2 = 2.0, g_4 = 1.0.$$

Thus, the normalized low-pass filter has the form shown in Figure 5-38.

Filter Implementation

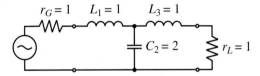

Figure 5-38 Normalized third-order low-pass filter.

Step 2 The inductances and capacitances in Figure 5-38 are replaced by $\lambda_0/4$ open- and short-circuited series and shunt stubs, as depicted in Figure 5-39. The line impedances and the admittance are multiplied by the bandwidth factor (5.67).

$$Z_1 = Z_3 = \text{bf} \cdot g_1, \quad Y_2 = \text{bf} \cdot g_2$$

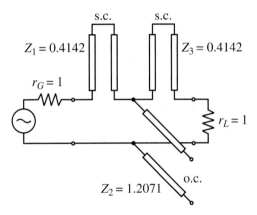

Figure 5-39 Replacing inductors and capacitors by series and shunt stubs.

Step 3 Unit elements of $\lambda_0/4$ line length are inserted and Kuroda's identity is used to convert all series stubs into shunt stubs as seen in Figure 5-40.

Step 4 Denormalization of the unit elements and explicit computation of the individual line lengths can now be conducted. Using the phase velocity $v_p = 0.6c = 1.8 \times 10^8$ m/s, the length is computed to be $l = \lambda_0/4 = v_p/(4f_0) = 15$ mm. Thus, the resulting design in microstrip line implementation is as shown in Figure 5-41.

Finally, for this bandstop filter we can also utilize a commercial simulation package to simulate the filter response of the microstrip line configuration shown in Figure 5-41. The attenuation profile is given in Figure 5-42, and shows that the filter specifications are met.

256   **Chapter 5 • An Overview of RF Filter Design**

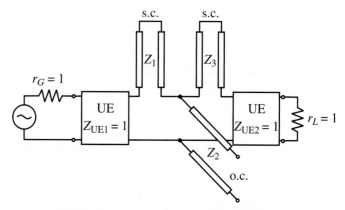

(a) Unit elements at source and load sides

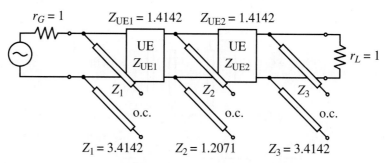

(b) Conversion from series to shunt stubs

Figure 5-40   Introducing unit elements and converting series stubs to shunt stubs.

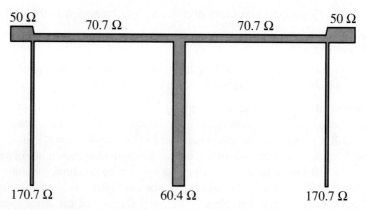

Figure 5-41   Characteristic impedances of final microstrip line implementation of bandstop filter design.

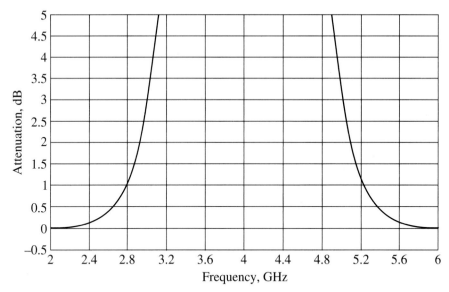

Figure 5-42 Attenuation versus frequency response for third-order bandstop filter.

5.4 Coupled Filter

The literature dealing with coupled filter designs and analyses is extensive. For our cursory treatment we will introduce only the most salient points and refer the reader to the references listed at the end of this chapter.

Our discussion briefly covers the odd and even wave coupling of transmission lines that share a common groundplane, which results in odd and even characteristic line impedances. This sets the stage for an understanding of the coupling between two microstrip lines and their input-output impedances as part of a two-port chain matrix representation. Cascading these elements gives rise to bandpass filter structures that are most easily designed with the aid of RF circuit simulation packages.

5.4.1 Odd and Even Mode Excitation

A simple modeling approach of coupled microstrip line interaction is established when considering the geometry depicted in Figure 5-43. The configuration consists of two lines separated over a distance S and attached to a dielectric medium of thickness d and dielectric constant ε_r. The microstrip lines are W wide, and the thickness is negligible when compared with d. The capacitive and inductive coupling phenomena between the lines and ground are schematically given in Figure 5-44. Here, equal indices denote self-capacitances and inductances, whereas index 12 stands for coupling between line 1 and line 2.

We can now define an even mode voltage V_e and current I_e and an odd mode voltage V_{od} and current I_{od} in terms of the total voltages and currents at terminals 1 and 2 such that

$$V_e = \frac{1}{2}(V_1 + V_2), \quad I_e = \frac{1}{2}(I_1 + I_2) \tag{5.68a}$$

and

$$V_{od} = \frac{1}{2}(V_1 - V_2), \quad I_{od} = \frac{1}{2}(I_1 - I_2). \tag{5.68b}$$

This is consistent with the voltage and current convention shown in Figure 5-44. For the even mode of operation (V_e, I_e), voltages are additive and currents flow in the same direction. However, for the odd mode of operation (V_{od}, I_{od}), the terminal voltages are subtractive and currents flow in opposite directions.

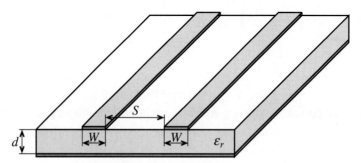

Figure 5-43 Coupled microstrip lines.

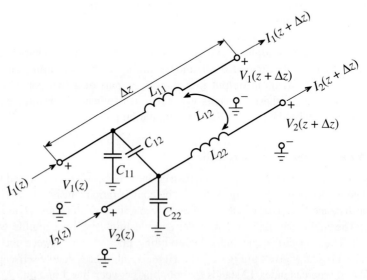

Figure 5-44 Equivalent circuit diagram and appropriate voltage and current definitions for a system of two lossless coupled transmission lines.

Coupled Filter

The benefit of introducing odd and even modes of operation is seen when establishing the fundamental equations. It can be shown that for two lines we obtain a set of first-order, coupled ordinary differential equations similar in form to the transmission line equations in Chapter 2:

$$-\frac{dV_e}{dz} = j\omega(L_{11} + L_{12})I_e \quad (5.69a)$$

$$-\frac{dI_e}{dz} = j\omega C_{11} V_e \quad (5.69b)$$

and

$$-\frac{dV_{od}}{dz} = j\omega(L_{11} - L_{12})I_{od} \quad (5.70a)$$

$$-\frac{dI_{od}}{dz} = j\omega(C_{11} + 2C_{12})V_{od}. \quad (5.70b)$$

What is important to notice is the fact that even and odd modes allow us to decouple the governing equations. The characteristic line impedances Z_{0e} and Z_{0o} for the even and odd modes can be defined in terms of even and odd mode capacitances C_e, C_{od}, and the respective phase velocities as follows:

$$Z_{0e} = \frac{1}{v_{pe} C_e}, \quad Z_{0o} = \frac{1}{v_{po} C_{od}}. \quad (5.71)$$

If both conductors are equal in size, we can conclude for the even mode

$$C_e = C_{11} = C_{22} \quad (5.72a)$$

and for the odd mode

$$C_{od} = C_{11} + 2C_{12} = C_{22} + 2C_{12}. \quad (5.72b)$$

The capacitances are, in general, difficult to find since fringing fields and different media have to be taken into account. For instance, even the parameters of the microstrip line conductor over a dielectric substrate cannot be computed based on the simple capacitance per unit length formula $C_{11} = \varepsilon_0 \varepsilon_r (w/d)$ because the width-to-thickness ratio is not sufficiently large for this formula to apply. Moreover, the cross-coupling capacitance C_{12} requires very intricate treatment. For this reason, it is common practice to resort to a numerically computed impedance grid, such as the one shown in Figure 5-45.

COUPLING EFFECTS

In microstrip line designs, the phase velocities between the even and odd modes differ due to differences in effective dielectric constants. This difference becomes more prominent as the proximity between the strips increases. However, for weak coupling, the condition $v_{pe} \cong v_{po} = v_p$ approximately holds true, resulting in $\beta_e \cong \beta_o = \beta$.

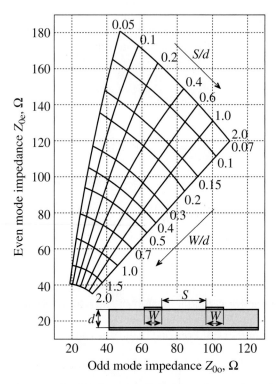

Figure 5-45 Even and odd characteristic impedance for microstrip lines.

COMBLINE FILTER

For frequencies below 1 GHz (for instance, between 200 to 600 MHz), microstrip bandpass filters that have a small footprint are difficult to design. This problem can be overcome by the design of a combline filter structure where the microstrip lines are stacked and terminated on one side to ground and at the other side to a capacitor. For instance, five element bandpass filters at 500 MHz with rejection of 40 dB and a footprint of 1 inch by 1 inch have been reported.

5.4.2 Bandpass Filter Section

We turn our attention to coupled microstrip lines as the main building block of a bandpass filter shown in Figure 5-46. This figure depicts the geometric arrangement with input and output ports, open-circuit conditions, and the corresponding transmission line representation.

Without delving into details of the rather complicated treatment (see Gupta in Further Reading), this configuration has the impedance matrix coefficients for open-circuited transmission line segments in the form

$$Z_{11} = -j\frac{1}{2}(Z_{0e} + Z_{0o})\cot(\beta l) = Z_{22} \quad (5.73a)$$

$$Z_{12} = -j\frac{1}{2}(Z_{0e} - Z_{0o})\frac{1}{\sin(\beta l)} = Z_{21}. \quad (5.73b)$$

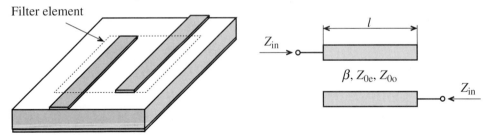

(a) Arrangement of two coupled microstrip lines (b) Transmission line representation

Figure 5-46 Bandpass filter element.

When cascading these building blocks into multiblock filter configurations, our desire is to match both ports of this segment to the adjacent elements. This is also known as finding the **image impedance**. For the input impedance at port 1 we can write

$$Z_{in} = \frac{V_1}{I_1} = \frac{AZ_L + B}{CZ_L + D} \tag{5.74a}$$

and the load impedance at port 2

$$Z_L = \frac{-V_2}{I_2} = \frac{DZ_{in} + B}{CZ_{in} + A}. \tag{5.74b}$$

Since we require $Z_{in} = Z_L$, we find from (5.74) that $A = D$ and

$$Z_{in} = \sqrt{\frac{B}{C}}. \tag{5.75}$$

If (5.73) is converted into a chain matrix form, the coefficients A, B, C, D can be determined. Inserting B and C into (5.75), one finds for the input, or image impedance

$$Z_{in} = \frac{1}{2\sin(\beta l)} \sqrt{(Z_{0e} - Z_{0o})^2 - (Z_{0e} + Z_{0o})^2 \cos^2(\beta l)}. \tag{5.76}$$

The bandpass filter behavior of (5.76) becomes apparent when plotting the real part of the input impedance response as a function of the electric length in the range $0 \leq \beta l \leq 2\pi$, as depicted in Figure 5-47.

According to Figure 5-47, the characteristic bandpass filter performance is obtained when the length is selected to be $\lambda/4$ or $\beta l = \pi/2$. For this case, the upper and lower cutoff frequencies are found as

$$(\beta l)_{1,2} = \theta_{1,2} = \pm \cos^{-1}\left[\frac{Z_{0e} - Z_{0o}}{Z_{0e} + Z_{0o}}\right]. \tag{5.77}$$

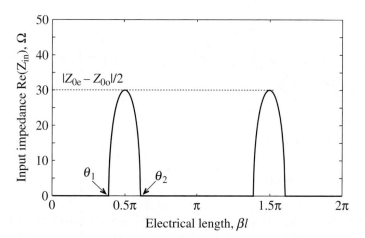

Figure 5-47 Input impedance behavior of equation (5.76). Z_{0e} and Z_{0o} are arbitrarily set to 120 Ω and 60 Ω, respectively.

Also noticeable is the periodic impedance response in Figure 5-47, which indicates that the upper operating frequency has to be band-limited to avoid multiple bandpass filter responses at higher frequencies.

5.4.3 Cascading Bandpass Filter Elements

A single bandpass element as discussed in the previous section does not result in good filter performance with steep passband to stopband transitions. However, the ability to cascade these building blocks ultimately results in high-performance filters. Figure 5-48 depicts a generic multielement design.

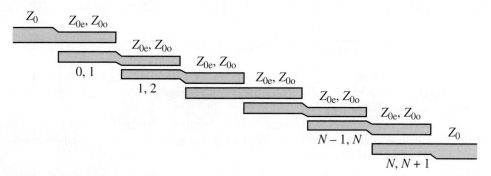

Figure 5-48 Multielement configuration of a fifth-order coupled-line bandpass filter ($N = 5$).

To design such a structure that meets a particular bandpass filter specification, a number of computations have to be performed. The following sequence of steps is needed to translate a set of design requirements into a practical filter realization (see Matthaei et al. in Further Reading).

- *Selection of standard low-pass filter coefficients.* Depending on whether a Butterworth or Chebyshev design with desired rejection and ripple is needed, the designer can directly select the appropriate standard low-pass filter coefficients $g_0, g_1, \ldots, g_N, g_{N+1}$ listed in Tables 5-2 to 5-6.
- *Identification of normalized bandwidth, upper, and lower frequencies.* From the desired filter specifications for lower and upper frequencies ω_L, ω_U and the center frequency $\omega_0 = (\omega_U + \omega_L)/2$, we define the normalized bandwidth of the filter as

> **DIPLEX FILTER**
> Connecting two radios operating at different frequency bands to the same antenna can be accomplished with a so-called diplex filter, which consists of a complementary low-pass and high-pass filter connected to a single point junction, or tee-junction. As a result, the first radio would operate below the 3 dB cutoff of the low-pass filter where the high-pass filter attenuates the second radio and vice versa.

$$BW = \frac{\omega_U - \omega_L}{\omega_0}. \tag{5.78}$$

This factor allow us to compute the following parameters:

$$J_{0,1} = \frac{1}{Z_0}\sqrt{\frac{\pi BW}{2g_0 g_1}} \tag{5.79a}$$

$$J_{i,i+1} = \frac{1}{Z_0} \frac{\pi BW}{2\sqrt{g_i g_{i+1}}} \tag{5.79b}$$

$$J_{N,N+1} = \frac{1}{Z_0}\sqrt{\frac{\pi BW}{2g_N g_{N+1}}} \tag{5.79c}$$

which in turn permit us to determine the odd and even characteristic line impedances:

$$Z_{0o}\big|_{i,i+1} = Z_0[1 - Z_0 J_{i,i+1} + (Z_0 J_{i,i+1})^2] \tag{5.80a}$$

and

$$Z_{0e}\big|_{i,i+1} = Z_0[1 + Z_0 J_{i,i+1} + (Z_0 J_{i,i+1})^2] \tag{5.80b}$$

where the indices $i, i+1$ refer to the overlapping elements seen in Figure 5-48. Here, Z_0 is the characteristic line impedance at the beginning and the end of the filter structure.
- *Selection of actual microstrip dimensions.* Based on Figure 5-45, the individual odd and even line impedances can be converted into microstrip dimensions. For instance,

if the dielectric material and the thickness of the PCB board are given, we can determine the separation S and the width W of the copper strips. The length of each coupled line segment has to be equal to $\lambda/4$ at the center frequency, as described in Section 5.4.2.

The preceding steps result in a first and often crude design, which can be made more precise by introducing length and width corrections to account for fringing field effects. In addition, the use of simulation packages often allows further adjustments and fine tuning to ensure a design that actually performs according to the specifications.

5.4.4 Design Example

In the following example, we go through the steps outlined in the previous section by designing a particular bandpass filter.

Example 5-6. Bandpass filter design with coupled line transmission line segments

A coupled-line bandpass filter with 3 dB ripples in the passband is to be designed for a center frequency of 5 GHz and lower and upper cut-off frequencies of 4.8 and 5.2 GHz, respectively. The attenuation should be at least 30 dB at 5.3 GHz. Select the number of elements and find odd and even mode characteristic impedances of the coupled transmission lines.

Solution. According to Section 5.4.3, the first step in the design of this filter is to choose an appropriate low-pass filter prototype. The order of the filter can be selected from the requirement of 30 dB attenuation at 5.3 GHz. Using frequency conversion for the bandpass filter (5.52), we find that for 5.3 GHz the normalized frequency of the low-pass filter prototype is

$$\Omega = \frac{\omega_c}{\omega_U - \omega_L}\left(\frac{\omega}{\omega_c} - \frac{\omega_c}{\omega}\right) = 1.4764.$$

From Figure 5-21 we determine that the order of the filter should be at least $N = 5$ to achieve 30 dB attenuation at $\Omega = 1.4764$. The coefficients for an $N = 5$ Chebyshev filter with 3 dB ripples are

$$g_1 = g_5 = 3.4817, g_2 = g_4 = 0.7618, g_3 = 4.5381, g_6 = 1.$$

The next step in the design is to find the even and odd excitation mode characteristic impedances of the coupled transmission lines as described by (5.80). The results of these computations are listed in the following table.

i	$Z_0 J_{i,i+1}$	$Z_{0o}(\Omega)$	$Z_{0e}(\Omega)$
0	0.1900	42.3056	61.3037
1	0.0772	46.4397	54.1557
2	0.0676	46.8491	53.6077
3	0.0676	46.8491	53.6077
4	0.0772	46.4397	54.1557
5	0.1900	42.3056	61.3037

To confirm the validity of our theoretical design, we can use a simulation package to analyze the performance of the bandpass filter just designed. The result of the simulation is shown in Figure 5-49.

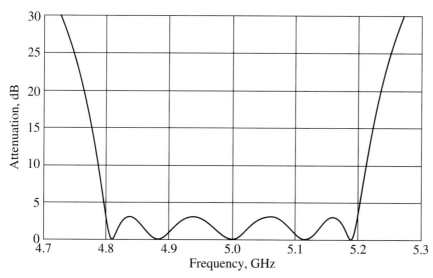

Figure 5-49 Simulations of the fifth-order coupled-line Chebyshev bandpass filter with 3 dB ripple in the passband. The lower cutoff frequency is 4.8 GHz and the upper cutoff frequency is 5.2 GHz.

The filter response in Figure 5-49 confirms that the specifications are met for f_L and f_U, and the attenuation at 5.3 GHz exceeds the 30 dB requirement.

Often, the theoretical filter design leads to coefficients whose validity must be double checked against an RF circuit simulator to test the actual performance.

Another reason for resorting to a simulation package is the need to verify the design methodology independently and to test the filter performance over a range of parameter variations in terms of geometry and dielectric properties. Most of these parametric studies can be accomplished with little effort on a personal computer. After performing the initial theoretical design, computer simulations typically precede the actual board construction and testing.

Practically Speaking. **A low-pass microstrip filter implementation**

In this section, we examine a microstrip low-pass filter, comparing the predictions of a simple MATLAB code developed by the authors and a rigorous ADS CAD simulator with the actually measured filter performance.

The fifth-order low-pass filter in Figure 5-37 was adjusted for a cutoff frequency of 1.5 GHz. All characteristic line impedances and electrical lengths remained the same, only the physical lengths of the microstrips were changed. Figure 5-50(a) plots the theoretical S-parameters (magnitude in dB) of the low-pass filter implemented with generic lossless transmission lines. The forward voltage gain $|S_{21}|$ exhibits several 0.5 dB ripples in the passband from 0 to 1.5 GHz, as originally designed. We notice that these ripples appear in the reflection coefficient $|S_{11}|$ as fluctuations between almost no reflection to −9.6 dB reflection. Beyond the 1.5 GHz cutoff, we see a relatively sharp fifth-order $|S_{21}|$ rolloff, slightly faster near the cutoff, consistent with a Chebyshev filter behavior. We also observe that $|S_{11}|$ approaches 1 in the stopband, meaning that the incoming signal is completely reflected back to the generator. This underscores the importance of proper port impedance matching for such a filter design.

Figure 5-50(b) shows the ADS simulation results for a microstrip realization of this filter. The microstrip layout is shown in Figure 5-51(a). The simulated substrate is 0.062 inch thick FR4 with $\varepsilon_r = 4.6$ and a loss tangent of $\tan \Delta = 0.02$ at 1 GHz. The performance of the simulated filter differs considerably from the ideal case: we notice increasing insertion loss towards the upper end of the passband due to losses in the FR4 dielectric and the copper conductor. The cutoff corner is not as sharp, and the rolloff is not as steep as in the ideal case. These discrepancies are attributed to nonideal behavior of the relatively short microstrips as well as the addition of copper patches to interconnect the strips.

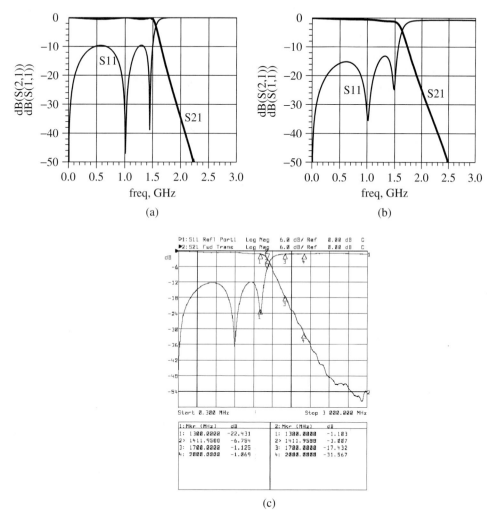

Figure 5-50 *S*-parameters of the 1.5 GHz microstrip low-pass filter: (a) Matlab simulation, (b) ADS simulation, (c) network analyzer mesurement.

Figure 5-50(c) shows a network analyzer measurement obtained with an actual PCB implementation of this filter. Figure 5-51(b) presents a photograph of this PCB. The implemented microstrip pattern is the same as the simulated pattern in Figure 5-51(a), with the addition of

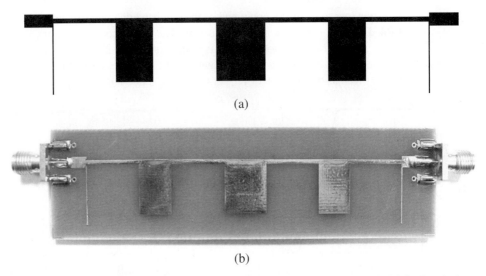

Figure 5-51 Microstrip pattern for the 1.5 GHz low-pass filter: (a) ADS simulation, (b) PCB implementation.

two SMA edge connectors for cable attachment. The PCB measures 4.05 by 1.1 inch and includes a continuous ground plane on the opposite side. The measured $|S_{21}|$ response is similar to the simulation, except that the cutoff frequency is slightly lower, around 1.4 GHz. This indicates that the dielectric constant of the FR4 material was underestimated in the simulation. Consistent with the simulation, the insertion loss increases at the upper end of the passband, reaching 1 dB at 1.3 GHz, and the rolloff is slightly slower than in an ideal fifth-order filter. However, we notice that the attenuation in the stopband stalls at around 50 dB above 2.3 GHz instead of continuing to increase. This is most likely due to an imperfect ground plane. In general, any physical filter structure exhibits signal leakage and thus high stopband attenuation is not easy to achieve. Another factor to consider is the physical size of the circuit. The 4 by 1 inch structure is too large to be practical in most cases. We could reduce the size almost in half by using a high dielectric substrate, but realistically, this type of filter only becomes practical in applications above approximately 5 GHz.

5.5 Summary

Our emphasis in this chapter has been on exposure to filter design concepts that are ubiquitous in many RF/MW circuit designs. Rather than going into detailed derivations, the intent

Summary

of this chapter is to present a generic discussion of some of the key issues facing a design engineer in the construction of practical filters.

Beginning with a general classification of high-pass, low-pass, bandpass, and bandstop filters, we introduce a common terminology that is needed when developing filter specifications. Terms that are often used such as cut-off, lower, upper, and center frequencies, shape factor, bandwidth, insertion loss, and rejection, are defined and placed in context with simple first-order high- and low-pass filters as well as series and parallel resonant circuits. Since the resonator circuits permit the realization of bandpass and bandstop designs, the sharpness of the impedance or admittance behavior is quantified through the so-called quality factor

$$Q = 2\pi \frac{\text{averaged stored energy}}{\text{energy loss per cycle}}\bigg|_{\omega = \omega_c}$$

a measure that can be further broken down into the filter Q_F and external Q_E quality factors. Specifically, the notation of insertion loss

$$\text{IL [dB]} = 10 \log \frac{P_A}{P_L} = -20 \log |S_{21}|$$

which defines the amount of power lost by inserting the filter between the source and load ports, is of central important in the design of high-frequency filters. Depending on the attenuation profile necessary to realize the various filter types, the loss factor

$$\text{LF} = \frac{1}{|S_{21}|^2} = 1 + \varepsilon^2 Q_{LD}^2$$

is employed to realize a particular response. However, our definition assumes that no power is dissipated in the filter network.

To enable a more comprehensive approach, the low-pass filter design based on a normalized frequency scale is chosen as the standard type. Through frequency scaling and shifting, all filter types can then be readily realized. The benefit of this approach is that only a few sets of standard low-pass filter coefficients have to be derived depending on whether a Butterworth filter with a maximally flat profile or Chebyshev filter with an equiripple attenuation profile is desired.

One practical distributed implementation is achieved through the Richards transformation:

$$S = j\tan\left(\frac{\pi}{4}\Omega\right).$$

This transformation is central in establishing a link between lumped capacitive and inductive elements and distributed transmission line theory. The various series and shunt transmission line segments can be spatially separated through unit elements before applying Kuroda's identities to convert some of the transmission sections into easy-to-implement segmental elements. In particular, shunt microstrip stubs are often easier to implement than series stubs. With the aid of Kuroda's identities, this can be accomplished elegantly.

The fact that the proximity of microstrip lines causes electromagnetic coupling is exploited to design bandpass and bandstop filters. Without delving into the theoretical explanations too deeply, two coupled line segments are used as the basic building block of a two-port network representation. Through odd and even mode impedance analysis, we can find the image impedance

$$Z_{in} = \frac{1}{2\sin(\beta l)}\sqrt{(Z_{0e} - Z_{0o})^2 - (Z_{0e} + Z_{0o})^2 \cos^2(\beta l)}$$

as the characteristic bandpass response. This single element can be cascaded into multiple section filters to fulfill various design requirements. By using an RF/MW simulation package, the same example is revisited and the coupled filter response is computed as a function of various element numbers, and geometric dimensions of the microstrip lines.

Although the topic of filter design could only be covered briefly, Chapter 5 conveys the basic engineering steps needed to arrive at a functional high-frequency filter realization. We attempted to make the process of choosing the appropriate filter coefficients, scaling the results to actual frequencies, and implementing the process in microstrip lines as much of a cookbook approach as possible. However, Chapter 5 should also make clear the usefulness of commercial simulation packages in carrying out a detailed numerical analysis. Indeed, for most modern filter design examples, an RF/MW simulation package is an indispensable tool for predicting the filter performance. Moreover, from the circuit schematic it is relatively straightforward to use special layout programs to generate the actual PCB layout file that becomes the basis for the physical board construction.

Further Reading

R. Levy and S. B. Cohn, "A Brief History of Microwave Filter Research, Design and Development," *IEEE Trans. on Microwave Theory and Techniques,* MTT-32, pp. 1055–1067, 1982.

S. Kobajashi and K. Saito, "A Miniaturized Ceramic Bandpass Filter for Cordless Phones," *IEEE International Microwave Symposium Dig.,* pp. 249–252, 1995.

C. C. You, C. L. Huang, and K. I. Sawamoto, "A Direct Coupled Lambda/4 Coaxial Resonator Bandpass Filter for Land Mobile Communications," *IEEE Trans. on Microwave Theory and Techniques,* MTT-34, pp. 972–976, 1986.

M. Korber, "New Microstrip Filter Topologies," *Microwave Journal,* 40, pp. 138–144, 1997.

B. Rawat, R. Miller, and B. E. Pontius, "Bandpass Filters for Mobile Communications," *Microwave Journal,* 27, pp. 146–152, 1984.

R. W. Rhea, *HF Filter Design and Computer Simulations,* Nobel Publishing, Atlanta, GA, 1994.

S. Butterworth, "On the Theory of Filter Amplifiers," *Wireless Eng.*, Vol. 7, pp. 536–541, 1930.

S. Darlington, "A History of Network Synthesis and Filter Theory for Circuits Composed of Resistors, Inductors, and Capacitors," *IEEE Transactions on Circuits and Systems,* Vol. 46, pp. 4–13, 1999.

B. Mayer and M. H. Vogel, "Design Chebyshev Bandpass Filters Efficiently," *RF Design,* pp. 50–56, September 2002.

D. Bradly, "The Design, Fabrication and Measurement of Microstrip Filter and Coupler Circuits," *High Frequency Electronics,* pp. 22–30, July 2002.

C. A. Corral, C. S. Lindquist, and P. B. Aronhtme, "Sensitivity of the Band-Edge Selectivity of Various Classical Filters," Proceedings of the 40th Midwest Symposium of Circuits and Systems, p. 324, 1997.

K. C. Gupta, R. Garg, and I. J. Bahl, *Microstrip Lines and Slot Lines,* Artech House, Dedham, MA, 1979.

G. L. Matthaei, et al., *Microwave Filters, Impedance-Matching Networks, and Coupling Structures,* McGraw-Hill, New York, 1964.

E. H. Bradley, "Design and Development of Stripline Filters," *IEEE Trans. on Microwave Theory and Techniques,* Vol. 4, No. 2, pp. 86–93, 1956.

A. Bhargava, "Combline Filter Design Simplified," RF Design, pp. 42–48, January 2004.

C. G. Montgomery, R. H Dicke, and E. M. Purcell, *Principles of Microwave Circuits,* MIT Radiation Laboratory Series, Vol. 8, McGraw-Hill, New York, 1948.

D. M. Pozar, *Microwave Engineering,* 2nd ed., John Wiley, New York, 1998.

P. A. Rizzi, *Microwave Engineering: Passive Circuits,* Prentice Hall, Englewood Cliffs, NJ, 1988.

L. Weinberg, *Network Analysis and Synthesis,* McGraw-Hill, New York, 1962.

A. Zverev, *Handbook of Filter Synthesis*, John Wiley, New York, 1967.

Problems

5.1 For the simple filter circuit shown,

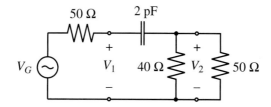

determine the following quantities:

- Transfer function $S_{21}(\omega)$.
- Insertion loss versus frequency behavior
- Phase versus frequency behavior $\varphi(\omega)$
- Group delay $t_g(\omega)$

Plot these factors for the frequency range from DC to 1 GHz.

5.2 Derive expressions for internal, external, and loaded quality factors for the standard series and parallel resonance circuits discussed in Section 5.1.4.

5.3 In Section 5.1.5, the admittance of the parallel resonant circuit is expressed in terms of a quality factor. Prove the resulting equation (5.29).

5.4 A transmission line can be used as a resonator if it is driven by a source and terminated by an open or short circuit. The figure below, for instance, shows a short-circuit termination.

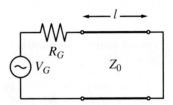

Assume that losses are low, i.e., $\alpha l < 0.1 Np$, (a) show that the input impedance of this configuration can be written as $Z_{in} = Z_0 \alpha l + j Z_0 \tan(\beta l)$, where l is the length, $\beta = \omega/v_p$ is the phase constant, v_p is the phase velocity. (b) Find equations for the unloaded and loaded quality factors for the lowest series resonance, and provide numerical results if the the resonance frequency is 5 GHz, $Z_0 = R_G = 50\ \Omega$, the phase velocity is $0.7\ c$, and the attenuation is 5 mNp/m.

5.5 A single-stage bandstop filter is inserted between source and load resistances of $R_G = R_L = 50\ \Omega$. (a) Find the insertion loss for this filter and plot its frequency response in dB if $L = 20$ nH, $C = 80$ pF, and $R = 50\ \Omega$, and (b) Determine the loaded quality factor from the 3 dB bandwidth and center frequency.

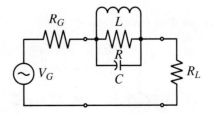

5.6 For the filter circuit shown,

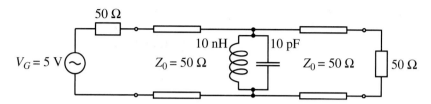

find the loaded, unloaded, and external quality factors. In addition, determine the power generated by the source as well as the power absorbed by the load at resonance. Furthermore, plot the insertion loss as a function of frequency in the range of ±50% of the resonance frequency.

5.7 Repeat Problem 5.4 for the following filter circuit:

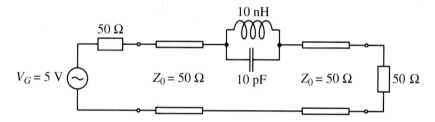

5.8 The insertion loss method mentioned at the beginning of this chapter can be generalized (see book by Rizzi) to the case where the input power of the network is represented by the input power and load power in the form

$$P_{in} = (1 - |\Gamma_{in}|^2)P_A \text{ and } P_L = (1 - |\Gamma_{in}|^2)P_A - P_d$$

where P_A and P_d are, respectively, the power available from the matched source and the power dissipated by the filter network. Find (a) the insertion loss of the network, and (b) show that for a lossless network ($P_d = 0$) one obtains IL [dB] $= 10 \log(1 - |\Gamma_{in}|^2)$.

5.9 You are required to build a low-pass Butterworth filter that provides an attenuation value of at least 50 dB at $f = 1.5 f_{3dB}$. Which filter order is required? How many components (inductors and capacitors) do you need to realize this filter?

5.10 Convert the filter your designed in the previous example into a high-pass filter of the same type and cutoff frequency.

5.11 Your task is to design a three-port diplex Butterworth filter with a 3 dB cutoff at 440 MHz. The low-pass (port 2) and high-pass (port 3) filters should each be of third order and connected to a single output (port 1). Draw the filter configuration and plot the magnitude of the S-parameters S_{11}, S_{21}, S_{31} in dB over a suitably chosen frequency range.

5.12 Design a prototype low-pass Butterworth filter that will provide at least 20 dB attenuation at the frequency of $f = 2 f_{3dB}$.

5.13 Plot the insertion loss of a low-pass Chebyshev filter that has 6 dB ripple in the passband and at least 50 dB attenuation at $f = 2f_{cutoff}$.

5.14 Using the low-pass prototype developed in Problem 5.12, design a high-pass filter with cutoff frequency of 1 GHz. Plot the attenuation profile.

5.15 To suppress noise in a digital communication system, a bandpass RF filter is required with a passband from 1.9 GHz to 2 GHz. The minimum attenuation of the filter at 2.1 GHz and 1.8 GHz should be 30 dB. Assuming that a 0.5 dB ripple in the passband can be tolerated, design a filter that will use a minimum number of components.

5.16 In the design of an amplifier for cellular phone applications, it is discovered that the circuit exhibits excessive noise at 3 GHz. Develop a bandstop filter with a center frequency of $f_c = 3$ GHz, 30 dB stopband width of 10% around f_c, 3 dB stopband width of 36.7%, and passband ripple of 3 dB.

5.17 In previous chapters we examined the input impedance behavior of an open-circuited stub assuming that open-circuit conditions at the end are ideal. In practical realizations, due to fringing fields, leakage occurs. This can be modeled as an additional parasitic capacitance, as shown below:

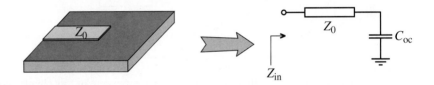

Using your favorite mathematical program, find the input impedance of the 50 Ω open-circuited stub of length $l = 1$ cm for frequencies ranging from 10 MHz to 100 GHz. In your computations assume that an equivalent load capacitance is $C_{oc} = 0.1$ pF and the phase velocity of the line is $v_p = 1.5 \times 10^8$ m/s. Compare your results to the input impedance behavior of the ideal open-circuited stub.

5.18 Assuming all physical parameters of the open-circuited stub to be the same as in Problem 5.17, find the effective fringing capacitance C_{oc} if the lowest frequency at which the input impedance of the stub equals zero is 3.3 GHz.

5.19 After reconsidering the design in Problem 5.17, it is decided to use an open-circuited stub of half of the length (i.e., $l = 5$ mm). Since the board is already manufactured with a 1 cm stub, you cut a slit in the middle so that the length of the resulting stub is 5 mm, as shown below.

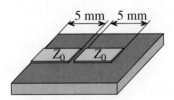

Due to proximity effects, the equivalent circuit in this case is as follows:

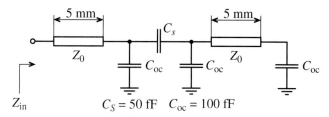

Using a mathematical spreadsheet, compute the input impedance of this configuration for frequencies ranging from 10 MHz to 20 GHz, assuming that the characteristic line impedance is 50 Ω and the phase velocity is $v_p = 1.5 \times 10^8$ m/s. Compare the results with the input impedance behavior of the 5 mm open-circuited stub, taking into account a fringing capacitance $C_{oc} = 100$ fF ($f = 10^{-15}$).

5.20 In Chapter 2 we introduced a quarter-wave transformer that is able to transform any real load impedance into any other real value. In our analysis, we always assumed that there are no parasitic elements involved. In reality, the connection of two transmission lines with different impedances leads to a discontinuity in the line width as follows:

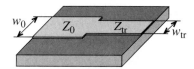

Due to this discontinuity, additional parasitic elements have to be taken into account. The equivalent circuit for the above configuration is:

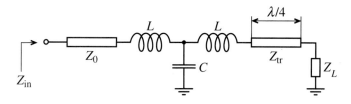

For a load impedance of $Z_L = 25$ Ω and a $Z_0 = 100$ Ω line impedance, find a characteristic impedance Z_{tr} of the quarter-wave transformer and compute the input impedance Z_{in} of the entire system for a frequency range from 10 MHz to 20 GHz, assuming that the transmission line is a quarter wavelength at 10 GHz and parasitic elements have the following values: $L = 10$ pH, $C = 100$ fF.

5.21 Prove the first three Kuroda's identities given in Table 5-6 by computing the appropriate **ABCD**-matrices.

5.22 Develop a low-pass filter with a cutoff frequency of 200 MHz, attenuation of 50 dB at 250 MHz and 3 dB ripple in the passband. Choose the filter implementation that requires the least number of components.

5.23 Design a three-element bandpass filter with 3 dB ripples in the passband. The center frequency is 900 MHz and the bandwidth is 30 MHz. Use a mathematical spreadsheet and plot the insertion loss of the filter.

5.24 In Project I of Section 5.3.3, we designed a microstrip realization of the Chebyshev-type low-pass filter with 3 GHz cutoff frequency. Repeat this design using an FR4 substrate with dielectric constant of $\varepsilon_r = 4.6$ and thickness of $h = 25$ mil. In addition, obtain the physical width and length of each microstrip line.

5.25 Design a five-element bandstop filter having a maximally flat response. The 3 dB bandwidth of the filter should be 15% with a center frequency of 2.4 GHz. The filter has to be matched to 75 Ω impedance at both sides.

5.26 Design a fifth-order low-pass filter with linear phase response. The cutoff frequency of the filter is 5 GHz. Provide two designs: the first one using lumped elements and the second design using microstrip lines. In both cases, assume that an FR4 substrate is used ($\varepsilon_r = 4.6$, $h = 20$ mil).

5.27 As a part of a satellite communication link, a bandpass filter for image rejection in the down conversion stage has to be designed. The bandwidth of a signal is 300 MHz and the center frequency is 10 GHz. It is essential to provide maximally flat response in the passband and obtain at least 40 dB attenuation at 10.4 GHz.

5.28 Prove equations (5.74a) and (5.74b) and show how equation (5.76) results.

5.29 Design a maximally flat bandpass filter with coupled transmission line segments that is fitted in a 50 Ω line impedance system and that meets the following requirements: attenuation of at least 30 dB at 5.5 GHz and 3 dB at 4 GHz and 5 GHz. Specifically, determine (a) number of elements, (b) standardized filter coefficients, (c) odd and even mode impedances, and (d) sketch the layout.

6

Active RF Components

6.1	Semiconductor Basics	278
	6.1.1 Physical Properties of Semiconductors	278
	6.1.2 The *pn*-Junction	285
	6.1.3 Schottky Contact	295
6.2	RF Diodes	298
	6.2.1 Schottky Diode	299
	6.2.2 PIN Diode	301
	6.2.3 Varactor Diode	307
	6.2.4 IMPATT Diode	310
	6.2.5 Tunnel Diode	312
	6.2.6 TRAPATT, BARRITT, and Gunn Diodes	313
6.3	Bipolar-Junction Transistor	313
	6.3.1 Construction	314
	6.3.2 Functionality	316
	6.3.3 Frequency Response	322
	6.3.4 Temperature Behavior	324
	6.3.5 Limiting Values	328
	6.3.6 Noise Performance	329
6.4	RF Field Effect Transistors	330
	6.4.1 Construction	330
	6.4.2 Functionality	331
	6.4.3 Frequency Response	338
	6.4.4 Limiting Values	339
6.5	Metal Oxide Semiconductor Transistors	339
	6.5.1 Construction	340
	6.5.2 Functionality	341
6.6	High Electron Mobility Transistors	342
	6.6.1 Construction	343
	6.6.2 Functionality	343
	6.6.3 Frequency Response	346
6.7	Semiconductor Technology Trends	347
6.8	Summary	352

Our focus in the first five chapters has been primarily geared toward passive RF devices and their electric circuit behavior. In this chapter, we extend and broaden our scope to include an investigation of various active circuit elements. Of specific interest for the design of amplifiers, mixers, and oscillators are solid-state devices such as diodes and transistors. What complicates a unified treatment is the wealth of special purpose components developed and marketed by a range of companies for a wide host of industrial applications. We cannot adequately address the multitude of technological advances currently shaping the RF/MW commercial markets. This is not the intent of this text; rather, we emphasize a number of key concepts driving the technological RF/MW evolution. These concepts are utilized later for the design of amplifiers, mixers, oscillators, and other circuits developed in subsequent chapters. Our goal is to enable the reader to formulate and develop his or her own network descriptions as part of an integrated strategy to construct suitable models of analog RF circuits.

Before developing appropriate network models for active devices, a short discussion of solid-state physics involving *pn* and metal-semiconductor junctions is presented. The aim is to provide a solid-state perspective of the electric circuit representations derived from the physical device level. This is needed because

- at high-frequencies additional capacitive and inductive effects enter the solid-state devices and affect their performance
- the high-frequency behavior of many active devices markedly departs from that of low-frequency components and therefore requires special treatment
- to utilize simulation tools such as SPICE or more specialized RF CAD programs, a working knowledge of the physical parameters that directly or indirectly influence the circuit behavior must be obtained.

Chapter 6 provides a concise summary of the most important semiconductor fundamentals that are encountered at high frequencies. By analyzing the *pn*-junction and the Schottky contact, we gain a more complete picture of electronic circuit functions that form the foundation of rectifier, amplifier, tuning, and switching systems. In particular, the metal-semiconductor interface is shown to be especially useful for high-frequency operation. The RF industry has seen many specialized diode developments. Chief among them are the Schottky, PIN, and tunnel diode, to name but a few. Next, our attention is turned toward the bipolar and field effect transistors, which are more complex implementations of the previously investigated *pn*-junction and Schottky contact. We learn about the construction, functionality, and temperature as well as noise performance of the bipolar and the metal-semiconductor field effect transistors.

6.1 Semiconductor Basics

6.1.1 Physical Properties of Semiconductors

The operation of semiconductor devices is naturally dependent on the physical behavior of the semiconductors themselves. This section presents a brief introduction to the basic building blocks of semiconductor device modeling, particularly the operation of the ***pn*-junction**.

Semiconductor Basics

In our discussion, we will concentrate on the three most commonly used semiconductors: germanium (Ge), silicon (Si), and gallium arsenide (GaAs). Figure 6-1(a) schematically shows the bonding structure of pure silicon. Each silicon atom shares its four valence electrons with the four neighboring atoms, forming four covalent bonds.

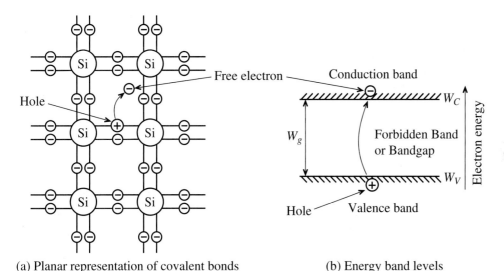

(a) Planar representation of covalent bonds (b) Energy band levels

Figure 6-1 Lattice structure and energy levels of silicon. (a) schematic planar crystal arrangement with thermal breakup of one valent bond resulting in a hole and a moving electron for $T > 0$ K. (b) equivalent energy band level representation whereby a hole is created in the valence band W_V and an electron is produced in the conduction band W_C. The energy gap between both bands is indicated by W_g.

In the absence of thermal energy, i.e., when the temperature is equal to zero degree Kelvin ($T = 0$ K $= -273.15°$C, where T [K] $= 273.15 + T$ [°C]), all electrons are bonded to the corresponding atoms and the semiconductor is not conductive. However, when the temperature increases, some of the electrons obtain sufficient energy to break up the covalent bond and cross the energy gap $W_g = W_C - W_V$, as shown in Figure 6-1(b) (at room temperature $T = 300$ K, the **bandgap energy** is equal to 1.12 eV for Si, 0.62 eV for Ge, and 1.42 eV for GaAs). These free electrons form negative charge carriers that allow electric current conduction. The concentration of the conduction electrons in the semiconductor is denoted as n. When an electron breaks the covalent bond, it leaves behind a positively charged vacancy, which can be occupied by another free electron. These types of vacancies are called **holes** and their concentration is denoted by p.

Electrons and holes undergo random motion through the semiconductor lattice as a result of the presence of thermal energy ($T > 0$ K). If an electron happens to meet a hole, they recombine and both charge carriers disappear. In thermal equilibrium, we have equal

RF SEMICONDUCTOR TECHNOLOGY

The implementation of RF system designs can be accomplished in a number of different semiconductor technologies. While standard silicon is preferred at low frequency, due to its widespread use in integrated circuits, materials based on SiGe, GaAs, InP are used for high to extremely high operating frequencies. For instance, InP is employed in optical communication systems up to 200 GHz and beyond.

number of recombinations and generations of holes and electrons. The concentrations obey the Fermi statistics according to

$$n = N_C \exp\left[-\frac{W_C - W_F}{kT}\right] \quad (6.1a)$$

$$p = N_V \exp\left[-\frac{W_F - W_V}{kT}\right] \quad (6.1b)$$

where

$$N_{C,V} = 2(2m^*_{n,p}\pi kT/h^2)^{3/2} \quad (6.2)$$

are the **effective carrier concentrations** in the conduction (N_C) and valence (N_V) bands, respectively. The terms W_C and W_V denote the energy levels associated with the conduction and valence bands and W_F is the **Fermi energy level**, which indicates the energy level that has a 50% probability of being occupied by an electron. For **intrinsic** (i.e., pure) semiconductors at room temperature, the Fermi level is very close to the middle of the bandgap. In (6.2), m^*_n and m^*_p refer to the effective mass of electrons and holes in the semiconductor (different from the free electron rest mass due to interaction with the crystal lattice), k is Boltzmann's constant, h is Planck's constant, and T is the absolute temperature measured in Kelvin.

In an intrinsic semiconductor, the number of free electrons produced by thermal excitation is equal to the number of holes (i.e., $n = p = n_i$). The electron and hole concentrations are described by the concentration law

$$np = n_i^2 \quad (6.3)$$

where n_i is the intrinsic concentration. Equation (6.3) is true not only for intrinsic but also for doped semiconductors, which are discussed later in this section.

Substitution of (6.1) into (6.3) results in the expression for the intrinsic carrier concentration:

$$n_i = \sqrt{N_C N_V} \exp\left[-\frac{W_C - W_V}{2kT}\right] = \sqrt{N_C N_V} \exp\left[-\frac{W_g}{2kT}\right]. \quad (6.4)$$

The effective electron and hole masses as well as the concentrations N_C, N_V, and n_i for $T = 300$ K are summarized in Table 6-1 and are listed in Table E-1 in Appendix E.

Classical electromagnetic theory specifies the electrical conductivity in a material to be $\sigma = J/E$, where J is the current density and E is the applied electric field. The conductivity in the classical model (Drude model) can be found through the carrier concentration N, the associated elementary charge q, the drift velocity v_d, and the applied electric field E:

$$\sigma = qNv_d/E. \quad (6.5)$$

Semiconductor Basics

Table 6-1 Effective concentrations and effective mass values at $T = 300$ K

Semiconductor	m_n^*/m_0	m_p^*/m_0	N_C, cm^{-3}	N_V, cm^{-3}	n_i, cm^{-3}
Silicon (Si)	1.08	0.56	2.8×10^{19}	1.04×10^{19}	1.45×10^{10}
Germanium (Ge)	0.55	0.37	1.04×10^{19}	6.0×10^{18}	2.4×10^{13}
Gallium Arsenide (GaAs)	0.067	0.48	4.7×10^{17}	7.0×10^{18}	1.79×10^6

In semiconductors, we have both electrons and holes contributing to the conductivity of the material. At low electric fields, the drift velocity v_d of the carriers is proportional to the applied field strength through a proportionality constant known as **carrier mobility** μ. Thus, for semiconductors we can rewrite (6.5) as

$$\sigma = qn\mu_n + qp\mu_p \qquad (6.6)$$

where μ_n, μ_p are the mobilities of electrons and holes, respectively. For intrinsic semiconductors, we can simplify (6.6) further by recalling that $n = p = n_i$, that is,

$$\sigma = qn_i(\mu_n + \mu_p) = q\sqrt{N_C N_V}\exp\left[-\frac{W_g}{2kT}\right](\mu_n + \mu_p). \qquad (6.7)$$

RF&MW→

Example 6-1. Computation of the temperature dependence of the intrinsic semiconductor conductivity

It is desired to find the conductivities for the intrinsic materials of Si, Ge, and GaAs as functions of temperature To make the computations easier, we assume that the bandgap energy and the mobilities for holes and electrons are temperature independent over the range of interest $-50°C \leq T \leq 200°C$.

Solution. As a first step, it is convenient to combine into one parameter $\sigma_0(T)$ the coefficient of the exponential term in (6.7); that is,

$$\sigma_0(T) = q\sqrt{N_C N_V}(\mu_n + \mu_p)$$

where electron and hole mobilities are found from Table E-1:

$$\mu_n = 1350(\text{Si}), 3900(\text{Ge}), 8500(\text{GaAs})$$

$$\mu_p = 480(\text{Si}), 1900(\text{Ge}), 400(\text{GaAs}).$$

All values are given in units of cm^2/(V · s). N_C, N_V are computed according to (6.2) as

$$N_{C,V}(T) = N_{C,V}^{300}\left(\frac{T}{300}\right)^{3/2}.$$

This leads to the form

$$\sigma = \sigma_0(T)\exp\left(-\frac{W_g}{2kT}\right)$$

$$= q(\mu_n + \mu_p)\sqrt{N_C^{300} N_V^{300}}\left(\frac{T}{300\,\text{K}}\right)^{3/2}\exp\left(-\frac{W_g}{2kT}\right)$$

where the bandgap energy $W_g = W_C - W_V$ is, respectively, 1.12 eV (Si), 0.62 eV (Ge), and 1.42 eV (GaAs). The three conductivities are plotted in Figure 6-2.

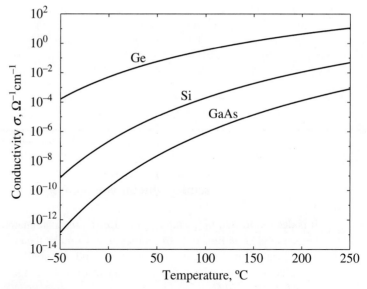

Figure 6-2 Conductivity of Si, Ge, GaAs in the range from –50°C to 250°C.

The electric properties of semiconductors are strongly influenced by the ambient temperature. In this example, we have neglected the temperature dependence of the bandgap energy, which is discussed in Chapter 7. Knowledge of the temperature behavior of active devices is an important design consideration where internal heating, due to power dissipation, can easily result in temperatures exceeding 100–150°C.

Semiconductor Basics

A major change in the electrical properties of a semiconductor can be initiated by introducing impurity atoms. This process is called **doping**. To achieve ***n*-type** doping (which supplies additional electrons to the conduction band), we introduce atoms with a larger number of valence electrons than the atoms in the intrinsic semiconductor lattice that they substitute. For instance, the implantation of phosphorus (P) atoms into Si introduces loosely bound electrons into the neutral crystal lattice, as shown in Figure 6-3(b).

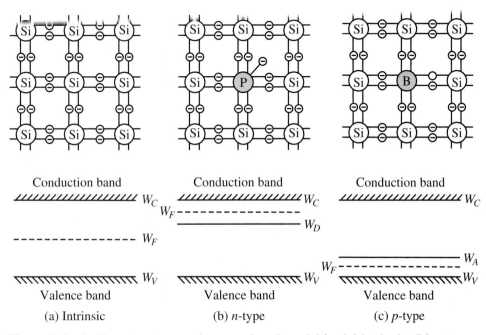

Figure 6-3 Lattice structure and energy band model for (a) intrinsic, (b) *n*-type, and (c) *p*-type semiconductors at no thermal energy. W_D and W_A are donor and acceptor energy levels.

It is intuitively apparent that the energy level of this "extra" electron is closer to the conduction band than the energy of the remaining four valence electrons. When the temperature is increased above absolute zero, the loosely bound electron separates from the atom, forming a free negative charge and leaving behind the fixed positive ion of phosphorus. Thus, while still maintaining charge neutrality, the atom has donated an electron to the conduction band without creating a hole in the valence band. This results in an increase in the Fermi level since more electrons are located in the conduction band. Contrary to the intrinsic semiconductor (n_i, p_i), we now have an ***n*-type** semiconductor in which the electron concentration is related to the hole concentration as

$$n_n = N_D + p_n \tag{6.8}$$

where N_D is the donor concentration and p_n represents the minority hole concentration. To find n_n and p_n, we have to solve (6.8) in conjunction with (6.3). The result is

$$n_n = \frac{N_D + \sqrt{N_D^2 + 4n_i^2}}{2} \qquad (6.9a)$$

$$p_n = \frac{-N_D + \sqrt{N_D^2 + 4n_i^2}}{2}. \qquad (6.9b)$$

If the donor concentration N_D is much greater than the intrinsic electron concentration n_i, then

$$n_n \approx N_D \qquad (6.10a)$$

$$p_n \approx \frac{-N_D + N_D(1 + 2n_i^2/N_D^2)}{2} = \frac{n_i^2}{N_D}. \qquad (6.10b)$$

Let us now consider adding impurity atoms with fewer valence electrons than the atoms forming the intrinsic semiconductor lattice. These types of elements are called **acceptors**, and an example of such an element for the Si lattice is boron (B). As seen in Figure 6-3(c), one of the covalent bonds appears to be empty. This empty bond introduces additional energy states in the bandgap that are closely situated to the valence band. Again, when the temperature is increased from absolute zero, some electrons gain extra energy to occupy empty bonds but do not possess sufficient energy to cross the bandgap. Thus, impurity atoms will accept additional electrons, forming negative net charges. At the sites where the electrons are removed, holes will be created. These holes are free to migrate and will contribute to the conduction current of the semiconductor. By doping the semiconductor with acceptor atoms, we have created a *p*-type semiconductor with

$$p_p = N_A + n_p \qquad (6.11)$$

where N_A, n_p are the acceptor and minority electron concentrations. Solving (6.11) together with (6.3), we find hole p_p and electron n_p concentrations in the *p*-type semiconductor:

$$p_p = \frac{N_A + \sqrt{N_A^2 + 4n_i^2}}{2} \qquad (6.12a)$$

$$n_p = \frac{-N_A + \sqrt{N_A^2 + 4n_i^2}}{2}. \qquad (6.12b)$$

Similarly to (6.9), for high doping levels when $N_A \gg n_i$, we observe

$$p_p \approx N_A \qquad (6.13a)$$

Semiconductor Basics

$$n_p \approx \frac{-N_A + N_A(1 + 2n_i^2/N_A^2)}{2} = \frac{n_i^2}{N_A}. \qquad (6.13b)$$

Minority and majority concentrations play key roles in establishing the current flow characteristics in the semiconductor materials.

6.1.2 The pn-Junction

The physical contact of a *p*-type with an *n*-type semiconductor leads to one of the most important concepts when dealing with active semiconductor devices: the **pn-junction**. Because of the difference in the carrier concentrations between the two types of semiconductors, a current flow will be initiated across the interface. This current is commonly known as a **diffusion current,** and is composed of electrons and holes. To simplify our discussion, we consider a one-dimensional model of the *pn*-junction as seen in Figure 6-4.

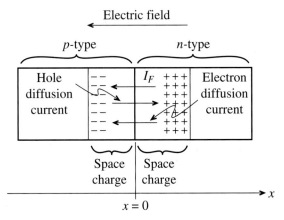

Figure 6-4 Current flows in the *pn*-junction.

The diffusion current is composed of $I_{n_{\text{diff}}}$ and $I_{p_{\text{diff}}}$ components:

$$I_{\text{diff}} = I_{n_{\text{diff}}} + I_{p_{\text{diff}}} = qA\left(D_n \frac{dn}{dx} + D_p \frac{dp}{dx}\right) \qquad (6.14)$$

where A is the semiconductor cross-sectional area orthogonal to the *x*-axis, and D_n, D_p are the diffusion constants for electrons and holes in the form (Einstein relation)

$$D_{n,p} = \mu_{n,p}\frac{kT}{q} = \mu_{n,p} V_T. \qquad (6.15)$$

The thermal potential $V_T = kT/q$ is approximately 26 mV at room temperature of 300 K.

Since the *p*-type semiconductor was initially neutral, the diffusion current of holes is going to leave behind a negative space charge. Similarly, the electron current flow from the *n*-semiconductor will leave behind positive space charges. As the diffusion current flow takes place, an electric field E is created between the net positive charge in the *n*-semiconductor and the net negative charge in the *p*-semiconductor. This field, in turn, induces a current $I_F = \sigma A E$ which opposes the diffusion current such that $I_F + I_{\text{diff}} = 0$. Substituting (6.6) for the conductivity, we find

$$I_F = qA(n\mu_n + p\mu_p)E = I_{n_F} + I_{p_F}. \tag{6.16}$$

Since the total current is equal to zero, the electron portion of the current is also equal to zero; that is,

$$I_{n_{\text{diff}}} + I_{n_F} = qD_n A \frac{dn}{dx} + qn\mu_n AE = q\mu_n A\left(V_T \frac{dn}{dx} - n\frac{dV}{dx}\right) = 0 \tag{6.17}$$

where the electric field E has been replaced by the derivative of the potential $E = -dV/dx$. Integrating (6.17), we obtain the **diffusion barrier voltage** or, as it is often called, the **built-in potential**:

$$\int_0^{V_{\text{diff}}} dV = V_{\text{diff}} = V_T \int_{n_p}^{n_n} n^{-1} dn = V_T \ln\left(\frac{n_n}{n_p}\right) \tag{6.18}$$

where again n_n is the electron concentration in the *n*-type and n_p is the electron concentration in the *p*-type semiconductor. The same diffusion barrier voltage could have been found had we considered the hole current flow from the *p* to the *n*-semiconductor and the corresponding balancing field-induced current flow. The resulting equation describing the barrier voltage is

$$V_{\text{diff}} = V_T \ln\left(\frac{p_p}{p_n}\right). \tag{6.19}$$

If the concentration of acceptors in the *p*-semiconductor is $N_A \gg n_i$, and the concentration of donors in the *n*-semiconductor is $N_D \gg n_i$, then $n_n \approx N_D$, $n_p \approx n_i^2/N_A$. By using (6.13b) and (6.18), we obtain

$$V_{\text{diff}} \approx V_T \ln\left(\frac{N_A N_D}{n_i^2}\right) \tag{6.20}$$

Exactly the same result will be obtained from (6.19), if we substitute $p_p \approx N_A$ and $p_n \approx n_i^2/N_D$.

Semiconductor Basics

——————————————————————RF&MW→

Example 6-2. Determining the diffusion barrier or built-in voltage of a *pn*-junction

For a particular Si *pn*-junction, the doping concentrations are given as $N_A = 10^{18} \text{cm}^{-3}$ and $N_D = 5 \times 10^{15} \text{cm}^{-3}$, with an intrinsic concentration of $n_i = 1.5 \times 10^{10} \text{cm}^{-3}$. Find the barrier voltages for $T = 300$ K.

Solution. The barrier voltage is directly determined from (6.20):

$$V_{\text{diff}} = V_T \ln\left(\frac{N_A N_D}{n_i^2}\right) = \frac{kT}{q} \ln\left(\frac{N_A N_D}{n_i^2}\right) = 0.796 \text{ (V)}.$$

We note that the built-in potential is strongly dependent on the doping concentrations and temperature.

For different semiconductor materials such as GaAs, Si, Ge, the built-in voltage will be different even if the doping densities are the same. This is due to significantly different intrinsic carrier concentrations.

If we desire to determine the potential distribution along the *x*-axis, we can employ Poisson's equation, which for a one-dimensional analysis is written as

$$\frac{d^2 V(x)}{dx^2} = -\frac{\rho(x)}{\varepsilon_r \varepsilon_0} = -\frac{dE}{dx} \qquad (6.21)$$

where $\rho(x)$ is the charge density and ε_r is the relative dielectric constant of the semiconductor. Assuming uniform doping and the **abrupt junction approximation**, as shown in Figure 6-5(b), the charge density in each material is

$$\rho(x) = -qN_A, \quad \text{for } -d_p \leq x \leq 0 \qquad (6.22a)$$

$$\rho(x) = qN_D, \quad \text{for } 0 \leq x \leq d_n \qquad (6.22b)$$

where d_p and d_n are the extents of the space charges in the *p*- and *n*-type semiconductors, respectively.

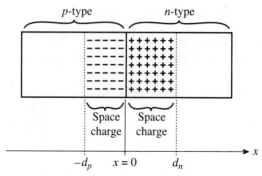

(a) *pn*-junction with space charge extent

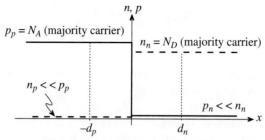

(b) Acceptor and donor concentrations

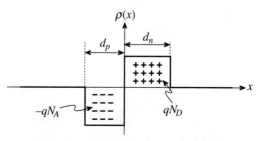

(c) Polarity of charge density distribution

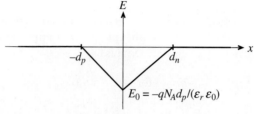

(d) Electric field distribution

Figure 6-5 The *pn*-junction with abrupt charge carrier transition in the absence of an externally applied voltage.

Semiconductor Basics

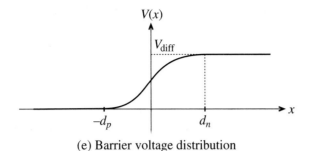

(e) Barrier voltage distribution

Figure 6-5 The *pn*-junction with abrupt charge carrier transition in the absence of an externally applied voltage. (Continued)

The electric field in the semiconductor is found by integrating (6.21) with the spatial limits $-d_p \leq x \leq d_n$ such that

$$E(x) = \int_{-d_p}^{x} \frac{\rho(x)}{\varepsilon_r \varepsilon_0} dx = \begin{cases} -\dfrac{qN_A}{\varepsilon_r \varepsilon_0}(x + d_p), & \text{for } -d_p \leq x \leq 0 \\ -\dfrac{qN_D}{\varepsilon_r \varepsilon_0}(d_n - x), & \text{for } 0 \leq x \leq d_n. \end{cases} \quad (6.23)$$

The resulting electric field profile is depicted in Figure 6-5(d). In deriving (6.23), we used the fact that the charge balance law demands that the total space charge in the semiconductor equals zero, which for highly doped semiconductors is equivalent to the condition

$$N_A \cdot d_p = N_D \cdot d_n. \quad (6.24)$$

To obtain the voltage distribution profile, we now carry out the integration of (6.23) as follows:

$$V(x) = -\int_{-d_p}^{x} E(x) dx = \begin{cases} \dfrac{qN_A}{2\varepsilon_r \varepsilon_0}(x + d_p)^2, & \text{for } -d_p \leq x \leq 0 \\ \dfrac{q}{2\varepsilon_r \varepsilon_0}(N_A d_p^2 + N_D d_n^2) - \dfrac{qN_D}{2\varepsilon_r \varepsilon_0}(d_n - x)^2, & \text{for } 0 \leq x \leq d_n. \end{cases} \quad (6.25)$$

Since the total voltage drop must be equal to the diffusion voltage V_{diff}, it is found that

$$V(d_n) = V_{\text{diff}} = \frac{qN_A d_p^2}{2\varepsilon_r \varepsilon_0} + \frac{qN_D d_n^2}{2\varepsilon_r \varepsilon_0}. \quad (6.26)$$

Substituting $d_p = d_n N_D/N_A$ and solving (6.26) for d_n, we obtain the extent of the positive space charge domain into the n-semiconductor:

$$d_n = \left[\frac{2\varepsilon V_{\text{diff}} N_A}{q N_D}\left(\frac{1}{N_A + N_D}\right)\right]^{1/2} \quad (6.27)$$

where $\varepsilon = \varepsilon_0 \varepsilon_r$. An identical derivation involving $d_n = d_p N_A/N_D$ gives us the space charge extent into the p-semiconductor:

$$d_p = \left[\frac{2\varepsilon V_{\text{diff}} N_D}{q N_A}\left(\frac{1}{N_A + N_D}\right)\right]^{1/2}. \quad (6.28)$$

The entire length is then the addition of (6.27) and (6.28):

$$d_S = d_n + d_p = \left[\frac{2\varepsilon V_{\text{diff}}}{q}\left(\frac{1}{N_A} + \frac{1}{N_D}\right)\right]^{1/2}. \quad (6.29)$$

We next turn our attention to the computation of the **junction capacitance**. This is an important parameter for RF devices, since low capacitances imply rapid switching speeds and suitability for high-frequency operation. The junction capacitance can be found via the well-known one-dimensional capacitor formula

$$C = \frac{\varepsilon A}{d_S}.$$

Substituting (6.29) for the distance d_S, we express the capacitance as

$$C = A\left[\frac{q\varepsilon}{2V_{\text{diff}}}\frac{N_A N_D}{N_A + N_D}\right]^{1/2}. \quad (6.30)$$

If an external voltage V_A is applied across the junction, two situations arise that explain the rectifier action of the diode, as shown in Figure 6-6. The reverse polarity [Figure 6-6(a)] increases the space charge domain and prevents the flow of current, except for a small leakage current involving the minority carrier concentration (holes in the n-semiconductor, and electrons in the p-semiconductor). In contrast, the forward polarity reduces the space charge domain by injecting excess electrons into the n-type and holes into the p-type semiconductor. To describe these situations, the previously given equations (6.27) and (6.28) have to be modified by replacing the barrier voltage V_{diff} with $V_{\text{diff}} - V_A$:

$$d_p = \left[\frac{2\varepsilon(V_{\text{diff}} - V_A) N_D}{q N_A}\left(\frac{1}{N_A + N_D}\right)\right]^{1/2} \quad (6.31)$$

Semiconductor Basics

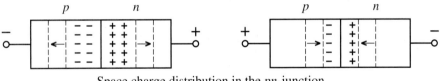

Space charge distribution in the *pn*-junction

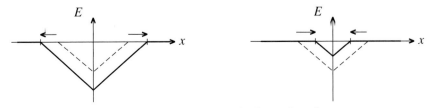

Electric field distribution in the *pn*-junction

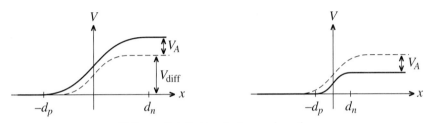

Voltage distribution in the *pn*-junction

(a) Reverse biasing ($V_A < 0$) (b) Forward biasing ($V_A > 0$)

Figure 6-6 External voltage applied to the *pn*-junction in reverse and forward directions.

$$d_n = \left[\frac{2\varepsilon(V_{\text{diff}} - V_A)N_A}{q}\left(\frac{1}{N_D(N_A + N_D)}\right)\right]^{1/2}. \quad (6.32)$$

This leads to a total length of the space charge or depletion domain

$$d_S = \left[\frac{2\varepsilon(V_{\text{diff}} - V_A)}{q}\left(\frac{1}{N_A} + \frac{1}{N_D}\right)\right]^{1/2}. \quad (6.33)$$

Depending on the polarity of V_A, we notice from (6.31)–(6.33) that either the space charge domain is enlarged or diminished.

RF&MW→

Example 6-3. Computation of the junction capacitance and the space charge region length of a *pn*-junction

For an abrupt *pn*-junction Si semiconductor at room temperature ($\varepsilon_r = 11.9$, $n_i = 1.5 \times 10^{10}$ cm^{-3}) with donor and acceptor concentrations equal to $N_D = 5 \times 10^{15}$ cm^{-3} and $N_A = 10^{15}$ cm^{-3}, we desire to find the space charge regions d_p, d_n and the junction capacitance at zero biasing voltage. Show that the depletion-layer capacitance of a *pn*-junction can be cast into the form

$$C_J = C_{J0}\left(1 - \frac{V_A}{V_{\text{diff}}}\right)^{-1/2}$$

and determine C_{J0}. Sketch the depletion capacitance as a function of applied voltage. Assume that the cross-sectional area of the *pn*-junction is $A = 10^{-4}$ cm^2.

Solution. We return to the capacitance expression (6.30) where we introduce the applied voltage V_A. Thus,

$$C_J = A\left[\frac{q\varepsilon}{2V_{\text{diff}}(1 - V_A/V_{\text{diff}})}\frac{N_A N_D}{N_A + N_D}\right]^{1/2}$$

This is immediately recognized as the preceding formula, if we set

$$C_{J0} = A\left[\frac{q\varepsilon}{2V_{\text{diff}}}\frac{N_A N_D}{N_A + N_D}\right]^{1/2}.$$

Substituting $V_{\text{diff}} = V_T \ln(N_A N_D / n_i^2) = 0.616$ V, it is found that $C_{J0} = 1.07$ pF.

For the space charge extents we use (6.28) and (6.29):

$$d_n = \left[\frac{2\varepsilon V_{\text{diff}} N_A}{q}\frac{N_A}{N_D}\left(\frac{1}{N_A + N_D}\right)\right]^{1/2} = 0.164 \ \mu\text{m}$$

$$d_p = \left[\frac{2\varepsilon V_{\text{diff}} N_D}{q}\frac{N_D}{N_A}\left(\frac{1}{N_A + N_D}\right)\right]^{1/2} = 0.821 \ \mu\text{m}.$$

The dependence of the junction capacitance on the applied voltage is depicted in Figure 6-7.

Semiconductor Basics

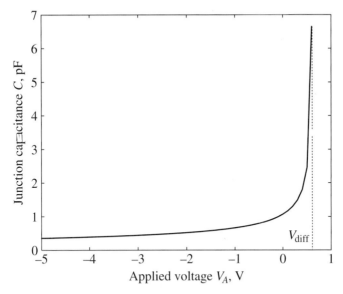

Figure 6-7 The *pn*-junction capacitance as a function of applied voltage.

In Figure 6-7 the junction capacitance for applied voltages near the built-in potential will approach infinity. However, in reality the value begins to saturate, as further discussed in Chapter 7.

For the current flow through the diode, we list the Shockley diode equation, which is derived in Appendix F:

$$I = I_0(e^{V_A/V_T} - 1) \tag{6.34}$$

where I_0 is the **reverse saturation** or **leakage current**. The current-voltage characteristic, often called the ***I-V* curve**, is generically depicted in Figure 6-8.

This curve reveals that for negative voltages, a small, voltage-independent current $-I_0$ will flow, whereas for positive voltages, an exponentially increasing current is observed. The function shown in Figure 6-8 is an idealization since it does not take into account breakdown phenomena. Nonetheless, (6.34) reveals the rectifier property of the *pn*-junction when an alternating voltage is applied.

The existence of the depletion layer, or junction capacitance, requires $V_A < V_{\text{diff}}$, see Example 6-3. However, under forward bias condition, we encounter an additional **diffusion capacitance** due to the presence of diffusion charges Q_d (minority carriers) stored in the

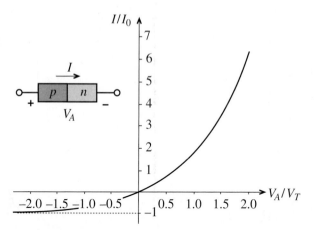

Figure 6-8 Current-voltage behavior of *pn*-junction based on Shockley equation.

semiconductor layers. This charge is quantified by realizing that the charge Q_d can be computed as diode current I multiplied by the mean transit time of carriers through the diode τ_T, or

$$Q_d = I\tau_T = \tau_T I_0(e^{V_A/V_T} - 1) \tag{6.35}$$

It is apparent that the diffusion capacitance assumes a nonlinear relation with the applied voltage and the junction temperature. The diffusion capacitance is computed as

$$C_d = \frac{dQ_d}{dV_A} = \frac{I_0 \tau_T}{V_T} e^{V_A/V_T} \tag{6.36}$$

and is seen to be strongly dependent on the operating voltage.

In general, the total capacitance C of a *pn* diode can roughly be divided into three regions:

1. $V_A < 0$: only the depletion capacitance is significant: $C = C_J$
2. $0 < V_A < V_{\text{diff}}$: depletion and diffusion capacitances combine: $C = C_J + C_d$
3. $V_A > V_{\text{diff}}$: only the diffusion capacitance is significant: $C = C_d$.

The influence of the diffusion capacitance is appreciated if we consider a diode that is operated at $V_A = 1$ V and that has an assumed mean transit time of $\tau_T = 100$ ps $= 10^{-10}$ s, and a reverse saturation current of $I_0 = 1$ fA $= 10^{-15}$ A measured at room temperature of 300 K (i.e., $V_T = 26$ mV). Substituting these values into (6.36), we find $C = C_d = 194$ nF. This is a rather large value that is significant for reverse recovery, and restricts the high-frequency use of conventional *pn*-junction diodes. The reverse recovery relates to the diffusion charge; it keeps the diode conducting in the reverse direction until all charges are removed.

6.1.3 Schottky Contact

W. Schottky analyzed the physical phenomena involved when a metallic electrode is contacting a semiconductor. For instance, if a *p*-semiconductor is in contact with a copper or aluminum electrode, there is a tendency for the electrons to diffuse into the metal, leaving behind an increased concentration of holes in the semiconductor. The consequences of this effect are modified valence and conduction band energy levels near the interface. This can be displayed by a local change in the energy band structure, depicted in Figure 6-9(a).

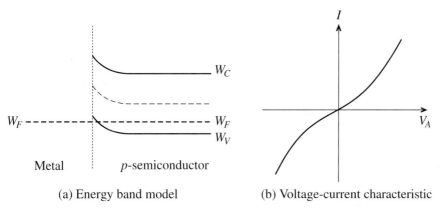

(a) Energy band model   (b) Voltage-current characteristic

Figure 6-9 Metal electrode in contact with *p*-semiconductor.

Because of the higher concentration of holes, the valence band bends toward the Fermi level. The conduction band, as the result of a lower electron concentration, bends away from the Fermi level. For such a configuration we always obtain a low resistance contact (see Figure 6-9(b)), irrespective of the polarity of the applied voltage.

The situation becomes more complicated, but technologically much more interesting, when a metallic electrode is brought in contact with an *n*-semiconductor. Here, the more familiar behavior of a *pn*-junction emerges: a small positive volume charge density is created in the semiconductor due to electron migration from the semiconductor to the metal. This mechanism is due to the fact that the Fermi level is higher in the semiconductor (lower work function) than in the metal (higher work function) when the two materials are apart. However, as both materials are contacted, the Fermi level again has to be the same and band distortions are created. Electrons diffuse from the *n*-semiconductor and leave behind positive space charge. The depletion zone grows until the electrostatic repulsion of the space charges prevents further electron diffusion. To clarify the issues associated with a metal *n*-semiconductor contact, Figure 6-10 shows the two materials before and after bonding.

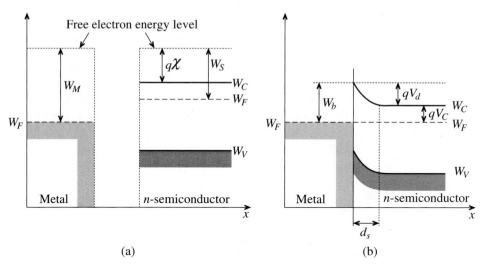

Figure 6-10 Energy band diagram of Schottky contact, (a) before and (b) after contact.

The energy $W_b = qV_b$ is related to the metal work function $W_M = qV_M$ (V_M is measured from the Fermi level to the reference level where the electron becomes a detached free particle; values of V_M for some commonly used metals are summarized in Table 6-2) and the electron affinity $q\chi$. Here, χ is 4.05 V for Si, 4.0 V for Ge, and 4.07 V for GaAs and is measured from the conduction band to the same reference level where the electron becomes a free carrier, according to

$$W_b = q(V_M - \chi). \tag{6.37}$$

Table 6-2 Work function potentials of some metals

Material	Work Function Potential, V_M
Silver (Ag)	4.26 V
Aluminum (Al)	4.28 V
Gold (Au)	5.10 V
Chromium (Cr)	4.50 V
Molybdenum (Mo)	4.60 V
Nickel (Ni)	5.15 V
Palladium (Pd)	5.12 V
Platinum (Pt)	5.65 V
Titanium (Ti)	4.33 V

Semiconductor Basics

An expression for the built-in Schottky barrier voltage V_d is established just as in the pn-junction, which involves (6.37) and the additional voltage V_C between conduction and Fermi levels:

$$V_d = (V_M - \chi) - V_C \tag{6.38}$$

where V_C is dependent on the doping N_D and the concentration of states in the conduction band N_C based on $N_C = N_D \exp(V_C/V_T)$. Solving for the voltage gives $V_C = V_T \ln(N_C/N_D)$. Although real metal-semiconductor interfaces usually involve an additional very narrow isolation layer, we will neglect the influence of this layer and only deal with the length of the space charge in the semiconductor:

$$d_S = \left[\frac{2\varepsilon(V_d - V_A)}{q} \frac{1}{N_D}\right]^{1/2}. \tag{6.39}$$

Therefore, it is found that the junction capacitance of the Schottky contact

$$C_J = A\frac{\varepsilon}{d_S} = A\left[\frac{q\varepsilon}{2(V_d - V_A)} N_D\right]^{1/2} \tag{6.40}$$

is almost identical to (6.30). A simple computation now can predict a typical value for V_d, as illustrated in the following example.

Example 6-4. Computation of the barrier voltage, depletion capacitance, and space charge region width for a Schottky diode

A Schottky diode is created as an interface between a gold contact material and an n-type silicon semiconductor, see Table 6-1. The semiconductor is doped to $N_D = 10^{16} \text{cm}^{-3}$ and the work function V_M for gold is 5.1 V. Also, as mentioned above, the affinity for Si is $\chi = 4.05$ V. Find the Schottky barrier V_d, space charge width d_S, and capacitance C_J if the dielectric constant of silicon is $\varepsilon_r = 11.9$. Assume the cross-sectional diode area to be $A = 10^{-4} \text{cm}^2$, and the temperature is 300 K.

Solution. Since the concentration of states in the conduction band of silicon is $N_C = 2.8 \times 10^{19} \text{cm}^{-3}$, we can compute the conduction band potential as

$$V_C = V_T \ln\left(\frac{N_C}{N_D}\right) = 0.21 \text{V}.$$

Substituting the obtained value for V_C into (6.38), we find the built-in barrier voltage

$$V_d = (V_M - \chi) - V_C = 0.84 \text{ V}.$$

The space charge width is obtained from (6.39)

$$d_S = \sqrt{\frac{2\varepsilon_0 \varepsilon_r}{q} \frac{V_d}{N_D}} = 0.332 \text{ }\mu\text{m}.$$

Finally, the junction capacitance according to the formula for the parallel-plate capacitor, see (6.40), gives us

$$C_J = A\frac{\varepsilon_0 \varepsilon_r}{d_S} = 3.2 \text{ pF}.$$

The metal-semiconductor junction diode for similar size and doping has a junction capacitance comparable to that of a pn-junction. However, the absence of diffusion capacitance permits higher frequency operation.

6.2 RF Diodes

In this section we will review some practical realizations of diodes that are most commonly used in RF and MW circuits. As presented in the previous section, a classical *pn*-junction diode is not very suitable for high-frequency applications because of the diffusion capacitance. Today, Schottky diodes find widespread applications in RF detectors, mixers, attenuators, oscillators, and amplifiers.

After discussing the Schottky diode in Section 6.2.1, we will continue investigating a number of special RF diodes. In Section 6.2.2, the PIN diode is analyzed and placed in context with its primary uses as a variable resistor and high-frequency switch. Besides relying on the rectifier property of diodes, we can also exploit the dependence of the junction capacitance on the applied voltage to construct voltage-controlled tuning circuits, where diodes are used as variable capacitors. An example of such a specialized diode is the varactor diode, covered in Section 6.2.3. At the end of this section, we will discuss a few more exotic diode configurations, such as IMPATT, tunnel, TRAPATT, BARRITT, and Gunn diodes, which are less frequently used, but which are still of interest due to their unique electric properties.

6.2.1 Schottky Diode

Compared with the conventional *pn*-junction, the Schottky barrier diode has a different reverse-saturation current mechanism; it is determined by the thermionic emission of the majority carriers across the potential barrier. This current is orders of magnitude larger than the diffusion-driven minority carriers constituting the reverse-saturation current of the ideal *pn*-junction diode. For instance, the Schottky diode has a typical reverse-saturation current density on the order of 10^{-6} A/cm² compared with 10^{-11} A/cm² of a conventional Si-based *pn*-junction diode. The schematic diagram of a cross-sectional view of the Schottky diode with the corresponding circuit elements is given in Figure 6-11.

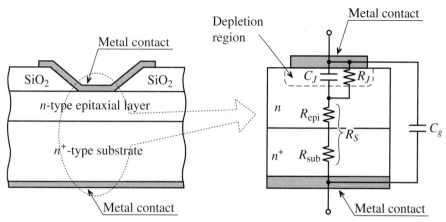

Figure 6-11 Cross-sectional view of Si Schottky diode.

The metal electrode (tungsten, aluminum, gold, etc.) is in contact with a weakly doped *n*-semiconductor layer, epitaxially grown on a highly doped n^+ substrate. The dielectric is assumed to be ideal; that is, the conductance is zero. The current-voltage characteristic is described by the following equation:

$$I = I_S(e^{(V_A - IR_S)/V_T} - 1) \tag{6.41}$$

where $V_T = q/(kT)$ is the thermal voltage. The reverse-saturation current is given by

$$I_S = A\left(R^* T^2 \exp\left[\frac{-qV_b}{kT}\right]\right) \tag{6.42}$$

and R^* is the so-called **Richardson constant** for thermionic emission of the majority carrier across the potential barrier. A typical value of R^* for Si is 100 A/cm²K².

The corresponding **small-signal** equivalent circuit model is illustrated in Figure 6-12. In this circuit, we note that the junction resistance R_J is dependent on the bias current, just as is the diode series resistance, which is composed of epitaxial and substrate resistances $R_S = R_{epi} + R_{sub}$. The bond wire inductance is fixed, and its value is approximately on the order of $L_S = 0.1$ nH. As discussed above, the junction capacitance C_J is given by (6.40). Because of the resistance R_S, the actual junction voltage is equal to the applied voltage minus the voltage drop over the diode series resistance, resulting in the modified exponential expression (6.41).

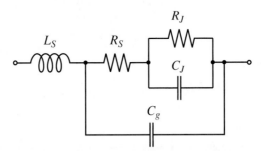

Figure 6-12 Circuit model of typical Schottky diode under forward bias.

Typical component values for Schottky diodes are $R_S \approx 2 \ldots 5\ \Omega$, $C_g = 0.1 \ldots 0.2$ pF, and $R_J = 0.2 \ldots 2$ kΩ. Often, the additional IR_S term in (6.41) is neglected for small bias currents below 0.1 mA. However, for certain applications, the series resistance may form a feedback loop, which means the resistance is multiplied by a gain factor of potentially large magnitude. For this situation, the IR_S term has to be taken into account.

In circuit realizations of high-frequency Schottky diodes, the planar configuration in Figure 6-11 gives rise to relatively large parasitic capacitances for very small metal contacts of typically 10 μm diameter and less. The stray capacitances can be somewhat minimized through the addition of an isolation ring, as depicted in Figure 6-13.

The small-signal junction capacitance and junction resistance can be found by expanding the electric current expression (6.41) around the **quiescent** or **operating point** V_Q. The total diode voltage is written as a DC bias V_Q and a small AC signal carrier frequency component v_d:

$$V = V_Q + v_d. \tag{6.43}$$

The substitution of (6.43) in (6.41) for a negligible IR_S term yields

$$I = I_S(e^{V/V_T} - 1) = I_S(e^{V_Q/V_T} e^{v_d/V_T} - 1). \tag{6.44}$$

Expanding this equation in a Taylor series about the Q-point and retaining the first two terms gives

$$I(V) \cong I_Q + \left.\frac{dI}{dV}\right|_{V_Q} v_d = I_Q + \frac{I_S v_d}{V_T} e^{V_Q/V_T} = I_Q + (I_Q + I_S)\frac{v_d}{V_T} = I_Q + \frac{v_d}{R_J}. \tag{6.45}$$

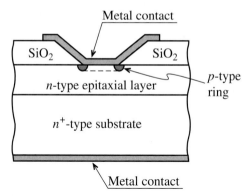

Figure 6-13 Schottky diode with additional isolation ring suitable for very high frequency applications.

Here, the dynamic junction resistance $R_J(V_Q)$ is identified as

$$R_J(V_Q) = \frac{V_T}{I_Q + I_S} \qquad (6.46)$$

and the junction capacitance is given by (6.40), with V_Q replacing V_A.

6.2.2 PIN Diode

PIN diodes find applications as high-frequency switches and variable resistors (attenuators) in the range from 10 kΩ to less than 1 Ω for RF signals up to 50 GHz. They contain an additional layer of an intrinsicx (*I*-layer) or lightly doped semiconductor sandwiched between highly doped p^+ and n^+ layers. Depending upon application and frequency range, the thickness of the middle layer ranges from 1 to 200 µm, sometimes even more. Under forward bias, the diode behaves as if it possesses a variable resistance controlled by the applied current. However, under reverse bias, the lightly doped inner layer creates space charges, whose extent reaches the highly doped outer layers. This effect takes place even for small reverse voltages and remains essentially constant up to high voltages, with the consequence that the diode behaves similarly to a dual plate capacitor. For instance, a Si-based PIN diode with an internal *I*-layer of 20 µm and a surface area of 200 by 200 µm has a diffusion capacitance on the order of 0.2 pF.

A generic PIN diode and its practical implementation in **mesa processing technology** is presented in Figure 6-14. The advantage of the mesa configuration over the conventional planar construction is a significant reduction in fringing capacitance.

The mathematical representation of the *I-V* characteristic depends on the level and polarity of the applied voltage. To keep things simple, we will rely to a large extent on discussions already outlined for the *pn*-junction.

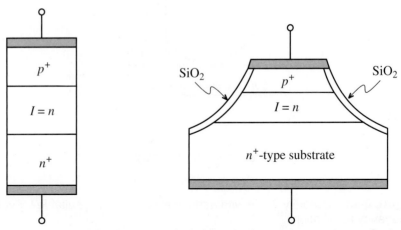

(a) Simplified structure of a PIN diode   (b) Fabrication in mesa processing technology

Figure 6-14 PIN diode construction.

In the forward direction, and for a weakly doped n-type intrinsic layer, the current through the diode is

$$I = A\left(\frac{qn_i^2 W}{N_D \tau}\right)(e^{V_A/(2V_T)} - 1) \tag{6.47}$$

where W is the width of the intrinsic layer, τ is the **excess minority carrier lifetime**, which can be on the order of up to $\tau = 10$ μs, and N_D is the doping concentration in the middle layer of the lightly doped n-semiconductor. The factor 2 in the exponent takes into account the presence of two junctions. For a pure intrinsic layer $N_D = n_i$, (6.47) leads to the form

$$I = A\left(\frac{qn_i W}{\tau}\right)\left(e^{V_A/(2V_T)} - 1\right). \tag{6.48}$$

The total charge can be calculated from the relation $Q = I\tau$. This allows us to find the diffusion capacitance:

$$C_d = \frac{dQ}{dV_A} = \tau\left(\frac{dI}{dV_A}\right) = \frac{I\tau}{2V_T}. \tag{6.49a}$$

In the reverse direction, the capacitance is dominated by the parallel plate capacitance of the depleted I-layer; the capacitance C_J is approximately

$$C_J = \varepsilon_I\left(\frac{A}{W}\right) \tag{6.49b}$$

where ε_I is the dielectric constant of the intrinsic layer.

RF Diodes

The RF resistance of a PIN diode is found by treating the I-layer as a cylindrical conductor with cross-sectional area A and length W. The result is

$$R_J(I_Q) = \frac{W}{\sigma A} = \frac{W}{qp(\mu_n + \mu_p)A} = \frac{W^2}{(\mu_n + \mu_p)\tau I_Q} \quad (6.50)$$

where I_Q is the bias current and $p \approx n$.

Based on the PIN diode's resistive behavior under forward bias ("switch on") and capacitive behavior under reverse bias ("switch off" or **isolation**) we can proceed to construct simple small-signal models. For the PIN diode in series connection, the electric circuit model is seen in Figure 6-15 terminated with source and load resistances. The junction resistance and diffusion capacitance, as derived in (6.49) and (6.50), may in practice model the PIN diode behavior only very approximately. More quantitative information is obtained through measurements or sophisticated computational modeling efforts.

> **HIGH-LEVEL CARRIER INJECTION**
>
> When an intrinsic layer is sandwiched between highly doped n^+ and p^+ layers, a forward bias current pushes a large carrier concentration into this layer. The result is a dramatic increase in conductivity, also known as conductivity modulation. Since charge neutrality is maintained ($n \approx p$), we can approximate the conductivity as $\sigma \approx qp(\mu_n + \mu_p)$.

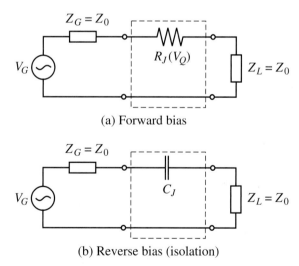

Figure 6-15 PIN diode in series connection.

The bias point setting required to operate the PIN diode has to be provided through a DC circuit that must be separated from the RF signal path. The DC isolation is achieved by a **radio frequency choke (RFC)**, representing a short circuit at DC and an open circuit at high frequency. Conversely, **blocking capacitors** (C_B) represent an open circuit at DC and a short circuit at RF. Figure 6-16 shows a typical attenuator circuit where the PIN diode is used either in series or shunt connection.

Although in the following discussion we will use a DC bias, a low-frequency AC bias can also be employed. In this case, the current through the diode consists of two components, $I = (dQ/dt) + Q/\tau_p$. The implication of this is deferred to the problem section.

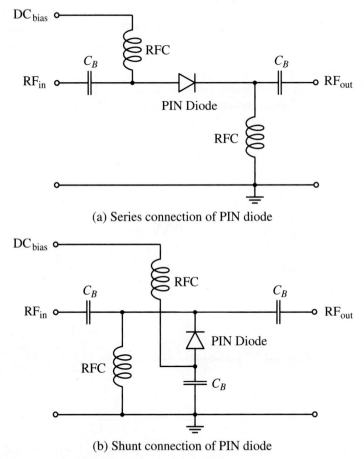

(a) Series connection of PIN diode

(b) Shunt connection of PIN diode

Figure 6-16 Attenuator circuit with biased PIN diode in series and shunt configurations.

For positive DC bias voltage, the series-connected PIN diode represents a low resistance to the RF signal. The shunt-connected PIN diode, however, creates a short-circuit condition, permitting only a negligibly small RF signal to appear at the output port. The shunt connection acts like a high attenuation device with high insertion loss. The situation is reversed for a negative bias condition, where the series-connected PIN diode behaves like a capacitor with high impedance or high insertion loss, whereas the shunt-connected diode with a high shunt impedance does not affect the RF signal appreciably.

A common notation found in datasheets is the **transducer loss** TL (identical to insertion loss) conveniently expressed in terms of the S parameter $|S_{21}|$ so that with (4.52)

$$\text{TL [dB]} = -20 \log|S_{21}|. \tag{6.51}$$

The following example computes the transducer loss for a PIN diode in series configuration.

Example 6-5. Computation of transducer loss of a PIN diode in series configuration for forward- and reverse-bias conditions

Find the transducer loss of a forward- and reverse-biased PIN diode in series connection ($Z_G = Z_L = Z_0 = 50\ \Omega$). Assume the junction resistance R_J under forward bias ranges between 1 and 20 Ω. Furthermore, assume that the reverse-bias operating conditions result in the junction capacitance being $C_J = 0.1, 0.3, 0.6, 1.3$, and 2.5 pF, and the frequency range of interest extends from 10 MHz to 50 GHz.

Solution. Based on (6.51) and Figure 6-15, the transducer loss is found with the aid of the voltage divider rule to be

$$\text{TL}_{\text{forward}} \text{[dB]} = -20 \log\left(\frac{100\ \Omega}{100\ \Omega + R_J}\right) = 20 \log\left(1 + \frac{R_J}{100\ \Omega}\right)$$

and

$$\text{TL}_{\text{reverse}} \text{[dB]} = -20 \log\left|\frac{100\ \Omega}{100\ \Omega - j1/(\omega C_P)}\right|$$

$$= 10 \log\left[1 + \left(\frac{1}{(100\ \Omega)\omega C_P}\right)^2\right].$$

Figure 6-17 plots the transducer loss in dB under forward bias condition for the given range of junction resistances. In contrast, Figure 6-18 graphs the reverse-bias condition where the PIN diode essentially has a purely capacitive response.

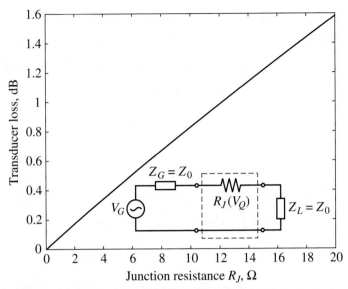

Figure 6-17 Transducer loss of series connected PIN diode under forward-bias condition. The diode behaves as a resistor.

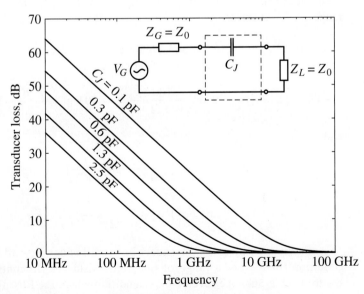

Figure 6-18 Transducer loss of series connected PIN diode under reverse-bias condition. The diode behaves as a capacitor.

6.2.3 Varactor Diode

The PIN diode with its capacitive behavior under reverse bias already suggests that a variable capacitance versus voltage characteristic can be created by a specific middle layer doping profile. A varactor diode accomplishes this task by a suitable choice of the intrinsic layer thickness W in addition to selecting a particular doping distribution $N_D(x)$.

Example 6-6. Determination of the required doping profile for a particular capacitance-voltage behavior

Find the appropriate doping concentration profile $N_D(x)$ that ensures that the varactor diode capacitance changes depending on the applied reverse-biasing voltage as $C(V_A) = C_0'/(V_A - V_{\text{diff}})$. Here, $C_0' = 5 \times 10^{-12}$ FV, and the cross-sectional diode area is $A = 10^{-4}$ cm^2.

Solution. The extent of the space charge length can be predicted based on (6.39):

$$x = \left[\frac{2\varepsilon_I(V_{\text{diff}} - V_A)}{q}\left(\frac{1}{N_D}\right)\right]^{1/2}$$

which determines the junction capacitance $C = \varepsilon_I A/x$. In the derivation of the preceding formula, we assumed that the doping concentration in the I-layer is much lower than the doping in the adjacent layers. If the space charge domain is increased by a differential increment ∂x, the charge is modified to

$$\partial Q = qN_D(x)A\partial x.$$

This differential increase in length can be expressed by a corresponding decrease in capacitance. Differentiating the capacitor formula, we obtain

$$\partial x = -\varepsilon_I A \partial C/C^2.$$

Upon substitution of ∂x into the expression for ∂Q and noting that $\partial Q = C\partial V_A$, we have

$$\partial Q \equiv C\partial V_A = -qN_D(x)A^2\varepsilon_I\partial C/C^2.$$

This gives us the desired expression for the doping profile:

$$N_D(x) = -\frac{C^3}{q\varepsilon_I A^2}\left(\frac{\partial V_A}{\partial C}\right).$$

For the desired capacitance, we find

$$N_D(x) = \frac{C_0'}{qAx} = \frac{2\times 10^{11}}{x}\ \text{cm}^{-2}.$$

Obviously, we cannot enforce the doping profile to reach infinity as x approaches the beginning of the I-layer. Nonetheless, by approximating a hyperbolic function, it is possible to ensure the desired capacitance-voltage behavior.

Figure 6-19 presents the simplified electric circuit model of the varactor diode consisting of a substrate resistance and voltage-dependent capacitance of the form $(V_{\text{diff}} - V_A)^{-1/2}$. This is the case when the doping profile is abrupt, as opposed to the hyper-abrupt profile of Example 6-6. Therefore, we have for the capacitance the generic representation

$$C_V = C_{V0}\left(1 - \frac{V_Q}{V_{\text{diff}}}\right)^{-1/2} \tag{6.52}$$

where V_Q is the reverse bias.

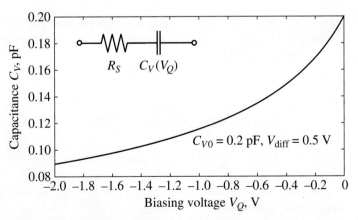

Figure 6-19 Simplified electric circuit model and capacitance behavior of varactor diode.

One of the main applications of this diode is frequency tuning of microwave circuits. This is due to the fact that the cutoff frequency f_V of the first-order varactor model

$$f_V = \frac{1}{2\pi R_S C_V(V_Q)} \qquad (6.53)$$

can be controlled through the reverse bias V_Q.

In addition, the varactor diode can be used to generate short pulses, as schematically explained in Figure 6-20. An applied voltage V_A across a series connection of resistor and diode creates a current flow I_V. This current is in phase with the voltage over the positive cycle. During the negative voltage cycle, the stored carriers in the middle layer contribute to the continued current flow until all carriers are removed. At this point the current drops abruptly to zero. A transformer can now couple out a voltage pulse, which is predicted by Faraday's law $V_{out} = L(dI_V/dt)$. The pulse width can be approximated based on the length of the middle layer W and the saturation drift velocity v_{dmax} of the injected carrier concentration.

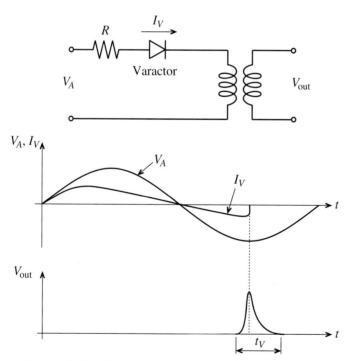

Figure 6-20 Pulse generation with a varactor diode.

If we assume $W = 10 \ \mu m$ and $v_{dmax} \approx 10^6$ cm/s, we obtain a transit time that is equivalent to a pulse width of

$$t_v = \frac{W}{v_{dmax}} = \frac{10 \ \mu m}{10^4 \ m/s} = 1 \ ns. \qquad (6.54)$$

6.2.4 IMPATT Diode

IMPATT stands for IMPact Avalanche and Transit Time diode and exploits the avalanche effect as originally proposed by Read. The principle of this diode construction, which is very similar to the PIN diode, is depicted in Figure 6-21. The key difference is the high electric field strength that is generated at the interface between the n^+ and p layer, resulting in an avalanche of carriers through impact ionization.

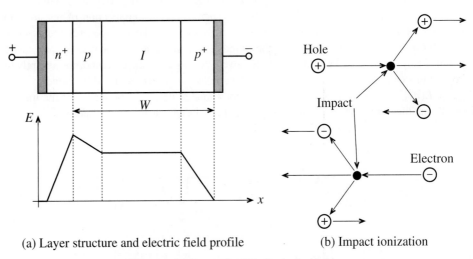

(a) Layer structure and electric field profile   (b) Impact ionization

Figure 6-21 IMPATT diode behavior.

The additional ionization current I_{ion} that is generated when the applied RF voltage V_A produces an electric field that exceeds the critical threshold level is seen in Figure 6-22. The current slowly decreases during the negative voltage cycle as the excess carriers are removed. The phase shift between this ionization current and the applied voltage can be tailored so as to reach 90°. The total diode current suffers an additional delay since the excess carriers must travel through the intrinsic layer to the p^+ layer. The time constant is dependent on the length and drift velocity. Choosing the intrinsic layer length appropriately, in conjunction with a suitable doping concentration, can create an additional time delay of 90°.

The electric circuit diagram of an IMPATT device, shown in Figure 6-23, is more intricate than the PIN diode and the reactance reveals an inductive behavior below the diode's resonance frequency f_0 before turning capacitive above the resonance frequency. The total resistance is positive for $f < f_0$ and becomes negative for $f > f_0$.

The resonance frequency is determined based on the operating current I_Q, dielectric constant, saturation drift velocity $v_{d\max}$, and the differential change in the ionization coefficient α with respect to the differential change in electric field strength $\alpha' = \partial\alpha/\partial E$. The resonance frequency is predicted as

$$f_0 = \frac{1}{2\pi}\sqrt{2I_Q \frac{v_{d\max}}{\varepsilon}\alpha'}. \tag{6.55}$$

RF Diodes

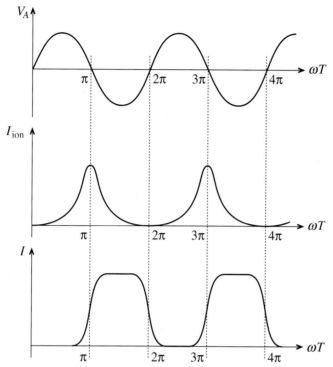

Figure 6-22 Applied voltage, ionization current, and total current of an IMPATT diode.

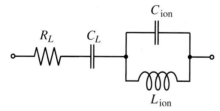

Figure 6-23 Electric circuit representation for the IMPATT diode.

The additional circuit parameters are specified as follows:

$$R = R_L + \frac{v_{d\max}}{2\pi^2 f_0^2 C_L W \left[1 - (f/f_0)^2\right]} \quad (6.56a)$$

$$C_L = \frac{\varepsilon A}{W} \quad (6.56b)$$

$$C_{ion} = \frac{\varepsilon A}{d} \quad (6.56c)$$

$$L_{ion} = \frac{1}{(2\pi f_0)^2 C_{ion}} \tag{6.56d}$$

where R_L is the combined resistance of the semiconductor layers, d is the length of the avalanche region of the p-layer, and W is the total length, as shown in Figure 6-21. The negative resistance of this diode above the resonance frequency can be understood in terms of returning electric energy to the RF or MW resonance circuit, which means the diode operates as an active device. Thus, the circuit attenuation can be substantially reduced to the point where additional power is transferred to the load impedance. Unfortunately, the 180-degree phase shift comes with a price. The **efficiency** of converting DC to RF power at operating frequencies of 5 to 10 GHz is very low, with typical values in the range of 10 to 15%.

6.2.5 Tunnel Diode

Tunnel diodes are *pn*-junction diodes that are made of n and p layers with extremely high doping (concentrations approach 10^{19}–10^{20} cm^{-3}) that create very narrow space charge zones. This can be seen immediately from equations (6.27) and (6.28). The result is that electrons and holes exceed the effective state concentrations in the conduction and valence bands. The Fermi level is shifted into the conduction band W_{Cn} of the n^+ layer and into the valence band W_{Vp} of the p^+ semiconductor. We notice from Figure 6-24 that the permissible electron states in either semiconductor layer are only separated through a very narrow potential barrier.

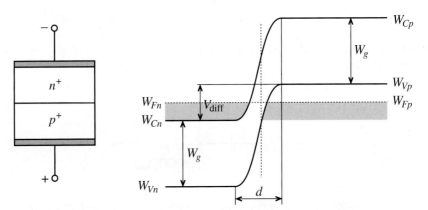

Figure 6-24 Tunnel diode and its band energy representation.

At small forward voltages, carriers more easily traverse the potential barrier, moving from the conduction to the valence band across the junction. This phenomenon is known as tunneling. As the voltage further increases, the band structure shifts apart, thereby reducing the ability of the carriers to tunnel across the junction. Above a critical voltage, $V_A \approx V_{diff}$, tunneling becomes impossible and the diode behaves like a conventional *pn*- junction. Interstingly, at low voltage the current rises quickly, but at higher voltages and reduced tunneling, the current drops, causing a negative slope. This can be exploited as a negative resistance in, for instance, an oscillator.

6.2.6 TRAPATT, BARRITT, and Gunn Diodes

For completeness, we briefly mention these additional three diode types without going into any details of their circuit representation and quantitative electric parameter derivations.

The **TRA**pped **P**lasma **A**valanche **T**riggered **T**ransit (TRAPATT) diode can be considered an enhancement of the IMPATT diode in that a higher efficiency (up to 75%) is realized through the use of bandgap traps. Such traps are energy levels that are situated inside the bandgap and allow the capture of electrons. External circuits ensure that during the positive cycle, a high barrier voltage is generated, resulting in carrier multiplication of the electron-hole plasma. The consequence is a breakdown in the rectifier properties of the diode during the negative cycle. The operating frequency is slightly lower than for the IMPATT diode. This is due to the fact that the buildup of the electron-hole plasma is slower than the transit time through the middle layer in an IMPATT diode.

For the **BARR**ier **I**njection **T**ransit **T**ime (BARRIT) diode, we are dealing with a transit time diode whose p^+np^+ doping profile acts like a transistor without base contact. The space charge domain extends from the cathode through the middle layer into the anode. The small-signal circuit model consists of a resistor and shunt capacitor whose values are dependent on the DC bias current. Unlike the IMPATT diode, this RC circuit can create a negative phase of up to –90 degrees at a relatively low efficiency of 5% and less. The BARRIT diode finds applications in radar mixer and detector circuits.

The **Gunn** diode is named after its inventor J. B. Gunn, who found in 1963 that in certain III-V semiconductors (GaAs, InP), a sufficiently high electric field can cause electrons to scatter into regions where the bandgap separation increases. As a result of this increase in bandgap energy, the electrons suffer a loss in mobility μ_n. This phenomenon is so dramatic that, for instance in GaAs, the drift velocity ($v_d = nq\mu_n$) can drop from 2×10^7 cm/s to less than 10^7 cm/s for electric field strengths growing from 5 kV/cm to 7 kV/cm. The negative differential mobility

$$\mu_n = \frac{dv_d}{dE} < 0$$

is used as a negative resistance in oscillator circuits. The Gunn diode consists of only one type of semiconductor with a doping profile of n^+nn^+. The lightly doped middle layer is the active layer that experiences most of the applied voltage. When a large enough voltage (and consequently electric field) is applied, the reduced mobility effect causes the conduction in the middle layer to become unstable. A thin slice of lower mobility (and conductivity) forms at the cathode and travels towards the anode. The reduced electric field in the rest of the middle layer prevents additional slices from forming. When the slice reaches the anode, it is absorbed, allowing the electric field in the middle layer to rise again, in turn triggering the formation of another slice at the cathode. The time the slice takes to travel across the middle layer largely determines the operating frequency of the Gunn diode.

6.3 Bipolar-Junction Transistor

The transistor was invented in 1948 by Bardeen and Brattain at the former AT&T Bell Laboratories and has over the past 50 years received a long list of improvements and refinements. Initially developed as a single point-contact device, the transistor has proliferated into a wide host of sophisticated devices ranging from the still popular **bipolar junction transistor**s (BJTs) over the modern **GaAs field effect transistor**s (GaAs FETs) to the most recent **high electron mobility transistor**s (HEMTs). Transistors are often arranged in the millions in digital integrated circuits (ICs) as part of microprocessor, memory, and peripheral chips. However, in RF and MW applications, the single discrete transistor has retained its importance. Many RF circuits still rely on discrete transistors in low-noise, linear, and high-power configurations. For this reason, we need to investigate both the DC and RF behavior of transistors in some detail.

The constituents of a bipolar transistor are three alternately doped semiconductor layers, in *npn* or *pnp* configuration. As the word *bi*polar implies, the internal current flow is due to both minority and majority carriers. In the following, we recapitulate some of the salient characteristics.

6.3.1 Construction

The BJT is one of the most widely used active RF elements due to its low-cost construction, relatively high operating frequency, low-noise performance, and high power handling capacity. The high-power capacity is achieved through a special interdigital emitter-base construction as part of a planar structure. Figure 6-25 shows both the cross-sectional planar construction and the top view of an interdigitated emitter-base connection.

Because of the interleaved construction shown in Figure 6-25(b), the base-emitter resistance is kept at a minimum while not compromising the gain performance. As we will see, a low base resistance directly improves the signal-to-noise ratio by reducing the current density through the base-emitter junction and by reducing the random thermal motion in the base (thermal noise); see Chapter 7 for more details.

For applications exceeding 1 GHz, it is important to reduce the emitter width to typically less than 1 μm, while increasing the doping to levels of 10^{20} to 10^{21} cm^{-3} to both reduce base resistance and increase current gain. Unfortunately, it becomes extremely difficult to ensure the tight tolerances, and self-aligning processes are required. Furthermore, the acceptor and donor doping concentrations quickly reach the solubility limits of the Si or GaAs semiconductor materials, providing a physical limitation of the achievable current gain. For these reasons, **heterojunction bipolar transistors** (HBTs) are becoming increasingly popular. HBTs achieve high current gains without having to dope the emitter excessively. Due to additional semiconductor layers (for instance, GaAlAs-GaAs sandwich structures), an enhanced electron injection into the base is achieved while the reverse hole injection into the emitter is suppressed. The result is a high **emitter efficiency** as defined by the ratio of electron current into the base to the sum of the same electron current and reverse emitter hole current. Figure 6-26 shows a cross-sectional view of such a structure.

Besides GaAs, heterojunctions have been accomplished with InP emitter and InGaAs base interfaces; even additional heterojunction interfaces between the GaInAs base and InP collector (double heterojunctions) have been fabricated. The material InP has the advantage of high breakdown voltage, larger carrier velocity, and higher thermal conductivity compared to GaAs. Operational frequencies exceeding 100 GHz, and a carrier transit time between base and collector of less than 0.5 ps have been achieved. Unfortunately, InP is a difficult material to handle and the manufacturing process has not yet matured to a level that allows it to compete with the Si and GaAs technologies.

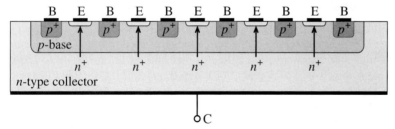

(a) Cross-sectional view of a multifinger bipolar junction transistor

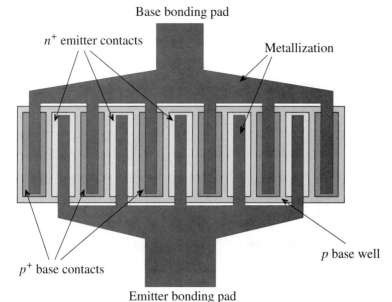

(b) Top view of a multifinger bipolar junction transistor

Figure 6-25 Interdigitated structure of high-frequency BJT.

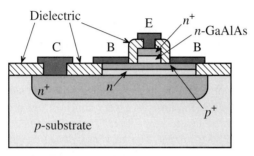

Figure 6-26 Cross-sectional view of a GaAs heterojunction bipolar transistor involving a GaAlAs-GaAs interface.

6.3.2 Functionality

In general, there are two types of BJTs: *npn* and *pnp* transistors. The difference between these two types lies in the doping of the semiconductor used to produce base, emitter, and collector. For an *npn* transistor, the collector and emitter are made of *n*-type semiconductor, whereas the base is of *p*-type. For a *pnp* transistor, the semiconductor types are reversed (*n*-type for base, and *p*-type for emitter and collector). Usually, the emitter has the highest, and the collector has the lowest, concentration of doping atoms. The BJT is a *current-controlled device* that is best explained by referring to Figure 6-27, which shows the structure, electrical symbol, and diode model with the associated voltage and current convention for the *npn* structure. We omit the discussion of the *pnp* transistor since it requires only a reversal of voltage polarity and diode directions.

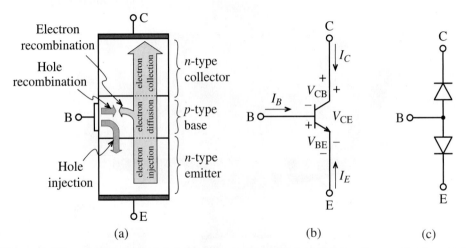

Figure 6-27 *npn* transistor: (a) structure with electrical charge flow under forward active mode of operation, (b) transistor symbol with voltage and current directions, and (c) diode model.

The first letter in the voltage designation always denotes the positive and the second letter the negative voltage reference points. Under normal mode of operation (i.e., the **forward active** mode), the emitter-base diode is operated in forward direction (with $V_{BE} \approx 0.7$ V) and the base-collector diode in reverse. Thus, the emitter injects electrons into the base, and, conversely, a hole current reaches the emitter from the base. If we maintain the collector-emitter voltage to be larger than the so-called **saturation voltage** (typically around 0.1 V), and since the base is a very thin (on the order of $d_B \leq 1$ μm) and relatively lightly doped *p*-type layer, only a small amount of electrons recombine with the holes supplied through the base current. The vast majority of electrons reach the base-collector junction and are collected by the applied reverse voltage V_{BC}.

For the **reverse active mode**, the collector-emitter voltage is negative (typically $V_{CE} < -0.1$ V) and the base-collector diode is forward biased, while the base-emitter diode is now operated in reverse direction. Unlike the forward active mode, it is now the electron flow from the collector that bridges the base and reaches the emitter.

Finally, the **saturation mode** involves the forward biasing of both the base-emitter and base-collector junctions. This mode typically plays an important role when dealing with switching circuits.

For a common-emitter configuration, Figure 6-28(a) depicts a generic biasing arrangement where the base current is fixed through an appropriate choice of biasing resistor R_B and voltage source V_{BB}, resulting in a suitable quiesent point, or Q-point. The base current versus base-emitter voltage, Figure 6-28(b), follows a typical diode I-V behavior, which constitutes the input characteristic of the transistor. The base current and base-emitter voltage at the intersection point between the load line and the transistor input characteristic are identified as I_B^Q and V_{BE}^Q. The collector current versus collector-emitter voltage behavior as part of the transistor output characteristic follows a more complicated pattern since the collector current must be treated as a parametric curve dependent on the base current ($I_{B1} < I_{B2} \ldots$), as seen in Figure 6-28(c).

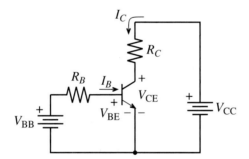

(a) Biasing circuit for *npn* BJT in common-emitter configuration

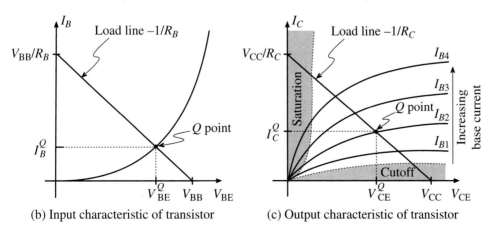

(b) Input characteristic of transistor   (c) Output characteristic of transistor

Figure 6-28 Biasing and input, output characteristics of an *npn* BJT.

The quantitative BJT behavior is analyzed by investigating the three modes of operation. This is accomplished by setting appropriate operating points and formulating the various current flows. For simplicity, we will neglect the spatial extent of the individual space charge domains and assume typical representative voltage and current conditions. To keep track of the different minority/majority and doping conditions in the three semiconductor layers, Table 6-3 summarizes the parameters and corresponding notation.

Table 6-3 BJT parameter nomenclature

Parameter description	Emitter (*n*-type)	Base (*p*-type)	Collector (*n*-type)
Doping level	N_D^E	N_A^B	N_D^C
Minority carrier concentration in thermal equilibrium	$p_{n_0}^E = n_i^2/N_D^E$	$n_{p_0}^B = n_i^2/N_A^B$	$p_{n_0}^C = n_i^2/N_D^C$
Majority carrier concentration in thermal equilibrium	$n_{n_0}^E$	$p_{p_0}^B$	$n_{n_0}^C$
Spatial extent	d_E	d_B	d_C

For the following BJT analysis, it is implicitly understood that the concentrations obey the inequality $p_{n0}^E \ll n_{p0}^B \ll p_{n0}^C$.

Forward Active Mode ($V_{CE} > V_{CEsat} = 0.1$ V, $I_B > 0$)

To find the minority carrier concentrations, we consider the configuration shown in Figure 6-29. Here, the concentration is plotted as a function of distance across the three semiconductor layers. For predicting the spatial minority carrier concentrations in the respective layer, we rely on the so-called **short diode** (see Appendix F) analysis, which approximates the exponentials as linear charge concentration gradients.

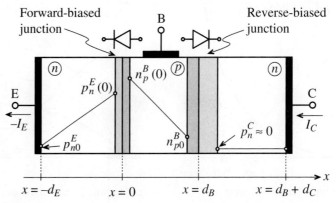

Figure 6-29 Minority carrier concentrations in forward active BJT.

Bipolar-Junction Transistor

The minority carrier concentrations in each layer are given as follows:

- Emitter: $p_n^E(-d_E) = p_{n0}^E$ and $p_n^E(0) = p_{n0}^E e^{V_{BE}/V_T}$
- Base: $n_p^B(0) = n_{p0}^B e^{V_{BE}/V_T}$ and $n_p^B(d_B) = n_{p0}^B e^{V_{BC}/V_T} \approx 0$
- Collector: $p_n^C(d_B) = p_{n0}^C e^{V_{BC}/V_T} \approx 0$.

The last two concentrations are zero because the base-collector voltage is negative (for instance, for typical transistor values of $V_{CE} = 2.5$ V and $V_{BE} = 0.7$ V, we find $V_{BC} = -1.8$ V, which yields $\exp(V_{BC}/V_T) = \exp(-1.8/0.026) \to 0$). Based on the aforementioned carrier concentrations, we can now predict the diffusion current density of holes $J_{p\,\text{diff}}^E$ in the emitter:

$$J_{p\,\text{diff}}^E = -qD_p^E \left[\frac{dp_n^E(x)}{dx}\right] = -\frac{qD_p^E}{d_E}[p_n^E(0) - p_n^E(-d_E)]$$

$$= -\frac{qD_p^E p_{n0}^E}{d_E}(e^{V_{BE}/V_T} - 1). \qquad (6.57)$$

For the diffusion current density of electrons in the base layer $J_{n\,\text{diff}}^B$, we similarly obtain

$$J_{n\,\text{diff}}^B = qD_n^B \left[\frac{dn_p^B(x)}{dx}\right] = \frac{qD_n^B}{d_B}[n_p^B(d_B) - n_p^B(0)] = -\frac{qD_n^B n_{p0}^B}{d_B} e^{V_{BE}/V_T}. \qquad (6.58)$$

From the preceding two equations, the collector and base currents are established as

$$I_{FC} = -J_{n\,\text{diff}}^B A = \frac{qD_n^B n_{p0}^B}{d_B} A e^{V_{BE}/V_T} = I_S e^{V_{BE}/V_T} \qquad (6.59)$$

and

$$I_{FB} = -J_{p\,\text{diff}}^E A = \frac{qD_p^E p_{n0}^E}{d_E} A(e^{V_{BE}/V_T} - 1) \qquad (6.60)$$

where index F denotes forward current, A is the junction cross-sectional area, and $I_S = (qD_n^B n_{p0}^B A)/d_B$ is the **saturation current**. The emitter current is directly found by adding (6.59) and (6.60). The forward current gain β_F is defined as

$$\beta_F = \frac{I_{FC}}{I_{FB}} = \frac{D_n^B n_{p0}^B d_E}{D_p^E p_{n0}^E d_B}. \qquad (6.61)$$

To arrive at (6.61), it is assumed that the exponential function in (6.60) is much larger than 1, allowing us to neglect the term -1. Moreover, the ratio between collector and emitter currents, or α_F, is expressed as

$$\alpha_F = \frac{I_{FC}}{(-I_{FE})} = \frac{\beta_F}{1+\beta_F}. \tag{6.62}$$

Example 6-7. Computation of the maximum forward current gain in a bipolar-junction transistor

Find the maximum forward current gain for a silicon-based BJT with the following parameters: donor concentration in the emitter, $N_D^E = 10^{19}$ cm^{-3}; acceptor concentration in the base, $N_A^B = 10^{17}$ cm^{-3}; spatial extent of the emitter, $d_E = 0.8$ µm, and of the base, $d_B = 1.2$ µm.

Solution. To apply (6.61), we need to determine the diffusion constants in base and emitter as described by the Einstein relation (6.15). Substituting this relation into (6.61), we obtain the forward current gain:

$$\beta_F = \frac{\mu_n n_{p0}^B d_E}{\mu_p p_{n0}^E d_B}.$$

Furthermore, using the expressions for the minority carrier concentrations in base and emitter from Table 6-3, we arrive at the final expression for β_F:

$$\beta_F = \frac{\mu_n N_D^E d_E}{\mu_p N_A^B d_B} = 187.5.$$

As discussed in Section 6.3.3 and in the following chapter, the current gain is only approximately constant. In general, it depends on the transistor operating conditions and temperature.

Bipolar-Junction Transistor

Reverse Active Mode ($V_{CE} < -0.1$ V, $I_B > 0$)

The minority carrier concentrations are shown in Figure 6-30 with the associated space charge domains (i.e., the base-emitter diode is reverse biased whereas the base-collector diode is forward biased).

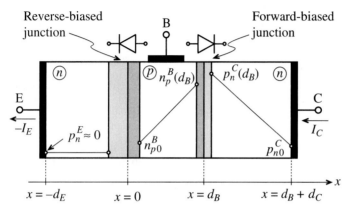

Figure 6-30 Reverse active mode of BJT.

The minority carrier concentrations in each layer are as follows:

- Emitter: $p_n^E(-d_E) \approx 0$ and $p_n^E(0) = p_{n0}^E e^{V_{BE}/V_T} \approx 0$
- Base: $n_p^B(0) = n_{p0}^B e^{V_{BE}/V_T} \approx 0$ and $n_p^B(d_B) = n_{p0}^B e^{V_{BC}/V_T}$
- Collector: $p_n^C(d_B) = p_{n0}^C e^{V_{BC}/V_T}$ and $p_n^C(d_B + d_C) \approx p_{n0}^C$.

From the diffusion current density, we find the reverse emitter current

$$I_{RE} = -J_{n\text{diff}}^B A = -qD_n^B \left(\frac{dn_p^B}{dx}\right) A = \frac{qD_n^B n_{p0}^B}{d_B} A e^{V_{BC}/V_T} = I_S e^{V_{BC}/V_T} \quad (6.63)$$

and the reverse base current

$$I_{RB} = -J_{p\text{diff}}^C A = -qD_p^C \left(\frac{dp_n^C}{dx}\right) A = \frac{qD_p^C p_{n0}^C A}{d_C} \left(e^{V_{BC}/V_T} - 1\right). \quad (6.64)$$

In a similar manner, as done for the forward current gain, we define the **reverse current gain** β_R

$$\beta_R = \frac{I_{RE}}{I_{RB}} = \frac{D_n^B n_{p0}^B d_C}{D_p^C p_{n0}^C d_B} \quad (6.65)$$

and the collector emitter current ratio α_R

$$\alpha_R = \left.\frac{I_{RC}}{(-I_{RE})}\right|_{V_{BC}} = \frac{\beta_R}{1+\beta_R}. \tag{6.66}$$

Saturation Mode $(V_{BE}, V_{BC} > V_T, I_B > 0)$
This mode of operation implies the forward bias of both diodes, so that the diffusion current density in the base becomes the combination of forward and reverse carrier flows; that is, with (6.59) and (6.63):

$$J^B_{n\text{diff}} = J_{RE} - J_{FC} = -\frac{I_S}{A}e^{V_{BE}/V_T} + \frac{I_S}{A}e^{V_{BC}/V_T}. \tag{6.67}$$

From (6.67), it is possible to find the emitter current by taking into account the forward base current. This forward base current (6.60) injects holes into the emitter and thus has to be taken with a negative sign to comply with our positive emitter current direction convention. Making the exponential expressions in (6.67) compatible with (6.60), we add and subtract unity and finally obtain

$$I_E = -I_S\left(e^{V_{BE}/V_T}-1\right)-\frac{I_S}{\beta_F}\left(e^{V_{BE}/V_T}-1\right)+I_S\left(e^{V_{BC}/V_T}-1\right). \tag{6.68}$$

Because the BJT can be treated as a symmetric device, the collector current is expressible in a similar manner as the contribution of three currents: the forward collector and reverse emitter currents, given by the negative of (6.67), and an additional hole diffusion contribution as a result of the reverse base current I_{RB}. The resulting equation is

$$I_C = I_S\left(e^{V_{BE}/V_T}-1\right)-\frac{I_S}{\beta_R}\left(e^{V_{BC}/V_T}-1\right)-I_S\left(e^{V_{BC}/V_T}-1\right). \tag{6.69}$$

Finally, the base current $I_B = -I_C - I_E$ is found from the preceding two equations:

$$I_B = I_S\left[\frac{1}{\beta_R}\left(e^{V_{BC}/V_T}-1\right)+\frac{1}{\beta_F}\left(e^{V_{BE}/V_T}-1\right)\right]. \tag{6.70}$$

Here, we should recall that the internal emitter current flow is denoted opposite in sign to the customary external circuit convention.

6.3.3 Frequency Response

The **transition frequency** f_T (also known as the **cutoff frequency**) of a microwave BJT is an important figure of merit since it determines the operating frequency at which the common-emitter, short-circuit current gain h_{fe} decreases to unity. The transition frequency f_T is related to the transit time τ that is required for carriers to travel through the emitter-collector structure:

$$f_T = \frac{1}{\tau}. \tag{6.71}$$

This transit time is generally composed of three delays:

$$\tau = \tau_E + \tau_B + \tau_C \tag{6.72}$$

where τ_E, τ_B, and τ_C are delays in emitter, base, and collector, respectively. The base-emitter depletion region charging time is given by

$$\tau_E = r_E C = \frac{V_T}{I_E}(C_E + C_C) \cong \frac{V_T}{I_C}(C_E + C_C) \tag{6.73a}$$

where C_E, C_C are emitter and collector junction capacitances, and r_E is the emitter resistance obtained by differentiation of the emitter current with respect to base-emitter voltage. The second delay in (6.72) is the base layer charging time, and its contribution is given as

$$\tau_B = \frac{d_B^2}{\eta D_n^B} \tag{6.73b}$$

where the factor η is doping profile dependent and ranges from $\eta = 2$ for uniformly doped base layers up to $\eta = 60$ for highly nonuniform layers. Finally, the transit time τ_C through the base-collector junction space charge zone w_C can be computed as

$$\tau_C = \frac{w_C}{v_S} \tag{6.73c}$$

with v_S representing the saturation drift velocity. In the preceding formulas, we have neglected the collector charging time $\tau_{CC} = r_C C_C$, which is typically very small when compared with τ_E.

As seen in (6.73a), the emitter charging time is inversely proportional to the collector current, resulting in higher transition frequencies for increasing collector currents. However, as the current reaches sufficiently high values, the concentration of charges injected into the base becomes comparable with the doping level of the base, which causes an increase of the effective base width and, in turn, reduces the transition frequency. Usually, BJT data sheets provide information about the dependence of the transition frequency on the collector current. For instance, Figure 6-31 shows the transition frequency as a function of collector current for the wideband *npn* transistor BFG403W measured at $V_{CE} = 2$ V, $f = 2$ GHz, and at an ambient temperature of 25°C.

Another aspect of the BJT operated at RF and MW frequencies is that, at high frequencies, the skin effect physically restricts current flow to the outer perimeter of the emitter (see also Section 1.4). To keep the charging time as low as possible, the emitter is constructed in a grid pattern of extremely narrow (less than 1 µm) strips. Unfortunately, the trade-off is a high current density over the small surface area, limiting the power handling capabilities. Additional ways to increase the cutoff frequency are to reduce the base transit time constant τ_B by high doping levels and concomitantly fabricate very short base layers of less than 100 nm. In addition, a small base thickness has the advantage of reducing power loss.

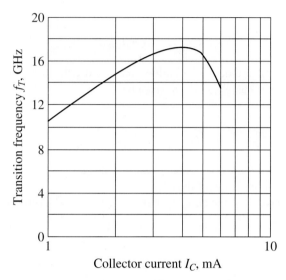

Figure 6-31 Transition frequency as a function of collector current for the 17 GHz *npn* wideband transistor BFG403W (courtesy of NXP).

Another useful quantify besides f_T is the maximum frequency of oscillation f_{max} whose derivation is not straightforward. This figure of merit denotes the upper limit of the device (either BJT or FET) operating as an amplifier when the power gain has reached unity. More detail about the relation between transition and maximum frequencies is found in the problem section.

6.3.4 Temperature Behavior

We have seen in this chapter that almost all parameters describing both the static and dynamic behavior of semiconductor devices are influenced by the junction temperature T_j. As an example of such a dependence, in Figure 6-32 the forward current gain β_F of a typical transistor for a given V_{CE} is plotted as a function of collector current I_C for various junction temperatures T_j. As we can see from this graph, the current gain rises from 40 at $I_C = 3.5$ mA and $T_j = -50°C$ to more than 80 at $T_j = 50°C$.

Another example that shows strong temperature influence is the dependence of the input characteristic of a transistor described by the base current as a function of base-emitter voltage, as depicted in Figure 6-33. Again, if we compare the behavior of the transistor at $T_j = -50°C$ and $T_j = 50°C$, we notice that at $T_j = -50°C$ and a base-emitter voltage of 1.25 V the transistor is in cutoff state, whereas at $T_j = 50°C$ the BJT already conducts 4 mA base current. These two examples underscore the importance of temperature considerations in the design of RF circuits. For instance, the design of a cellular phone for worldwide use must ensure that our circuit preforms according to specifications under all temperature conditions encountered by the operator. Standard specifications usually cover the temperature range from $-50°C$ to $80°C$.

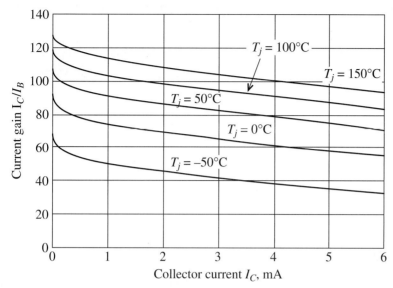

Figure 6-32 Typical current gain β_F as a function of collector current for various junction temperatures at a fixed V_{CE}.

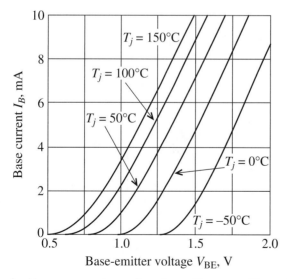

Figure 6-33 Typical base current as a function of base-emitter voltage for various junction temperatures at a fixed V_{CE}.

The junction temperature also plays an important role when dealing with the maximum power dissipation. In general, the manufacturer provides a **power derating curve** that specifies the temperature T_S up to which the transistor can be operated at the maximum available power P_{tot}. For junction temperatures T_j exceeding this value, the power has to be reduced to values dictated by the thermal resistance between the junction and the soldering point (or case) R_{thjs} according to

$$P = P_{tot} \frac{T_{jmax} - T_j}{T_{jmax} - T_S} = \frac{T_{jmax} - T_j}{R_{thjs}} \qquad (6.74)$$

where T_{jmax} is the maximum junction temperature. Typical values for Si BJTs vary between 150 and 200°C.

For the RF transistor BFG403W, the maximum total power P_{tot} of 16 mW can be maintained up to $T_S = 140°C$. For higher temperatures $T_S \leq T_j \leq T_{jmax}$, the power must be derated until the maximum junction temperature T_{jmax} of 150°C is reached. Manufacturers have to develop effective ways to dissipate the thermal energy generated by the transistor. Usually, this is done by employing heat sinks and using materials with high thermal conductivity. Instead of the thermal resistance at the soldering point R_{thjs}, the manufacturer may supply additional information involving heat resistances between junction-to-case (R_{thjc}), case-to-sink (R_{thcs}), and sink-to-air (R_{thha}) interfaces.

To simplify the thermal analysis, it is convenient to resort to a thermal equivalent circuit with the following correspondences:

- Thermal power dissipation = electric current
- Temperature = electric voltage.

A typical thermal circuit in equilibrium is shown in Figure 6-34. Here, the total electric power supplied to the device is balanced through a thermal circuit involving thermal resistances. Therefore,

$$R_{thjc} = \frac{T_j - T_s}{P_W} = \frac{1}{\gamma_{th} A_{BJT}} \qquad (6.75)$$

where the junction and soldering point temperatures T_j and T_s and thermal power P_W determine the thermal resistance in Kelvin per watt (K/W), and whose value can also be expressed in terms of the thermal conductivity γ_{th} and the surface area A_{BJT} of the BJT. The solder point temperature is affected by the transition between casing and heat sink. This constitutes a thermal resistance R_{thcs} with values up to 5 K/W. Finally, the heat sink represents a thermal resistance of

$$R_{thha} = \frac{1}{\delta_{hs} A_{hs}} \qquad (6.76)$$

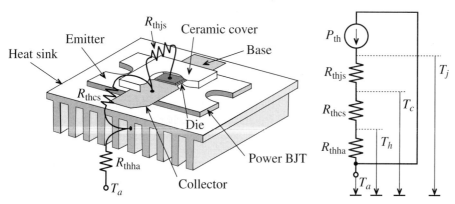

Figure 6-34 Thermal equivalent circuit of BJT.

where δ_{hs} is a convection coefficient that can vary widely between 10 W/(K·m^2) for still air, 100 W/(K·m^2) for forced air, up to 1000 W/(K·m^2) for water cooling, and A_{hs} is the total area of the heat sink.

The following example describes a common design problem.

RF&MW

Example 6-8. Thermal analysis involving a BJT mounted on a heat sink

An RF power BJT generates a total power P_W of 15 W at a case temperature of 25°C, while the junction temperature is 150°C (representing the maximum acceptable temperature). The maximum ambient operating temperature is specified by the user to be $T_a = 60°C$. What is the maximum dissipated power if the thermal resistances between case-to-sink and sink-to-air are 2 K/W and 10 K/W, respectively?

Solution. With reference to Figure 6-34, we are dealing with three thermal resistances: R_{thjs}, R_{thcs}, and R_{thha}. The junction-to-solder point resistance can be found based on equation (6.75):

$$R_{thjs} = \frac{T_j - T_s}{P_W} = \frac{150°C - 25°C}{15 \text{ W}} = 8.3 \text{ K/W}.$$

Adding up all resistances gives us a total thermal resistance of

$$R_{thtot} = R_{thjs} + R_{thca} + R_{thhs} = 20.3 \text{ K/W}.$$

The dissipated power P_{th} follows from the temperature drop (junction temperature T_j minus ambient temperature T_a) divided by the total thermal resistance:

$$P_{th} = \frac{T_j - T_a}{R_{thtot}} = \frac{150°C - 60°C}{20.3 \text{ K/W}} = 4.43 \text{ W}.$$

To operate the BJT in thermal equilibrium, we have to reduce the total electric power $P_{tot} = P_W$ to the point where it is in balance with the computed thermal power $P_{tot} = P_{th}$. Thus, a reduction from 15 W to 4.43 W is required.

While the design engineer cannot influence the junction-to-solder point heat resistance, it is the choice of casing and heat sink that typically allows major improvements in thermal performance.

6.3.5 Limiting Values

The total power dissipation capabilities at a particular temperature restrict the range of safe operation of the BJT. In our discussion, we will focus exclusively on the active mode in the common-emitter configuration and will neglect the switch-mode behavior whereby the BJT is operated either in saturation or cutoff mode. For a given maximum BJT power rating, we can either vary the collector-emitter voltage V_{CE} and plot the allowable collector current $I_C = P_{tot}/V_{CE}$ (here we assume that base current is negligibly small compared to the collector current due to high β), or vary I_C and plot the allowable collector-emitter voltage $V_{CE} = P_{tot}/I_C$. The result is the **maximum power hyperbola**. This does not mean that I_C and V_{CE} can be increased without bound. In fact, we need to ensure that $I_C \leq I_{Cmax}$ and $V_{CE} \leq V_{CEmax}$, as depicted in Figure 6-35. The **safe operating area** (SOAR) is defined as a

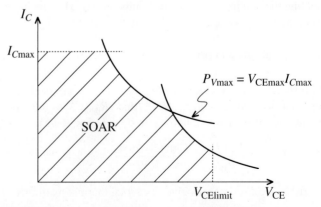

Figure 6-35 Operating domain of BJT in active mode with breakdown mechanisms.

set of biasing points where the transistor can be operated without risk of unrecoverable damage to the device. The SOAR domain, shown as a shaded region in Figure 6-35, is more restrictive than a subset bounded by the maximum power hyperbola, since we have to take into account two more breakdown mechanisms:

1. **Breakdown of the first kind**. Here, the collector current density exhibits a nonuniform distribution that results in a local temperature increase, which in turn lowers the resistance of a portion of the collector domain, creating a channel. The consequence is a further increase in current density through this channel until the positive feedback begins to destroy the crystal structure (**avalanche breakdown**), ultimately destroying the transistor itself.
2. **Breakdown of the second kind**. This breakdown mechanism can take place independently of the first mechanism and affects primarily power BJTs. Internal overheating may cause an abrupt increase in the collector current for constant V_{CE}. This breakdown mechanism usually occurs at the base-collector junction when the temperature increases to such high values that the intrinsic concentration is equal to the collector doping concentration. At this point, the resistance of the junction is abruptly reduced, resulting in a dramatic current increase and melting of the junction.

It is interesting to point out that the BJT can exceed the SOAR, indeed even the maximum power hyperbola, for a short time since the temperature response has a much larger time constant (on the order of microseconds) in comparison with the electric time constants.

Additional parameters of importance to a design engineer are the maximum voltage conditions for open emitter, base, and collector conditions; that is, V_{CBO} (collector-base voltage, open emitter), V_{CEO} (collector-emitter, open base), and V_{EBO} (emitter-base voltage, open collector). For instance, values for the BFG403W are as follows: $V_{CBO}|_{max} = 10$ V, $V_{CEO}|_{max} = 4.5$ V, and $V_{EBO}|_{max} = 1.0$ V.

6.3.6 Noise Performance

The noise performance of a transistor becomes an important design consideration when dealing with very weak signals such as those encountered in the receiver chain of a communication system, for instance in low-noise amplifiers and mixers. Furthermore, as we will see in Chapter 10, noise has a significant impact on oscillator performance. Main noise contributors in these circuits are passive (resistors, inductors, capacitors) and active devices (BJTs, FETs). In particular, the BJT generates noise through three major mechanisms: (a) thermal noise, (b) shot noise, and (c) $1/f$ noise. We discuss these qualitatively here, referring to Appendix H and additional literature at the end of this chapter for further details.

(a) Thermal or Johnson noise is a characteristic of all resistive components and is independent of current flow. It is the result of the random thermal motion of electric charges in the material. The power spectral density of thermal noise is constant with frequency (this is known as white noise) and is proportional to the device temperature. Although resistances are encountered in many different locations in the transistor, their noise contributions are usually converted into input-referred noise sources for modeling purposes. See Appendix H for a more detailed analysis of noise in a two-port network.

(b) Shot noise occurs when current flows through a *pn*-junction. The reason for this noise source is the discrete nature of charges that have to overcome a potential barrier at the junction. Since the base-emitter junction in a BJT is forward biased under normal operation, it generates a substantial amount of shot noise. The power spectral density of this type of noise is independent of frequency (white) and is proportional to the current through the junction (in this case the base and collector currents). Shot noise constitutes a major disadvantage of BJTs over FETs. The latter are not subject to this type of noise; they are therefore preferred in low-noise amplifiers.

(c) $1/f$ noise is also known as flicker or pink noise. The precise mechanisms of this noise source are not well understood, but it is thought to result from the interaction of charges with semiconductor surfaces. As the name implies, the power spectral density of this noise source decreases with frequency. A corner frequency is usually quoted where the $1/f$ noise power spectral density equals that of white (thermal and shot) noise. The empirically determined corner frequency typically falls in the low kHz range. Consequently, this noise source can be safely ignored in RF/microwave amplifier applications. However, in oscillators the low-frequency $1/f$ noise translates directly into phase noise. Moreover, this noise source is important in mixers because of their highly nonlinear nature, especially if the mixer performs direct, RF-to-baseband conversion. In these cases, BJTs hold an advantage over FETs since their $1/f$ noise is substantially lower due to a smaller influence of semiconductor surfaces on device operation. Interestingly, the *pnp* BJT is also known to possess lower $1/f$ noise than its *npn* counterpart.

6.4 RF Field Effect Transistors

Unlike BJTs, **field effect transistors** (FETs) are **monopolar devices,** meaning that only one carrier type, either holes or electrons, contributes to the current flow through the channel. If hole contributions are involved we speak of ***p*-channel**, otherwise of ***n*-channel** FETs. Moreover, the FET is a voltage-controlled device. A variable electric field controls the current flow from the **source** to the **drain** by changing the applied voltage on the **gate** electrode.

6.4.1 Construction

Traditionally, FETs are classified according to how the gate is connected to the conducting channel. Specifically, the following four types are used:

1. **Metal Insulator Semiconductor FET** (MISFET). Here, the gate is separated from the channel through an insulation layer. One of the most widely used types, the **Metal Oxide Semiconductor FET** (MOSFET), belongs to this class.
2. **Junction FET** (JFET). This type relies on a reverse-biased *pn*-junction that isolates the gate from the channel.
3. **Metal Semiconductor FET** (MESFET). If the reverse-biased *pn*-junction is replaced by a Schottky contact, the channel can be controlled just as in the JFET case.

4. **Hetero FET**. As the name implies (and unlike the previous three cases, whose constructions rely on a single semiconductor material such as Si, GaAs, SiGe, or InP), heterostructures utilize abrupt transitions between layers of different semiconductor materials. Examples are GaAlAs to GaAs or GaInAs to GaAlAs interfaces. The **High Electron Mobility Transistor** (HEMT) belongs to this class.

Figure 6-36 provides an overview of the first three types. In all cases, the current flow is directed from the source to the drain, with the gate controlling the current flow.

Due to the presence of a large capacitance formed by the gate electrode and the insulator or reverse-biased *pn*-junction, MISFETs and JFETs have a relatively low cutoff frequency and are usually operated in low and medium frequency ranges of typically up to 1 GHz. GaAs MESFETs find applications up to 60–70 GHz, and HEMT can operate beyond 100 GHz. Since our interest is geared toward RF applications, the emphasis will be on the last two types.

In addition to the above physical classification, it is customary to electrically classify FETs according to **enhancement** and **depletion** types. This implies that the channel either experiences an increase in carriers (for instance the *n*-type channel is injected with electrons) or a depletion in carriers (for instance the *n*-type channel is depleted of electrons) as the gate voltage increases. In Figure 6-36 (a), the FET is nonconducting, or **normally-off**, until a sufficiently positive gate voltage sets up a conduction channel. Normally-off FETs can only be operated in enhancement mode. Alternatively, **normally-on** FETs can be of both enhancement and depletion types.

6.4.2 Functionality

Because of its importance in RF and MW amplifier, mixer, and oscillator circuits, we focus our analysis on the MESFET, whose physical behavior is in many ways similar to the JFET. The analysis is based on the geometry shown in Figure 6-37 where the transistor is operated in depletion mode.

The Schottky contact builds up a channel space charge domain that affects the current flow from the source to the drain. The space charge extent d_S can be controlled via the gate voltage in accordance to our discussion in Section 6.1.3, where (6.39) is adjusted such that V_A is replaced by the gate source voltage V_{GS}:

$$d_S = \left(\frac{2\varepsilon}{q} \frac{V_d - V_{GS}}{N_D}\right)^{1/2}. \tag{6.77}$$

For instance, the barrier voltage V_d is approximately 0.9 V for a GaAs-Au interface. The resistance R between source and drain is predicted by

$$R = \frac{L}{\sigma(d - d_S)W} \tag{6.78}$$

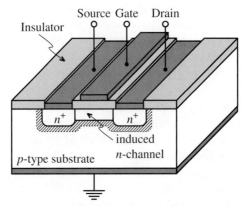

(a) Metal insulator semiconductor FET (MISFET)

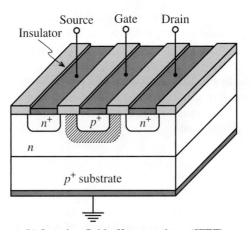

(b) Junction field effect transistor (JFET)

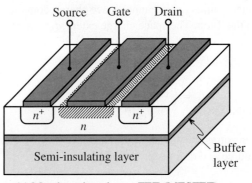

(c) Metal semiconductor FET (MESFET)

Figure 6-36 Construction of (a) MISFET, (b) JFET, and (c) MESFET. The shaded areas depict the space charge domains.

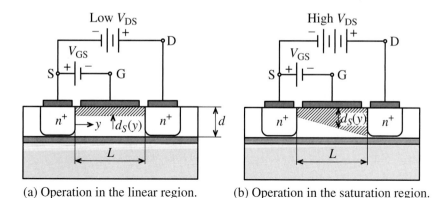

(a) Operation in the linear region.   (b) Operation in the saturation region.

Figure 6-37 Functionality of MESFET for different drain-source voltages.

with the conductivity given by $\sigma = q\mu_n N_D$, and W being the gate width. Substituting (6.77) into (6.78) yields the drain current equation

$$I_D = \frac{V_{DS}}{R} = G_0\left[1 - \left(\frac{2\varepsilon}{qd^2}\frac{V_d - V_{GS}}{N_D}\right)^{1/2}\right]V_{DS} \quad (6.79)$$

where we have defined the conductance $G_0 = \sigma W d/L$. This equation shows that the drain current depends linearly on the drain source voltage, a fact that is only true for small V_{DS}.

As the drain-source voltage increases, the space charge domain near the drain contact increases as well, resulting in a nonuniform distribution of the depletion region along the channel, see Figure 6-37(b). If we assume that the voltage along the channel changes from 0 at the source location to V_{DS} at the drain end, then we can compute the drain current for the nonuniform space charge region. This approach is also known as the **gradual-channel approximation**. The approximation rests primarily on the assumption that the cross-sectional area at a particular location y along the channel is given by $A(y) = (d - d_S(y))W$ and the electric field E is only y-directed. The channel current is thus

$$I_D = -\sigma E A(y) = \sigma \frac{dV(y)}{dy}(d - d_S(y))W \quad (6.80)$$

where the difference between V_d and V_{GS} in the expression for $d_S(y)$ has to be augmented by the additional drop in voltage $V(y)$ along the channel; that is, (6.77) becomes

$$d_S(y) = \left[\frac{2\varepsilon}{qN_D}(V_d - V_{GS} + V(y))\right]^{1/2}. \quad (6.81)$$

> **DUAL GATE FETs**
>
> For certain applications, it proves beneficial to combine a common drain FET with a common gate FET. This series connection of two FETs is known as a cascode circuit, which improves high frequency performance by reducing the Miller effect, as discussed in Chapter 7. The series combination of both devices can be achieved through wafer integration, and is then known as a dual-gate FET.

Substituting (6.81) into (6.80) and carrying out the integration on both sides of the equation yields

$$\int_0^L I_D dy = I_D L = \sigma W \int_0^{V_{DS}} \left(d - \left[\frac{2\varepsilon}{qN_D}(V + V_d - V_{GS})\right]^{1/2} \right) dV \tag{6.82}$$

The result is the **output characteristic** of the MESFET in terms of the drain current as a function of V_{DS} and V_{GS}, or

$$I_D = G_0 \left(V_{DS} - \frac{2}{3}\sqrt{\frac{2\varepsilon}{qN_D d^2}}[(V_{DS} + V_d - V_{GS})^{3/2} - (V_d - V_{GS})^{3/2}] \right). \tag{6.83}$$

We note that this equation reduces to (6.79) for small V_{DS}.

When the space charge extends over the entire channel depth d, the drain-source voltage for this situation is called **drain saturation voltage** V_{DSsat} and is given by

$$d_S(L) = d = \sqrt{\frac{2\varepsilon}{qN_D}(V_d - V_{GS} + V_{DSsat})} \tag{6.84}$$

or, explicitly,

$$V_{DSsat} = \frac{qN_D d^2}{2\varepsilon} - (V_d - V_{GS}) = V_P - V_d + V_{GS} = V_{GS} - V_{T0} \tag{6.85}$$

where we introduced the so-called **pinch-off voltage** $V_P = qN_D d^2/(2\varepsilon)$ and **threshold voltage** $V_{T0} = V_d - V_P$. The associated drain saturation current is found by inserting (6.85) into (6.83), with the result

$$I_{Dsat} = G_0 \left[\frac{V_P}{3} - (V_d - V_{GS}) + \frac{2}{3\sqrt{V_P}}(V_d - V_{GS})^{3/2} \right]. \tag{6.86}$$

The maximum saturation current in (6.86) is obtained when $V_{GS} = 0$, which we define as $I_{Dsat}(V_{GS} = 0) = I_{DSS}$. In Figure 6-38, the typical input-output transfer as well as the output characteristic behavior is shown.

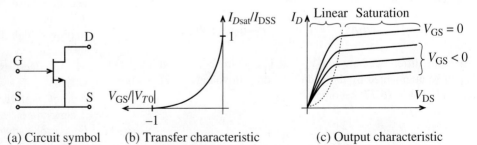

(a) Circuit symbol   (b) Transfer characteristic   (c) Output characteristic

Figure 6-38 Transfer and output characteristics of an *n*-channel MESFET.

The saturation drain current (6.86) is often approximated by the simple relation

$$I_{D\text{sat}} = I_{DSS}\left(1 - \frac{V_{GS}}{V_{T0}}\right)^2. \tag{6.87}$$

How well (6.87) approximates (6.86) is discussed in the following example.

Example 6-9. Drain saturation current in a MESFET

A GaAs MESFET has the following parameters: $N_D = 10^{16}\,\text{cm}^{-3}$, $d = 0.75\,\mu\text{m}$, $W = 10\,\mu\text{m}$, $L = 2\,\mu\text{m}$, $\varepsilon_r = 12.0$, $V_d = 0.8\,\text{V}$, and $\mu_n = 8500\,\text{cm}^2/(\text{Vs})$. Determine (a) the pinch-off voltage, (b) the threshold voltage, (c) the maximum saturation current I_{DSS}, and plot the drain saturation current based on (6.86) and (6.87) for V_{GS} ranging from –4 to 0 V.

Solution. The pinch-off voltage for the FET is independent of the gate-source voltage, and is computed as

$$V_P = \frac{qN_D d^2}{2\varepsilon} = 4.24\,\text{V}.$$

Knowing V_P and the barrier voltage $V_d = 0.8\,\text{V}$, we find the threshold voltage to be $V_{T0} = V_d - V_P = -3.44\,\text{V}$. The maximum saturation drain current is again independent of the applied drain-source voltage and, based on (6.86), is equal to

$$I_{DSS} = G_0\left[\frac{V_P}{3} - V_d + \frac{2}{3\sqrt{V_P}}V_d^{3/2}\right] = 6.89\,\text{A}$$

where $G_0 = \sigma q N_D W d / L = q^2 \mu_n N_D^2 W d / L = 8.16\,\text{S}$.

Figure 6-39 shows results for the saturation drain current computed using the exact formula (6.86) and the quadratic law approximation given by (6.87).

Because of the excellent agreement, the quadratic law approximation (6.87) is more widely used in the literature and data sheets than the exact equation.

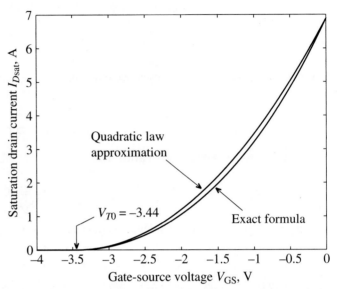

Figure 6-39 Drain current versus V_{GS} computed using the exact and the approximate equations (6.86) and (6.87).

If V_{DS} reaches the saturation voltage V_{DSsat} for a given V_{GS}, the space charges pinch off the channel. This implies that the drain current saturates. Interestingly, *pinch-off does not imply a zero I_D* since there is no charge barrier impeding the flow of carriers. It is the electric field as a result of the applied voltage V_{DS} that "pulls" the electrons across the depletion space charge domain. Any additional increase $V_{DS} > V_{DSsat}$ will result in a shortening of the channel length from the original length L to the new length $L' = L - \Delta L$. The result is that (6.86) must be modified to

$$I'_D = I_D\left(\frac{L}{L - \Delta L}\right) = I_D\left(\frac{L}{L'}\right). \tag{6.88}$$

The change in channel length as a function of V_{DS} is heuristically taken into account through the so-called **channel length modulation parameter** $\lambda = \Delta L/(L'V_{DS})$. This is particularly useful when expressing the drain current in the saturation region:

$$I'_{Dsat} = I_{Dsat}(1 + \lambda V_{DS}) \tag{6.89}$$

where measurements show a slight increase in drain current as V_{DS} is increased.

Example 6-10. I-V characteristic of a MESFET

For the following discrete gate-source voltages $V_{GS} = -1, -1.5, -2,$ and -2.5 V, plot the drain current I_D of a MESFET as a function of drain-source voltage V_{DS} in the range from 0 to 5 V. Assume that the device parameters are the same as in the previous example and that the channel length modulation parameter λ is 0.03 V^{-1}. Compare your results with the case where $\lambda = 0$.

Solution. In the analysis of the MESFET behavior, we have to be careful about choosing the appropriate formulas. At very low drain-source voltages, the drain current can be described by a simple linear relation (6.79). As the voltage increases, this approximation becomes invalid and a more complicated expression for I_D has to be employed, see (6.83). Further increase in V_{DS} ultimately leads to channel pinch-off, where $V_{DS} \geq V_{DSsat} = V_{GS} - V_{T0}$. In this case, the drain current is equal to the saturation current given by (6.86). Additional increases in V_{DS} beyond the saturation voltage result only in minor increases of the drain current due to a shortening of the channel. At this point, I_D is linearly dependent on V_{DS}. Substituting (6.86) into (6.89) for $V_{DS} \geq V_{DSsat}$, we obtain

> **DEVICE SPREAD**
> A handicap of FETs over BJTs is the manufacturing process related device characteristic fluctuation. This may be tolerable for digital CMOS technology where the switching involves only on/off states, but causes considerable problems in analog circuits. Drain current and gate-source voltage variations need to be compensated through appropriate bias circuits.

$$I_D = G_0 \left\{ \frac{V_P}{3} - (V_d - V_{GS}) + \frac{2}{3} \frac{(V_d - V_{GS})^{3/2}}{V_P^{1/2}} \right\} (1 + \lambda V_{DS}).$$

To provide a smooth transition from the linear to the saturation region for nonzero λ, we multiply (6.83) by $(1 + \lambda V_{DS})$. Thus, the final expression for the drain current for $V_{DS} \leq V_{DSsat}$ is

$$I_D = G_0 \left\{ V_{DS} - \frac{2}{3} \frac{(V_{DS} + V_d - V_{GS})^{3/2} - (V_d - V_{GS})^{3/2}}{V_P^{1/2}} \right\} (1 + \lambda V_{DS})$$

The results of applying these formulas to predict I_D for zero (dashed line) as well as nonzero λ (solid line) are shown in Figure 6-40.

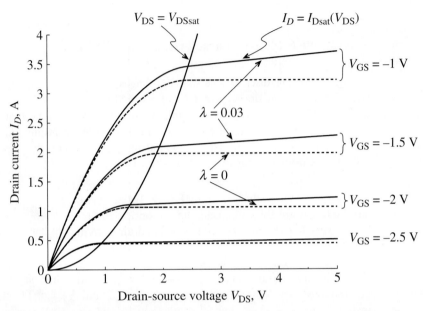

Figure 6-40 Drain current as a function of applied drain-source voltage for different gate-source biasing conditions.

> *Channel length modulation is similar to the Early Effect encountered in a BJT, where the collector current in the active region increases slightly for increasing collector-emitter voltage, as discussed in Chapter 7.*

6.4.3 Frequency Response

The high-frequency MESFET performance is determined by the transit time of charge carriers traveling between source and drain, and the *RC* time constant of the device. Here, we will focus our attention on the transit time only and defer the time constant computation, which requires knowledge of the channel capacitance, to Chapter 7. Since electrons in Si and GaAs have much higher mobility than holes, *n*-channel MESFETs are used in RF and MW applications almost exclusively. Furthermore, since the electron mobility of GaAs is roughly five times higher than that of Si, GaAs MESFETs are usually preferred over Si devices.

The transit time τ of electrons traveling through the channel of gate length L is computed as

$$\tau = \frac{L}{v_{\text{sat}}} \tag{6.90}$$

Metal Oxide Semiconductor Transistors

where we have assumed a fixed saturation velocity v_{sat}. As an example, the transition frequency $f_T = 1/(2\pi\tau)$ for a gate length of 1.0 μm and a saturation velocity of approximately 10^7 cm/s is 15 GHz.

6.4.4 Limiting Values

The MESFET must be operated in a domain limited by maximum drain current I_{Dmax}, maximum gate-source voltage V_{GSmax}, and maximum drain-source voltage V_{DSmax}. The maximum power P_{max} is dictated by the product of V_{DS} and I_D, or

$$P_{max} = V_{DS}I_D \tag{6.91}$$

which in turn is related to the channel temperature T_C, ambient temperature T_a, and the thermal resistance between channel and soldering point R_{thjs} as follows:

$$T_C = T_a + R_{thjs}P. \tag{6.92}$$

Figure 6-41 clarifies this point. Also shown in this figure are three possible operating points. Bias point 3 indicates low amplification and possible clipping of the output current. However, the power consumption is at a minimum. Bias point 2 reveals acceptable amplification at substantially increased power consumption. Finally, bias point 1 shows high amplification at high power consumption and low output current swing. Choosing appropriate bias points for specific applications will be investigated in depth in subsequent chapters.

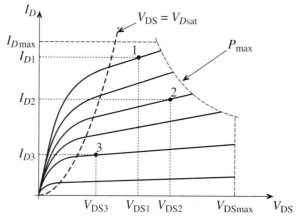

Figure 6-41 Typical maximum output characteristics and three operating points of MESFET.

6.5 Metal Oxide Semiconductor Transistors

Although widely used in digital microelectronic circuits, the MOSFET is also gaining importance in RF analog design, primarily because shrinking feature sizes, in particular gate lengths, have pushed the frequency response well into the GHz range. The metal gate is only of historical interest as modern process technology has replaced it with a highly conductive polysilicon. The thin insulation layer is formed out of silicon dioxide (SiO$_2$) and can range from less than 10 nm to approximately 50 nm. Depending on the bulk material, we have to differentiate between *n*-channel (*p*-type bulk) and *p*-channel (*n*-type bulk) transistors. When both transistor types are monolithically implemented, the process is known as Complementary MOS, or CMOS.

6.5.1 Construction

Because of the fact that electrons possess a much higher mobility, the *n*-channel enhancement transistor is preferred in RF analog circuits. The generic form is depicted in Figure 6-42. The word "enhancement" implies that a positive gate voltage is required to achieve a conduction channel between source and drain. It should be noted that the MOSFET is essentially a four-terminal device, where the bulk substrate constitutes the fourth connection.

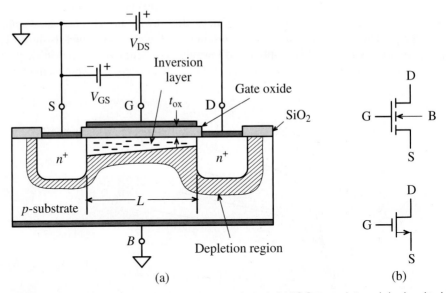

Figure 6-42 Cross-sectional view of an *n*-channel MOS transistor: (a) physical construction, (b) symbols.

As we see in Figure 6-42, the gate is isolated from the bulk semiconductor through a very thin insulating dielectric layer. This already allows the implementation of integrated capacitors as part of the microelectronic circuit. For instance, if the gate is negatively biased, the negative charges on the gate are balanced by positive charges that accumulate below the gate area in the p-type substrate. The bulk substrate becomes an accumulation channel, and its intrinsic capacitance C_{ox} per unit area can be predicted

$$C_{ox} = \frac{\varepsilon}{t_{ox}} = \frac{\varepsilon_0 \varepsilon_{rox}}{t_{ox}} \qquad (6.93)$$

where t_{ox} and ε_{rox} are, respectively, the thickness of the insulating oxide layer and its relative dielectric constant (about 3.9 for SiO_2). The total gate capacitance is then determined by the channel length L and width W

$$C_{GS} = C_{ox} WL. \qquad (6.94)$$

When a moderately positive voltage is applied to the gate, the situation begins to change in that the positive charges in the gate layer repel the positive carriers in the p-doped bulk channel, creating a depletion layer. Further increase in the gate voltage ultimately leads to the attraction of negative charges from the source and drain regions: the channel beneath the gate is inverted. The gate-source voltage V_{GS} at which channel inversion occurs is known as the threshold voltage V_{T0}. In general, channel inversion is a gradual process and is often divided into weak, moderate, and strong inversion. As a rule of thumb, if the difference between the gate-source voltage and threshold voltage, or the effective voltage $V_{eff} = V_{GS} - V_{T0}$, is more than 200 mV, we have strong inversion, a situation preferred by most circuit designers.

6.5.2 Functionality

In the previous section we have maintained the source and drain contacts at an equipotential level of 0 V, which implies that no current flows. We can now investigate two important situations dependent on the magnitude of V_{DS} with respect to V_{GS} (which in turn is assumed to be above V_{T0}).

1. Triode or linear region

Here V_{DS} is moderately positive in the sense that the drain current I_D increases linearly:

$$I_D = \mu_n C_{ox} \frac{W}{L} \left[(V_{GS} - V_{T0}) V_{DS} - \frac{1}{2} V_{DS}^2 \right] \qquad (6.95)$$

The channel between source and drain has a continuous charge carrier layer whose concentration decreases from a maximum at the source contact, where the gate-channel voltage is the highest, to a minimum at the drain contact, where it is the lowest. As V_{DS} increases, the squared term in (6.95) gains in influence, slowing the drain current rise.

2. Saturation region

In this mode of operation, V_{DS} is strongly positive, resulting in the drain current I_D reaching saturation. The current expression now takes on the form

$$I_D = \mu_n C_{ox} \frac{W(V_{GS} - V_{T0})^2}{L \quad 2}. \qquad (6.96)$$

This behavior is recorded due to the fact that the gate-drain voltage V_{GD} begins to drop below the threshold voltage ($V_{GD} < V_{T0}$) and the carrier inversion process is gradually reversed. In particular, beginning from the drain, negative carriers no longer exist: the channel is depleted of carriers, or pinched off. We have already encountered this phenomenon when discussing the MESFET. Again, this implies that the electrons move across the channel at saturation velocity. Interestingly, this saturation region is similar to the forward active region of a BJT; however, unlike the exponential current response, the MOSFET shows a square-law response. Furthermore, and consistent with our MESFET discussion, the decrease in effective channel length leads to channel length modulation

$$I_D = \mu_n C_{ox} \frac{W(V_{GS} - V_{T0})^2}{L\ 2}[1 + \lambda(V_{DS} - V_{eff})] \tag{6.97}$$

where λ is again the channel length modulation parameter.

Implicit in the above MOS treatment is the assumption that the bulk is kept at the same voltage as the source contact. Particularly for CMOS fabrication with high substrate doping, and where the active devices are created in a well process, this does not have to be the case. Known as the body effect, an additional source body voltage V_{SB} can be applied, which affects the threshold voltage V_{T0}.

6.6 High Electron Mobility Transistors

The **high electron mobility transistor** (HEMT), also known as **modulation-doped field effect transistor** (MODFET), exploits the differences in bandgap energy between dissimilar semiconductor materials such as GaAlAs and GaAs in an effort to substantially surpass the upper frequency limit of the MESFET, while maintaining low noise performance and high power rating. At present, transition frequencies of 100 GHz and above have been achieved. The high-frequency behavior is due to a separation of the carrier electrons from their donor sites at the interface between the doped GaAlAs and undoped GaAs layer (**quantum well**), where they are confined to a very narrow (about 10 nm thick) layer in which motion is possible only parallel to the interface. Here, we speak of a **two-dimensional electron gas** (2DEG) or plasma of very high mobility, up to 9000 cm^2/(V·s). This is a major improvement over GaAs MESFETs with $\mu_n \approx 4500$ cm^2/(V·s). Because of the thin layer, the carrier density is often specified in terms of a surface density, typically on the order of 10^{12}–10^{13} cm^{-2}.

To further reduce carrier scattering by impurities, it is customary to insert a spacer layer ranging between 20 and 100 nm of undoped GaAlAs. The layer is grown through a molecular beam epitaxial process and has to be sufficiently thin so as to allow the gate voltage V_{GS} to control the electron plasma through an electrostatic force mechanism. Besides single layer heterostructures (GaAlAs on GaAs), multilayer heterostructures involving several 2DEG channels have also been proposed. As can be expected, manufacturing an HEMT is significantly more expensive when compared with the GaAs MESFET, due to the precisely controlled thin-layer structures, steep doping gradients, and the use of more difficult-to-fabricate semiconductor materials.

6.6.1 Construction

The basic heterostructure is shown in Figure 6-43, where a GaAlAs *n*-doped semiconductor is followed by an undoped GaAlAs spacer layer, an undoped GaAs layer, and a highly resistive semi-insulating (s.i.) GaAs substrate.

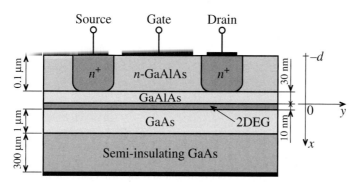

Figure 6-43 Generic heterostructure of a depletion-mode HEMT.

The 2DEG forms in the undoped GaAs layer at zero gate bias condition because the Fermi level is above the conduction band, so that electrons accumulate in this narrow potential well. As discussed later, the electron concentration can be depleted by applying an increasingly negative gate voltage.

HEMTs are primarily constructed of heterostructures with matching lattice constants to avoid mechanical tensions between layers. Specific examples are the GaAlAs-GaAs and InGaAs-InP interfaces. Research is also ongoing with mismatched lattices whereby, for instance, a larger InGaAs lattice is compressed onto a smaller GaAs lattice. Such device configurations are known as **pseudomorphic HEMTs**, or pHEMTs.

6.6.2 Functionality

The key issue that determines the drain current flow in a HEMT is the narrow interface between the GaAlAs and the GaAs layers. For simplicity, we neglect the spacer layer and concentrate our attention at the energy band model shown in Figure 6-44.

A mathematical model similar to (6.21) can be developed by writing down the one-dimensional Poisson equation in the form

$$\frac{d^2 V}{dx^2} = -\frac{qN_D}{\varepsilon_H} \qquad (6.98)$$

> **EVOLUTION OF HEMT**
>
> The first HEMT was reported in 1978 using lattice-matched layers of AlGaAs and GaAs. However, new fabrication techniques have recently resulted in a proliferation of devices, such as InP lattice-matched HEMT, as well as metamorphic HEMT, all geared toward improving the conduction band discontinuity. Due to their low noise, high frequency range, and superior large-signal behavior, HEMTs are well suited for high performance mixer circuits.

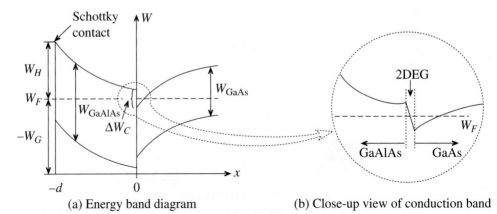

Figure 6-44 Energy band diagram of GaAlAs-GaAs interface for an HEMT.

where N_D and ε_H are the donor concentration and dielectric constant in the GaAlAs heterostructure. The boundary conditions for the potential are imposed such that $V(x=0)=0$ and at the metal-semiconductor side $V(x=-d) = -V_b + V_G + \Delta W_C/q$. Here, V_b is the barrier voltage (see (6.38)), ΔW_C is the energy difference in the conduction levels between the n-doped GaAlAs and GaAs, and V_G is composed of the gate-source voltage as well as the channel voltage drop $V_G = -V_{GS} + V(y)$. To find the potential, (6.98) is integrated twice. At the metal-semiconductor interface, we set

$$V(-d) = \frac{qN_D}{2\varepsilon_H}x^2 - E_y(0)d \qquad (6.99)$$

which yields

$$E(0) = \frac{1}{d}(V_{GS} - V(y) - V_{T0}) \qquad (6.100)$$

where we defined the HEMT threshold voltage V_{T0} as $V_{T0} = V_b - \Delta W_C/q - V_P$. Here, we have used the previously defined pinch-off voltage $V_P = qN_D d^2/(2\varepsilon_H)$, see (6.85). From the known electric field at the interface, we find the electron drain current

$$I_D = \sigma E_y A = -q\mu_n N_D E W d = q\mu_n N_D \left(\frac{dV}{dy}\right) W d. \qquad (6.101)$$

As mentioned previously, the current flow is restricted to a very thin layer so that it is appropriate to carry out the integration over a surface charge density Q_S at $x = 0$. The result is $\sigma = -\mu_n Q/(WLd) = -\mu_n Q_S/d$. For the surface charge density, we find with Gauss's law $Q_S = \varepsilon_H E(0)$. Inserted in (6.101), we obtain

$$\int_0^L I_D dy = \mu_n W \int_0^{V_{DS}} Q_S dV. \qquad (6.102a)$$

Upon using (6.100), it is seen that the drain current can be found

$$I_D L = \mu_n W \int_0^{V_{DS}} \frac{\varepsilon_H}{d}(V_{GS} - V - V_{T0})dV \qquad (6.102b)$$

or

$$I_D = \mu_n \frac{W\varepsilon_H}{Ld}\left[V_{DS}(V_{GS} - V_{T0}) - \frac{V_{DS}^2}{2}\right]. \qquad (6.102c)$$

Pinch-off occurs when the drain-source voltage is equal to or greater than the difference of gate-source and threshold voltages (i.e., $V_{DS} \geq V_{GS} - V_{T0}$). If the equality of this condition is substituted in (6.102c), it is seen that

$$I_D = \mu_n \frac{W\varepsilon_H}{2Ld}(V_{GS} - V_{T0})^2. \qquad (6.103)$$

The threshold voltage allows us to determine if the HEMT is operated as an enhancement or depletion type. For the depletion type, we require $V_{T0} < 0$, or $V_b - (\Delta W_C/q) - V_P < 0$. Substituting the pinch-off voltage $V_P = qN_D d^2/(2\varepsilon_H)$ and solving for d, this implies

$$d > \left[\frac{2\varepsilon_H}{qN_D}\left(V_b - \frac{\Delta W_C}{q}\right)\right]^{1/2} \qquad (6.104)$$

and if d is less than the preceding expression (i.e., $V_{T0} > 0$), we deal with an enhancement HEMT.

──────RF&MW──▶

Example 6-11. Computation of HEMT-related electric characteristics

Determine typical numerical values for a HEMT device such as pinch-off voltage, threshold voltage, and drain current for $V_{GS} = -1, -0.75, -0.5, -0.25,$ and 0 V as a function of drain-source voltage V_{DS}. Assume the following parameters: $N_D = 10^{18}$ cm^{-3}, $V_b = 0.81$ V, $\varepsilon_H = 12.5\varepsilon_0$, $d = 50$ nm, $\Delta W_C = 3.5 \times 10^{-20}$ W·s, $W = 10$ μm, $L = 0.5$ μm, and $\mu_n = 8500$ cm^2/(V·s).

Solution. The pinch-off voltage of a HEMT is evaluated as

$$V_P = qN_D d^2/(2\varepsilon_H) = 1.81 \text{ V}.$$

Knowing V_P we can find the threshold voltage as

$$V_{T0} = V_b - \Delta W_C/q - V_P = -1.22 \text{ V}.$$

Using these values, the drain current is computed by relying either on equation (6.102c) for $V_{DS} \leq V_{GS} - V_{T0}$ or equation (6.103) for $V_{DS} \geq V_{GS} - V_{T0}$. The results of these computations are plotted in Figure 6-45. We notice in this graph that unlike the GaAs MESFET in Figure 6-43, channel length modulation is not taken into account. In practical simulations, such a heuristic adjustment can be added.

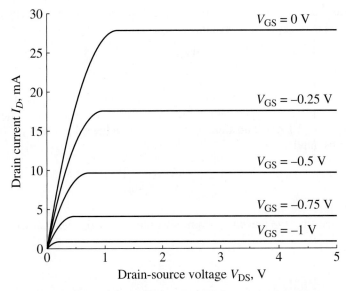

Figure 6-45 Drain current in a GaAs HEMT.

Both GaAs MESFET and HEMT exhibit similar output characteristics, and are thus represented by the same electric circuit model.

6.6.3 Frequency Response

The high-frequency performance of the HEMT is determined by the transit time similar to the MESFET. However, the transit time τ is expressed best through the electron mobility μ_n and the electric field E of the drain-source voltage according to

$$\tau = \frac{L}{\mu_n E_y} = \frac{L^2}{\mu_n V_{DS}}. \tag{6.105}$$

We therefore obtain a transition frequency $f_T = 1/(2\pi\tau)$ of approximately 190 GHz for the gate length of 1.0 μm and a mobility of $\mu_n = 8000 \text{ cm}^2/(\text{V·s})$ at a typical drain voltage V_{DS} of 1.5 V.

6.7 Semiconductor Technology Trends

Recent years have experienced explosive growth in the RF/microwave semiconductor industry owing to the proliferation of a host of applications. The major shares have been in the consumer markets of cellular communications and wireless networking, followed by satellite communications and cable television. For these applications, the operating frequencies typically do not exceed the relatively moderate limit of approximately 6 GHz. Interestingly, at these frequencies we no longer require special processing steps; today's standard digital semiconductor chip fabrication is sufficient. Consequently, the microelectronic scaling benefits of cost, size, and power can be fully utilized in an identical way as has been the case in digital circuits for the past 40 years.

The use of silicon technology in microwave circuits is relatively new. This particularly applies to CMOS (Complementary Metal Oxide Semiconductor), which is ubiquitous in digital integrated circuits. Due to aggressive gate length scaling, MOSFETs have recently attained adequate RF performance. MOSFET transition frequencies (f_T) and maximum oscillation frequencies (f_{max}) above 200 GHz have recently been reported, with more typical values in the range of 50 to 100 GHz. The f_{max} performance of MOSFETs has generally lagged behind f_T due to their high gate resistance. Through continuing process technology enhancements, the noise and power performance of MOSFETs is constantly improving, although they still have not caught up with other transistor technologies in use today. The advantage of silicon as the fundamental process technology is apparent: extremely low cost and the possibility of integrating entire RF/mixed signal circuits with digital logic on a single chip (a system-on-chip, SoC). Single-chip Bluetooth devices are already available and similar integration is likely to be achieved in cellular telephones and wireless networking (IEEE 802.11) in the near future. The frequency performance of modern scaled MOS devices can be approximated by the equation

FINFET

FETs are arranged as planar structures consisting of a source, a drain, and a channel with gate controlling the current flow. Shrinking the channel is limited to 10s of nm before source-drain leakage effects preclude further reduction. New devices attemt to exploit the vertical dimension where a thin, fin-like, layer establishes the channel between the source and the drain. The device is aptly called a FinFET.

$$2\pi f_T = \frac{v_{\text{eff}}}{L_G} \qquad (6.106)$$

where L_G is the gate length and v_{eff} is the effective saturation electron velocity ($v_{\text{eff}} \approx 5 \times 10^6 \text{cm/s}$ in Si). For instance, a device with gate length of 1 μm will result in a f_T of roughly 8 GHz, whereas today's process node sizes of 130 nm and below can easily achieve 80 GHz and more. Unfortunately, low bulk resistivity (~1–100 kΩ cm) and low breakdown electric field strength (~300 kV/cm) remain liabilities when it comes to low-noise and high-power applications.

In addition to MOSFETs, a high-performance bipolar transistor, the Silicon-Germanium (SiGe) HBT, has been available in silicon technology since the late 1980s. This type of transistor has generally outperformed the MOSFET in RF applications, but now displays nearly the same transition frequencies (100-200 GHz). The SiGe HBT is preferred when its higher power performance and higher reliability justify the higher cost. In recent years, the overall performance of SiGe HBTs have been improving to the point where they have begun displacing the more traditional microwave semiconductor technologies based on Gallium-Arsenide (GaAs) and Indium-Phosphide (InP) for applications up to approximately 40 GHz. SiGe HBTs can be used in a combined bipolar-CMOS process known as BiCMOS, which permits cost-effective integration of digital and RF circuits on a single chip. SiGe HBTs are most advantageous in power amplifiers and oscillators, while MOSFETs have superior noise characteristics and are therefore ideally suitable for low-noise amplifiers (LNAs).

The classical high-frequency (>10 GHz) and high-performance microwave transistor markets are still dominated by GaAs and InP semiconductor technologies. These compound semiconductors exhibit dramatically higher electron mobility than silicon (5000 cm^2/Vs and 3000 cm^2/Vs, respectively, versus 1900 cm^2/Vs for Si). Until recently, even the simple GaAs MESFET outperformed the best Si-based transistors. MESFETs (which dominated the microwave transistor market up to the 1990s) still find application in lower-cost microwave circuits, but the majority of recent developments have been in heterostructure devices, which include HEMTs and HBTs on both GaAs and InP substrates. State-of-the-art transistors with transition frequencies near 1 THz have been reported, with more common higher end commercial devices displaying f_T and f_{max} of 200 to 400 GHz. Of the two semiconductor substrates, GaAs is currently dominant, primarily due to its maturity and lower cost. InP is superior to GaAs in terms of overall performance; it reaches higher frequencies, but fabrication is less mature and tends to be expensive. InP currently finds applications in lasers, optical devices, and microwave circuits operating at the upper end of the spectrum, mainly above 100 GHz. However, it is expected to advance into other areas as fabrication technology continues to mature. Integrated circuits are common in GaAs and InP; nonetheless, they cannot achieve even a fraction of the density and scale of integration seen in modern Si-based chips.

Of the many experimental semiconductor technologies, the most promising appears to be based on wide bandgap semiconductors; it is intended for power amplifier applications. The most common wide bandgap materials are SiC (silicon carbide) and GaN (gallium nitride). SiC (4H or 6H polytype) MESFETs are emerging as the power amplifiers of choice at lower frequencies up to 12 GHz. The favorable characteristics of SiC are high electric field breakdown (>2000 kV/cm for SiC and >5000 kV/cm for GaN), excellent thermal conductivity, high temperature tolerance, and low dielectric constant. The good breakdown and thermal characteristics allow the fabrication of smaller device structures with higher bias voltages. The small device structure combined with a low dielectric constant leads to higher input and output impedances, thereby greatly simplifying impedance matching (which is a major problem in solid-state power amplifiers). Additional advantages are lower cost due to smaller die area and reduced heat sink requirements due to high temperature operation. SiC is comparable to Si in electron transport characteristics, achieving a f_{max} of up to 60 GHz. Both 4H and 6H SiC wafers are available (although high crystal defect density remains a major problem) and several SiC MESFETs have already entered the RF market.

The GaN HEMT (GaAlN/GaN heterostructure) is another promising wide bandgap technology device with electron mobility several times higher than that of Si, bulk GaN, or SiC. Moreover, GaN is highly temperature tolerant and possesses an extremely high resistivity (>10^{10} kΩ cm). Consequently, GaN HEMTs are expected to become the technology of choice for power amplifiers at frequencies up to 40 GHz and beyond, as fabrication technology matures (f_{max} of up to 120 GHz have been reported). However, one disadvantage is a lack of bulk GaN wafer substrates; most GaN devices are manufactured on either sapphire or SiC substrates. Sapphire is known to have poor thermal conductivity (a potentially significant disadvantage in power amplifiers), and both sapphire and SiC substrates present substantial lattice mismatch, adversely affecting reliability. Another shortcoming of SiC and GaN is the extremely poor hole mobility (for instance, GaN: $\mu_p \approx 40 \text{ cm}^2/(\text{Vs})$ at $N_D = 10^{16} \text{ cm}^{-3}$ and T = 300 K). This fact limits circuit designers to n-channel FETs, and precludes the use of bipolar or p-channel devices. In general, wide bandgap technologies still have a number of fabrication hurdles to overcome before they can gain market acceptance.

Practically Speaking. The internal structure of an RF power transistor

In this section, we take a look at the construction of a power transistor. This particular device is designed by NXP and carries the part number BLF4G22-100. It is a silicon LDMOS (Laterally Diffused Metal Oxide Semiconductor) power FET designed for W-CDMA (Wideband Code Division Multiple Access) cellular base station applications in the frequency range of 2.0–2.2 GHz. The device is rated for 24 W average output power and 150 W peak power. It operates at a nominal drain-source voltage of 28 V and delivers up to 13.5 dB of gain.

Figure 6-46 depicts the internal construction of the BLF4G22-100 in three levels of magnification under an optical microscope. Figure 6-46(a) shows the entire device which is approximately a 20 mm by 10 mm rectangle with wide copper strip-like leads on the broad sides. The strips leads are for the gate and drain of the device, which are also the RF input and output ports. The source is attached to the metal base, which is mounted on a grounded heat sink in a typical circuit. The transistor

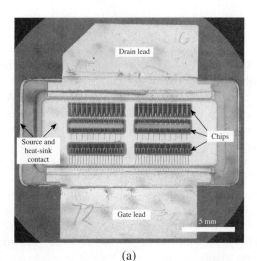

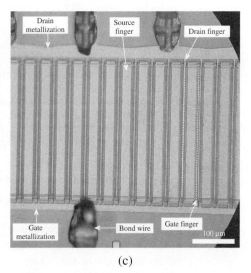

Figure 6-46 The internal structure of the LDMOS power transistor BLF4G22-100 at three levels of magnification: (a) overall package, (b) die interconnections, and (c) transistor die (courtesy of NXP).

needs a substantial heat sink to achieve its rated output power since its efficiency is quoted as 26% in W-CDMA applications, implying that about 3 times more power, or approximately 68 W, is dissipated as heat when outputting the maximum 24 W of RF power.

The metal base is designed to be directly soldered to the heat sink in order to minimize the package to heat sink thermal resistance and to provide a low-impedance connection to ground.

Figure 6-46(b) is a magnified view that highlights the interconnection of the silicon components. As one can see, the device consists of two identical transistors connected in parallel. The transistor die is a slender strip of silicon with an array-like metallization structure. We observe that other silicon devices are present besides the actual transistor. These are simply capacitors, which together with the bond wires acting as inductors, form matching-assist circuits. They do not match the transistor to the typical characteristic impedance of 50 Ω on the input or output, but only increase the device impedance slightly. This is helpful because the device is a parallel combination of many smaller transistors that results in very low input and output impedances, in general a few ohms or less. The *LC* matching-assist circuits lower the input and output reflection coefficients of the device, thereby greatly simplifying the task of building external matching networks (see Chapter 8). A possible drawback of this approach is that the simple *LC* matching circuits tend to be narrow-band, locking the transistor into a certain frequency range.

In Figure 6-46(c) we can examine the metallization pattern on the transistor die. The transistor is a finger structure repeating the source-gate-drain-gate-source pattern. Of these, the gate is the narrowest (0.6 µm wide according to the specifications) and barely identifiable even under such high magnification. The drain fingers are wider and are clearly visible in the image. The source fingers are the largest, to achieve a very low impedance contact to ground. Also visible in the image are the bond wire contact points, which appear quite large when compared to the other features on the die.

Figure 6-47 shows the power transistor installed on a circuit board containing biasing and matching circuits. The bias is supplied through a $\lambda/4$ microstrip line terminated by bypass capacitors on the DC side. The matching networks are constructed from microstrips of varying width and length ratios. The widest section of microstrip adjacent to the device has very low characteristic impedance; this is necessary to transform the even-lower device impedance to a reasonable level. Several free-floating copper patches are supplied around the matching circuit. They can be selectively connected to the microstrips to adjust the matching. Although not visible in the image, the copper base plate is attached to a water-cooling system.

Figure 6-47 The power transistor mounted on a demonstration board.

6.8 Summary

To understand the functionality and limitations of the most widely employed active RF solid-state devices, we commenced this chapter with a review of the key elements of semiconductor physics. The concepts of conduction, valence, and Fermi levels as part of the energy band model are used as the starting point to examine the various solid-state mechanisms.

We next turned our attention to the *pn*-junction, where we derived the barrier voltage

$$V_{\text{diff}} = V_T \ln\left(\frac{N_A N_D}{n_i^2}\right)$$

and the depletion and diffusion capacitances C_d and C_S in the forms

$$C_d = \frac{C_0}{\sqrt{1 - V_A/V_{\text{diff}}}} \quad \text{and} \quad C_S = \frac{\tau I_0}{V_T} e^{V_A/V_T}.$$

Both capacitances are of primary importance when dealing with the frequency response of a *pn*-diode, whose current is given by the Shockley equation

$$I = I_S\left(e^{V_A/V_T} - 1\right).$$

This equation underscores the nonlinear current-voltage diode characteristics.

Unlike the *pn*-junction, the Schottky contact involves an *n*-type semiconductor and a metal interface. The Schottky barrier potential V_d is now modified and requires the work

Summary

function of metal qV_M, semiconductor $q\chi$, and the conduction band potential V_C, expressed via

$$V_d = (V_M - \chi) - V_C.$$

Unlike the 0.7 V of a pn-junction, we obtain a typical value of 0.84 V for a Si-Au interface. Technologically, this contact is exploited in the Schottky diode, which has become ubiquitous in many RF applications such as modulators and mixers. The I-V characteristic remains the same as for the pn-junction diode, except that the reverse saturation current I_S is theoretically more intricate.

Additional special-purpose RF diodes are the PIN, varactor, and tunnel diodes. The PIN diode incorporates an additional intrinsic layer sandwiched between the p and n layers. This allows the switching between a low-resistance forward bias to a capacitive reverse, or isolation, bias. PIN diodes find applications in switchers and attenuators. The RF resistance of a forward biased PIN diode is approximately

$$R_J(I_Q) = \frac{W^2}{(\mu_n + \mu_p)\tau I_Q}.$$

The varactor diode incorporates the I-layer based on a special doping profile to achieve a particular capacitance-voltage behavior. Such a response is beneficial for frequency tuning and the generation of short pulses. The tunnel diode exhibits a negative slope during a particular portion of its I-V curve, thus making it suitable for oscillator circuits. Additional diodes of interest in the RF field are the IMPATT, TRAPATT, BARRITT, and Gunn diodes.

The BJT in many ways can be regarded as an extension of our previous diode discussion, since the npn structure constitutes a series connection of two diodes. The three transistor modes forward active, reverse active, and saturation are reflected in the emitter, collector, and base current expressions (6.69)–(6.71):

$$I_E = -I_S(e^{V_{BE}/V_T} - 1) - \frac{I_S}{\beta_F}(e^{V_{BE}/V_T} - 1) + I_S(e^{V_{BC}/V_T} - 1)$$

$$I_C = I_S(e^{V_{BE}/V_T} - 1) - \frac{I_S}{\beta_R}(e^{V_{BC}/V_T} - 1) - I_S(e^{V_{BC}/V_T} - 1)$$

$$I_B = I_S\left\{\frac{1}{\beta_R}(e^{V_{BC}/V_T} - 1) + \frac{1}{\beta_F}(e^{V_{BE}/V_T} - 1)\right\}.$$

The frequency response of a BJT is determined by the transition frequency $f_T = 1/(2\pi\tau)$ at which the short-circuit current gain equals unity. The time constant is composed of three delays $\tau = \tau_E + \tau_B + \tau_C$ associated with emitter, base, and collector domains.

Unlike the bipolar BJT, the FET is a monopolar device that displays superior high-frequency and low-noise performance. In particular, *n*-channel GaAs MESFETs are commonly found in many RF amplifiers, mixers, and oscillators. The key equation that determines the output characteristic of a MESFET is the drain current (6.83):

$$I_D = G_0 \left(V_{DS} - \frac{2}{3} \sqrt{\frac{2\varepsilon}{q N_D d^2}} [(V_{DS} + V_d - V_{GS})^{3/2} - (V_d - V_{GS})^{3/2}] \right).$$

Additional modifications to the drain current are required when the channel is pinched off and the FET is operated in the saturation domain with channel length modulation.

Finally, the HEMT device is almost identical in construction with the MESFET, but exploits the differences in bandgap energies between heterogeneous semiconductors. Here, the current flow is restricted to a very narrow quantum well layer, where the carrier mobility can attain twice the value of a MESFET. Because of carrier separation from the donor sites, extremely high operating frequencies have been reported (exceeding 100 GHz). The drain current representation is almost identical to the one discussed for the MESFET.

Further Reading

L. E. Larson, "Silicon technology tradeoffs for radio frequency/mixed-signal 'systems-on-a-chip,'" *IEEE Transactions on Electron Devices,* Vol. 50, pp. 683–699, 2003.

D. L. Harme, et al. "Design automation methodology and RF/analog modeling for RF CMOS and SiGe BiCMOS technologies," *IBM J. Res. and Dev.,* Vol. 47, No. 2/3, pp. 139–175, 2003.

D. Ueda, "Review of compound semiconductor devices for RF power applications," *IEEE Proceedings of the 14th International Symposium on Power Semiconductor Devices and ICs,* pp. 17–24, June 2002.

C. Nguyen and M. Micovic, "The state-of-the-art of GaAs and InP power devices and amplifiers," *IEEE Transactions on Electron Devices,* Vol. 48, pp. 472–478, 2001.

R. J. Trew, "Wide bandgap semiconductor transistors for microwave power amplifiers," *IEEE Microwave Magazine,* Vol. 1, pp. 46–54, 2000.

F. Schwierz, "Microwave Transistors–the last 20 years," *IEEE Proceedings of the 2000 International Conference on Devices, Circuits and Systems,* D28/1–D28/7, 2000.

J. J. Liou, "Semiconductor device physics and modelling. 1. Overview of fundamental theories and equations," *IEE Proceedings G,* Vol. 139, pp. 646–654, 1992.

R. Dingle et al., "Electron mobilities in modulation doped semiconductor heterojunction superlattices," *Appl. Phys. Lett.,* 33, pp. 665–667, 1978.

R. S. Cobbold, *Theory and Applications of Field-Effect Transistors,* John Wiley, New York, 1970.

A. M. Cowley and S. M. Sze, "Surface States and Barrier Height of Semiconductor Systems," *J. Appl. Physics,* Vol. 36, pp. 3212–3220, 1965.

M. B. Das, "Millimeter-Wave Performance of Ultra-Submicrometer Gate Field-Effect Transistors. A Comparison of MODFET, MESFET, and HBT-Structures," *IEEE Trans. on Electron Devices,* Vol. 34, pp. 1429–1440, 1987.

A. S. Grove, *Physics and Technology of Semiconductor Devices,* John Wiley, New York, 1967.

G. Massobrio and P. Antognetti, *Semiconductor Device Modeling with SPICE,* McGraw-Hill, New York, 1993.

J. L. Moll, *Physics of Semiconductors,* McGraw-Hill, New York, 1964.

D. V. Morgan and N. Parkman, *Physics and Technology of Heterojunction Devices,* P. Peregrinus Ltd., London, UK, 1991.

M. H. Norwood and E. Schatz, "Voltage Variable Capacitor Tuning—A Review," *Proceed. IEEE,* Vol. 56, pp. 788–798, 1968.

R.S. Pengelly, *Microwave Field-effect Transistors—Theory, Design and Applications,* Research Studies Press, London, UK, 1982.

C. T. Sah, "Characteristics of the Metal-Oxide Semiconductor Field-Effect Transistor," *IEEE Trans. on Electron Devices,* Vol. 11, pp. 324–345, 1964.

W. Shockley, *Electrons and Holes in Semiconductors,* Van Nostrand Reinhold, New York, 1950.

M. Shur, *GaAs Devices and Circuits,* Plenum Press, New York, 1987.

S. M. Sze, *Physics of Semiconductor Devices,* John Wiley, New York, 1981.

C. Weisbuch. *Physics and Fabrication of Microstructures and Microdevices,* Springer-Verlag, New York, 1986.

Problems

6.1 To appreciate the large number of atoms in semiconductors, let us consider the following simple calculation: silicon is a diamond structure cubic semiconductor (8 corner, 6 face, and 4 internal atoms) with a lattice constant of $a = 5.43 \times 10^{-8}$ cm. The atom arrangement is such that a corner atom contributes one-eighth and the face atom one-half. Find the density of atoms per cubic centimeter.

6.2 The conduction and valence band carrier concentration is determined by integration of the density of states based on the Fermi statistics:

$$N = \int g(E)dE.$$

For effective electron mass m_n^*, quantum mechanical considerations lead to the density function

$$g(E) = 4\pi(2m_n^*)^{3/2}\sqrt{E}/h^3.$$

(a) Determine the generic electron concentration of states N for energy values up to 1.5 eV.
(b) For an effective electron mass of $1.08 m_n$ or $1.08 \times 9.11 \times 10^{-31}$ kg, explicitly find the number of states.

6.3 Let us consider a p-type Si semiconductor, whose doping concentration at room temperature contains $N_A = 5 \times 10^{16}$ boron atoms per cubic centimeter ($n_i = 1.5 \times 10^{10}$ cm^{-3}). Find the minority and majority carrier concentrations as well as the conductivity of the semiconductor.

6.4 The Fermi-Dirac probability for indistinguishable particles is the underlying statistical theory describing the quantum mechanical distribution of particles per unit volume and per unit energy $N(E)$ normalized with respect to the number of quantum states per unit voltage and per unit energy $g(E)$ according to

$$f(E) = \frac{N(E)}{g(E)} = \frac{1}{1 + \exp((E - E_F)/(kT))}.$$

(a) Plot both the probability of states being occupied, $f(E)$, and the probability of states being empty, $1 - f(E)$, at room temperature and for $E_F = 5$ eV.
(b) Determine the temperature at which we have a 5% probability of encountering an empty state.

6.5 The intrinsic carrier concentration is typically recorded at room temperature. For GaAs, we find at $T = 300$ K the effective densities of state $N_C = 4.7 \times 10^{17}$ cm^{-3}, $N_V = 7.0 \times 10^{18}$ cm^{-3}. Assuming that the bandgap energy of 1.42 eV remains constant,
(a) find the intrinsic carrier concentration at room temperature,
(b) compute n_i at $T = 400$ K, and
(c) compute n_i at $T = 450$ K.

6.6 It is interesting to observe that a significant diffusion current density can be created even for moderate carrier concentration gradients. We can assume for a p-type Si semiconductor a linear hole concentration changing from 5×10^{17} cm^{-3} to 10^{18} cm^{-3} over a distance of 100 μm. Find the current density if the diffusion coefficient is given at $T = 300$ K to be $D_p = 12.4$ cm^2/s.

6.7 In Section 6.1.2 we derived the expressions for the electric field and potential distributions in the pn diode with an abrupt junction. Repeat these computations for the case of a gradual junction where the charge density changes linearly according to the following relation:

$$\rho(x) = \begin{cases} qN_A(x/d_p), -d_p \leq x \leq 0 \\ qN_D(x/d_n), 0 \leq x \leq d_n. \end{cases}$$

6.8 The built-in potential barrier of a *pn*-junction remains relatively constant even though the doping concentration may change over several orders of magnitude. We recall that the typical barrier potential in solid state circuits is assumed to be 0.5–0.9 V. In this problem we intend to show how one arrives at this voltage. Let us assume a *p*-type semiconductor with $N_A = 10^{18}$ cm^{-3} joined with an *n*-type semiconductor of concentration $N_D = 5 \times 10^{15}$ cm^{-3}.
(a) Find the barrier voltage at room temperature ($n_i = 1.45 \times 10^{10}$ cm^{-3}).
(b) Recompute the barrier voltage if N_A is reduced to $N_A = 5 \times 10^{16}$ cm^{-3}.

6.9 An abrupt *pn*-junction made of Si has the acceptor and donor concentrations of $N_A = 10^{18}$ cm^{-3} and $N_D = 5 \times 10^{15}$ cm^{-3}, respectively. Assuming that the device operates at room temperature, determine
(a) the barrier voltage,
(b) the space charge width in the *p*- and *n*-type semiconductors,
(c) the peak electric field across the junction, and
(d) the junction capacitance for a cross sectional area of 10^{-4} cm^2 and a relative dielectric constant of $\varepsilon_r = 11.7$.

6.10 For two *pn* diodes with abrupt junctions, one of which is made of Si and another is made of GaAs, with $N_A = 10^{17}$ cm^{-3} and $N_D = 2 \times 10^{14}$ cm^{-3} in both cases:
(a) find the barrier voltage,
(b) find the maximum electric field and the space charge region width, and
(c) plot the space charge, potential, and electric field distribution along the diode axis.

6.11 A silicon *pn*-junction has a conductivity of 10 S/cm and 4 S/cm for *p* and *n* layers, respectively. Using the necessary properties of silicon, calculate the built-in voltage of the junction at room temperature.

6.12 A Schottky contact between a metal and a semiconductor can be made of various materials. For both Si and GaAs, we would like to investigate the barrier voltage if the metal is either aluminum or gold. Use Table 6-2 and Table E-1 to find the four barrier voltages and associated depletion layer thicknesses at room temperature. Assume $N_D = 10^{16}$ cm^{-3}.

6.13 Consider a Schottky diode formed by the contact between *n*-type GaAs and silver. The diode is operated at a forward biasing current of 1 mA. The Richardson constant is $R^* = 4$ A/(cm^{2}K^2), the parasitic series resistance is 15 Ω, and the device cross section is $A = 10^{-2}$ mm^2. Compute the barrier voltage V_d and plot the magnitude and phase of the diode impedance versus frequency ranging from 1 MHz to 100 GHz for two doping densities N_D of 10^{15} and 10^{17} cm^{-3}. Assume that the device is operated at the temperature of 300 K.

6.14 It is often of significant practical interest to investigate the nonlinear current behavior of a Schottky diode for a given applied voltage. We recall

$$I = I_S\left(e^{(V_A - IR_S)/V_T} - 1\right)$$

with the reverse saturation current given as $I_S = 2 \times 10^{-11}$ A. For a substrate resistance $R_S = 1.8 \, \Omega$ write a computer program to predict the current if the applied voltage is allowed to vary within $0 \leq V_A \leq 1$ V.

6.15 A PIN diode is a semiconductor device with a lightly doped layer sandwiched between two highly doped n- and p-type materials. In the intrinsic layer, the minority and majority carriers possess a finite life time τ_p before recombination takes place. On the basis of the recombination lifetime, a simple PIN model can be constructed involving the diode current I and the stored charge Q:

$$I = \frac{Q}{\tau_p} + \frac{dQ}{dt}.$$

(a) Establish the frequency domain response $Q(\omega)/I$ of this first-order system.
(b) Plot the normalized charge response $20 \log[Q(\omega)/(I\tau_p)]$ versus angular frequency for τ_p of 10 ps, 1 ns, and 1 µs.

Note: For frequencies well below the cutoff frequency $f_p = 1/\tau_p$, the PIN diode behaves like a normal pn-junction diode. However, at frequencies above f_p, the PIN diode becomes a pure linear resistor whose value is controlled by the biasing current.

6.16 The fabrication of two different types of varactor diodes calls for the following two capacitance-voltage behaviors:
(a) $C = 5 \text{ pF} \sqrt{V_A/(V_A - V_{\text{diff}})}$
(b) $C = 5 \text{ pF}(V_A/(V_A - V_{\text{diff}}))^{1/3}$.

Determine the necessary donor doping profile $N_D(x)$ for the intrinsic layer. Assume the cross-sectional area of the varactor diode is 10^{-4} cm^2.

6.17 Consider a Si bipolar-junction transistor, whose emitter, base, and collector are uniformly doped with the following concentrations: $N_D^E = 10^{21}$ cm^{-3}, $N_A^B = 2 \times 10^{17}$ cm^{-3}, $N_D^C = 10^{19}$ cm^{-3}. Assume that the base-emitter voltage is 0.75 V and the collector-emitter potential is set to 2 V. The cross-sectional area of both junctions is 10^{-4} cm^2 and the emitter, base, and collector thicknesses are $d_E = 0.8$ µm, $d_B = 1.2$ µm, and $d_C = 2$ µm, respectively. Assuming that the device is operated at room temperature,
(a) Find the space charge region extents for both junctions,
(b) Compute the base, emitter, and collector currents, and
(c) Calculate the forward and reverse current gains β_F and β_R.

6.18 For a GaAs BJT, the maximum junction temperature is 420°C (which far exceeds the maximum junction temperature of Si, 200°C). The supplied power is 90 W. The thermal resistance between the BJT and the heat sink is estimated to be 1.5°C/W.

(a) Determine the maximum thermal resistance of the heat sink, if the ambient operating temperature does not exceed 50°C.

(b) For a heat convection coefficient of 100 W/°C · m², find the required surface area.

6.19 A BJT is encapsulated in a plastic housing and mounted on a heat sink ($R_{thha} = 3.75°C/W$). Under these conditions, the total power dissipation is supposed to be 20 W at an ambient temperature of 20°C. What rating has the engineer to choose for the BJT casing if the maximum junction temperature should not exceed 175°C?

6.20 Prove that the drain current equation (6.83) for a MESFET under the gradual-channel approximation reduces to (6.79) for small V_{DS}.

6.21 Derive the saturation drain current equation (6.86).

6.22 A MESFET with an *n*-type channel has the following parameters: $W/L = 10$, $\mu_n = 1000$ m²/(V·s), $d = 2$ μm, $\varepsilon_r = 11.7$, and $V_{T0} = -3$ V. Compute the saturation drain current at $V_{GS} = -1$ V.

6.23 Compute the output current I_D versus V_{DS} characteristics of the transistor from Problem 6.22 for drain-source voltage ranging from 0 to 5 V. First, assume that channel length modulation effect is negligible (i.e., $\lambda = 0$), and then repeat your computation for the case when $\lambda = 0.01$.

7

Active RF Component Modeling

7.1 Diode Models...362
    7.1.1 Nonlinear Diode Model...362
    7.1.2 Linear Diode Model..364
7.2 Transistor Models..367
    7.2.1 Large-Signal BJT Models...367
    7.2.2 Small-Signal BJT Models...376
    7.2.3 Large-Signal FET Models...388
    7.2.4 Small-Signal FET Models...391
    7.2.5 Transistor Amplifier Topologies......................................395
7.3 Measurement of Active Devices..397
    7.3.1 DC Characterization of Bipolar Transistor............................397
    7.3.2 Measurements of AC Parameters of Bipolar Transistors.................398
    7.3.3 Measurements of Field Effect Transistor Parameters...................403
7.4 Scattering Parameter Device Characterization.................................404
7.5 Summary..413

Almost all circuits of any complexity have to be modeled as part of computer aided design (CAD) programs prior to their practical realizations to assess quantitatively whether or not they meet design specifications. For the purpose of electric circuit simulation, a large number of software analysis packages offer a host of equivalent circuit models attempting to replicate the electric performance of the various discrete elements. Special electric circuit models have been developed to address such important design requirements as low- or high-frequency operation, linear or nonlinear system behavior, and normal or reverse mode of operation to name but a few.

    It is the purpose of this chapter to examine several equivalent circuit representations for diodes as well as mono- and bipolar transistors. The physical foundation of these devices is reviewed in Chapter 6. By developing a close link with the previous chapter, we will be able to observe how a basic understanding of solid-state device physics naturally leads to large-signal (nonlinear) circuit models. Subsequent discussions will focus on modifications that can be made to linearize these models and to refine them for high-frequency operation.

Considering the various BJT models, we restrict our discussion to only the most popular types such as the Ebers-Moll and Gummel-Poon models. Both types, and a number of linear derivatives, find widespread applications in such simulation tools as SPICE, ADS, MMICAD, and others. Often, the situation arises where the device manufacturer may not be able to specify all the required electric parameters, since they can easily exceed 40 independent parameters, and a so-called SPICE model representation is unattainable. Under those circumstances, the S-parameters are recorded for various bias conditions and operating frequencies to characterize the high-frequency behavior. In most cases, these S-parameters may provide the design engineer with sufficient information to complete the simulation task.

7.1 Diode Models

7.1.1 Nonlinear Diode Model

The typical large-scale circuit model treats both the *pn* and Schottky diode in the same fashion, as shown in Figure 7-1.

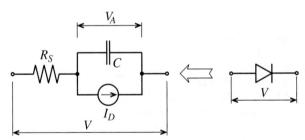

Figure 7-1 Large-scale diode model.

This model takes into account the nonlinear *I-V* characteristics of the Shockley diode equation (6.34) in slightly modified form

$$I_D = I_S(e^{V_A/(nV_T)} - 1) \tag{7.1}$$

where the **emission coefficient** n is chosen as an additional parameter aimed at bringing the model in closer agreement with actual measurements. This coefficient for most applications is close to 1.0. Furthermore, in Section 6.1.2 the diffusion and junction (or depletion layer) capacitances C_d and C_J are discussed. Both effects are combined in a single capacitance C, but in a more general form. Specifically, for the junction capacitance, we have to consider the space charge Q_J, which is differentiated with respect to the applied voltage, leading to

$$C_J = \frac{dQ_J}{dV_A} = \frac{C_{J0}}{(1 - V_A/V_{\text{diff}})^m} \tag{7.2}$$

where m is known as the **junction grading coefficient**. It assumes a value of 0.5 for the abrupt junction that is subject of our analysis in Section 6.1.2. For the more realistic case of a gradual transition, m lies in the range $0.2 \leq m \leq 0.5$. As mentioned in Chapter 6, the formula given in (7.2) is applicable only for certain positive applied voltages. If the applied voltage V_A approaches the built-in potential V_{diff}, the junction capacitance computed using (7.2) approaches infinity, which is obviously physically impossible. In practice, the junction capacitance becomes almost linearly dependent on the applied voltage once it exceeds a threshold potential V_m, which is usually equal to half of the built-in potential, $V_m \approx 0.5 V_{\text{diff}}$. Therefore, the approximate formula capable of describing the junction capacitance over the entire range of applied voltages is given by

$$C_J = \begin{cases} \dfrac{C_{J0}}{(1 - V_A/V_{\text{diff}})^m}, & V_A \leq V_m \\ \dfrac{C_{J0}}{(1 - V_m/V_{\text{diff}})^m}\left(1 + m\dfrac{V_A - V_m}{V_{\text{diff}} - V_m}\right), & V_A \geq V_m. \end{cases} \tag{7.3}$$

We also observe that the diffusion capacitance C_d is dependent on V_A:

$$C_d = \frac{dQ_d}{dV_A} = \frac{I_S \tau_T}{nV_T} e^{V_A/(nV_T)} \approx \frac{I_D \tau_T}{nV_T} \tag{7.4}$$

with the **transit time** τ_T.

In a realistic diode, the injection and extraction of charges is accomplished by the electric field that constitutes a voltage drop in the neutral-charge domains. This voltage drop is modeled as a series resistance R_S. Thus, the total voltage in Figure 7-1 is composed of two contributions:

$$V = R_S I_D + nV_T \ln(1 + I_D/I_S). \tag{7.5}$$

Temperature dependencies can also be introduced into this model. Besides the obvious thermal voltage, $V_T = kT/q$, the reverse saturation current I_S is found to be strongly influenced by temperature according to

$$I_S(T) = I_S(T_0)\left(\frac{T}{T_0}\right)^{p_t/n} \exp\left[-\frac{W_g(T)}{V_T}\left(1 - \frac{T}{T_0}\right)\right] \tag{7.6}$$

where T_0 is a reference temperature at which the saturation current is recorded. The literature primarily uses $T_0 = 300$ K (or 27°C). The **reverse saturation current temperature coefficient** p_t is either 3 or 2 depending on whether a pn or Schottky diode is modeled. The model parameter can thus account for the difference in temperature behavior between the two diode types. Also, the **bandgap energy** $W_g(T)$ is considered. As the temperature increases, this bandgap decreases, making it easier for charge carriers to transition from the valence into the conduction band. The semi-empirical formula assumes a specific

SPICE

The word SPICE stands for Simulation Program with Integrated Circuit Emphasis. The program was originally developed at Berkley to simulate interconnected (via a so-called netlist) passive and active circuit elements. It is capable of conducting nonlinear DC and transient, as well as linear AC analyses. SPICE became the starting point for many derivatives (PSPICE, HSPICE) and extensions, including Advanced Design System (ADS).

bandgap energy $W_g(0)$ recorded at $T = 0$ K and then adjusts this value as follows

$$W_g(T) = W_g(0) - \frac{\alpha_T T^2}{\beta_T + T}. \quad (7.7)$$

For instance, the experimentally determined parameters for Si are $W_g(0) = 1.16$ eV, $\alpha_T = 7.02 \times 10^{-4}$ eV/K and $\beta_T = 1108$ K. Additional temperature dependencies affecting the capacitances are usually small and are neglected.

Perhaps the most popular circuit simulation program in industry and academia is **SPICE**, which is capable of taking into account the nonlinear diode model depicted in Figure 7-1. This simulation program incorporates a range of physical model parameters; some of them are so specialized that they are beyond the scope of our textbook. The most important ones are summarized in Table 7-1. Also listed are the differences between the standard *pn* and Schottky diode.

Table 7-1 Diode model parameters and their corresponding SPICE parameters

Symbol	SPICE	Description	Typical values
I_S	IS	saturation current	1 fA–10 µA
n	N	emission coefficient	1
τ_T	TT	transit time	5 ps–500 µs
R_S	RS	ohmic resistance	0.1–20 Ω
V_{diff}	VJ	barrier voltage	0.6–0.8 V (*pn*) 0.5–0.6 V (Schottky)
C_{J0}	CJ0	zero-bias junction capacitance	5–50 pF (*pn*) 0.2–5 pF (Schottky)
m	M	grading coefficient	0.2–0.5
W_g	EG	bandgap energy	1.11 eV (Si) 0.69 eV (Si-Schottky)
p_t	XTI	saturation current temperature coefficient	3 (*pn*) 2 (Schottky)

7.1.2 Linear Diode Model

The nonlinear model is based on the device physics developed in Chapter 6. As such, this model can be used for static and dynamic analyses under practically any circuit conditions. However, if the diode is operated at a particular DC bias point and the signal variations about this point are small, we can develop a linear or **small-signal model**. The concept

Diode Models

of linearization implies the approximation of the exponential *I-V* characteristic through a tangent at the bias or *Q*-point. The tangent at this *Q*-point is the differential conductance G_d, which we can find as

$$G_d = \frac{1}{R_d} = \left.\frac{dI_D}{dV_A}\right|_Q = \frac{I_Q + I_S}{nV_T} \approx \frac{I_Q}{nV_T}. \qquad (7.8)$$

The tangent approximation is shown in Figure 7-2 along with the simplified linear circuit model. It should be emphasized that the differential capacitance is now the diffusion capacitance at bias point V_Q, or

$$C_d = \frac{I_S \tau_T}{nV_T} e^{V_Q/(nV_T)} \approx \frac{I_D \tau_T}{nV_T}. \qquad (7.9)$$

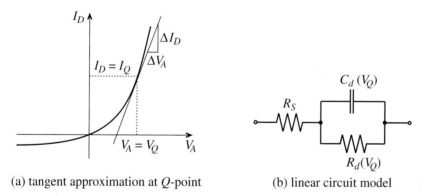

(a) tangent approximation at *Q*-point (b) linear circuit model

Figure 7-2 Small-signal forward biased diode model.

An apparent benefit of such a linearized circuit model is the ability to decouple the RF diode operation from the DC bias condition, as the following design example demonstrates.

―RF&MW→

Example 7-1. Derivation of the small-signal *pn* diode model

A conventional Si-based *pn* diode is operated at 300°K and has the following electric parameters at this temperature: $\tau_T = 500$ ps, $I_S = 5 \times 10^{-15}$ A, $R_S = 1.5\ \Omega$, $n = 1.16$. The DC operating conditions are chosen such that $I_Q = 50$ mA. To characterize the performance of a particular RF system in which this diode is used, we need to obtain

(a) the impedance behavior of the diode in the frequency range 10 MHz $\leq f \leq$ 1 GHz at 300 K, and

(b) the impedance response of the diode in the same frequency range, but for temperatures of 250 K, 350 K, and 400 K.

Solution. At a temperature of 300 K, we first determine from $I_Q = 50$ mA the corresponding V_Q, which is found from (7.1)

$$V_Q = nV_T \ln(1 + I_Q/I_S) = 0.898 \text{ V}.$$

Next, we can compute the differential resistance and capacitance as

$$R_d = \frac{nV_T}{I_Q} = 0.6 \text{ }\Omega \text{ and } C_d = \frac{I_S \tau_T}{nV_T} e^{V_Q/(nV_T)} = 832.9 \text{ pF}.$$

Knowing these parameters, we can find the impedance of the diode as a resistor R_S connected in series with the parallel combination of R_d and C_d:

$$Z = R_S + \frac{R_d}{1 + j\omega C_d R_d}.$$

The resulting frequency behavior is shown in Figure 7-3.

As temperature changes and the biasing current I_Q is maintained constant, the biasing voltage V_Q should change due to the temperature dependence of the thermal potential $V_T = kT/q$, bandgap energy W_g given by (7.7), and saturation current I_S described in (7.6). Results of these computations are presented in Table 7-2, and the corresponding frequency behavior of the diode impedance is shown in Figure 7-3.

Table 7-2 Diode model parameters for different temperatures

T, K	250	300	350	400
$W_g(T)$, eV	1.128	1.115	1.101	1.086
$I_s(T)$, A	5.1×10^{-19}	5.0×10^{-15}	3.3×10^{-12}	3.8×10^{-10}
V_Q, V	0.979	0.898	0.821	0.748
R_d, Ω	0.5	0.6	0.7	0.8
C_d, pF	999.5	832.9	713.9	624.7

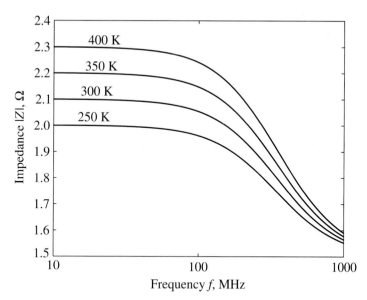

Figure 7-3 Frequency behavior of the diode impedance for different junction temperatures.

> We observe how the physical parameters developed for the pn-junction in Chapter 6 directly translate into the small-signal circuit model. The DC bias conditions influence the AC behavior because they affect the differential capacitance and resistance.

7.2 Transistor Models

Over the years, a number of large- and small-signal bipolar and monopolar transistor models have been developed. Perhaps the best-known one is the Ebers-Moll BJT model, which was initially introduced to characterize static and low-frequency modes of operation. The need to expand into RF/MW frequencies and high power applications required taking into account many second-order effects, such as low-current and high-injection phenomena. This has resulted in the Gummel-Poon model as a more refined BJT circuit representation.

7.2.1 Large-Signal BJT Models

We begin our discussion with the static **Ebers-Moll model**, which is one of the most popular large-signal models. Although this model was first introduced in 1954, it still is indispensable to understanding the basic model requirements and its extensions to more sophisticated large-signal models, as well as the derivation of most small-signal models. Figure 7-4 shows the generic *npn* transistor with the associated Ebers-Moll circuit model in the so-called **injection version**.

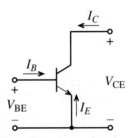

(a) Voltage and current convention for *npn* transistor

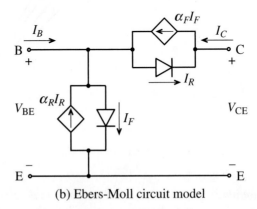

(b) Ebers-Moll circuit model

Figure 7-4 Large-signal Ebers-Moll circuit model.

In Figure 7-4, we encounter two diodes connected in forward and reverse polarity, as already seen in Chapter 6. In addition, two current-controlled current sources permit the mutual coupling of the two diodes as part of the base contact. The forward and reverse current gains (in common-base configuration) α_F and α_R possess typical values of $\alpha_F = 0.95 \ldots 0.99$ and $\alpha_R = 0.02 \ldots 0.05$. As a direct extension of the previously discussed single-diode model, the dual-diode Ebers-Moll equations take on the form

$$I_E = \alpha_R I_R - I_F \qquad (7.10)$$

$$I_C = \alpha_F I_F - I_R \qquad (7.11)$$

with the diode currents

$$I_R = I_{CS}(e^{V_{BC}/V_T} - 1) \qquad (7.12)$$

$$I_F = I_{ES}(e^{V_{BE}/V_T} - 1) \qquad (7.13)$$

where the **reverse collector and emitter saturation currents** I_{CS} and I_{ES} (whose numerical values range from 10^{-9} A to 10^{-18} A) can be related to the transistor saturation current I_S as follows:

$$\alpha_F I_{ES} = \alpha_R I_{CS} = I_S. \tag{7.14}$$

Despite their simplicity, the Ebers-Moll equations are capable of describing all major physical phenomena developed in Chapter 6. For the important cases of forward and reverse active modes, the circuit model can be simplified. The following two situations arise:

- Forward Active Mode ($V_{CE} > V_{CEsat} = 0.1$ V, $V_{BE} \approx 0.7$ V). With the base-emitter diode I_F conducting, and the base-collector diode reverse biased, we conclude that $I_R \approx 0$, and also $\alpha_R I_R \approx 0$. The base-collector diode and the base-emitter current source can thus be neglected.
- Reverse Active Mode ($V_{CE} < -0.1$ V, $V_{BC} \approx 0.7$ V). Here, the base-collector diode I_R is conducting, and the base-emitter diode is biased in reverse direction (i.e., $V_{BE} < 0$ V), which results in $I_F \approx 0$ and $\alpha_F I_F \approx 0$.

Figure 7-5 summarizes these two modes of operation when the emitter is chosen as a common reference point.

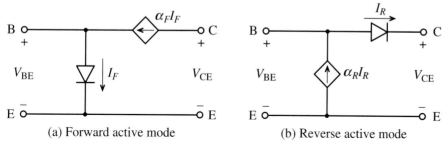

(a) Forward active mode   (b) Reverse active mode

Figure 7-5 Simplified Ebers-Moll equations for forward and reverse active modes.

This model can be modified to account for dynamic operation by including the familiar base-emitter and base-collector diffusion (C_{de}, C_{dc}) and junction (C_{je}, C_{jc}) diode capacitances. Unlike the simple charge analysis presented for the single-diode model, a more elaborate treatment is required for the BJT. For instance, the charge accounting for the emitter diffusion capacitance is composed of minority charges stored in (a) the neutral emitter zone, (b) the emitter-base space charge region, (c) the collector-base space charge region, and (d) the neutral base zone. An identical analysis applies to the collector diffusion capacitance. Figure 7-6 depicts the dynamic Ebers-Moll chip model. Further refinements for RF operation are often made by including the resistive and inductive properties of the lead wires as well as parasitic capacitances between the terminal points, see Figure 7-6(b).

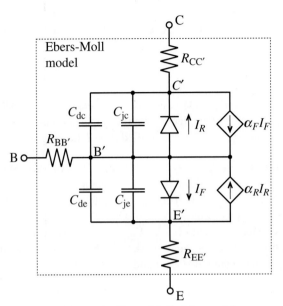

(a) Dynamic Ebers-Moll chip model

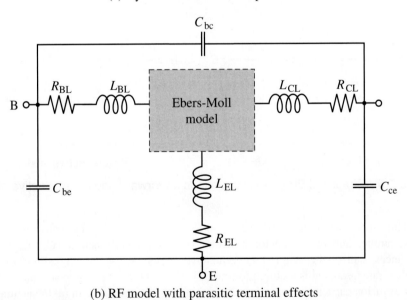

(b) RF model with parasitic terminal effects

Figure 7-6 Dynamic Ebers-Moll model and parasitic element refinements.

Transistor Models

Example 7-2. Transport versus injection form of the Ebers-Moll large-signal model

Instead of the injection model, it is the transport model that typically finds use in SPICE simulations. Let us go through the qualitative steps to arrive at this important representation.

Solution. We begin our discussion with the static BJT model, since the diffusion and junction capacitances can be added later in the derivation. First, we can show that the injection model in Figure 7-4 is equivalent to the transport model in Figure 7-7.

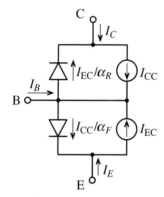

Figure 7-7 Transport representation of static Ebers-Moll model.

The equivalence of both models can be established if we re-express the collector and emitter currents as follows:

$$I_C = I_{CC} - I_{EC}/\alpha_R$$

$$I_E = -I_{CC}/\alpha_F + I_{EC}$$

with the current controlled sources now given as

$$I_{CC} = I_S(e^{V_{BE}/V_T} - 1)$$

$$I_{EC} = I_S(e^{V_{BC}/V_T} - 1).$$

A slightly different form can be obtained if both current sources are combined to a single source $I_{com} = I_{CC} - I_{EC}$, and the diode currents are re-expressed as

$$\frac{I_{EC}}{\alpha_R} \rightarrow \frac{1-\alpha_R}{\alpha_R} I_{EC} = \frac{I_{EC}}{\beta_R}$$

$$\frac{I_{CC}}{\alpha_F} \rightarrow \frac{1-\alpha_F}{\alpha_F} I_{CC} = \frac{I_{CC}}{\beta_F}.$$

This model configuration is shown in Figure 7-8 with base, collector, and emitter resistances. Also shown are the combined diffusion and junction capacitances C_{be} and C_{bc} associated with the base-emitter and base-collector diodes.

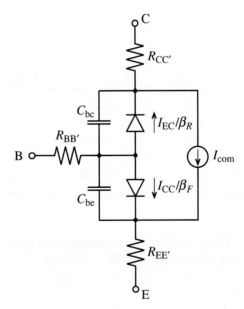

Figure 7-8 Dynamic Ebers-Moll transport model with single current source.

The Figure 7-8 configuration is useful since it leads directly to the large-signal BJT model under forward active bias condition. This bias allows us to neglect the base-collector diode current, but not its capacitative effect. Renaming the electric parameters, we arrive at the circuit depicted in Figure 7-9, where we replaced the forward-biased diode with an equivalent current source.

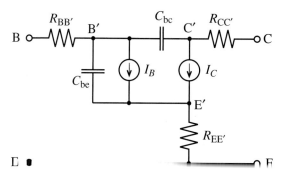

Figure 7-9 Large-signal BJT model in forward active mode.

This final form can be found in the SPICE library as a nonlinear representation of the standard BJT.

We notice how the dynamic transport model of the Ebers-Moll equations naturally lead to the SPICE large-signal model. An inherent difficulty for all circuit models is the unique determination of the model parameters through appropriate measurement strategies.

The Ebers-Moll model was one of the first BJT circuit representations and has retained its popularity and wide acceptance. However, shortly after its introduction, it became apparent that a number of physical phenomena could not be taken into account by this original model. Specifically, research has shown that (a) β_F and β_R are current dependent, and (b) the saturation current I_S is affected by the base-collector voltage (**Early effect**). Both effects significantly influence the overall BJT performance. For this reason, a number of refinements have been introduced to the original Ebers-Moll model, culminating in the Gummel-Poon model shown in Figure 7-10.

In this model, we notice the addition of two extra diodes to deal with the I_C-dependent forward and reverse current gains $\beta_F(I_C)$ and $\beta_R(I_C)$. Figure 7-11 depicts a typical curve for β_F. The two **leakage diodes** $L1$, $L2$ provide four new design parameters: coefficients I_{S1}, n_{EL} in $I_{L1} = I_{S1}(\exp[V_{BE}/(n_{EL}V_T)] - 1)$ for low-current normal mode operation, and I_{S2}, n_{CL} in $I_{L2} = I_{S2}(\exp[V_{BC}/(n_{CL}V_T)] - 1)$ for low-current inverse mode operation. Additionally, the Gummel-Poon model can handle the Early effect, whereby with increasing collector-emitter voltage, the space charge domain begins to extend far into the base region. The result is an increase in collector current for a fixed base current. If one draws tangents to each collector current curve (see Figure 7-12), they all converge approximately at a single voltage point $-V_{AN}$ known as the **forward Early voltage**. An identical analysis can be conducted if the BJT is operated in the reverse active mode, resulting in a voltage point V_{BN} known as the **inverse Early voltage**.

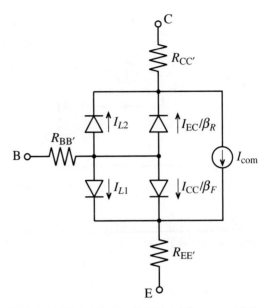

Figure 7-10 Static Gummel-Poon model.

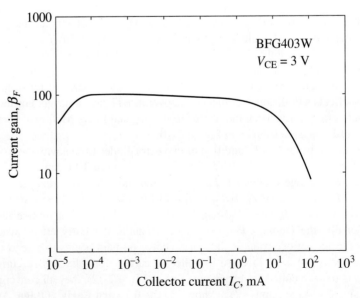

Figure 7-11 Typical dependence of β_F on the collector current I_C for a fixed collector-emitter voltage V_{CE}.

Transistor Models

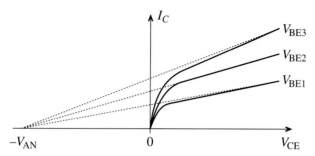

Figure 7-12 Collector current dependence on V_{CE} and its approximation through the Early voltage V_{AN}.

Both voltages are incorporated as additional factors in the model. Moreover, Gummel-Poon also permits the specification of a current-dependent base resistance and a distributed base-collector junction capacitance C_{jbc}. We will not go into details of the various underlying physical reasonings leading to the requirement of these additional model parameters. The interested reader is referred to the sources listed at the end of this chapter. Converting the static Gummel-Poon model (Figure 7-10) into dynamic form by including the diode capacitances and C_{jbc} leads to the equivalent circuit shown in Figure 7-13.

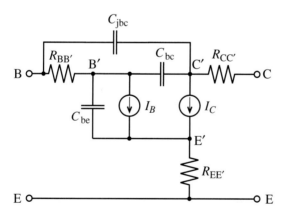

Figure 7-13 Large-signal Gummel-Poon model in forward active mode.

This circuit is similar to the large-signal Ebers-Moll form (see Figure 7-9), but with the differences that the base resistance $R_{BB'}$ is current dependent, the collector current takes into account the Early effect, and a distributed base-collector junction capacitance C_{jbc} enters the model.

In SPICE, both BJT models can be invoked, with Ebers-Moll requiring the specification of 26 circuit parameters, and up to 41 parameters for Gummel-Poon. Generally, the

BJT manufactures supply these parameters in their datasheets. Unfortunately, one increasingly encounters the situation where instead of the generally applicable SPICE model parameters, only the measured S-parameters are given. Since these measurements are recorded for particular operating frequencies and under certain bias conditions, it is then left to the circuit design engineer to interpolate the data for operating conditions not found in the datasheet.

7.2.2 Small-Signal BJT Models

From the large-signal Ebers-Moll equations, it is now easy to derive a small-signal model in the forward active mode. To this end, the large signal model (Figure 7-9) is converted into the linear hybrid-π model shown in Figure 7-14.

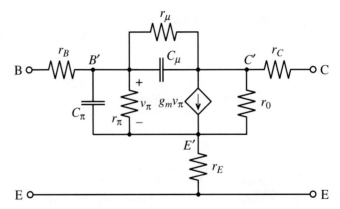

Figure 7-14 Small-signal hybrid-π Ebers-Moll BJT model.

We see that the base-emitter diode is replaced by a small-signal diode model and the collector current source is substituted by a voltage-controlled current source. To make the model more realistic, a resistor r_μ is connected in parallel to the feedback capacitor C_μ. For this model, we can directly establish the small-signal parameters by expanding the input voltage V_{BE} and output current I_C about the biasing or Q-point in terms of small AC voltage v_{be} and current i_c as follows:

$$V_{BE} = V_{BE}^Q + v_{be} \tag{7.15a}$$

$$I_C = I_C^Q + i_c = I_S \exp[(V_{BE}^Q + v_{be})/V_T]$$

$$= I_C^Q \left[1 + \left(\frac{v_{be}}{V_T}\right) + \frac{1}{2}\left(\frac{v_{be}}{V_T}\right)^2 + \ldots \right]. \tag{7.15b}$$

Transistor Models

Truncating the series expansion of the exponential expression after the linear term, we find for the small-signal collector current

$$i_c = \left(\frac{I_C^Q}{V_T}\right) v_{be} = g_m v_{be} \tag{7.16}$$

where we identify the **transconductance**

$$g_m = \left.\frac{dI_C}{dV_{BE}}\right|_Q = \left.\frac{d}{dV_{BE}} I_s e^{(V_{BE}/V_T)}\right|_Q \approx \frac{I_C^Q}{V_T} \tag{7.17}$$

and the **small-signal current gain** at the operating point

$$\left.\beta_F\right|_Q = \left.\frac{dI_C}{dI_B}\right|_Q = \beta_0. \tag{7.18}$$

The **input resistance** is determined through the chain rule:

$$r_\pi = \left.\frac{dV_{BE}}{dI_B}\right|_Q = \left.\frac{dI_C}{dI_B}\right|_Q \left.\frac{dV_{BE}}{dI_C}\right|_Q = \frac{\beta_0}{g_m}. \tag{7.19}$$

For the **output conductance**, we have

$$\frac{1}{r_0} = \left.\frac{dI_C}{dV_{CE}}\right|_Q = \left.\frac{d}{dV_{CE}}\left(I_s e^{V_{BE}/V_T}\left[1+\frac{V_{CE}}{V_{AN}}\right]\right)\right|_Q \approx \frac{I_C^Q}{V_{AN}} \tag{7.20}$$

which includes the Early effect, also known as **base-width modulation** because of the increased depletion layer extent into the base.

It is directly seen that this model in its simplest form at the terminals B'-C'-E' reduces for the low frequence case, and upon neglecting the collector-emitter resistance, to our familiar low-frequency transistor model. Here, the output current can simply be expressed in term of the input voltage v_{be} as

$$i_c = g_m v_{be} = g_m r_\pi \frac{v_{be}}{r_\pi} = \beta_0 \frac{v_{be}}{r_\pi}. \tag{7.21}$$

Additional small-signal BJT circuit models can be developed on the basis of the *h*-parameter network representation. For instance, if we recall the definitions of the *h*-parameters and apply them to a BJT in common-emitter configuration, we obtain

$$v_{be} = h_{11} i_b + h_{12} v_{ce} \tag{7.22}$$

$$i_c = h_{21} i_b + h_{22} v_{ce} \tag{7.23}$$

which is encoded in generic form in Figure 7-15.

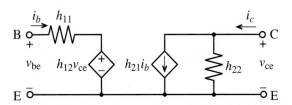

Figure 7-15 Generic h-parameter BJT representation with two sources.

In this notation, the indices denote $11 \Rightarrow$ input, $12 \Rightarrow$ reverse, $21 \Rightarrow$ forward, and $22 \Rightarrow$ output. The individual parameters can be computed via the following relations:

$$h_{11} = \left.\frac{v_{be}}{i_b}\right|_{v_{ce}=0} \quad \text{input impedance} \tag{7.24a}$$

$$h_{21} = \left.\frac{i_c}{i_b}\right|_{v_{ce}=0} \quad \text{foward current gain } \beta_F \tag{7.24b}$$

$$h_{12} = \left.\frac{v_{be}}{v_{ce}}\right|_{i_b=0} \quad \text{reverse voltage gain} \tag{7.24c}$$

$$h_{22} = \left.\frac{i_c}{v_{ce}}\right|_{i_b=0} \quad \text{output admittance.} \tag{7.24d}$$

It is observed that h_{12} represents the influence of the output voltage "fed back" to the input as part of a voltage-controlled voltage source. Conversely, h_{21} models the influence of the input "fed forward" to the output, or gain, as part of a current-controlled current source. The output-to-input feedback is modeled by the reverse-biased collector-base junction capacitance C_{cb}, which is generally on the order of 0.1 to 0.5 pF and a resistor r_{cb}, with values ranging in the low MΩ. Therefore, for low and intermediate frequencies up to approximately 50 MHz, this feedback can safely be neglected. However, in the GHz range, it may profoundly affect the BJT operation.

If the feedback resistor r_{bc} is neglected, a high-frequency circuit model results, as displayed in Figure 7-16. Also shown in this figure is a converted circuit such that the feedback capacitance C_{cb} appears as the Miller capacitance on the input and output sides. The Miller effect allows us to decouple the input from the output port by redistributing the feedback capacitance, as the following example shows.

Transistor Models

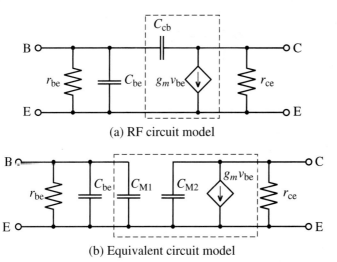

Figure 7-16 RF small-signal circuit model and converted circuit model using the Miller effect.

Example 7-3. The Miller effect

Show that the feedback capacitance C_{cb} can be expressed as $C_{M1} = C_{cb}(1 - v_{ce}/v_{be})$ at the input port, and as $C_{M2} = C_{cb}(1 - v_{be}/v_{ce})$ at the output port. Assume that the input and output voltages are approximately constant, and keep in mind that v_{ce}/v_{be} is negative under common-emitter configuration.

Solution. We need to convince ourselves that the two generic circuits shown in Figure 7-17, and high-lighted in Figure 7-16, are equivalent.

The current I_p is found by taking the voltage difference between output and input, divided by the feedback impedance

$$I_p = (V_1 - V_2)/Z_{12}$$

and for the equivalent input and output impedances Z_{11}, Z_{22}

$$Z_{11} = \frac{V_1}{I_p} = \frac{Z_{12}V_1}{(V_1 - V_2)} = Z_{12}(1 - V_2/V_1)^{-1}$$

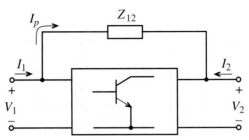

(a) Circuit with feedback impedance Z_{12}

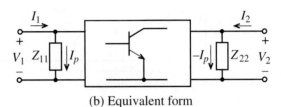

(b) Equivalent form

Figure 7-17 Miller transformation of feedback impedance.

and

$$Z_{22} = \frac{V_2}{(-I_p)} = \frac{Z_{12}V_2}{(V_2 - V_1)} = Z_{12}(1 - V_1/V_2)^{-1}.$$

With the assignments $Z_{12} = 1/(j\omega C_{cb})$, $Z_{11} = 1/(j\omega C_{M1})$, $Z_{22} = 1/(j\omega C_{M2})$ and $V_1 = v_{be}$, $V_2 = v_{ce}$, we find the equivalent capacitances

$$C_{M1} = C_{cb}(1 - v_{ce}/v_{be}) \qquad (7.25)$$

and

$$C_{M2} = C_{cb}(1 - v_{be}/v_{ce}). \qquad (7.26)$$

Decoupling of the input from the output port is accomplished by computing an equivalent capacitance that depends on a constant voltage amplification factor v_{ce}/v_{be}.

Another factor that is directly related to the BJT frequency behavior is the short-circuit current gain $h_{fe}(\omega)$, which implies the connection of the collector with the emitter, as depicted in Figure 7-18. Since the output is short-circuited ($v_{ce} = 0$), the Miller effect does not enter the analysis. We find $h_{fe}(\omega)$ by computing the ratio of collector to base currents

Transistor Models

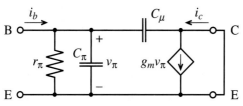

(a) Short-circuited hybrid-π model

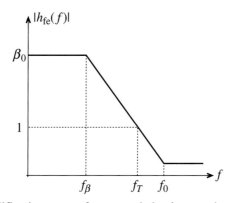

(b) Amplification versus frequency behavior on a log-log scale

Figure 7-18 Short-circuit current gain of BJT model.

$$h_{fe}(\omega) = \frac{i_c}{i_b} = \frac{g_m Z_{in}(1 - j\omega C_\mu/g_m)}{1 + j\omega C_\mu Z_{in}} \qquad (7.27)$$

where $Z_{in} = r_\pi/(1 + j\omega r_\pi C_\pi)$. Substituting Z_{in} into (7.27) and using (7.19) results in

$$h_{fe}(\omega) = \frac{\beta_0(1 - j\omega C_\mu/g_m)}{1 + j\omega r_\pi(C_\pi + C_\mu)} = \frac{\beta_0[1 - j(f/f_0)]}{1 + j(f/f_\beta)} \qquad (7.28)$$

with the maximum frequency f_0 and the beta cutoff frequency f_β:

$$f_0 = \frac{g_m}{2\pi C_\mu} \quad \text{and} \quad f_\beta = \frac{1}{2\pi r_\pi(C_\pi + C_\mu)}. \qquad (7.29)$$

The transition frequency f_T denotes the point where the magnitude of the current gain is unity (or 0 dB) under a short-circuit output condition. Setting the absolute value of (7.28) equal unity, we find

$$f_T = \frac{1}{2\pi}\sqrt{\frac{\beta_0^2 - 1}{r_\pi^2(C_\pi^2 + 2C_\pi C_\mu)}}. \qquad (7.30)$$

Since usually $\beta_0 \gg 1$ and $C_\pi \gg C_\mu$, we can simplify (7.30) as

$$f_T \approx \frac{\beta_0}{2\pi r_\pi C_\pi} = \frac{g_m}{2\pi C_\pi}. \qquad (7.31)$$

As already seen in Chapter 6, this frequency is related to the emitter-collector time delay, which is composed of the delays associated with base, emitter, and collector. Another name for f_T is the **gain-bandwidth product**, which is specified in data sheets for a particular collector-emitter voltage and collector current bias condition. An additional figures of merit can be established when one considers the power gain of the transistor. This will be further investigated in Chapter 9.

Finally, let us discuss a design project involving the BJT. In this project, we will go through the steps of deciding upon bias conditions, determining the input and output impedances as a function of frequency, and converting the impedance values to the relevant S-parameters. The transistor SPICE parameters used for this example are summarized in Table 7-3. The Matlab routine ex7_4.m provides computational details.

Table 7-3 SPICE parameters of the BJT

Symbol	Description	Typical value
β_F	forward current gain	145
I_S	saturation current	5.5 fA
V_{AN}	forward Early voltage	30 V
τ_F	forward transit time	4 ps
C_{JC0}	base-collector junction capacitance at zero applied junction voltage	16 fF
C_{JE0}	base-emitter junction capacitance at zero applied junction voltage	37 fF
m_C	collector capacitance grading coefficient	0.2
m_E	emitter capacitance grading coefficient	0.35
$V_{\text{diff}_{BE}}$	base-emitter diffusion potential	0.9 V
$V_{\text{diff}_{BC}}$	base-collector diffusion potential	0.6 V
r_B	base body resistance	125 Ω
r_C	collector body resistance	15 Ω
r_E	emitter body resistance	1.5 Ω
L_B	base lead inductance	1.1 nH
L_C	collector lead inductance	1.1 nH
L_E	emitter lead inductance	0.5 nH

Transistor Models

Example 7-4. Setting bias conditions, determining input/ output impedances, and computing the S-parameters for a BJT

Our task is to design an amplifier for a portable communication system. The system is supposed to operate from a 3.6 V battery source. Taking into considerations the maximum available current and battery lifetime, we demand that the current for the amplifier should not exceed approximately 10 mA. Assuming $V_{CE} = 2$ V and $I_C = 10$ mA as bias conditions for this transistor, and the BJT parameters given in Table 7-3, we need to determine the hybrid-π model. In addition, the resulting input/output impedances and the corresponding S-parameters for the frequency range of 1 MHz $< f <$ 100 GHz have to be found.

Solution. We begin this design by developing a simple voltage divider biasing network, as shown in Figure 7-19.

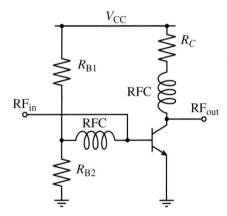

Figure 7-19 Biasing a BJT in common-emitter configuration.

With the power supply voltage $V_{CC} = 3.6$ V, desired collector-emitter voltage $V_{CE} = 2$ V, and collector current $I_C = 10$ mA, we can find the value for the collector resistor R_C as follows:

$$R_C = \frac{V_{CC} - V_{CE}}{I_C} = 160 \, \Omega.$$

Based on the current gain of $\beta_0 = 145$ and collector current of $I_C = 10$ mA, we find the base current to be $I_B = I_C/\beta_0 = 69 \, \mu A$.

The current through the resistor R_{B1} is equal to the sum of the current flowing through resistor R_{B2} and I_B. In practice, the values of R_{B1} and R_{B2} are selected to make the magnitude of I_B equal to 10% of the current through resistor R_{B2}. Keeping this in mind and realizing that the base-emitter voltage drop V_{BE} is approximately equal to the base-emitter built-in potential $V_{\text{diff}_{BE}}$, we find

$$R_{B2} = \frac{V_{\text{diff}_{BE}}}{10 I_B} = 1300 \, \Omega$$

and

$$R_{B1} = \frac{V_{CC} - V_{\text{diff}_{BE}}}{11 I_B} = 3560 \, \Omega.$$

This bias network is simple in the sense that it does not take into account temperature variations and device tolerances. More sophisticated networks are discussed in Chapter 8.

We are now ready to compute the hybrid-π model parameters. From equations (7.17)–(7.20), we obtain $g_m = I_C/V_T = 386$ mS, $r_\pi = \beta_0/g_m = 375 \, \Omega$, and $r_0 = V_{AN}/I_C = 3$ kΩ. To find C_μ and C_π we have to resort to the *pn*-junction analysis. Since the base-collector voltage is negative, the base-collector capacitance is determined by only the junction capacitance. From (7.3), we find

$$C_\mu = \frac{C_{JC0}}{(1 - V_{BC}/V_{\text{diff}_{BC}})^{m_C}} = 13 \, \text{fF}.$$

Because the base-emitter voltage is positive, C_π is a combination of both the junction and diffusion capacitances. From (7.3), and by assuming $V_{m_E} = 0.5 V_{\text{diff}_{BE}}$, we have

$$C_{\pi_{\text{junct}}} = \frac{C_{JE0}}{0.5^{m_E}} \left(1 + m_E \frac{V_{BE} - 0.5 V_{\text{diff}_{BE}}}{0.5} \right) = 55 \, \text{fF}$$

and

$$C_{\pi_{\text{diff}}} = \frac{I_S \tau_T}{V_T} e^{V_{BE}/V_T} = 1.085 \, \text{pF}.$$

Thus, the total base-emitter capacitance is

$$C_\pi = C_{\pi_{\text{junct}}} + C_{\pi_{\text{diff}}} = 1.14 \, \text{pF}.$$

After establishing all parameters of the hybrid-π model, we can compute the corresponding h-parameter matrix as described in (7.24). The result takes into account only the transistor die hybrid-π parameters without incorporating base, collector, and emitter resistances and parasitic inductances.

To consider the influence of the body-resistance and lead inductance, we can employ network analysis, as described in Chapter 4. Specifically, we can partition the equivalent transistor circuit into four two-port networks, as shown in Figure 7-20.

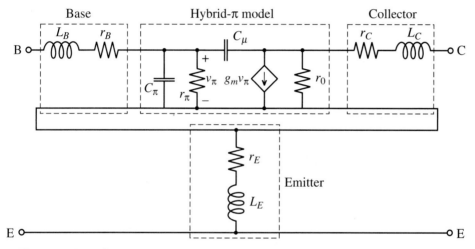

Figure 7-20 Complete transistor model divided into four two-port networks.

Relying on this network partitioning, we proceed as follows. To obtain the Z-parameters of the entire transistor, we first convert the h-parameters of the hybrid-π model into the *ABCD* representation. Next, this converted hybrid-π model is multiplied by the **ABCD**-matrix representations for base and collector leads:

$$\begin{bmatrix} A & B \\ C & D \end{bmatrix}_{tr} = \begin{bmatrix} 1 & r_B + j\omega L_B \\ 0 & 1 \end{bmatrix}_{base} \begin{bmatrix} A & B \\ C & D \end{bmatrix}_{h\text{-model}} \begin{bmatrix} 1 & r_C + j\omega L_C \\ 0 & 1 \end{bmatrix}_{collector}.$$

Finally, we convert the *ABCD* representation of the transistor with the attached base and collector leads into Z-parameter form and add the resulting matrix to the **Z**-matrix of the emitter lead.

$$\begin{bmatrix} Z_{11} & Z_{12} \\ Z_{21} & Z_{22} \end{bmatrix}_{\text{trans}} = \begin{bmatrix} Z_{11} & Z_{12} \\ Z_{21} & Z_{22} \end{bmatrix}_{\text{tr}} + \begin{bmatrix} r_E + j\omega L_E & r_E + j\omega L_E \\ r_E + j\omega L_E & r_E + j\omega L_E \end{bmatrix}_{\text{emitter}}.$$

The frequency responses of coefficients Z_{11} and Z_{22} are shown in Figure 7-21.

As we see from Figure 7-21(a), the addition of the body resistance to the basic hybrid-π model at low frequencies results in a significant increase in the input impedance due to the large base resistance. At high frequencies, the effect of base and emitter lead inductances becomes noticeable in terms of a sharp rise in the impedance.

For the output impedance, the situation is quite different. Since the base resistance does not have any effect on Z_{22}, the output impedance remains virtually unaffected by the addition of the leads and is dominated by the resistance r_0 up to very high frequencies. At that point, the inductive effect of the leads become dominant.

From the known Z-parameter representation of the transistor, we can easily compute the S-parameters using the conversion described in Chapter 4. The resulting input reflection coefficient S_{11} and gain S_{21} of the transistor are shown in Figure 7-22 as part of the Smith Chart and a polar plot, respectively.

As we notice in Figure 7-22(b), even though the emitter resistance and inductance seem to be negligible compared to the values of the other components in the model, their addition results in a significant drop in gain over the entire frequency range. This shows, once again, the influence of parasitic elements in RF circuits.

We have demonstrated an approach for computing the small-signal parameters of the transistor from known operating conditions of the underlying SPICE model. Even though a simple topology is investigated, this method can be directly applied to more complicated internal structures by breaking them down into a set of interconnected two-port networks.

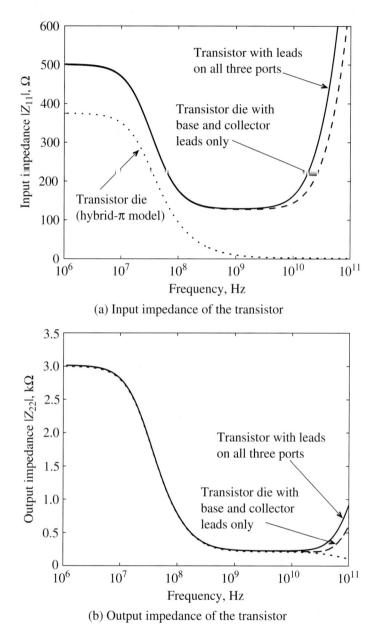

Figure 7-21 Input and output impedances as a function of frequency.

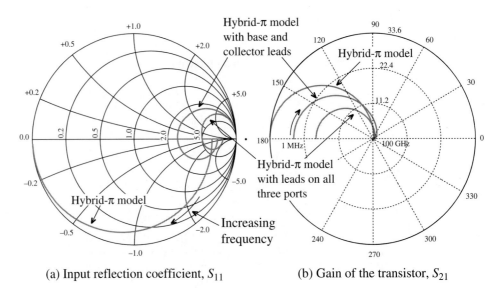

(a) Input reflection coefficient, S_{11} (b) Gain of the transistor, S_{21}

Figure 7-22 S_{11} and S_{21} responses of a BJT for various model configurations.

7.2.3 Large-Signal FET Models

FETs offer a number of advantages, but also suffer some disadvantages over BJTs. In choosing the appropriate active device for a particular circuit, one should take into consideration the following FET-related benefits:

- FETs exhibit a better temperature behavior.
- The noise performance of a FET is, in general, superior.
- The input impedance of FETs is normally very high.
- The drain current of a FET shows a quadratic functional behavior compared with the exponential collector current curve of a BJT.
- The upper frequency limit exceeds, often by a substantial margin, that of a BJT.
- The power consumption of a FET is lower.

In terms of the disadvantages one often hears:

- FETs generally possess lower gain.
- Because of the high input impedance, matching networks are more difficult to construct.
- The power handling capabilities tend to be inferior compared with BJTs.

The preceding list is debatable, since new device concepts and fabrication improvements continuously affect various transistor performance aspects.

For our FET modeling purposes, we will focus on the **noninsulated gate FET**. To this group we count the MESFET, often identified as GaAs FET (pronounced "gasfet"), and the HEMT. Both types are discussed in Chapter 6. In Figure 7-23, the basic n-channel,

Transistor Models

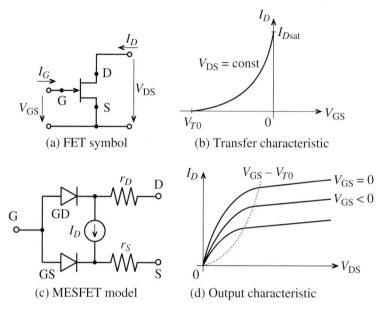

Figure 7-23 Static *n*-channel MESFET model.

depletion mode MESFET model (with negative threshold voltage) is shown along with the transfer and output characteristics.

The key equations for the drain current in forward, or normal, mode of operation follow from the analysis developed in Section 6.4. There, we obtained the drain current for both the linear and saturation regions. These current expressions constitute the starting point for deriving the model for the FET.

(a) Saturation region $(V_{DS} \geq V_{GS} - V_{T0} > 0)$

The saturation drain current given by (6.94) is repeated here for convenience

$$I_{Dsat} = G_0 \left(\frac{V_P}{3} - (V_d - V_{GS}) + \frac{2}{3\sqrt{V_P}} (V_d - V_{GS})^{3/2} \right). \qquad (7.32)$$

If we substitute in (7.32) the combination of threshold voltage V_{T0} and pinch-off voltage V_P (in other words, $V_d = V_{T0} + V_P$), an alternate form is obtained:

$$I_{Dsat} = G_0 \frac{V_P}{3} \left(1 - 3\left(1 - \frac{V_{GS} - V_{T0}}{V_P}\right) + \left(21 - \frac{V_{GS} - V_{T0}}{V_P}\right)^{3/2} \right). \qquad (7.33)$$

Making a binomial expansion of the square bracketed expression up to the second term allows us to write (7.33) as

$$I_{D\text{sat}} = G_0 \frac{V_P}{3}\left(\frac{3}{4}\right)\left(\frac{V_{GS} - V_{T0}}{V_P}\right)^2. \tag{7.34}$$

The constant factors in front of the square term in (7.34) are combined to form the **conduction parameter** β_n

$$\beta_n = \frac{1}{4}\left(\frac{G_0}{V_P}\right) = \frac{\mu_n \varepsilon Z}{2Ld} \tag{7.35}$$

where the definitions for the conductance $G_0 = \sigma Z d/L = \mu_n N_D q Z d/L$ and the pinch-off voltage $V_P = (qN_D d^2)/(2\varepsilon)$ from Section 6.4 are used. If the channel modulation effect is included, we arrive at

$$I_D = \beta_n (V_{GS} - V_{T0})^2 (1 + \lambda V_{DS}). \tag{7.36}$$

Here, the parameter $\lambda \approx 0.01 - 0.1 \text{ V}^{-1}$ models the slight increase in drain current for increasing drain-source voltage in the saturation region, see Figure 7-23(d).

(b) Linear or triode region $(0 < V_{DS} < V_{GS} - V_{T0})$

Identical steps, as outlined for the saturation region, can be invoked to manipulate the drain current expression (6.91) to yield

$$I_D = \beta_n [2(V_{GS} - V_{T0})V_{DS} - V_{DS}^2](1 + \lambda V_{DS}) \tag{7.37}$$

where, again, the channel modulation is considered to achieve a smooth transition from the linear into the saturation region. For instance, if $V_{DS} = V_{GS} - V_{T0}$ (that is, the transition from linear to saturation region) both drain currents are identical.

The FET can also be operated in reverse, or inverted, mode if $V_{DS} < 0$. For completeness, the two drain current relations are given without further comments.

(c) Reverse saturation region $(-V_{DS} \geq V_{GD} - V_{T0} > 0)$

$$I_D = -\beta_n (V_{GD} - V_{T0})^2 (1 - \lambda V_{DS}). \tag{7.38}$$

(d) Reverse linear or triode region $(0 < -V_{DS} < V_{GD} - V_{T0})$

$$I_D = \beta_n [2(V_{GD} - V_{T0})V_{DS} - V_{DS}^2](1 + \lambda V_{DS}). \tag{7.39}$$

Making the transition from the static to the dynamic FET model requires only the addition of gate-drain and gate-source capacitances, as illustrated in Figure 7-24. Also shown in this model are source and drain resistors associated with source-gate and drain-gate channel resistances. A gate resistor is typically not included because the metallic gate connection represents low resistance.

Transistor Models

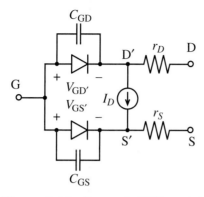

Figure 7-24 Dynamic FET model.

A summary of the most relevant SPICE modeling parameters for a MESFET is presented in Table 7-4.

Table 7-4 SPICE modeling parameters for a MESFET

Symbol	SPICE	Description
V_{T0}	VTO	Threshold voltage
λ	LAMBDA	Channel length modulation coefficient
β	BETA	Conduction parameter
C_{GD}	CGD	Zero-bias gate-to-drain capacitance
C_{GS}	CGS	Zero-bias gate-to-source capacitance
r_D	RD	Drain resistance
r_S	RS	Source resistance

7.2.4 Small-Signal FET Models

A small-signal FET circuit can directly be derived from the large-signal FET model (Figure 7-24). In this model, we simply replace the gate-drain and the gate-source diodes with their small-signal representations, derived in Section 7.1. In addition, the voltage-controlled current source is modeled via a transconductance g_m and a shunt conductance $g_0 = 1/r_{ds}$. The model can be tied in with a physical device representation, as Figure 7-25 shows.

This model can be described by a two-port Y-parameter network in the form

$$i_g = y_{11} v_{gs} + y_{12} v_{ds} \qquad (7.40a)$$

$$i_d = y_{21} v_{gs} + y_{22} v_{ds}. \qquad (7.40b)$$

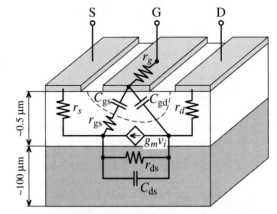

(a) Idealized MESFET device structure

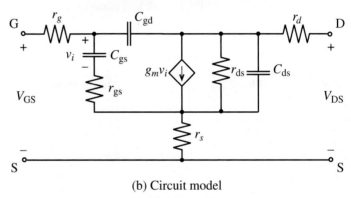

(b) Circuit model

Figure 7-25 Small-signal MESFET model.

Under low-frequency conditions, the input conductance of y_{11} and the feedback conductance of y_{12} are very small and can thus be neglected. This is consistent with the fact that the gate current is too small to be of practical consequence. However, for high-frequency operation, the capacitances are typically included, resulting in the circuit model shown in Figure 7-26.

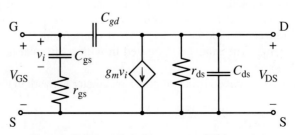

Figure 7-26 High-frequency FET model.

Transistor Models

For DC and low-frequency operation, the model in Figure 7-26 simplifies to the condition where the input is completely decoupled from the output. Transconductance g_m and output conductance g_0 can be readily computed for the forward saturation region from the drain current equation (7.36):

$$y_{21} = g_m = \left.\frac{dI_D}{dV_{GS}}\right|_Q = 2\beta_n(V_{GS}^Q - V_{T0})(1 + \lambda V_{DS}^Q) \tag{7.41}$$

$$y_{22} = \frac{1}{r_{ds}} = \left.\frac{dI_D}{dV_{DS}}\right|_Q = \beta_n \lambda (V_{GS}^Q - V_{T0})^2 \tag{7.42}$$

with the operating point, or Q-point, denoted by V_{DS}^Q and V_{GS}^Q.

The gate-source and gate-drain capacitances play a crucial role in determining the frequency performance. For the transition frequency f_T, we again have to consider the short-circuit current gain for the situation where the magnitude of the input current I_G is equal to the magnitude of the output current I_D, or specifically

$$|I_G| = \omega_T(C_{gs} + C_{gd})|V_{GS}| = |I_D| = g_m|V_{GS}| \tag{7.43}$$

which gives us

$$f_T = \frac{g_m}{2\pi(C_{gs} + C_{gd})}. \tag{7.44}$$

For low-frequency FET applications, it is primarily the charging time defined by these capacitances that severely limits the FET frequency response. This is in contrast to the channel transit time, as defined in Section 6.4.3, which for high-frequency applications limits the FET's operation as the following example illustrates.

> **TRANSITION FREQUENCIES OF FETs**
>
> Commercially available high performance GaAs FETs list f_T numbers of 200 GHz and higher, whereas experimental devices are reported above 1 THz.

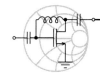

Example 7-5. Approximate determination of the cutoff frequency of a GaAs MESFET

A GaAs MESFET with a gold gate is fabricated to be 1.0 μm in length and 200 μm in width, and $d = 0.5$ μm in depth. The following electric characteristics are known: $\varepsilon_r = 13.1$, $N_D = 10^{16}$ cm^{-3}, and $\mu_n = 8500$ cm^2/V·s. Under suitably chosen approximations, we would like to find the cutoff frequency at room temperature.

Solution. To apply (7.44), it is necessary to find an approximate expression for the transconductance and capacitances. The transconductance can be found by knowing that the drain saturation current (7.33) is maximal for $V_{GS} = 0$, which gives

$$g_m = \left.\frac{dI_{Dsat}}{dV_{GS}}\right|_{V_{GS}=0} = G_0(1 - \sqrt{V_d/V_P})$$

where the built-in voltage V_d for the Schottky contact is found from (6.39) to be

$$V_d = (V_M - \chi) - V_C$$

with $V_C = V_T \ln(N_C/N_D) = 0.1$ V, $V_M = 5.1$ V, and $\chi = 4.07$ V. Substituting these values yields $V_d = 0.93$ V. The pinch-off voltage and the conductance are, respectively,

$$V_P = \frac{qN_D d^2}{2\varepsilon_0 \varepsilon_r} = 1.724 \text{ V} \quad \text{and} \quad G_0 = \frac{q\mu_n N_D W d}{L} = 136 \text{ mS}.$$

Thus, $g_m \approx 36.126$ mS. For the capacitance, we can approximately compute the surface area of the channel times the dielectric constant divided by the channel thickness:

$$C_{gs} + C_{gd} = \varepsilon_0 \varepsilon_r \left(\frac{WL}{d}\right) = 0.046 \text{ pF}.$$

From these values, we estimate f_T:

$$f_T = \frac{g_m}{2\pi(C_{gs} + C_{gd})} = 123.93 \text{ GHz}.$$

In contrast to an approximate channel transit time corresponding to 15 GHz, see Section 6.4.3, we now have the situation that the RC time constant is smaller. In other words, the channel transit time is the limiting factor in the high-speed performance of this MESFET.

An often used approximate formula for (7.44) can be derived if we set $g_m \approx G_0$. The explicit result is

$$f_T \approx \frac{q\mu_n N_D d^2}{2\pi \varepsilon L^2}. \tag{7.45}$$

This expression applied to the above example would have yielded 466.5 GHz.

7.2.5 Transistor Amplifier Topologies

The common single-transistor amplifier topologies are shown in Figure 7-27. The important performance characteristics of each topology are listed in Table 7-5 with approximate numerical representations intended to provide the reader with an intuitive high-level design guide.

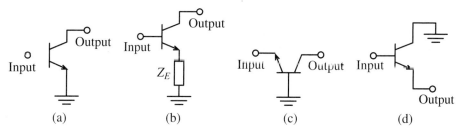

Figure 7-27 Single-transistor amplifier topologies: (a) common emitter (CE), (b) CE with emitter degeneration, (c) common base (CB), and (d) common collector (CC).

The common emitter (CE), or common source (CS) for FET devices, topology in Figure 7-27(a) is the standard configuration in low-noise amplifiers and power amplifiers since it provides the highest power gain as well as low-noise performance. Moreover, this topology has favorable, albeit frequency-dependent, input and output impedances. The stability and linearity of this stage are only fair, due to the Miller feedback and the direct feeding of the input signal into the base junction (which is less of a problem with FETs).

Some of the properties of the CE stage can be enhanced with emitter degeneration (Figure 7-27(b)). Here, an additional impedance in series with the emitter improves linearity and modifies the input impedance in a controlled way. Inductive emitter degeneration has the advantageous property of adding a resistive component to the input impedance without adding noise. Emitter degeneration also constitutes negative feedback that has a gain-leveling effect (with a resistive Z_E) and may improve stability. The disadvantages of emitter degeneration are (a) reduced gain, (b) increased noise due to any resistive component of the emitter impedance, and (c) degraded reverse isolation, which can lead to instability at high frequencies.

The distinctive features of the common base (CB), or common gate (CG) for FET devices, topology (Figure 7-27(c)) are the predictable low input impedance, very high output impedance and high bandwidth due to the absence of the Miller effect. The CB stage also possesses good reverse isolation and consequently better stability, at least at lower frequencies (although any parasitic feedback is usually positive, resulting in a destabilizing effect). One key drawback of the CB stage is reduced power gain, since it has no current gain. Also, care must be taken to properly ground the base (or gate), since the stage can be easily destabilized even by a small series impedance at that port, which may include the impedance internal to the device.

The CB stage often follows a CE stage in a configuration known as **cascode**. The CB provides a low impedance load to the CE, nearly eliminating the Miller effect. The dramatically improved reverse isolation helps with the stability and gain of the cascode circuit. The cascode is common in low-noise amplifiers, where it can provide higher gain with minimal negative impact on noise performance. The CB transistor is often smaller in area or width, which helps reduce the parasitic capacitive component of the output impedance.

Table 7-5 Performance matrix of various single-transistor amplifier topologies

Topology	CE and CS	CE with emitter degeneration	CB and CG	CC and CD
Input impedance	High	$Z_{inCE} + (1 + h_{21})Z_E$ (inductive Z_E best)	$1/g_m$ (low)	$h_{21}Z_L$ (highest)
Voltage gain	$-g_m Z_L$ (high)	$-g_m Z_L/(1 + g_m Z_E)$ (lower than CE)	$g_m Z_L$ (high)	Near 1
Current gain	$-h_{21}$ (high)	$-h_{21}$ (high)	Near 1	h_{21} (high)
Output impedance	High	High	$\propto h_{21}Z_S$ (highest)	$1/g_m$ (low)
Reverse isolation	Fair (Miller effect reduces bandwidth)	Worse than CE (since feedback is present)	Good	Poor (but increases bandwidth)
Noise	Low	Higher than CE if resistive Z_E	Same as CE	High (output noise fed back into input)
Stability	Fair (Miller effect)	Somewhat better than CE	Good (but destabilized by feedback)	Poor (especially with capacitive load)
Linearity	Fair for BJT, good for FET	Better than CE (esp. for BJT)	Same as CE	Good if $Z_L > 1/g_m$

The common collector (CC), or common drain (CD) for FET devices, topology (Figure 7-27(d)) is also known as the **emitter follower**. The distinctive features of this stage are the predictable low output impedance, very high input impedance, and good linearity. The drawbacks include a lack of voltage gain, resulting in lower power gain, and poor reverse isolation. The CC stage is usually employed in the output stages of an amplifier block, where driving a low load impedance with good linearity is required. The poor reverse isolation of the CC stage impairs its noise performance; as a result, it is not optimal as the first stage of a low-noise amplifier. Although the parasitic feedback can actually improve the bandwidth of the CC stage (in contrast to the CE stage), it has a destabilizing effect, especially with a capacitive load. This destabilizing effect can be exploited as part of an oscillator circuit.

7.3 Measurement of Active Devices

7.3.1 DC Characterization of Bipolar Transistor

We commence our analysis with the Ebers-Moll equations (7.10) and (7.11), re-expressed as collector and base currents:

$$I_C = I_S(e^{V_{BE}/V_T} - e^{V_{BC}/V_T}) - \frac{I_S}{\beta_R}(e^{V_{BC}/V_T} - 1) \tag{7.46a}$$

$$I_B = \frac{I_S}{\beta_F}(e^{V_{BE}/V_T} - 1) + \frac{I_S}{\beta_R}(e^{V_{BC}/V_T} - 1). \tag{7.46b}$$

The unknown coefficients to be determined through measurements are I_S, β_R, and β_F. In addition, forward and reverse Early voltages V_{AN} and V_{BN} become important when the BJT is operated with large V_{CE}. To separate forward and reverse current gain measurements, we resort to two measurement protocols, shown in Figure 7-28.

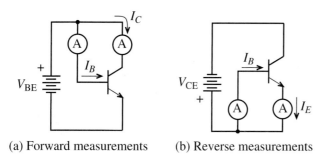

(a) Forward measurements (b) Reverse measurements

Figure 7-28 Forward and reverse measurements to determine Ebers-Moll BJT model parameters.

Under the forward measurement condition, the base-collector is short-circuited ($V_{BC} = 0$), simplifying (7.46) to

$$I_C = I_S(e^{V_{BE}/V_T} - 1) \tag{7.47a}$$

$$I_B = \frac{I_S}{\beta_F}(e^{V_{BE}/V_T} - 1) \tag{7.47b}$$

Monitoring the base and collector currents as a function of V_{BE} results in the graph shown in Figure 7-29.

Both currents are logarithmically plotted and shown for sufficiently large V_{BE} values, where the exponential terms dominates over the factor 1. A linear slope of $1/V_T$ for both currents is obtained, since

$$\ln I_C = \ln I_S + \frac{V_{BE}}{V_T} \tag{7.48a}$$

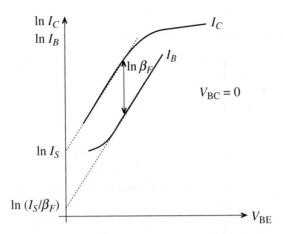

Figure 7-29 I_C and I_B versus V_{BE}.

$$\ln I_B = \ln I_S - \ln \beta_F + \frac{V_{BE}}{V_T}. \qquad (7.48b)$$

From these two curves, we can first extrapolate the collector current to obtain $\ln I_S$, and thus I_S. Extrapolating the base current yields a value for $\ln I_S - \ln \beta_F$, from which we can determine β_F. In general, the current gain is constant only over a narrow base-emitter voltage range. For low and high current injections, significant deviations occur. The Early effect is expressed as a linear gradient of the collector current:

$$I_C = I_S(e^{V_{CE}/V_T} - 1)\left(1 + \frac{V_{CE}}{V_{AN}}\right) \approx I_S e^{V_{CE}/V_T}\left(1 + \frac{V_{CE}}{V_{AN}}\right). \qquad (7.49)$$

This allows us to find V_{AN} by projecting a tangent, applied to the collector current in the active region, to the intercept point with the V_{CE}-axis in the second quadrant. The intercept point is the same for various base currents, as shown in Figure 7-12. The determination of the reverse mode parameters β_R, V_{BN} is carried out by interchanging the collector with the emitter terminal (see Figure 7-28(b)), and then following the identical procedure as done in the forward direction.

7.3.2 Measurements of AC Parameters of Bipolar Transistors

The determination of the AC parameters is more of a challenge depending on the model involved and the details required. Extracting the large-signal Ebers-Moll or Gummel-Poon circuit elements analytically is an actively pursued research endeavor. For our purposes, we concentrate on the small-signal low-frequency circuit model shown in Figure 7-30.

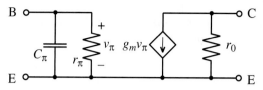

Figure 7-30 Small-signal, low-frequency model for parameter extraction.

This model is related to the hybrid-π model presented in Figure 7-14, but without the output feedback ($h_{12} = 0$) and ohmic contributions $r_B \approx r_E \approx r_C \approx 0$. For a Q-point in the forward active region, and consistent with (7.15)–(7.20), we can derive the following parameters:

Transconductance

$$g_m = \left.\frac{dI_C}{dV_{BE}}\right|_{V_{CE}=0} = \frac{I_C^Q}{V_T} \qquad (7.50a)$$

Input capacitance

$$C_\pi = \tau_{be}\frac{I_S}{V_T}e^{V_{BE}^Q/V_T} = \tau_{be}\frac{I_C^Q}{V_T} = g_m \tau_{be} \qquad (7.50b)$$

Input resistance

$$r_\pi = \left.\frac{dV_{BE}}{dI_B}\right|_{V_{CE}^Q} = \left.\frac{v_{be}}{i_b}\right|_{v_{ce}=0} = \frac{\beta_0}{g_m} \qquad (7.50c)$$

Output conductance

$$\frac{1}{r_0} = \left.\frac{dI_C}{dV_{CE}}\right|_{V_{BE}^Q} = \frac{I_C^Q}{V_{AN}} \qquad (7.50d)$$

where it is understood that the collector current in the presence of the Early effect is given by $I_C = \beta_0 I_B (1 + V_{CE}/V_{AN})$. Furthermore, since we decided to operate in the forward active mode, C_π denotes the diffusion capacitance, with the forward transit time τ_{be} of the base-emitter diode.

The parameter extraction for this simplified hybrid-π model begins with the process of setting the desired Q-point, resulting in known I_C^Q, I_B^Q. V_{AN} is extracted from the slope of the I-V curve. Thus, a measurement protocol would sequence through the following steps:

- Transconductance $g_m = I_C^Q / V_T$ for a given junction temperature
- DC current gain $\beta_0 = I_C^Q / I_B^Q$
- Input resistance $r_\pi = \beta_0 / g_m$

- Output resistance $r_0 = V_{AN}/I_C^Q$
- Input impedance $Z_{in} = (1/r_\pi + j\omega C_\pi)^{-1}$ recorded at a particular angular frequency and then solved for the capacitance C_π.

Instead of recording the input impedance and indirectly determining C_π, we can more elegantly find the transition frequency and thus C_π. This is accomplished by noting that the AC current gain at the transition frequency f_T is unity:

$$\frac{|i_c|}{|i_b|} = \left|\frac{\beta_0}{1+j\omega_T r_\pi C_\pi}\right| \equiv 1. \tag{7.51}$$

Knowing that $\beta_0 \gg 1$ leads to $f_T \approx \beta_0/(2\pi C_\pi r_\pi)$, it follows that

$$C_\pi \approx \frac{\beta_0}{2\pi f_T r_\pi}. \tag{7.52}$$

This approach can be implemented quite easily with a network analyzer. Sweeping the frequency until $|h_{fe}(\omega)|$ is equal to unity would allow us to enforce (7.51). The resulting transition frequency can then be substituted into (7.52) to find C_π. It is typically sufficient to record a single frequency point f_{meas} (less than f_T), and evaluate f_T as the gain-bandwidth product: $f_T = |h_{fe}(f_{meas})|f_{meas}$.

Example 7-6. Small-signal hybrid-π parameter extraction without Miller effect

An *npn* transistor is operated under DC bias of $I_C^Q = 6$ mA, $I_B^Q = 40$ μA, and the Early voltage is recorded to be $V_{AN} = 30$ V. Through a network analyzer measurement, the transition frequency is determined to be $f_T = 37$ GHz at room temperature. It is required to determine the hybrid-π parameters: β_0, r_π, C_π, r_0, and g_m.

Solution. Neglecting feedback from the output to the input, we can use the preceding equations directly and find

$$g_m = \frac{I_C^Q}{V_T} = 232 \text{ mS}.$$

The forward DC current gain β_0 of the transistor can be found simply as a ratio of the collector current to the base current:

$$\beta_0 = I_C^Q/I_B^Q = 150.$$

From the known β_0 and transconductance g_m, we find the input resistance as $r_\pi = \beta_0/g_m = 647\ \Omega$. The output resistance is a ratio of the forward Early voltage to the collector current $r_0 = V_{AN}/I_C^Q = 5\ \text{k}\Omega$. Finally, the capacitance is found from (7.52):

$$C_\pi = \frac{\beta_0}{2\pi f_T r_\pi} = 1.00\ \text{pF}.$$

The small-signal parameter determination is almost a cookbook design process. However, constant forward current gain may not always reflect realistic transistor behavior.

While Example 7-6 is applicable for low- and medium-range frequencies, the situation becomes more complicated for values approaching 1 GHz and beyond. Here, we cannot neglect the Miller effect, and our attempt must be directed toward finding a strategy to obtain C_μ. As discussed in Chapter 4, electric measurements at high frequencies cannot rely on impedance, admittance, or h-parameter determination because of the difficulties associated with enforcing short- and open-circuit conditions. At these frequencies, we must resort to S-parameter measurements. The use of S-parameters to find the feedback capacitor C_μ is explained in the following example.

Example 7-7. Small-signal hybrid-π parameter extraction with Miller effect included

We re-examine the previous example, but this time use the network analyzer to record the following S-parameters based on the characteristic impedance of 50 Ω at 500 MHz:

$$[\mathbf{S}] = \begin{bmatrix} 0.86 e^{-j19.5°} & 0.002 e^{j80.9°} \\ 21.1 e^{j170.9°} & 0.98 e^{-j1.5°} \end{bmatrix}.$$

Our goal is to find the feedback capacitance C_μ. In addition, we would like to observe how the input and output impedances are affected if C_μ is excluded.

Solution. Since the DC measurements do not change, we will not repeat them. For given S-parameters, we can compute the input impedance of the transistor (under open-circuited output) using matrix transformations described in Chapter 4:

$$Z_{in} = Z_{11} = Z_0 \frac{(1+S_{11})(1-S_{22})+S_{12}S_{21}}{(1-S_{11})(1-S_{22})-S_{12}S_{21}}$$

$$= R_{in} + jX_{in} = (6.1 - j29.4) \; \Omega$$

Setting the input impedance equal to the circuit model yields

$$Z_{in} = \frac{1}{1/r_\pi + j\omega(C_\pi + C_{M1})}$$

where C_{M1} is the Miller-transformed capacitance. Rearranging this equation leads to the form

$$C_{M1} = \frac{1}{\omega} \text{Im}\left(\frac{1}{Z_{in}}\right) - C_\pi$$

where $\omega = 2\pi f$ is the angular frequency at which the S-parameters are recorded. Explicitly, we find $C_{M1} = 10.4 \text{ pF} - 1.00 \text{ pF} = 9.4 \text{ pF}$. To compute the actual feedback capacitance C_μ, we can use (7.25), where the ratio of collector-emitter to base-emitter voltage under open-circuited output is equal to $(h_{12} - h_{11}h_{22}/h_{21})^{-1}$. This yields finally $C_\mu = 6.9 \text{ fF}$.

To compute the frequency behavior of the input and output impedances, we can first calculate the h-parameters of the transistor as given by (7.24) and then convert them into Z-parameter representation. Both input and output impedances are plotted with and without the feedback ($C_\mu = 0$) in Figure 7-31.

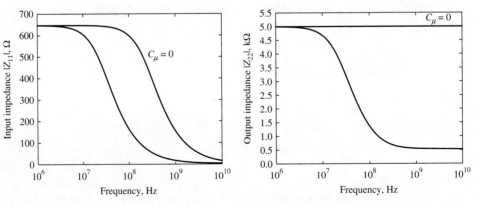

Figure 7-31 Input and output impedances with and without feedback.

This example underscores the importance of including the feedback effect once the frequency begins to exceeds 100 MHz.

Although the preceding examples are simple extraction cases, they convey an appreciation of how difficult a realistic situation can become if the entire SPICE parameter set is to be extracted. For the nonlinear large-signal circuit models, this is a research task with no clear solution methodology. Many manufacturers have therefore resorted to S-parameter characterization alone. This approach greatly simplifies BJT characterization by utilizing an appropriate test fixture, or jig, and relying on a network analyzer to measure the S-parameters at certain bias conditions and operation frequencies.

7.3.3 Measurements of Field Effect Transistor Parameters

Because the GaAs MESFET has gained such prominence in many RF circuits, it is important to take a closer look at its parameter extraction. Since the circuit model is the same for the HEMT, we can treat both cases in parallel. The fundamental equation for the drain current in the linear, or triode, region is derived in Chapter 6 and is repeated here for convenience:

$$I_D = \mu_n \frac{\varepsilon W}{d L} \left\{ (V_{GS} - V_{T0})V_{DS} - \frac{1}{2}V_{DS}^2 \right\} \approx \beta(V_{GS} - V_{T0})V_{DS}. \quad (7.53)$$

The only difference between MESFET and HEMT lies in the definition of the threshold voltage V_{T0}. Specifically, with the Schottky barrier voltage V_d, pinch-off voltage V_P, and energy difference ΔW_c between the conduction bands of the heterostructure in a HEMT, we obtain the following two expressions:

$$V_{T0} = V_d - V_P \quad \text{(MESFET)} \quad (7.54a)$$

$$V_{T0} = V_d - \Delta W_c/q - V_P \quad \text{(HEMT)}. \quad (7.54b)$$

In the saturation region, when $V_{DS} \geq V_{GS} - V_{T0}$, (7.53) becomes

$$I_D = I_{Dsat} = \beta(V_{GS} - V_{T0})^2. \quad (7.55)$$

Using (7.55), we can easily extract values for the conduction parameter β and the threshold voltage V_{T0} by plotting the square root of the drain current versus the applied gate-source voltage V_{GS}. A measurement arrangement of a MESFET for obtaining V_{T0} and β is shown in Figure 7-32.

The threshold voltage is determined indirectly by setting two different gate-source voltages V_{GS1} and V_{GS2}, while maintaining a constant drain-source voltage

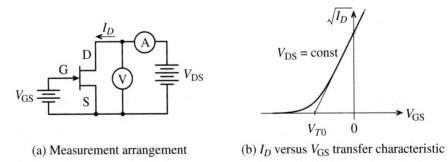

(a) Measurement arrangement (b) I_D versus V_{GS} transfer characteristic

Figure 7-32 Generic measurement arrangement and transfer characteristics in saturation region.

$V_{DS} = \text{const} \geq V_{GS} - V_{T0}$, such that the transistor is operated in the saturation region. The result of these two measurements gives

$$\sqrt{I_{D1}} = \sqrt{\beta}(V_{GS1} - V_{T0}) \tag{7.56a}$$

$$\sqrt{I_{D2}} = \sqrt{\beta}(V_{GS2} - V_{T0}). \tag{7.56b}$$

Here, we assume that the channel length modulation effect is negligible. Therefore, the measured current is close to the saturation drain current as given by (7.55). Taking the ratio of (7.56a) to (7.56b) and solving for V_{T0}, we obtain

$$V_{T0} = \frac{V_{GS1} - (\sqrt{I_{D1}}/\sqrt{I_{D2}})V_{GS2}}{1 - \sqrt{I_{D1}}/\sqrt{I_{D2}}}. \tag{7.57}$$

Next, we substitute (7.57) into (7.56a) and solve this equation for β. The extraction process can further be simplified if we choose $I_{D2} = 4I_{D1}$ so that (7.57) becomes $V_{T0} = 2V_{GS1} - V_{GS2}$. Upon substituting this expression in (7.56a), we see that $\beta = I_{D1}/(V_{GS2} - V_{GS1})^2$. The reactive components are typically extracted from the S-parameters, as is done for the BJT.

7.4 Scattering Parameter Device Characterization

The S-parameter measurement approach greatly simplifies the device-under-test (DUT) characterization by utilizing an appropriate test fixture, or jig, and relying on a vector voltmeter or network analyzer to record the four frequency- and bias-dependent S-parameters.

Although nowadays a vector voltmeter is seldom used for recording the S-parameters, it nonetheless allows us to gain valuable insight into the basic measurement procedure that is also at the heart of a network analyzer. We will, therefore, investigate this approach first. It is generically depicted in Figure 7-33, and requires an RF signal generator, two dual directional couplers, transistor biasing networks, the actual transistor fixture, and calibration kit to create short-circuit and through-line conditions.

Scattering Parameter Device Characterization

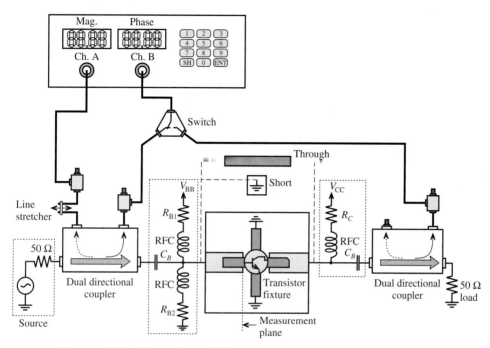

Figure 7-33 Recording of S-parameters with a vector voltmeter.

The function of a **dual directional coupler** in Figure 7-33 is to isolate the incident from the reflected power wave. This is accomplished as shown in Figure 7-34, where a cross-sectional view of a coaxial coupler is depicted. For incident power coming from the left through the main arm, two slots spaced $\lambda/4$ apart couple the energy into an auxiliary path to port 4. The incident wave does not produce any coupling toward port 3, since there is a 180° phase delay between signals coming from slot B and slot A, essentially canceling the entire wave. However, a reflected wave from the DUT will enter the coupler at port 2 coming from the right and subsequently couple out the wave energy through the auxiliary path to port 3, canceling any wave leaving port 4. Therefore, port 3 provides an output for the reflected power, whereas port 4 records the incident power. The two figures of merit for a directional coupler are the **coupling factor** cf and its **directivity factor** df. The factor cf is defined as

$$cf\,[\text{dB}] = 10\,\log\left(\frac{P_i}{P_n}\right) \qquad (7.58a)$$

and denotes the logarithmic ratio of the power in the main port, either 1 or 2 ($i = 1, 2$), to the power in the auxiliary port, either 3 or 4 ($n = 3, 4$). The directivity

$$df\,[\text{dB}] = 10\,\log\left(\frac{P_3}{P_4}\right) \qquad (7.58b)$$

specifies the ratio of the powers in the auxiliary arms for the condition of forward power applied to the main port 1 and port 2 being terminated into its characteristic impedance. For high signal discrimination we expect to see a large directivity value.

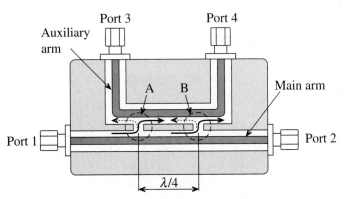

Figure 7-34 Cross-sectional view of directional coupler and signal path adjustment.

The actual signal propagation paths are observed in Figure 7-33. Here, the vector voltmeter records with channels A and B the incident and reflected powers from the input port of the active device. Taking the ratio of the voltage magnitudes yields $|S_{11}|$. For recording the phase angle, it is important to obtain an appropriate phase reference. For this reason, the DUT is removed and a short circuit is inserted for phase reference. To ensure equal path length (i.e., from the signal source to channel A, and from the short to channel B), a line stretcher is used to perform the necessary adjustment to achieve a zero phase difference.

The same test setup can also be utilized to find the forward gain S_{21}. Switching channel B to the directional coupler situated on the output side of the DUT yields the ratio between the output and input voltages, or $|S_{21}|$. The phase adjustment now calls for replacing the DUT with a through section element and, again, equalizing the signal paths with the line stretcher.

The remaining two S-parameters, S_{22} and S_{12}, are measured by reversing the DUT jig and exchanging the biasing networks. As Figure 7-33 implies, the S-parameter measurements depend on the setting of an appropriate bias or Q-point and the signal source frequency. Thus, it is possible to generate, a wide range of parametric curves.

Instead of employing a vector voltmeter, the more popular approach involves the use of the network analyzer. This instrument is capable of processing magnitude and phase information of a single or dual-port RF network. A simplified block diagram highlighting the functionality is shown in Figure 7-35.

Scattering Parameter Device Characterization

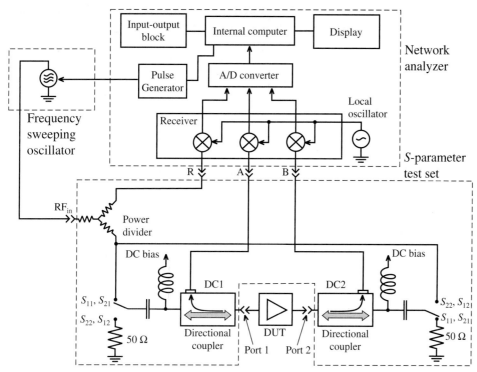

Figure 7-35 Block diagram of a network analyzer with S-parameter test set.

The advantage of a network analyzer lies in the fact that all the separate functional units associated with the vector voltmeter based measurement procedures are incorporated into one instrument for entirely automated testing of the RF or MW device. The operation is such that a sweeping RF generator applies the RF signal to the directional couplers. In the forward direction, the reference channel R records the incident power wave and channel A provides the S_{11} parameter via directional coupler 1 (DC1). At the same time, parameter S_{21} is recorded via directional coupler 2 (DC2). Switching to the reverse direction, the reference channel R records the incident power launched into port 2 of the device under test, while channel B records S_{22} and channel A then yields S_{12}. This arrangement allows electronic switching between calibration and testing conditions, permitting the recording of the entire S-parameter set without changing the test jig. An interfrequency mixing and amplification stage feeds the signal into an analog-to-digital conversion unit and subsequently into a microcomputer and display system. The computer provides the user with the computed S-parameters (in magnitude and phase) as well as such postprocessed parameters as group delay, return and insertion losses, voltage standing wave ratio, input and output impedances, and many additional features.

The computer system allows for software compensation of many imperfections introduced by the test arrangement. As a case in point, we recall the recording of the S-parameters in Section 4.4.7 via the through-reflect-line (TRL) technique. This is only one of a number of calibration schemes proposed to compensate for the various error sources introduced by the measurement process.

Practically Speaking. Modeling an RF transistor using circuit simulation software

In this section, we examine how a modern circuit simulator can be used to analyze a typical RF transistor. The software program that is employed is Advanced Design Systems (ADS) of Agilent Technologies, although similar circuit simulators can be found by other vendors. We create a schematic of a simple amplifier using the manufacturer-supplied model of a commercially available HEMT, simulate its S-parameters and compare the results to the S-parameter measurements published in the datasheet. We also take a closer look at how the transistor is actually modeled.

Figure 7-36 presents the schematic of the modeled amplifier. The device chosen for the simulation is the Agilent ATF551M4, a so-called Enhancement Mode Pseudomorphic HEMT (E-pHEMT) designed for low-noise amplifier applications in the 450 MHz– 10 GHz range. The manufacturer has supplied a nonlinear ADS model of this transistor (available at Agilent's website www.agilent.com). The transistor package has two source connections, each of which is attached to ground through two via connections, as recommended in the datasheet. The HEMT is biased using an active biasing circuit with two matched low-frequency *pnp* transistors. Details of biasing circuits are deferred to Chapter 8. It is sufficient, however, to mention that the supplied biasing circuit (suggested in an Agilent application note) keeps the HEMT drain voltage and current reasonably independent of temperature and device-to-device variations. The biasing conditions are $V_{DS} = 3$ V and $I_D = 20$ mA. If we examine the datasheet, this biasing point requires $V_{GS} \approx +0.53$ V, permitting the use of a simple single-ended power supply (represented by the 5 V DC source in the schematic). The biasing point was chosen so that the transistor achieves close to maximum gain, lowest distortion and lowest noise. It is also one of the few bias points with *S*-parameters listed in the datasheet.

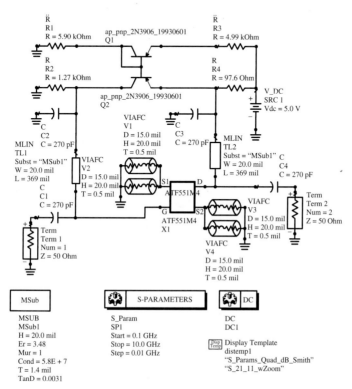

Figure 7-36 Circuit schematic of a 5 GHz amplifier with a HEMT device mounted on a simulated PCB with a biasing network and an *S*-parameter measurement setup.

The RF signal is blocked from entering the DC biasing circuit using a filter consisting of a microstrip transmission line and a bypass capacitor. The transmission line length was chosen to be $\lambda/4$ in length at 5 GHz. This ensures that the DC biasing circuit behaves like an open circuit. The 5 GHz value is the center frequency for our amplifier, where the simulated and measured *S*-parameters should match. As we know from Chapter 2, for other frequencies, the transmission line no longer behaves like an open circuit, thus influencing the *S*-parameter coefficients. The schematic feature MSub1 specifies the PCB substrate parameters necessary for the simulation of the microstrips. For this particular circuit, a Rogers 4350 substrate is used, with parameters derived from the datasheet.

The gate and drain of the HEMT are connected through DC blocking capacitors to S-parameter measurement ports (similar to a network analyzer). Other features on the schematic include the S-parameters setup block (SP1) specifying a frequency sweep from 0.1 to 10 GHz, the DC simulation block (DC1) and a display template for S-parameters.

Before looking at the simulation results, let us examine the HEMT model in greater detail. Figure 7-37 presents the transistor model, whose circuit details are revealed by employing the "Push Into Hierarchy" command of the ADS schematic editor. It is a subcircuit that includes the transistor die model and the package parasitics modeled as capacitors, inductors, and microstrip sections. The transistor model is represented by the device symbol FET1, which is linked to the model block MESFETM1 that specifies all the model parameters. Here, the built-in "Advanced Curtice Quadratic" GaAs FET model is used. This is one of the simplest MESFET models in use today, and is similar to the model depicted in Figure 7-24. Figure 7-38 shows the basic circuit topology of this model, which is virtually identical in most circuit-based MESFET models. Most of the circuit elements in this model are simple resistors/inductors/capacitors, with the exception of the bias-dependent (nonlinear) current source I_D with the associated R_{DS}, and the junction capacitances C_{GS} and C_{GD}. In the Curtice Quadratic model, we have the following current representation

$$I_D = \beta(V_{GS} - V_{TO})^2 (1 + \lambda V_{DS}) \tanh(\alpha V_{DS})$$

and

$$C_{GS} = \frac{C_{GS0}}{\left(1 - \frac{V_{GS}}{V_{BI}}\right)^{0.5}}$$

where the parameters are defined in the MESFETM1 block: α = Alpha = hyperbolic tangent parameter, β = Beta = transconduction parameter, λ = Lambda = channel length modulation, V_{TO} is the threshold voltage, C_{GS0} is the zero-bias gate-source capacitance (C_{GS} in the parameter listing) and V_{BI} is the built-in junction potential.

Scattering Parameter Device Characterization

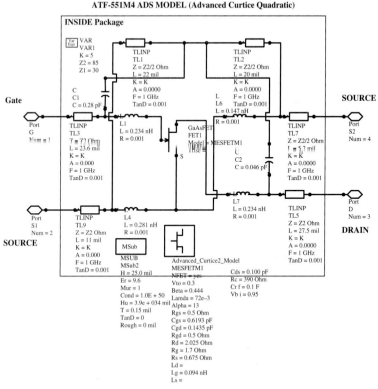

Figure 7-37 ADS circuit model for the HEMT including the transistor model and package parasitics.

We point out a small difference in the formulation of I_D: the inclusion of a hyperbolic tangent factor. This allows us to rely on a single continuous formula in both the linear, or triode, and the saturation regions at the expense of some accuracy. This is often convenient, as continuous functions simplify the parameter fitting for individual transistors.

Running the ADS simulation produces the S-parameter plots seen in Figure 7-39. We can observe that without any additional impedance matching, the transistor provides about 13 dB of gain (S_{21}) at 5 GHz and the gain generally decreases with frequency, as expected. However, the gain also drops substantially at low frequencies and near 10 GHz because the biasing feeds look more like short

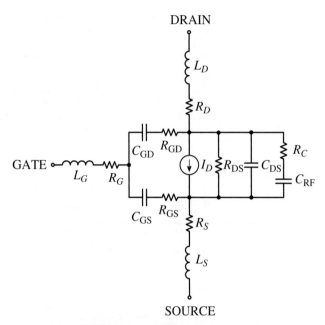

Figure 7-38 Basic circuit topology of the Curtice Quadratic GaAs MESFET model employed for the transistor die.

circuits at these frequencies. The reverse gain (S_{12}) is below −20 dB, which can be regarded as good reverse isolation. The input and output ports are not matched to 50 Ω, as seen by observing the S_{11} and S_{22} plotted in the Smith Charts. Thus, additional matching circuits may be needed in a final LNA design; we will take on these issues in Chapters 8 and 9.

A marker has been placed in each plot to examine the numerical S-parameter values at 5 GHz. This is also seen in the table below the graphs. We can now compare the model predictions with the datasheet values: $S_{11} = 0.662\angle 175.3°$, $S_{21} = 4.099\angle 59.4°$ (12.25 dB), $S_{12} = 0.076\angle 19.2°$ (−22.4 dB), $S_{22} = 0.270\angle -71.5°$. The agreement is not perfect, but this is understandable in light of the fact that accuracy sacrifices had to be made in order to produce a universally applicable model. Also, some of the observed discrepancy is due to the different ground-source connections that are made when measuring the datasheet's S-parameters.

The simulation results furthermore allow examining the gain, stability, and noise performance of the amplifier in detail. These topics will be discussed in Chapter 9.

Summary

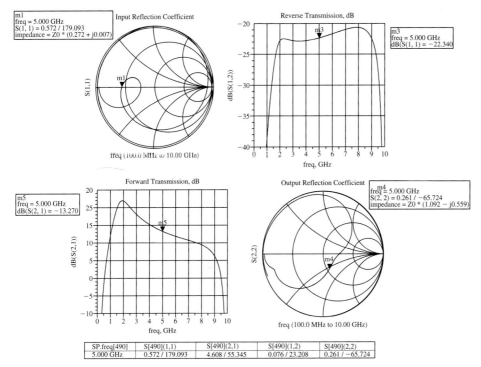

Figure 7-39 ADS S-parameter sweep results for the 5 GHz amplifier.

7.5 Summary

Electric circuit models for active devices form the backbone of most CAD software packages. These circuits range from simple linear models to very sophisticated large-signal models. Specifically, a large-scale BJT SPICE model that takes into account temperature influences can involve over 40 adjustable parameters, whose determination is a daunting task.

In this chapter, we reviewed the basic large-scale diode model that is used for both the conventional *pn*-junction diode and the Schottky diode. Junction and diffusion capacitances and the temperature-dependent saturation current are the key ingredients constituting this model. By identifying a bias or *Q*-point and considering only small-signal responses, we arrive at the linear diode model with the differential conductance and diffusion capacitance

$$G_d = \frac{1}{R_d} = \left.\frac{dI_D}{dV_A}\right|_{V_Q} = \frac{I_Q}{nV_T} \quad \text{and} \quad C_d = \frac{I_S \tau_T}{nV_T} e^{V_Q/(nV_T)} \approx \frac{I_Q \tau_T}{nV_T}.$$

The diode model is used as the basic building block to develop the static large-scale BJT model, as originally proposed by Ebers and Moll. Issues such as forward active and reverse active operating regions are explained by simplifying the basic Ebers-Moll equations. Starting from the injection model, we converted the Ebers-Moll BJT equations to the transport representation and subsequently to the large-scale BJT model in forward active mode. Additional refinements and modifications of the Ebers-Moll model have resulted in the more sophisticated Gummel-Poon model, whose large-signal forward active mode circuit is shown in Figure 7-13. For the small-signal representation, the hybrid-π is a popular linearization of the large-scale Ebers-Moll representation. The hybrid-π parameters are computed for a given collector current operating point:

$$g_m = I_C^Q/V_T, \quad r_\pi = \beta_0/g_m, \quad \beta_F|_Q = \beta_0, \text{ and } 1/r_0 = I_C^Q/V_{\text{AN}}.$$

For high-frequency operation, the capacitive coupling between input and output ports significantly influences the transistor operation. By taking into account the so-called Miller effect, the collector-base capacitance is transformed into input and output capacitances, thus permitting us, again, to separate the two ports if the voltage gain is known. Since lead inductances and resistances also influence the high-frequency performance, we go through a detailed design project to investigate, among other topics, how the input and output impedances are affected as the frequency increases.

Attention is next directed toward FET circuit models, specifically the relevant high-frequency types of MESFET and HEMT. Saturation, linear (triode), reverse saturation, and reverse linear (triode) regions are defined in close relation with Chapter 6. Specifically, the drain currents in the saturation region

$$I_D = \beta_n(V_{\text{GS}} - V_{T0})^2(1 + \lambda V_{\text{DS}})$$

and in the linear region

$$I_D = \beta_n[2(V_{\text{GS}} - V_{T0})V_{\text{DS}} - V_{\text{DS}}^2](1 + \lambda V_{\text{DS}})$$

form the basis of the static and dynamic circuit models. Of particular interest are the small-signal low- and high-frequency FET models. The cutoff frequency allows us to quantify the frequency limitations of the device. For low to medium frequency devices, the charging time of the device's capacitances determines the frequency performance, whereas in high speed devices, the channel transit time becomes the limiting factor.

Finally, we discuss how to extract the electric parameters of the active devices. For the DC characterization of the BJT, we can primarily rely on the collector and base currents as functions of base-emitter voltage. From these curves, the saturation current, current gain, and Early voltage are obtained. Measurement of the AC parameters is more of a challenge, and only the linear hybrid-π model allows a cookbook approach, as outlined by equations (7.50). FET model characterization follows a similar path as outlined for the DC BJT model, and involves recording the drain current versus gate-source voltage.

In many cases, both for BJT and FET, the S-parameter representation is the most common way to characterize an active device for a given bias and operating frequency. For this purpose, either a vector voltmeter or network analyzer is used to record the input/output power waves of the device under test. Measurements with the vector voltmeter require directional couplers, signal sources, switches, and a forward and reverse measurement protocol. This is all automated by connecting an S-parameter test set to the three channels of a network analyzer. The recording of S_{11}, S_{22}, S_{21}, and S_{12} for particular bias conditions and operating frequencies generally provides sufficient information for the circuit designer to characterize the device.

Further Reading

P. Antognetti and G. Massobrio, *Semiconductor Device Modeling with SPICE,* McGraw-Hill, New York, 1988.

J. J. Ebers and J. L. Moll, "Large-Scale Behaviour of Junction Transistors," *Proc. of IRE,* Vol. 42, pp. 1761–1778, December 1954.

H. K. Gummel and H. C. Poon, "An Integral Charge Control Model of Bipolar Transistors," *Bell System Tech. Journal,* Vol. 49, pp. 827–851, 1970.

T.-H. Hsu and C. P. Snapp, "Low-Noise Microwave Bipolar Transistor with Sub-Half-Micrometer Emitter Width," *IEEE Trans. on Electron Devices,* Vol. ED-25, No. 6, June 1978.

E. S. Yang, *Microelectronic Devices,* McGraw-Hill, NY, 1988.

B. Razavi, *RF Microelectronics,* Prentice Hall, 1998.

Y. Tsividis, *Operation and Modeling of the MOS Transistor,* McGraw-Hill, 2nd edition, 1999.

R. Jaeger and T. Blalock, *Microelectronic Circuit Design,* McGraw-Hill, 2nd edition, 2003.

R. T. Howe and C. G. Sodini, *Microelectronics: An Integrated Approach,* Prentice-Hall, 1997.

F. Ellinger, *Radio Frequency Integrated Circuits and Technologies,* Springer-Verlag, 2007.

Problems

7.1 A silicon *pn*-junction diode has the following parameters at $T = 300$ K: $I_S = 5 \times 10^{-15}$ A, $n = 1.2$, $\tau_T = 100$ ps, and $R_S = 10\ \Omega$. Assuming that the diode is operated under such biasing conditions that the applied junction voltage is maintained at 0.7 V, find the differential resistance and the diode capacitance for temperatures ranging from 200 to 450 K.

7.2 The reverse saturation current of a *pn* diode is $I_S = 0.01$ pA at $T_j = 25°C$, and it has an emission coefficient of 1.6. For a junction temperature of 120°C, find the reverse saturation current and the diode current I_D at an applied diode voltage of $V_A = 0.8$ V.

7.3 The task for a process engineer is to obtain the model parameters for a Schottky diode. From measurements, it is determined that the saturation current is equal to $I_S = 2$ pA. To obtain the remaining parameters (n and τ_T), the engineer decides to use the differential capacitance of the diode. It is assumed that the electric measurements at room temperature indicate a diffusion capacitance of $C_d = 0.329$ pF at an applied junction voltage of $V_A = 0.5$ V, and $C_d = 0.371$ nF at $V_A = 0.7$ V. Find the emission coefficient n and the transit time τ_T.

7.4 A GaAs Schottky diode with a gold contact is operated at 80 mA. The following parameters are given at 300 K: $\tau_T = 40$ ps, $R_S = 3\,\Omega$, $n = 1.2$, $I_S = 10^{-14}$ A. (a) Plot the magnitude of the small-signal impedance behavior in the frequency range from 1 MHz to 5 GHz. (b) Repeat the calculations for a temperature of 400 K.

7.5 For the diode configuration shown below, compute the S-parameters of the circuit when the control voltage equals either +5 V or –5 V, and the frequency ranges from 1 MHz to 10 GHz. The diode model parameters are $I_S = 5 \times 10^{-15}$ A, $n = 1.2$, $\tau_T = 100$ ps, $m = 0.5$, $C_{J0} = 10$ pF, $V_{\text{diff}} = 0.7$ V, and $R_S = 10\,\Omega$. The ambient operating temperature is $T = 300$ K, and we set infinite values for the blocking capacitors and RFCs.

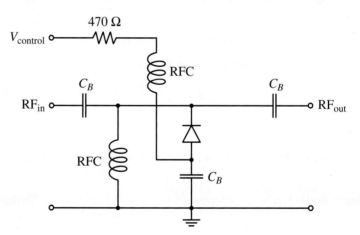

7.6 Determine the change in the forward-bias voltage of an ideal Si pn-junction diode with a change in temperature from –20°C to 80°C. Assume that current is kept constant and the initial bias voltage was 0.7 V at $T = 300$ K.

7.7 Find the maximum operating frequency of the ideal pn-junction diode whose parameters are given in Example 7-1. Assume $I_Q = 1$ mA. The maximum frequency can be estimated based on the RC time constant of the diode.

7.8 Consider three ideal pn-junction diodes whose parameters are identical, except for the bandgap energy. Find the ratio of the forward-biased currents for these diodes, if the applied voltage is the same in each case and the diodes are made of Ge, Si, and GaAs, respectively.

7.9 The base terminal current is constrained to be zero in an *npn* BJT (open-circuit condition). Assuming that the device is operated at room temperature and has $\alpha_F = 0.99$ and $\alpha_R = 0.05$, use the large-signal Ebers-Moll model to find the base-emitter voltage as a function of the applied collector-emitter voltage V_{CE}.

7.10 Express the transconductance g_m of a bipolar junction transistor in terms of its collector current. Compare this expression with the expression for the differential resistance of a *pn* diode.

7.11 Show that for a small-signal transistor model, as depicted in Figure 7-16, under open-circuited output condition the input Miller capacitance can be written as $C_{M1} = (1 + g_m r_{ce})C_\mu$. In addition, obtain an upper frequency limit for which this formula is applicable.

7.12 For a hybrid-π BJT model, plot the short-circuit current gain h_{fe} in the frequency range from 10 MHz to 10 GHz. Assume the following parameters are given at a collector bias point of 20 mA and $T = 300°K$: $\beta_0 = 140$, $C_\mu = 0.1$ pF, and $C_\pi = 5$ pF.

7.13 In Example 7-4 we discussed the relatively complicated case of a microwave transistor analysis where we have taken into account effects associated with parasitic elements such as lead inductances and resistances. In most practical applications, the situation is even more complicated due to the presence of internal matching and stability networks incorporated into the transistor housing by the manufacturer.

For the internal circuit shown, compute the *S*-parameters in the frequency range of 100 MHz to 20 GHz.

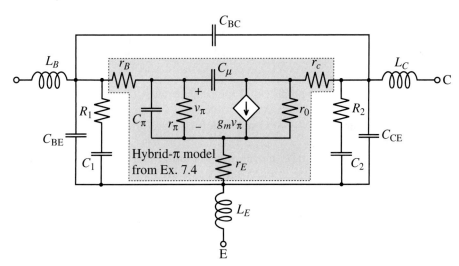

The following component values are given: $R_1 = 25\ \Omega$, $R_2 = 20\ \Omega$, $C_1 = C_2 = 0.2$ pF, $C_{BE} = C_{CE} = 0.1$ pF, and $C_{BC} = 10$ fF. Assume that the biasing conditions and the values for all inductances and components in the hybrid-π model are the same as in Example 7-4.

7.14 An easy way to determine the capacitance C_μ in the hybrid-π BJT model is to make a capacitance measurement between base and collector, as follows.

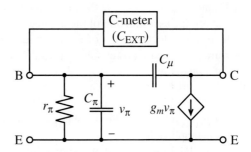

If the frequency is sufficiently low so that $1/(\omega C_\mu) \gg r_B$, we can directly relate the externally recorded capacitance to the feedback capacitance C_μ. Show that this is true by proving that the voltage v_π is zero and that r_π, C_π, and g_m do not influence the measurement. If a precision instrument measures an external capacitance $C_{ext} = 0.6$ pF at 1 MHz, can r_B, which typically ranges between 25 and 200 Ω, be neglected?

7.15 For the hybrid-π model, it is required to find the parameters r_π, r_B, and g_m from low-frequency measurements (which allow us to neglect C_μ and C_π).

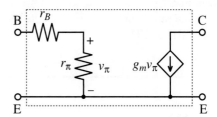

At the operating point and at room temperature (25°C), we record a DC base current of $I_B = 100$ µA at a base-emitter voltage of $V_{BE} = 1.0$ V, and a collector current of $I_C = 25$ mA. We furthermore record a low-frequency input impedance of 356 Ω.

7.16 A small-signal BJT model has the following parameters: $g_m = 40$ mS, $f_T = 600$ MHz, $r_{ce} = 2.5$ kΩ, $r_{bb'} = 125$ Ω, and $C_{b'c} = 2$ pF. A load $R_L = 50$ Ω is attached as shown.

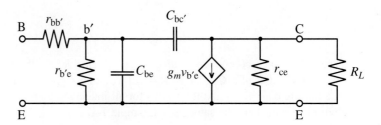

Under the assumption that $V_L \approx -g_m V_{b'e} R_L$, find the Miller capacitance C_{M1} such that the input circuit can be approximated as

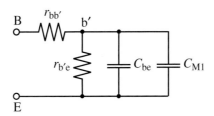

7.17 Neglecting all parasitic elements, including base, emitter, and collector resistances in the transistor described in Example 7-4, find the maximum frequency f_0, the beta cutoff frequency f_β, and the transition frequency f_T.

7.18 Obtain the *h*-parameter representation for a BJT in common-base configuration, neglecting base, emitter, and collector resistances (r_B, r_E, and r_C).

7.19 Derive the *h*-parameter representation for the following high-frequency FET model:

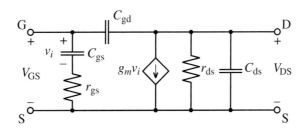

7.20 Using the equivalent circuit shown in Problem 7.19, obtain the *h*-parameter representation for a FET in common-gate configuration.

7.21 For the FET circuit model in Problem 7.19, find the equivalent input and output impedances by replacing C_{gd} with its equivalent Miller capacitances. Assume open circuit load.

7.22 For the simplified FET model shown, design a procedure to determine the capacitances C_{gs} and C_{gd}, as well as g_m.

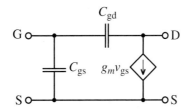

Show that for low-frequency operation, it is sufficient to record the drain current and gate-source voltage under a short circuit output condition.

7.23 FET models are often given in terms of Y-parameters, as the following generic figure shows:

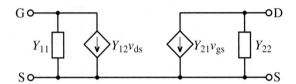

Convert this model into a P_i-network and determine its coefficients A, B, C, and D.

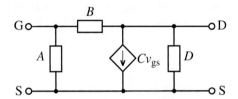

7.24 For the model parameters in Problem 7.16, plot the cutoff frequency f_T as a function of load resistance in the range $10 \ \Omega \leq R_L \leq 200 \ \Omega$.

7.25 The frequency dependence of the transistor's forward current gain h_{fe} is detailed in (7.28) and (7.29). For the transistor analyzed in Examples 7-6 and 7-7, find a frequency range where we may approximately state $|h_{fe}(\omega)| \propto 1/f$.

7.26 Given the S-parameters in Example 7-7, estimate f_T using the approximation in Problem 7.25.

8

Matching and Biasing Networks

8.1 Impedance Matching Using Discrete Components 422
    8.1.1   Two-Component Matching Networks 422
    8.1.2   Forbidden Regions, Frequency Response, and Quality Factor 431
    8.1.3   T and Pi Matching Networks 442
8.2 Microstrip Line Matching Networks 446
    8.2.1   From Discrete Components to Microstrip Lines 446
    8.2.2   Single-Stub Matching Networks 450
    8.2.3   Double-Stub Matching Networks 454
8.3 Amplifier Classes of Operation and Biasing Networks 458
    8.3.1   Classes of Operation and Efficiency of Amplifiers 458
    8.3.2   Bipolar Transistor Biasing Networks 463
    8.3.3   Field Effect Transistor Biasing Networks 469
8.4 Summary 478

As pointed out in Chapter 2, to achieve maximum power transfer, we need to match the impedance of the load to that of the source. Usually this is accomplished by incorporating additional passive networks connected between source and load. These networks are generically referred to as matching networks. However, their functionality is not simply limited to matching source and load impedances for optimal power flow. In fact, for many practical circuits, matching networks are designed not only to meet the requirement of minimum power loss, but are also based on additional constraints, such as minimizing the noise influence, maximizing power handling capabilities, and linearizing the frequency response. In a more general context, the purpose of a matching network can be defined as a transformation that converts a given impedance value to another over a frequency range.

In this chapter, we restrict our coverage to the techniques of performing impedance transformation using passive networks. The emphasis is to ensure minimum reflections between source and load. All remaining considerations, such as noise figure and linearity, are left for discussion in Chapter 9.

We commence with a study of networks based on discrete components. These circuits are easy to analyze and can be used up to frequencies in the low GHz range. Next, we continue with the analysis and design of matching networks using distributed elements, such as microstrip lines and stub sections. These networks are more suitable for operating frequencies exceeding 1 GHz, or for cases where vertical circuit dimensions are of importance, as required in RF integrated circuit designs.

To simplify our treatment and to gain clarity in the design methodology, the Smith Chart will be utilized extensively throughout as a primary design tool.

8.1 Impedance Matching Using Discrete Components

8.1.1 Two-Component Matching Networks

In a generic sense, our engineering efforts primarily strive for two main goals: first, to meet system specifications, and second, to find the most inexpensive and reliable way to accomplish this first task. The cheapest and most reliable matching networks are usually those that contain the fewest number of components.

The topic of this section is to analyze and design the simplest possible type of matching networks: so-called **two-component networks**, also known as **L-sections** or L-type networks due to their element arrangement. These networks use two reactive components to transform the load impedance Z_L to the desired input impedance Z_{in}. In conjunction with the load and source impedances, the components are alternately connected in series and shunt configurations, as shown in Figure 8-1, which depicts eight possible arrangements of capacitors and inductors.

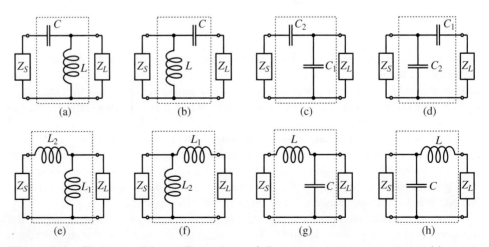

Figure 8-1 Eight possible configurations of discrete two-component matching networks.

Impedance Matching Using Discrete Components

In designing an appropriate matching network, we have two broad approaches at our disposal:

1. Derive the values of the elements analytically
2. Rely on the Smith Chart as a graphical design tool.

The first approach yields very precise results and is suitable for computer synthesis. Alternatively, the second approach is more intuitive, easier to verify, and faster for an initial design, since it does not require complicated computations. The example below details the use of the analytical approach to design a particular L-type matching network.

Example 8-1. Analytical approach to the design of an L-type matching network

The output impedance of a transmitter operating at a frequency of 2 GHz is $Z_T = (150 + j75)\Omega$. Design an L-type matching network, as shown in Figure 8-2, such that maximum power is delivered to the antenna whose input impedance is $Z_A = (75 + j15)\Omega$.

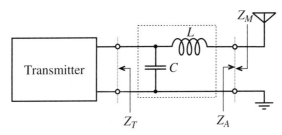

Figure 8-2 Transmitter to antenna matching circuit design.

Solution. The condition of maximum power transfer from the source to the load requires the source impedance to be equal to the complex conjugate of the load impedance. In our case this implies that the output impedance Z_M of the matching network has to be equal to the complex conjugate of Z_A, i.e., $Z_M = Z_A^* = (75 - j15)\Omega$.

The impedance Z_M can be computed as a series connection of an inductor L and a parallel combination of C and Z_T:

$$Z_M = \frac{1}{Z_T^{-1} + jB_C} + jX_L = Z_A^* \qquad (8.1)$$

where $B_C = \omega C$ is the susceptance of the capacitor and $X_L = \omega L$ is the reactance of the inductor. Expressing transmitter and antenna impedances in terms of their real and imaginary parts (i.e., $Z_T = R_T + jX_T$ and $Z_A = R_A + jX_A$), we can rewrite (8.1) as

$$\frac{R_T + jX_T}{1 + jB_C(R_T + jX_T)} + jX_L = R_A - jX_A. \tag{8.2}$$

Separating real and imaginary parts in (8.2), a system of two equations is found:

$$R_T = R_A(1 - B_C X_T) + (X_A + X_L)B_C R_T \tag{8.3a}$$

$$X_T = R_T R_A B_C - (1 - B_C X_T)(X_A + X_L). \tag{8.3b}$$

Solving (8.3a) for X_L and substituting into (8.3b) results in a quadratic equation for B_C, whose solution is

$$B_C = \frac{X_T \pm \sqrt{\frac{R_T}{R_A}(R_T^2 + X_T^2) - R_T^2}}{R_T^2 + X_T^2}. \tag{8.4}$$

Since $R_T > R_A$, the argument of the square root is positive and greater than X_T^2. Therefore, to ensure a positive B_C, we must choose the plus sign in (8.4). Substituting (8.4) into (8.3a) yields X_L as

$$X_L = \frac{1}{B_C} - \frac{R_A(1 - B_C X_T)}{B_C R_T} - X_A. \tag{8.5}$$

Inserting numerical values into (8.4) and (8.5), we find

$$B_C = 9.2 \text{ mS} \Rightarrow C = B_C/\omega = 0.73 \text{ pF}$$

$$X_L = 76.9 \text{ }\Omega \Rightarrow L = X_L/\omega = 6.1 \text{ nH}.$$

This example shows the analytical approach of designing an L-type matching network by solving a quadratic equation for C and then a linear equation for L. The process is tedious, but can be easily implemented via a mathematical spreadsheet.

As we may anticipate from Example 8-1, the analytical approach to designing matching networks can become very complicated and computationally intensive even for simple L-sections. Instead of the preceding method, we can use the Smith Chart for rapid and relatively precise designs of the matching circuits. The appeal of this approach is that its complexity remains almost the same independent of the number of components in the network. Moreover, by observing the impedance transformation on the Smith Chart, we obtain a "feel" of how the individual circuit elements contribute to achieving a particular matching condition. Any errors in component selection and value assignment are observed immediately and the design engineer can directly intervene. With the help of a personal computer, this process is carried out in real time. That is, the parameter choice (L or C) and its value can be instantaneously displayed as part of the Smith Chart on the computer screen.

The effect of connecting a single reactive component (either capacitor or inductor) to a complex load is described in considerable detail in Section 3.4. Here, we point out the following:

- The addition of a reactance connected in series with a complex impedance results in motion along a constant-resistance circle in the combined Smith Chart
- A shunt connection produces motion along a constant-conductance circle.

This is indicated in Figure 8-3 for the combined ZY-Smith Chart. Concerning the direction of rotation, the general rule of thumb is that whenever an inductor is involved, we rotate in

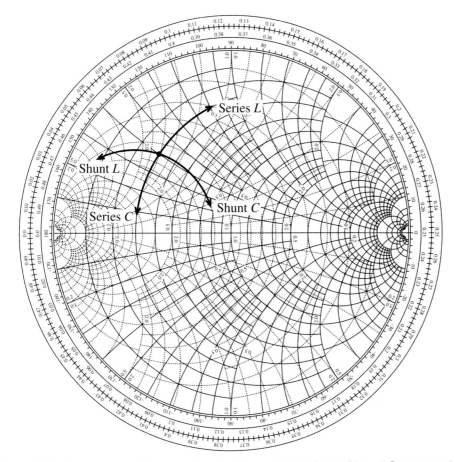

Figure 8-3 Impedance effect of series and shunt connections of L and C to a complex load in the Smith Chart.

the direction that moves the impedance into the upper half of the Smith Chart. In contrast, a capacitance results in movement toward the lower half.

Having established the effect of connecting a single component to the load, we can now develop suitable two-component matching networks that perform the transformation from any load impedance to any specified input impedance. In general, designing an L-type matching network, or for that matter any passive network, in the ZY Smith Chart consists of moving along either constant-resistance or constant-conductance circles.

In the following example, we illustrate this graphical design technique as an alternative to the analytical approach discussed in Example 8-1. Most modern CAD programs allow us to conduct this graphical approach interactively on the computer screen. In fact, simulation packages such as ADS permit the direct placement of components with the corresponding impedance behavior displayed on the Smith Chart.

Example 8-2. Graphical approach to the design of the L-type matching network

Design the L-type reactive matching network discussed in Example 8-1 using the Smith Chart as a graphical design tool.

Solution. The first step is to compute normalized transmitter and antenna impedances. Since no characteristic impedance Z_0 is given, we arbitrarily select $Z_0 = 75 \ \Omega$. Therefore, the normalized transmitter and antenna impedances are $z_T = Z_T/Z_0 = 2 + j1$ and $z_A = Z_A/Z_0 = 1 + j0.2$, respectively. Since the first component connected to the transmitter is a shunt capacitor, the total impedance of this parallel combination is positioned somewhere on the circle of constant conductance that passes through the point z_T in the combined Smith Chart (see Figure 8-4).

Next, an inductor is added in series with the parallel combination of transmitter z_T and capacitor; the resulting impedance will move along the circle of constant resistance. For maximum power gain, we require an output impedance of the matching network connected to the transmitter to be equal to the complex conjugate of the antenna impedance. This circle has to pass through $z_M = z_A^* = 1 - j0.2$, as shown in Figure 8-4.

The intersection of the two circles in the Smith Chart determines the normalized impedance formed by the shunt connection of transmitter and capacitor. Reading from the Smith Chart, we find that this impedance is approximately $z_{TC} = 1 - j1.22$ with the corresponding

Impedance Matching Using Discrete Components

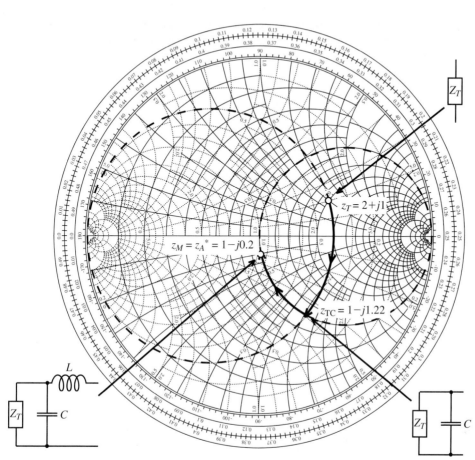

Figure 8-4 Design of the two-element matching network as part of the ZY Smith Chart.

admittance of $y_{TC} = 0.4 + j0.49$. Therefore, the normalized susceptance of the shunt capacitor is $jb_C = y_{TC} - y_T = j0.69$ and the normalized reactance of the inductor is $jx_L = z_A^* - z_{TC} = j1.02$. Finally, the actual values for the inductor and capacitor are

$$L = (x_L Z_0)/\omega = 6.09 \text{ nH}$$

$$C = b_C/(\omega Z_0) = 0.73 \text{ pF}.$$

This example presents a simple, yet precise graphical approach to designing L-type matching networks. The method can be readily extended to more complicated systems.

The design procedure described in Example 8-2 can be applied to any L-type matching network shown in Figure 8-1. The generic solution procedure for optimal power transfer includes the following six steps:

1. Find the normalized source and load impedances.
2. In the Smith Chart, plot circles of constant resistance and conductance that pass through the point denoting the source impedance.
3. Plot circles of constant resistance and conductance that pass through the point of the *complex conjugate* of the load impedance.
4. Identify the intersection points between the circles in steps 2 and 3. The number of intersection points determines the number of possible L-type matching networks.
5. Find the values of the normalized reactances and susceptances of the inductors and capacitors by tracing a path along the circles from the source impedance to the intersection point and then to the complex conjugate of the load impedance.
6. Determine the actual values of inductors and capacitors for a given frequency.

In the preceding steps, it is not necessary to move from the source to the complex conjugate load impedance. As a matter of fact, we can transform the load to the complex conjugate source impedance. The following example illustrates the first approach, whereas Section 8.1.2 discusses the second method.

Example 8-3. Design of general two-component matching networks

Using the Smith Chart, design all possible configurations of discrete two-element matching networks that match the source impedance $Z_S = (50 + j25)\,\Omega$ to the load $Z_L = (25 - j50)\,\Omega$. Assume a characteristic impedance of $Z_0 = 50\,\Omega$ and an operating frequency of $f = 2$ GHz.

Solution. We follow the six steps listed previously.

1. The normalized load and source impedances are:

$$z_S = Z_S/Z_0 = 1 + j0.5 \text{ or } y_S = 0.8 - j0.4$$

$$z_L = Z_L/Z_0 = 0.5 - j1 \text{ or } y_L = 3 + j0.8.$$

2. We plot circles of constant resistance and constant conductance that pass through the points of the normalized source impedance (dashed line circles in Figure 8-5), and

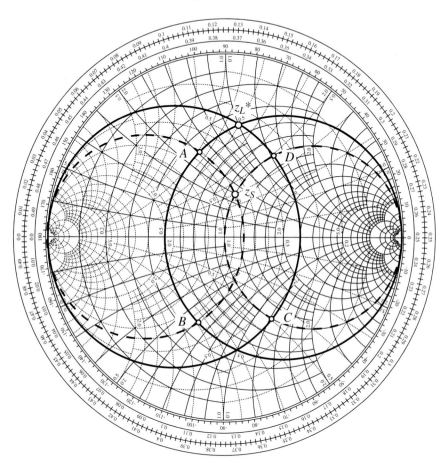

Figure 8-5 Design of a matching network using the Smith Chart

3. Complex conjugate of the load impedance (solid line circles in Figure 8-5).
4. These circles intersect in four points denoted as A, B, C, and D, with the normalized impedances and admittances being as follows:

$$z_A = 0.5 + j0.6, \quad y_A = 0.8 - j1$$
$$z_B = 0.5 - j0.6, \quad y_B = 0.8 + j1$$
$$z_C = 1 - j1.2, \quad y_C = 3 + j0.5$$
$$z_D = 1 + j1.2, \quad y_D = 3 - j0.5.$$

5. Since there are four intersection points, we expect four possible configurations of L-type matching networks. Indeed, if we move along the $z_S \rightarrow z_A \rightarrow z_L^*$ path, we see that from point z_S to z_A, the impedance is transformed along the circle of constant conductance, indicating a shunt connection. Moreover, we move toward the upper half of the Smith Chart (see Figure 8-3), which indicates that the first component connected to the source should be a shunt inductor. From points z_A to z_L^*, the impedance is transformed along the circle of constant resistance, with movement toward the upper half of the chart indicating series connection of an inductance. Therefore, the $z_S \rightarrow z_A \rightarrow z_L^*$ path results in a **shunt L, series L** matching network, as shown in Figure 8-1(f). If the $z_S \rightarrow z_B \rightarrow z_L^*$ path is chosen, we obtain a **shunt C, series L** network (Figure 8-1(h)). For $z_S \rightarrow z_C \rightarrow z_L^*$, the matching network is **series C, shunt L** (Figure 8-1(a)). Finally, for the $z_S \rightarrow z_D \rightarrow z_L^*$ path, a matching network is constructed by a **series L, shunt L** combination, which is shown in Figure 8-1(e).

6. We finally have to find the actual component values for the matching networks identified in the previous step. If we direct our attention again to the $z_S \rightarrow z_A \rightarrow z_L^*$ path, we see that from the source impedance to the point z_A, the normalized admittance of the circuit is changed by

$$jb_{L_2} = y_A - y_S = (0.8 - j1) - (0.8 - j0.4) = -j0.6.$$

From here, the value of the shunt inductor is:

$$L_2 = -\frac{Z_0}{b_{L_2}\omega} = 6.63 \text{ nH}.$$

Transformation from point z_A to z_L^* is done by adding an inductor connected in series to the impedance z_A. Therefore,

$$jx_{L_1} = z_L^* - z_A = (0.5 + j1) - (0.5 + j0.6) = j0.4$$

and the value of this inductor is

$$L_1 = \frac{x_{L_1} Z_0}{\omega} = 1.59 \text{ nH}.$$

Quantifying the components for the remaining three matching networks in the same way, the results are shown in Figure 8-6.

The Smith Chart allows immediately observing whether or not a particular impedance transformation is capable of achieving the

Impedance Matching Using Discrete Components

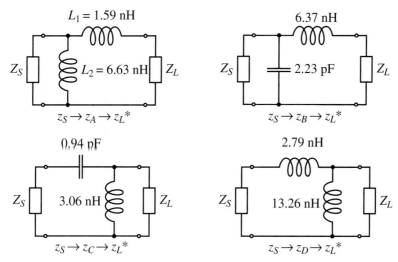

Figure 8-6 Matching networks for four different paths in the Smith Chart.

desired matching. Moreover, the total number of possible network configurations can readily be seen.

8.1.2 Forbidden Regions, Frequency Response, and Quality Factor

Before continuing with the frequency analysis of L-type matching networks, let us first note that not every network topology depicted in Figure 8-1 can perform the required matching between arbitrary load and source impedances. For example, if the source is $Z_S = Z_0 = 50 \, \Omega$, and if we use a matching network shown in Figure 8-1(h), then the addition of the capacitor in parallel with the source produces motion in the clockwise direction away from the circle of constant resistance that passes through the origin. This implies that all load impedances that reside within the shaded region in Figure 8-7(a) cannot be matched to the 50 Ω source by this particular network.

Similar "forbidden regions" can be developed for all L-type matching network topologies depicted in Figure 8-1. Examples of such regions for several other networks based on a 50 Ω source impedance are shown in Figure 8-7. Here, the shaded areas denote values of the load impedance that cannot be matched to the 50 Ω source. It is emphasized that the forbidden regions in Figure 8-7 are applicable only when dealing with a $Z_S = Z_0 = 50 \, \Omega$ source impedance. The regions take on totally different shapes for other source impedance values.

As explained in Example 8-3 and displayed in Figure 8-7, for any given load and input impedances, there are at least two possible configurations of L-type networks that

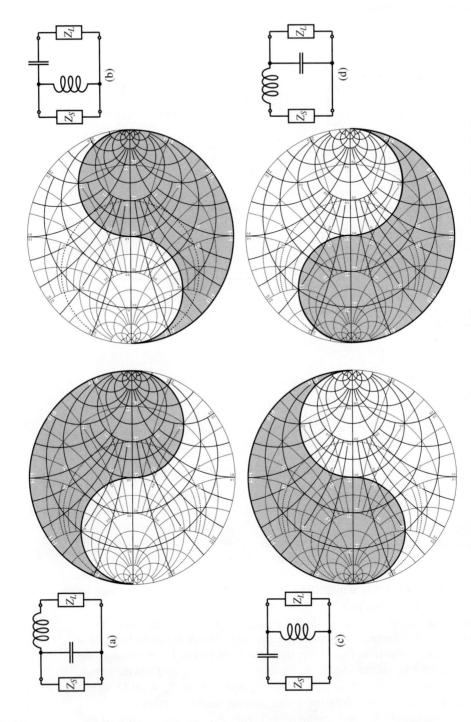

Figure 8-7 Forbidden regions for L-type matching networks with $Z_S = Z_0 = 50\ \Omega$.

432

accomplish the required match. The question now is, what is the difference between these realizations, and which network should ultimately be chosen?

Besides the obvious reasons for selecting one network over another (for instance, availability of components with required values), there are key technical considerations, including DC biasing, stability, and frequency response. In the remainder of this section, we concentrate primarily on the frequency response and quality factor of the L-type matching networks, whereas DC biasing issues are covered later in Section 8.3. Stability is deferred to Chapter 9.

Since any L-type matching network consists of series and shunt combinations of capacitors and/or inductors, the frequency response of these networks can be classified as either low-pass, high-pass, or bandpass filters. To demonstrate such behavior, let us consider a matching network that transforms a complex load, consisting of resistance $R_L = 80\ \Omega$ connected in series with capacitor $C_L = 2.65$ pF, into a 50 Ω input impedance. Let us further assume that the operating frequency for this circuit is $f_0 = 1$ GHz.

At 1 GHz, the normalized load impedance is $z_L = 1.6 - j1.2$, and, according to Figure 8-7, we can use either one of the matching networks shown in Figure 8-7(c) or Figure 8-7(d), following a similar design procedure as described in Example 8-2. However, because the source impedance z_S is real ($z_S = 1$), it is easier to transform from the load to the source impedance since $z_S^* = z_S = 1$. This is shown in Figure 8-8(a). The corresponding matching networks are shown in Figures 8-8(b) and 8-8(c).

The frequency responses of these two networks in terms of the input reflection coefficient $\Gamma_{in} = (Z_{in} - Z_S)/(Z_{in} + Z_S)$ and the transfer function $H = V_{out}/V_S$ (where the output voltage V_{out} is measured across the load resistance $R_L = 80\ \Omega$) are shown in Figures 8-9(a) and (b), respectively. It is apparent that both networks exhibit perfect matching only at a particular frequency $f_0 = 1$ GHz, and begin to deviate quickly when moving away from f_0.

The previously developed matching networks can also be viewed as resonance circuits with f_0 being the resonance frequency. As discussed in Section 5.1.1, these networks may be described by a loaded quality factor, Q_L, which is equal to the ratio of the resonance frequency f_0 to the 3 dB bandwidth BW

$$Q_L = \frac{f_0}{BW}. \qquad (8.6)$$

The question now is how to find the bandwidth of the matching network. To answer this, we will exploit the similarity between the bell-shaped response of the matching network's transfer function near f_0 (see Figure 8-9(b)) and the frequency response of a bandpass filter.

For frequencies close to f_0, the matching network in Figure 8-8(c) can be redrawn as a bandpass filter with a loaded quality factor calculated based on (8.6). The equivalent bandpass filter is shown in Figure 8-10(a). The equivalent capacitance C_T in this circuit is obtained by replacing the series combination of R_L and C_L in Figure 8-8(c) with an equivalent parallel connection of R_{LP} and C_{LP} and then adding the capacitances C and C_{LP}: $C_T = C + C_{LP}$. The equivalent shunt inductance L_{LN} is obtained by first replacing the series connection of the voltage source V_S, resistance R_S, and inductance L with the

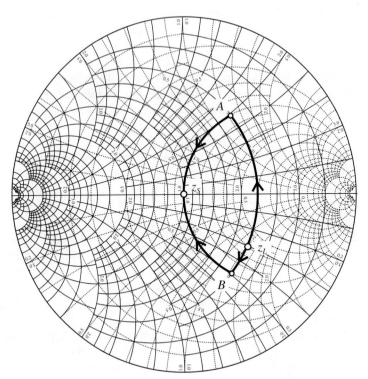

(a) Impedance transformations displayed in Smith Chart

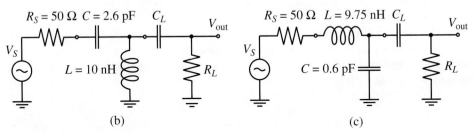

Figure 8-8 Two design realizations of an L-type matching network.

Norton equivalent current source $I_N = V_S/(R_S + j\omega_0 L)$ connected to the parallel combination of conductance G_{SN} and inductance L_N, where the admittance is given as follows: $G_{SN} + (j\omega_0 L_N)^{-1} = (R_S + j\omega_0 L)^{-1}$. Next, the current source I_N and conductance G_{SN} are converted back into a Thévenin equivalent voltage source

$$V_T = I_N/G_{SN} = V_S \frac{R_S - j\omega_0 L}{R_S} = V_S(1 - j1.22) \tag{8.7}$$

and series resistance

Impedance Matching Using Discrete Components

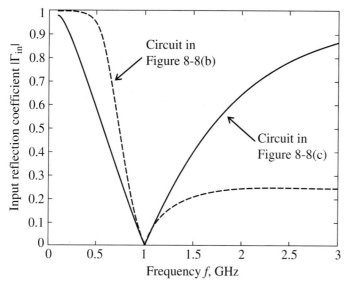

(a) Frequency response of input reflection coefficient

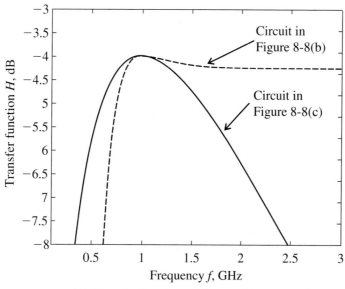

(b) Transfer function of the matching networks

Figure 8-9 Frequency response of the two matching network realizations.

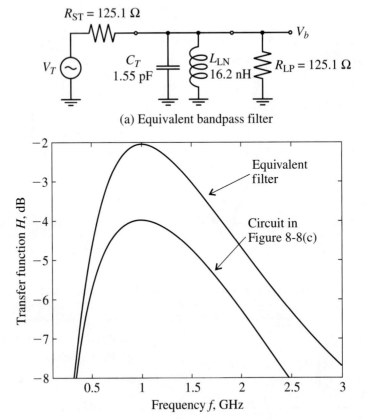

(a) Equivalent bandpass filter

(b) Frequency response of the matching network compared to the equivalent filter response

Figure 8-10 Comparison of the frequency response of the L-type matching network and an equivalent bandpass filter.

$$R_{ST} = G_{SN}^{-1} = \frac{R_S^2 + (\omega_0 L)^2}{R_S}. \tag{8.8}$$

The resonance circuit in Figure 8-10 is loaded by the combined resistance $R_T = R_L \| R_{ST} = 62.54\ \Omega$. Thus, the loaded quality factor Q_L of the equivalent bandpass filter is given by

$$Q_L = \frac{f_0}{BW} = \omega_0 R_T C = \frac{R_T}{|X_C|} = 0.61. \tag{8.9}$$

It is noticed that the maximum gain of the equivalent bandpass filter is higher than the gain of the original matching network. This is explained by the fact that for the matching network we measure the output voltage on the load R_L, while for the equivalent filter we measure the output voltage at the equivalent load resistance R_{LP}, which is connected in parallel with the capacitance C_T. Therefore, the conversion from V_b to V_{out} at the resonance frequency can be found through the voltage divider rule:

$$|V_{out}| = |V_b| \frac{R_L}{\left|R_L + \frac{1}{j\omega_0 C_L}\right|} = 0.7908 |V_b|$$

which gives us

$$20 \log \frac{|V_{out}|}{|V_S|} = -2.0382 + 20 \log \frac{|V_b|}{|V_S|} = -3.9794 \text{ dB}$$

a result that agrees very well with Figure 8-9(b).

From the known Q_L, we can directly find the bandwidth of the filter: $BW = f_0/Q_L = 1.63$ GHz. The frequency response in Figure 8-9(b) shows that the 3 dB point for $f < f_0$ occurs at $f_{min} = 0.40$ GHz and for $f > f_0$ at $f_{max} = 2.19$ GHz. Thus, the bandwidth of the matching network is $BW = f_{max} - f_{min} = 1.79$ GHz, which agrees reasonably well with the result obtained for the equivalent bandpass filter.

The equivalent bandpass filter analysis allows us to explain the bell-shaped response of the matching network in the neighborhood of f_0, and provides us with a good estimation of the bandwidth of the circuit. The drawbacks to this approach are its complexity and approximate nature. It would be desirable to develop a simpler method of estimating the quality factor of the matching network without first having to develop an equivalent bandpass filter, or even computing the frequency response of the network. This is accomplished through the use of a so-called **nodal quality factor** Q_n.

Let us go back to Figure 8-8(a), where we illustrate the impedance transformation as we move from one node of the circuit to another. We note that at each node of the matching network, the impedance can be expressed in terms of an equivalent series impedance $Z_S = R_S + jX_S$ or admittance $Y_P = G_P + jB_P$. Hence, at each node we can find Q_n as the ratio of the absolute value of the reactance X_S to the corresponding resistance R_S

$$Q_n = \frac{|X_S|}{R_S} \tag{8.10}$$

or as the ratio of the absolute value of susceptance B_P to the conductance G_P

$$Q_n = \frac{|B_P|}{G_P}. \tag{8.11}$$

Using (8.10) and (8.11) and the impedance transformations in Figure 8-8(a), we can deduce that for the matching network shown in Figure 8-8(c), the maximum nodal quality factor is obtained at point B, where the normalized impedance is $1 - j1.23$, resulting in

$$Q_n = |1.23|/1 = 1.23. \tag{8.12}$$

To relate the nodal quality factor Q_n to Q_L, we compare the result of (8.12) with (8.9) and find

$$Q_L = \frac{Q_n}{2}. \tag{8.13}$$

This result is true for any L-type matching network. For more complicated configurations, the loaded quality factor of the matching network is usually estimated as simply the maximum nodal quality factor. Even though this approach does not yield a very accurate estimate of the circuit bandwidth, it nonetheless allows us to compare networks and to select a network with higher or lower bandwidth.

To simplify the matching network design process even further, we can draw constant-Q_n contours in the Smith Chart. Figure 8-11 shows such contours for Q_n valued 0.3, 1, 3, and 10.

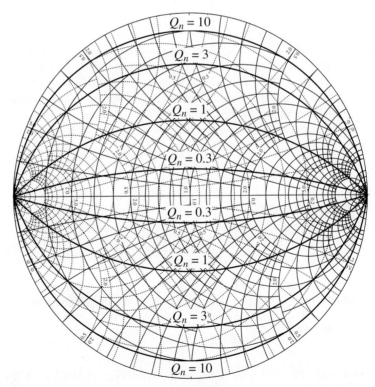

Figure 8-11 Constant Q_n contours displayed in the Smith Chart.

Impedance Matching Using Discrete Components

To obtain the equations for these contours, we refer back to the general derivation of the Smith Chart in Chapter 3. There, it is shown in (3.6) and (3.7) that the normalized impedance can be written as

$$z = r + jx = \frac{1 - \Gamma_r^2 - \Gamma_i^2}{(1-\Gamma_r)^2 + \Gamma_i^2} + j\frac{2\Gamma_i}{(1-\Gamma_r)^2 + \Gamma_i^2}. \quad (8.14)$$

Thus, the nodal quality factor can be written as

$$Q_n = \frac{|x|}{r} = \frac{2|\Gamma_i|}{1 - \Gamma_r^2 - \Gamma_i^2}. \quad (8.15)$$

Rearranging terms in (8.15), a circle equation can be formulated

$$\Gamma_i^2 + \left(\Gamma_r \pm \frac{1}{Q_n}\right)^2 = 1 + \frac{1}{Q_n^2} \quad (8.16)$$

where the plus sign is taken for positive reactance x and the minus sign for negative x.

With these constant-Q_n circles in the Smith Chart, it is possible to find the loaded quality factor of an L-type matching network by simply reading the corresponding Q_n and dividing it by 2. This procedure is discussed in Example 8-4.

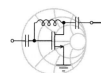

Example 8-4. Design of a narrowband matching network

Using the forbidden regions in Figure 8-7, design two L-type networks that match a $Z_L = (25 + j20)\Omega$ load impedance to a 50 Ω source at 1 GHz. Determine the loaded quality factors of these networks from the Smith Chart and compare them to the bandwidth obtained from their frequency response. Assume that the load consists of a resistance and inductance connected in series.

Solution. As we see from Figure 8-7, the normalized load impedance $z_L = 0.5 + j0.4$ lies inside of the constant conductance circle $g = 1$. There are two L-type matching networks that satisfy our requirements. The first consists of a series inductor and shunt capacitor, as shown in Figure 8-7(a), and the second is a series capacitor with shunt inductor, as shown in Figure 8-7(b). Following the same procedure as described in Example 8-2, we obtain the two matching networks shown in Figure 8-12.

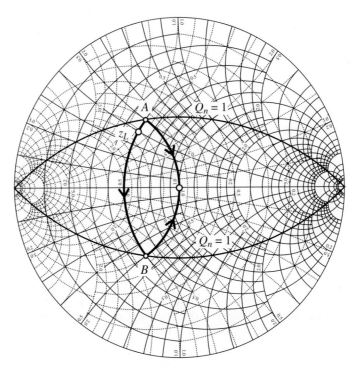

(a) Impedance transformation in the Smith Chart

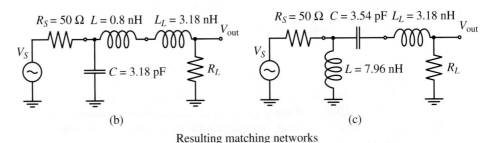

(b)   (c)

Resulting matching networks

Figure 8-12 Two L-type matching networks for a 50 Ω source and a $Z_L = (25 + j20)\Omega$ load impedance operated at a frequency of 1 GHz.

According to Figure 8-12(a), the nodal quality factor for both networks is equal to $Q_n = 1$. Thus, we can expect that the bandwidth should be equal to $\text{BW} = f_0/Q_L = 2f_0/Q_n = 2$ GHz. This is checked by plotting the corresponding frequency responses for the designed matching networks, as depicted in Figure 8-13.

We observe that the bandwidth for the network corresponding to Figure 8-12(c) is approximately $\text{BW}_c = 2.4$ GHz. Interestingly, the

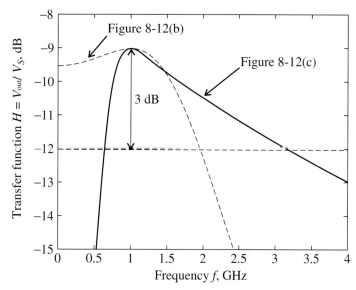

Figure 8-13 Frequency responses for the two matching networks.

matching network corresponding to Figure 8-12(b) does not possess a lower cutoff frequency. However, if we assume that the frequency response is symmetric around the resonance frequency $f_0 = 1$ GHz, then the bandwidth will be $BW_b = 2(f_{max} - f_0) = 1.9$ GHz, with the upper cutoff frequency being $f_{max} = 1.95$ GHz.

Despite their design for the same resonance frequency, certain matching network configurations exhibit better high- or low-frequency rejection, as Figure 8-13 exemplifies.

In many practical applications, the quality factor of the matching network is of importance. For example, if we design a broadband amplifier, we would like to utilize networks with low Q in order to increase the bandwidth. However, for oscillator design it is desirable to achieve high-Q networks to eliminate undesired harmonics in the output signal. Unfortunately, as we have seen in the previous example, L-type matching networks provide no control over the value of Q_n, and we must either accept or reject the resulting quality factor. To gain the freedom of choosing the values of Q and thus affect the bandwidth behavior of the circuit, we can introduce a third element in the matching network. The addition of this third element results in either a T- or Pi-network, both of which are discussed next.

8.1.3 T and Pi Matching Networks

As already pointed out, the loaded quality factor of the matching network can be estimated from the maximum nodal Q_n. The addition of the third element into the matching network introduces an additional degree of freedom in the circuit, and allows us to control the value of Q_L by choosing an appropriate intermediate impedance.

The following two examples illustrate the design of T- and Pi-type matching networks with specified Q_n factor.

Example 8-5. Design of a T matching network

Design a T-type matching network that transforms a load impedance $Z_L = (60 - j30)\Omega$ into a $Z_{in} = (10 + j20)\Omega$ input impedance and that has a maximum nodal quality factor of 3. Compute the values for the matching network components, assuming that matching is required at $f = 1$ GHz.

Solution. There are several possible solutions that satisfy the design specifications. In this example, we investigate only one design since the rest can easily be obtained by using the same approach.

The general topology of the T-type matching network is shown in Figure 8-14.

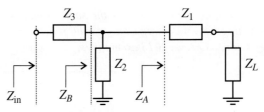

Figure 8-14 General topology of a T-type matching network.

The first element in this network is connected in series with the load impedance. Because Z_1 is purely reactive, the combined impedance Z_A will reside somewhere on the constant resistance circle described by $r = r_L$. Similarly, Z_3 is connected in series with the input, so that the combined impedance Z_B (consisting of Z_L, Z_1, and

Z_2) is positioned somewhere on the constant resistance circle with $r = r_{in}$. Because the network should have a nodal quality factor $Q_n = 3$, we can choose the impedance values in such a way that Z_B is located on the intersection of the constant resistance circle $r = r_{in}$ and the $Q_n = 3$ circle (see point B in Figure 8-15).

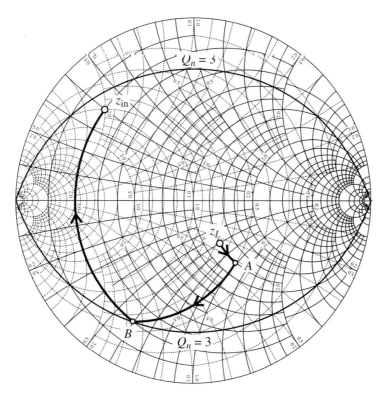

Figure 8-15 Design of a T-type matching network for a specified $Q_n = 3$.

We next find the intersection point A of the constant conductance circle that passes through the point B obtained from the previous step. The circle of constant resistance $r = r_L$ now allows us to determine the required value of the remaining component of the network to reach the point z_{in}.

The complete T-type matching network with the actual component values is illustrated in Figure 8-16. The computed elements are based on the required matching frequency of $f = 1$ GHz.

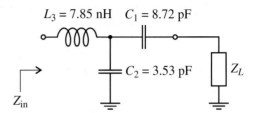

Figure 8-16 T-type matching network circuit schematics.

The extra degree of freedom to adjust the quality factor (bandwidth) of a matching network comes at the expense of an additional circuit element.

In the following example, the design of a Pi-type matching network is developed with the intent to achieve a minimum nodal quality factor. A low quality factor design directly translates into a wider bandwidth of the network, as required, for instance, in broadband FET and BJT amplifiers.

Example 8-6. Design of a Pi-type matching network

For a broadband amplifier, it is required to develop a Pi-type matching network that transforms a load impedance of $Z_L = (10 - j10)\Omega$ into an input impedance of $Z_{\text{in}} = (20 + j40)\Omega$. The design should involve the lowest possible nodal quality factor. Find the component values, assuming that matching should be achieved at a frequency of $f = 2.4$ GHz.

Solution. Since the load and input impedances are fixed, we cannot produce a matching network that has a quality factor lower than the highest Q_n computed at the locations Z_L and Z_{in}. Therefore, the minimum value for Q_n is determined at the input impedance location as $Q_n = |X_{\text{in}}|/R_{\text{in}} = 40/20 = 2$. The Smith Chart design of the Pi-type matching network based on $Q_n = 2$ is depicted in Figure 8-17.

In the design, we employ a method very similar to the one used in Example 8-5. First, we plot a constant-conductance circle $g = g_{\text{in}}$ and find its intersection with the $Q_n = 2$ contour in the Smith Chart. This intersection is denoted as point B. Next, we find the intersection point of the constant-conductance circle $g = g_L$ with the constant-resistance circle that passes through the point B. The resulting point is labelled A in Figure 8-17.

Impedance Matching Using Discrete Components

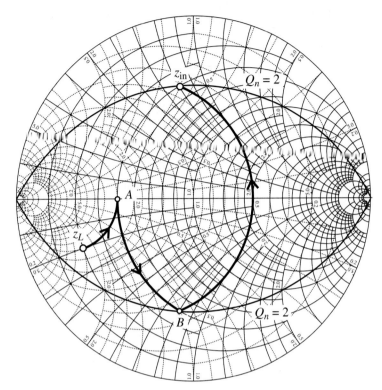

Figure 8-17 Design of a Pi-type matching network using a minimal Q_n.

The network components can be determined based on converting the Smith Chart points into actual capacitances and inductances, as detailed in Example 8-2. The resulting circuit configuration is shown in Figure 8-18.

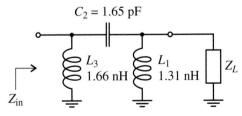

Figure 8-18 Pi-type matching network configuration.

It is interesting to note that, unlike the situation discussed in Example 8-5, the relative positions of Z_L and Z_{in} are such that only one possible Pi-type network configuration with $Q_n = 2$ exists. All other realizations of the Pi-type network will result in an increased nodal quality factor. Furthermore, if we had a lower load resistance, we would not be able to implement this Pi-type network for the given Q_n.

As this example shows, the bandwidth cannot be increased arbitrarily by reducing the nodal quality factor. The limits are set by the desired complex input and output impedances. In case of real impedances, very wide bandwidths can be achieved through more sophisticated techniques.

8.2 Microstrip Line Matching Networks

> **TAPERED MATCHING**
>
> The $\lambda/4$ transformer matching of two dissimilar resistances can be modified to handle a finite bandwidth. Multiple transmission lines of fixed lengths and stepped line impedances, or gradually varying line impedances, so-called tapers, find use if real (or approximately real) impedances are to be matched over a wide BW.

In the previous sections, we have discussed the design of matching networks involving discrete components. However, with increasing frequency and correspondingly reduced wavelength, the influence of parasitics in the discrete elements becomes more noticeable. The design now requires us to take these parasitics into account, thus significantly complicating the component value computations. This, along with the fact that discrete components are only available for certain values, limits their use in high-frequency circuit applications. As an alternative to lumped elements, distributed components are widely used when the wavelength becomes sufficiently small compared with the circuit dimensions, a topic already discussed in Chapter 2.

8.2.1 From Discrete Components to Microstrip Lines

In the mid-GHz range, design engineers often employ a mixed approach by combining lumped and distributed elements. These types of matching networks usually contain a number of transmission lines connected in series and capacitors spaced in a shunt configuration, as illustrated in Figure 8-19. The reader is also referred to Figure 1-2(a) for a practical example.

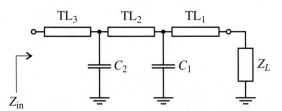

Figure 8-19 Mixed design of matching network involving transmission line sections (TL) and discrete capacitive elements.

Inductors are usually avoided in such designs because they tend to have higher resistive losses than capacitors. In general, only one shunt capacitor with two transmission lines connected in series on both sides is sufficient to transform any given load impedance to any

Microstrip Line Matching Networks

input impedance. Similarly to the L-type matching networks, such configurations may involve the additional requirement of a fixed Q_n, necessitating additional components to control the quality factor of the circuit.

The arrangement of components shown in Figure 8-19 is very attractive in practice, since it permits tuning the circuit after it has been manufactured. Changing the values of the capacitors as well as placing them at different locations along the transmission lines offers a wide range of flexibility. The tuning capability makes these types of matching networks popular for prototyping. Usually, all transmission lines have the same width (i.e., the same characteristic impedance) to simplify the actual tuning.

Example 8-7 discusses the Smith Chart approach to the design of a matching network containing two 50 Ω transmission lines connected in series and a single shunt capacitor placed between them.

Example 8-7. Design of a matching network with lumped and distributed components

Design a matching network that transforms the load $Z_L = (30 + j10)\Omega$ to an input impedance $Z_{in} = (60 + j80)\Omega$. The matching network should contain only two series transmission lines and a shunt capacitance. Both transmission lines have a 50 Ω characteristic line impedance, and the frequency at which matching is desired is $f = 1.5$ GHz.

Solution. The first step involves identifying the normalized load impedance $z_L = 0.6 + j0.2$ as a point in the Smith Chart. We can then draw a SWR circle that indicates the combined impedance of the load connected to the 50 Ω transmission line. The position on the SWR circle is determined by the length of the transmission line, as investigated in Chapter 3.

The second step requires plotting a SWR circle that passes through the normalized input impedance point $z_{in} = 1.2 + j1.6$ shown in Figure 8-20.

The choice of the point from which we transition from the load SWR circle to the input SWR circle can be made arbitrarily. In Figure 8-20, the point A is chosen, which approximately corresponds to a normalized admittance value of $y_A = 1 - j0.6$. The addition of the parallel capacitor results in the movement along the circle of constant conductance $g = 1$ and transforms the impedance from point A to point B on the input SWR circle of the Smith Chart. From point B, an impedance transformation is required along the constant SWR circle by adding a series connected transmission line.

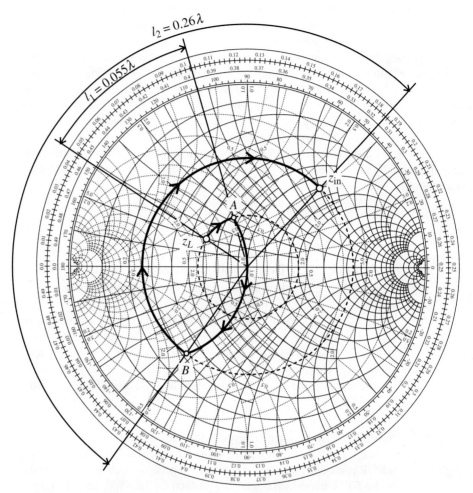

Figure 8-20 Design of the distributed matching network for Example 8-7.

As a final step, the electrical length of the transmission lines must be determined. This can be done by reading the two lengths l_1, l_2 from the so-called **WTG** (wavelengths toward generator) scale displayed on the outer perimeter of the Smith Chart (see Figure 8-20). The resulting circuit schematic for the matching network is shown in Figure 8-21.

It is interesting to investigate the tuning capability range for this circuit configuration. Figure 8-22 shows the dependence of the real r_{in} and imaginary x_{in} parts of the input impedance as functions of the distance l between the load and the capacitor location. In other words, the total length $l_1 + l_2$ is kept fixed, and the placement of the capacitor

Microstrip Line Matching Networks

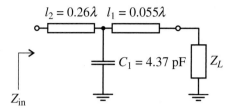

Figure 8-21 Matching network combining series transmission lines and shunt capacitance.

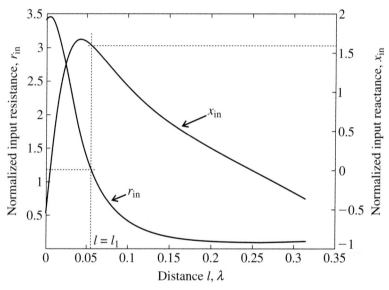

Figure 8-22 Input impedance as a function of the position of the shunt capacitor in Example 8-7.

is varied from the load end to the beginning of the network (i.e., $0 \leq l \leq l_1 + l_2$). The dashed lines indicate the original design. It is noticed that x_{in} undergoes the expected inductive (positive values) to capacitive (negative values) transition.

In this example, we have designed a combined matching network that involves both distributed (transmission lines) and a lumped (capacitor) element. These types of networks have rather large tuning capabilities, but are very sensitive to the placement of the capacitor along the transmission line. Even small deviations from the target location result in drastic changes in the input impedance.

8.2.2 Single-Stub Matching Networks

The next logical step in the transition from lumped to distributed element networks is the complete elimination of all lumped components. This is accomplished by employing open- and/or short-circuited stub lines.

In this section, we consider matching networks that consist of a series transmission line connected to a shunt open-circuited or short-circuited stub. Let us investigate two topologies: the first one involves a series transmission line connected to the parallel combination of load and stub, as shown in Figure 8-23(a), and the second involves a shunt stub connected to the series combination of the load and transmission line, as depicted in Figure 8-23(b).

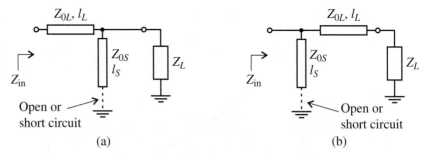

Figure 8-23 Two topologies of single-stub matching networks.

The matching networks in Figure 8-23 possess four adjustable parameters: length l_S and characteristic impedance Z_{0S} of the stub, and length l_L and characteristic impedance Z_{0L} of the transmission line.

Example 8-8 demonstrates the design procedure for the matching network topology shown in Figure 8-23(a), with the characteristic impedances of both stub Z_{0S} and transmission line Z_{0L} fixed to the same arbitrary value Z_0, and their electrical lengths variable to meet the particular input impedance requirement.

Example 8-8. Single-stub matching network design with fixed characteristic impedances

For a load impedance of $Z_L = (60 - j45)\Omega$, design two single-stub matching networks that transform the load to a $Z_{in} = (75 + j90)\Omega$ input impedance. Assume that both stub and transmission line in Figure 8-23(a) have a characteristic impedance of $Z_0 = 75\ \Omega$.

Solution. The basic concept is to select the length l_S of the stub such that it produces a susceptance B_S sufficient to move the load

Microstrip Line Matching Networks

admittance $y_L = 0.8 + j0.6$ to the SWR circle that passes through the normalized input impedance point $z_{in} = 1 + j1.2$, as illustrated in Figure 8-24.

We notice that the input SWR circle associated with $z_{in} = 1 + j1.2$ intersects the constant conductance circle $g = 0.8$ at two points ($y_A = 0.8 + j1.05$ and $y_B = 0.8 - j1.05$), suggesting two possible solutions. The corresponding susceptance values for the stub are $jb_{SA} = y_A - y_L = j0.45$ and $jb_{SB} = y_B - y_L = -j1.65$, respectively. In the first case, the length of an open-circuit stub can be found in the Smith Chart by measuring the length l_{SA}, starting from the $y = 0$ point (open circuit) and moving along the outer perimeter of the Smith Chart $g = 0$ toward the generator (clockwise) to the point

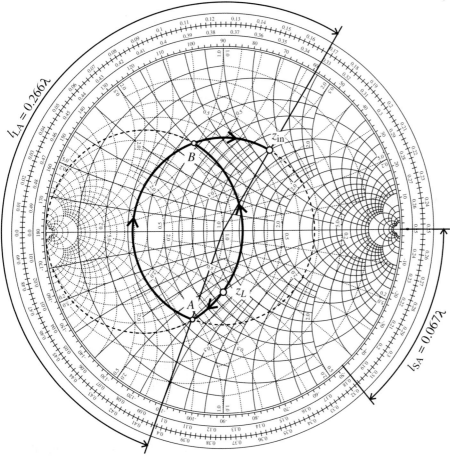

Figure 8-24 Smith Chart design for the single-stub matching network based on Example 8-8.

where $y = j0.45$. The length for this case is $l_{SA} = 0.067\lambda$. The open-circuited stub can be replaced by a short-circuited stub if its length is increased by a quarter wavelength. Such a substitution may become necessary if a coaxial cable is used because of excessive radiation losses due to the large cross-sectional area. In printed circuit design, open-circuited stubs are sometimes preferred, since they eliminate the deployment of a via, which is otherwise necessary to obtain the ground connection for a short-circuited stub.

Similar to the first solution, b_{SB} yields the length $l_{SB} = 0.337\lambda$ for the open-circuited stub, and $l_{SB} = 0.087\lambda$ for the short-circuited stub. For this case, we also notice that creating a short-circuited stub requires a shorter length than an open-circuited stub. This is due to the fact that the open-circuited stub models a negative susceptance.

The length of the series transmission line segment is found in the same way as described in Example 8-7; it is equal to $l_{LA} = 0.266\lambda$ for the first solution, and $l_{LB} = 0.07\lambda$ for the second solution.

A circuit designer often has to minimize the size of the circuit board and, therefore, must be concerned about employing the shortest possible transmission line segments. Depending on the impedance requirements, this can either be an open- or short-circuited stub section.

In the next example, we illustrate the generic design procedure for the matching network topology shown in Figure 8-23(b). Unlike the previous example, we now fix the lengths of both the stub and the transmission line segment, but vary their characteristic impedances. In a microstrip line circuit design, this is typically accomplished by changing the width of the lines.

Example 8-9. Design of a single-stub matching network using transmission lines with different characteristic impedances

Using the matching network topology shown in Figure 8-23(b), choose the characteristic impedances of the stub and transmission line such that the load impedance $Z_L = (120 - j20)\Omega$ is transformed into the input impedance $Z_{in} = (40 + j30)\Omega$. Assume that the length of the transmission line is $l_L = 0.25\lambda$ and the stub has the length of

$l_S = 0.375\lambda$. Furthermore, determine whether a short-circuited or an open-circuited stub is necessary for this circuit.

Solution. The combined impedance Z_1 of the series connection of the load impedance with the transmission line can be computed using the formula for the quarter-wave transformer:

$$Z_1 = Z_{0L}^2/Z_L. \qquad (8.17)$$

The addition of the open-circuited stub results in an input admittance of

$$Y_{in} = Y_1 + jB_S \qquad (8.18)$$

where $Y_1 = Z_1^{-1}$ is the admittance of the previously computed series combination of load impedance and transmission line, and $jB_S = \pm jZ_{0S}^{-1}$ is the susceptance of the stub. The plus or minus signs correspond to either a short-circuited or an open-circuited stub.

Combining (8.17) and (8.18), we find

$$G_{in} = R_L/Z_{0L}^2 \qquad (8.19a)$$

$$B_{in} = X_L/Z_{0L}^2 \pm Z_{0S}^{-1} \qquad (8.19b)$$

where we have used the input admittance and load impedance representation in terms of their real and imaginary components: $Y_{in} = G_{in} + jB_{in}$, $Z_L = R_L + jX_L$.

Using (8.19a), we find the characteristic impedance of the transmission line to be

$$Z_{0L} = \sqrt{\frac{R_L}{G_{in}}} = \sqrt{\frac{120}{0.016}} = 86.6 \; \Omega.$$

Substituting the obtained value into (8.19b), we find that the minus sign should be used; that is, we need to implement an open-circuited stub with a characteristic impedance of

$$Z_{0S} = \frac{1}{X_L/Z_{0L}^2 - B_{in}} = 107.1 \; \Omega.$$

This design approach is very easy to implement as long as the characteristic impedance stays within reasonable limits ranging approximately from 20 to 200 Ω.

In practical realizations, single-sided unbalanced stubs are often replaced by the balanced design, as shown in Figure 8-25.

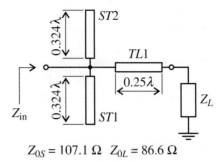

Figure 8-25 Balanced stub design for Example 8-9.

Naturally, the combined susceptance of the parallel connection of stubs $ST1$ and $ST2$ has to be equal to the susceptance of the unbalanced stub. Therefore, the susceptance of each side of the balanced stub must be equal to half of the susceptance of the unbalanced stub. We note that the length l_{SB} of each side does not scale linearly. In other words, the length of the balanced stub is not half of the length of the unbalanced stub l_S. Rather, it has to be computed as

$$l_{SB} = \frac{\lambda}{2\pi}\tan^{-1}\left(2\tan\frac{2\pi l_S}{\lambda}\right) \qquad (8.20)$$

for an open-circuited stub, or

$$l_{SB} = \frac{\lambda}{2\pi}\tan^{-1}\left(\frac{1}{2}\tan\frac{2\pi l_S}{\lambda}\right) \qquad (8.21)$$

for a short-circuited stub. This result can also be found graphically by using the Smith Chart.

8.2.3 Double-Stub Matching Networks

TUNER APPLICATIONS

Tuners with movable shorts as part of stub line adjustments allow a wide range of load impedance matching for high power applications. They are generally prefered over lumped element tuning in power amplifier designs.

The single-stub matching networks in the previous section are quite versatile and allow matching between any input and load impedances, so long as they have a nonzero real part. One of the main drawbacks of such matching networks is that they require a variable-length transmission line between the stub and the input port, or between the stub and the load impedance. Usually, this does not pose a problem for fixed networks, but it may create difficulties for variable tuners. In this section, we examine matching networks that overcome this drawback by incorporating a second stub. The general topology of such a network that matches an arbitrary load impedance to an input impedance $Z_{in} = Z_0$ is shown in Figure 8-26.

Microstrip Line Matching Networks

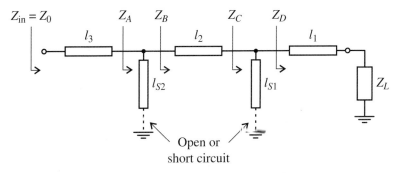

Figure 8-26 Double-stub matching network arrangement.

In double-stub matching networks, two short- or open-circuited stubs are connected in shunt with a fixed-length transmission line placed in between. The length l_2 of this line is usually chosen to be one-eighth, three-eighths, or five-eighths of a wavelength. The three-eighths and five-eighths wavelength spacings are typically employed in high-frequency applications to simplify the tuner construction.

Let us assume for our subsequent discussion that the length of the line segment between the two stubs is $l_2 = (3/8)\lambda$. To facilitate the analysis, we start from the input side of the tuner and work backward to the load end.

For a perfect match, it is required that $Z_{in} = Z_0$ and therefore $y_A = 1$. Since it is assumed that the lines are lossless, the normalized admittance $y_B = y_A - jb_{S2}$ is located somewhere on the constant conductance circle $g = 1$ in the Smith Chart. Here, b_{S2} is the susceptance of the stub and l_{S2} is the associated length. For an $l_2 = (3/8)\lambda$ line the $g = 1$ circle is rotated by $2\beta l_2 = 3(\pi/2)$ radians or 270° toward the load (i.e., in the counter-clockwise direction, as depicted in Figure 8-27). The admittance y_C (being the series connection of Z_L with line l_1 in parallel to stub l_{S1}) needs to reside on this rotated $g = 1$ circle (called the y_C circle) in order to ensure matching.

By varying the length of the l_{S1} stub, we can transform point y_D in such a way that the resulting y_C is indeed located on the rotated $g = 1$ circle. This procedure can be performed for any load impedance, except for the case when point y_D (i.e., the series connection of Z_L and line l_1) is located inside the $g = 2$ circle. This represents the forbidden region that has to be avoided. To overcome this problem in practical applications, commercial double-stub tuners usually have input and output transmission lines whose lengths are related according to $l_1 = l_3 \pm \lambda/4$. In this case, if a particular load impedance cannot be matched, one simply connects the load to the opposite end of the tuner, which moves y_D out of the forbidden region.

The following example demonstrates the computation of the stub lengths to achieve matching for a specific load impedance.

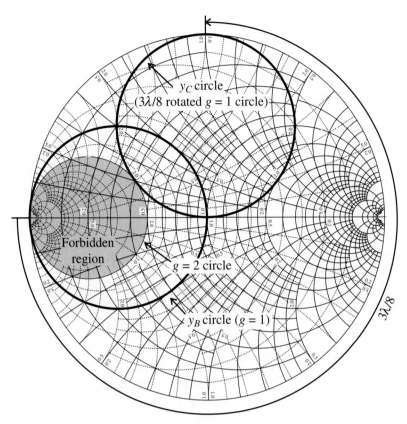

Figure 8-27 Smith Chart analysis of a double-stub matching network shown in Figure 8-26.

Example 8-10. Design of a double-stub matching network

It is assumed that in the double-stub matching network shown in Figure 8-26, the lengths of the transmission lines are $l_3 = l_2 = 3\lambda/8$ and $l_1 = \lambda/8$. Find the lengths of the short-circuited stubs that match the load impedance $Z_L = (50 + j50)\Omega$ to a 50 Ω input impedance. The characteristic line impedance for all components is $Z_0 = 50\ \Omega$.

Solution. First, the normalized admittance y_D has to be determined and checked that it does not fall inside the forbidden region.

Using the Smith Chart (see Figure 8-28), we find $y_D = 0.4 + j0.2$. Since $g_D < 2$, we are assured that the admittance y_D does not reside within the forbidden region. Next, we plot the rotated $g = 1$ circle as explained previously. This allows us to fix the intersection of the rotated $g = 1$ circle with the constant conductance circle that passes through the point y_D. The intersection point provides the value of y_C. In fact, there are two intersection points that yield two possible solutions. If we choose $y_C = 0.4 - j1.8$, then the susceptance of the first stub should be $jb_{S1} = y_C - y_D = -j2$, which permits us to determine the length of the first short-circuited stub: $l_{S1} = 0.074\lambda$.

Rotating y_C by $l_2 = 3\lambda/8$, we find $y_B = 1 + j3$. This implies that we have to make the susceptance of the second stub equal to $jb_{S2} = -j3$ so that $y_{in} = y_A = 1$. Using the Smith Chart, we find that the length of the second stub is $l_{S2} = 0.051\lambda$.

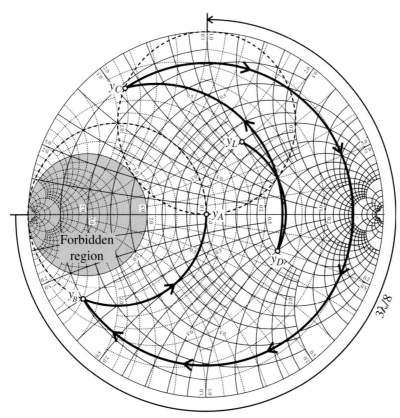

Figure 8-28 Double-stub matching network design for Example 8-10.

In certain practical realizations, the stubs are replaced by varactor diodes. This allows electronic tuning of the diode capacitances, and thus the shunt admittances.

8.3 Amplifier Classes of Operation and Biasing Networks

An indispensable building block in any RF circuit is the active or passive biasing network. The purpose of biasing is to provide the appropriate quiescent point for the active devices under specified operating conditions and maintain a constant setting, irrespective of transistor parameter variations and temperature fluctuations.

In the following section, we introduce a brief analysis of the different classes of amplifier operation. This will enable us to develop an understanding of how BJTs and FETs need to be appropriately biased.

8.3.1 Classes of Operation and Efficiency of Amplifiers

Depending on the application for which the amplifier is designed, specific bias conditions are required. There are several classes of amplifier operation that describe the biasing of an active device in an RF circuit.

In Figure 8-29, the transfer characteristic of an ideal transistor is displayed. It is assumed that the transistor does not reach saturation or breakdown regions, and in the linear operating region the output current is proportional to the input voltage. The voltage V^* corresponds either to the threshold voltage in case of FETs, or the base-emitter built-in potential in case of BJTs.

The distinction between different classes of operation is made based upon the so-called **conduction angle**, which indicates the portion of the signal cycle when the current flows through the load. As depicted in Figure 8-29(a), in **Class A** operation the current is present during the entire output signal cycle. This corresponds to a $\Theta_A = 360°$ conduction angle. If the transfer characteristic of the transistor in the linear region is close to that of a linear function, then the output signal is an amplified replica of the input signal without suffering any distortion. In practical circuits, however, there is always a certain degree of nonlinearity present, which results in a distorted output signal of the amplifier.

In **Class B** (Figure 8-29(b)), the current is present during only half of the cycle, corresponding to a $\Theta_B = 180°$ conduction angle. During the second half of the cycle, the transistor is in the cutoff region, and no current flows through the device. **Class AB** [Figure 8-29(c)] combines the properties of the classes A and B; it has a conduction angle Θ_{AB} ranging from $180°$ to $360°$. This type of amplifier is typically employed when high-power "linear" amplification of the RF signal is required.

In a **Class C** amplifier (Figure 8-29(d)), we have a nonzero current for less than half of the cycle (i.e., the conduction angle is $0 < \Theta_C < 180°$). This results in maximum distortion of the output signal.

Amplifier Classes of Operation and Biasing Networks

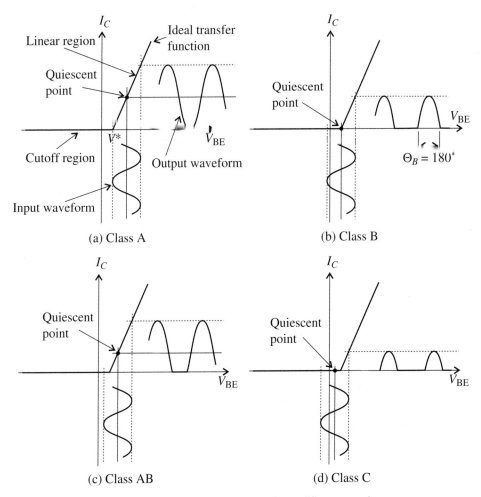

Figure 8-29 Various classes of amplifier operation.

A logical question that arises is why not all amplifiers are operated in Class A, since this mode delivers the least signal distortion. The answer is directly linked to the amplifier efficiency. Efficiency η is defined as the ratio of the average RF power P_{RF} delivered to the load over the average power P_S supplied by the DC source, and is usually measured in percent:

$$\eta = \frac{P_{RF}}{P_S} 100\%. \tag{8.22}$$

SWITCHED AMPLIFIERS

In an attempt to increase efficiency, switched transistor operation (so-called Class D, E, F) in conjunction with appropriate filtering of the harmonics, can convert up to 100% of the DC power into RF power. However, non-zero switching times and resistive losses typically limit the efficiency to less than 100%.

The theoretical maximum efficiency of the Class A amplifier is only 50%, but the efficiency of Class C can reach values close to 100%. Fifty percent efficiency of Class A amplifiers means that half of the power supplied by the DC source is dissipated as heat. This situation may not be acceptable in portable communication systems where most devices are battery operated. In practical applications, designers usually choose the class of operation that gives maximum efficiency but still preserves the informational content of the RF signal. Even though Class C amplifiers are efficient, the output power tends to be low for a given power rating of the transistor. For this reason, other classes of amplifiers (D, E, F) have been introduced. They all use the transistor in switched mode.

In the following example, we derive the maximum theoretical efficiency η of the amplifier as a function of conduction angle.

Example 8-11. Amplifier efficiency computation

Derive the general expression for the amplifier efficiency η as a function of conduction angle Θ_0. List the values of η for both Class A and Class B amplifiers.

Solution. The electrical current through the load for a conduction angle of Θ_0 has a waveform shown in Figure 8-30(a), where the cosine current amplitude is given by I_0.

Similarly, the power supply current I_S has a maximum value of I_0 plus the quiescent current I_Q:

$$I_S = I_Q + I_0 \cos\Theta. \qquad (8.23)$$

The value of the quiescent current necessary to ensure the specified conduction angle Θ_0 can be found from (8.23) by setting I_S to zero at $\Theta = \Theta_0/2$:

$$I_Q = -I_0 \cos(\Theta_0/2). \qquad (8.24)$$

The average power supply current is then computed as an integral over the conduction angle ranging between the limits of $\Theta = -\Theta_0/2$ and $\Theta = \Theta_0/2$; that is,

Amplifier Classes of Operation and Biasing Networks

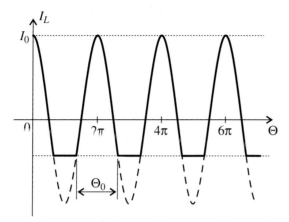

(a) Load current waveform at the output of the transistor

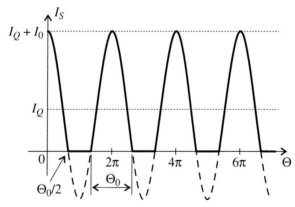

(b) Corresponding power supply current waveform

Figure 8-30 Load and power supply current waveforms as a function of conduction angle.

$$\langle I_S \rangle = \frac{1}{2\pi} \int_{-\Theta_0/2}^{\Theta_0/2} I_S d\Theta$$

$$= -\frac{I_0}{2\pi}\left[\Theta_0 \cos\left(\frac{\Theta_0}{2}\right) - 2\sin\left(\frac{\Theta_0}{2}\right)\right]. \quad (8.25)$$

Thus, the average power from the power supply is

$$P_S = V_{CC}\langle I_S \rangle = -\frac{I_0 V_{CC}}{2\pi}\left[\Theta_0 \cos\left(\frac{\Theta_0}{2}\right) - 2\sin\left(\frac{\Theta_0}{2}\right)\right] \quad (8.26)$$

where V_{CC} is the supply voltage.

The RF output current is the fundamental harmonic of I_S:

$$I_{RF} = \frac{1}{\pi}\int_{-\pi}^{\pi} I_S \cos\Theta \, d\Theta = \frac{1}{\pi}\int_{-\Theta_0/2}^{\Theta_0/2} (I_Q + I_0 \cos\Theta)\cos\Theta \, d\Theta$$

where we took advantage of the even symmetry of I_S. Evaluating the above integral:

$$I_{RF} = \frac{1}{2\pi}\left(4I_Q \sin\left(\frac{\Theta_0}{2}\right) + I_0(\Theta_0 + \sin\Theta_0)\right).$$

Substituting (8.24) for I_Q, we simplify this to:

$$I_{RF} = \frac{I_0}{2\pi}(\Theta_0 - \sin\Theta_0).$$

The RF output voltage can be a sinusoid swinging between $+V_{CC}$ and $-V_{CC}$, if all harmonics other than the fundamental are short-circuited by the matching network:

$$V_{RF} = V_{CC}.$$

The resulting RF output power is:

$$P_{RF} = \frac{1}{2}V_{RF}I_{RF} = \frac{I_0 V_{CC}}{4\pi}(\Theta_0 - \sin\Theta_0). \qquad (8.27)$$

Dividing (8.27) by (8.26), we find the amplifier efficiency

$$\eta = -\frac{\Theta_0 - \sin\Theta_0}{2[\Theta_0 \cos(\Theta_0/2) - 2\sin(\Theta_0/2)]} \qquad (8.28)$$

where the conduction angle Θ_0 is measured in radians.

The graph of η as a function of the conduction angle Θ_0 is shown in Figure 8-31.

Substituting $\Theta_0 = 2\pi$ into (8.28), we find that the efficiency of a Class A amplifier is indeed 50%. To determine the efficiency of a Class B amplifier, we simply use the conduction angle $\Theta_0 = \pi$ in (8.28), which yields

$$\eta_B = -\frac{\pi - \sin\pi}{2[\pi\cos(\pi/2) - 2\sin(\pi/2)]} = \frac{\pi}{4} = 0.785.$$

That is, Class B yields an efficiency of 78.5%.

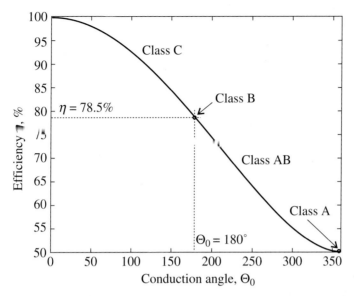

Figure 8-31 Maximum theoretical efficiency of an ideal amplifier as a function of conduction angle.

Efficiency is an important design consideration when dealing with low power consumption, as required, for instance, in mobile communication systems, where battery lifetime must be maximized.

8.3.2 Bipolar Transistor Biasing Networks

There are generally two types of biasing networks: passive and active. **Passive** (or **self-biased**) networks are the simplest type, and usually incorporate a resistive network, which provides the appropriate voltages and currents for the RF transistor. The main disadvantages of such networks are that they are sensitive to changes in transistor parameters and that they provide poor temperature stability. To compensate for these drawbacks, **active biasing** networks, are often employed.

In this section, we consider several network configurations for biasing RF BJTs. Two possible topologies are shown in Figure 8-32.

The combination of the blocking capacitor C_B and the RFC connected to the base and collector terminals of the transistor in Figure 8-32 serve the purpose of isolating the RF signal from the DC power source. At high frequencies, the RFCs are usually replaced by quarter-wave transmission lines that convert the short-circuit condition on the C_B side to an open-circuit condition on the transistor side.

EFFICIENCY MEASURES

To more precisely specify amplifier efficiency, we need to take the input RF power into account. Two popular efficiency measures exist for this purpose, power-added efficiency

$$\text{PAE} = \frac{P_{RFout} - P_{RFin}}{P_{DC}}$$

$$= \eta\left(1 - \frac{1}{G}\right)$$

and total, or overall, efficiency

$$\eta_{total} = \frac{P_{RFout}}{P_{DC} + P_{RFin}}$$

$$= \left(\frac{1}{\eta} + \frac{1}{G}\right)^{-1}$$

where G is the gain. Total efficiency is more accurate, but power-added efficiency retains greater popularity.

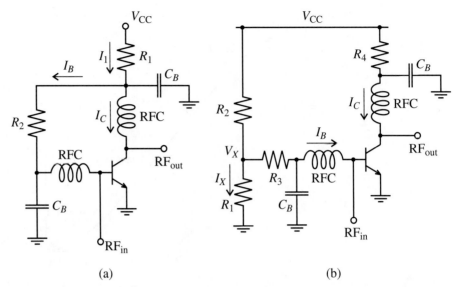

Figure 8-32 Passive biasing networks for an RF BJT in common-emitter configuration.

The following example discusses how to compute the resistors for the two biasing networks shown in Figure 8-32.

Example 8-12. Design of passive biasing networks for a BJT in common-emitter configuration

Design biasing networks according to Figures 8-32(a) and (b) for the BJT settings of $I_C = 10$ mA, $V_{CE} = 3$ V, and $V_{CC} = 5$ V. Assume that the transistor has a $\beta = 100$ and $V_{BE} = 0.8$ V.

Solution. As seen in Figure 8-32(a), the current I_1 through resistor R_1 is equal to the sum of the collector and base currents. Since $I_B = I_C/\beta$, we obtain

$$I_1 = I_C + I_B = I_C(1 + \beta^{-1}) = 10.1 \text{ mA}.$$

The value of R_1 can be found as

$$R_1 = \frac{V_{CC} - V_{CE}}{I_1} = 198 \; \Omega.$$

Similarly, the base resistor R_2 is computed as

$$R_2 = \frac{V_{CE} - V_{BE}}{I_B} = \frac{V_{CE} - V_{BE}}{I_C/\beta} = 22 \; k\Omega.$$

For the circuit in Figure 8-32(b), the situation is slightly more complicated. Here, we have the freedom of choosing the value of the voltage potential V_X and the current I_X through the voltage divider resistor R_2. Setting V_X to 1.5 V, we determine the base resistor R_3 to be

$$R_3 = \frac{V_X - V_{BE}}{I_B} = \frac{V_X - V_{BE}}{I_C/\beta} = 7 \; k\Omega.$$

The value of I_X is usually chosen to be 10 times larger than I_B. Therefore, $I_X = 10 I_B = 1$ mA and the values of the resistances for the voltage divider are computed as

$$R_1 = \frac{V_X}{I_X} = 1.5 \; k\Omega \text{ and } R_2 = \frac{V_{CC} - V_X}{I_X + I_B} = 3.18 \; k\Omega.$$

Finally, the collector resistor is found as

$$R_4 = (V_{CC} - V_{CE})/I_C = 200 \; \Omega.$$

Typical values for C_B and RFC at 1 GHz are 200 pF and 200 nH, respectively. The freedom to select particular voltages and currents is in practice restricted by standardized resistance values.

An example of an active biasing network for a BJT in common-emitter configuration is shown in Figure 8-33. Here we employ a low-frequency transistor Q_1 to provide the necessary base current for the RF transistor Q_2. If transistors Q_1 and Q_2 have the same thermal properties, then this biasing network also results in good temperature stability.

Example 8-13 illustrates the determination of the component values for the active biasing network depicted in Figure 8-33.

BIAS FOR PULSED MEASUREMENTS

Often, active devices must be measured over short (pulsed) durations so as to prevent thermal influences. Special bias networks with small blocking caps (~20pF) and RFCs (~50nH) are used for sufficiently high test frequencies $\left(\geq 1 \; GHz\right)$ and pulse durations ≥ 100 ns.

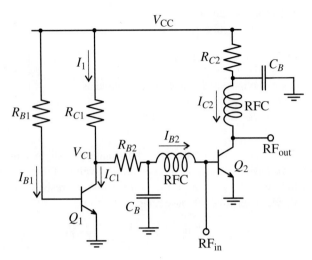

Figure 8-33 Active biasing network for a common-emitter RF BJT.

Example 8-13. Design of an active biasing network for a BJT transistor in common-emitter configuration

Design a biasing network as shown in Figure 8-33 for $I_{C2} = 10$ mA, $V_{CE2} = 3$ V, and $V_{CC} = 5$ V. Assume that both transistors have $V_{BE} = 0.8$ V, and $\beta = 100$ that can fluctuate between 50 and 150 as a function of temperature.

Solution. We need to determine the β-dependence of the Q_2 bias, specifically its collector current

$$I_{C2} = \beta I_{B2} = \beta \frac{V_{C1} - V_{BE}}{R_{B2}}.$$

To determine V_{C1}, we express the load to this transistor by its Thévenin equivalent voltage source V_{Th} and resistance R_{Th}:

$$V_{C1} = V_{Th} - R_{Th} I_{C1}$$

where $R_{Th} = R_{C1} \| R_{B2} = \dfrac{R_{C1}R_{B2}}{R_{C1}+R_{B2}}$, and $V_{Th} = \dfrac{V_{CC}R_{B2}+V_{BE}R_{C1}}{R_{C1}+R_{B2}}$.

This follows from the voltage divider formed by R_{C1} and R_{B2} between V_{CC} and $V_{B2} = V_{BE}$. The collector current of Q_1 is set by its base resistor:

$$I_{C1} = \beta I_{B1} = \beta \dfrac{V_{CC}-V_{BE}}{R_{B1}}.$$

Substituting the above equations into the expression for I_{C2} and simplifying, we obtain:

$$I_{C2} = \dfrac{\beta(R_{B1}-\beta R_{C1})}{R_{B1}(R_{C1}+R_{B2})}(V_{CC}-V_{BE}).$$

The next step is to set the derivative of I_{C2} to zero to minimize the β-dependence:

$$\left.\dfrac{dI_{C2}}{d\beta}\right|_{\beta=\beta_0} = \dfrac{R_{B1}-2\beta_0 R_{C1}}{R_{B1}(R_{C1}+R_{B2})}(V_{CC}-V_{BE}) = 0.$$

This reduces to the circuit constraint:

$$R_{C1} = \dfrac{R_{B1}}{2\beta_0}.$$

We can now compute all the component values. The base resistance of Q_1 is set by the base current, which in turn depends on the desired collector current according to:

$$R_{B1} = \dfrac{V_{CC}-V_{BE}}{I_{B1}} = \dfrac{(V_{CC}-V_{BE})\beta_0}{I_{C1}}.$$

Since we have some freedom in choosing I_{C1}, we pick a value of 10 times the Q_2 base current. We notice that reducing I_{C1} results in a reduction of V_{C1}. Therefore, we need to leave some swing room to account for the β variation. Using $I_{C1} = 10\dfrac{I_{C2}}{\beta_0} = 1\,\text{mA}$, we obtain $R_{B1} = 420\,\text{k}\Omega$. From the above derived constraint, $R_{C1} = 2.1\,\text{k}\Omega$. Since $I_{B2} = \dfrac{I_{C2}}{\beta_0} = 100\,\mu\text{A}$ and $V_{C1} = V_{CC}-(I_{C1}+I_{B2})R_{C1} = 2.69\,\text{V}$, we obtain

$$R_{B2} = \dfrac{V_{C1}-V_{BE}}{I_{B2}} = 18.9\,\text{k}\Omega.$$

Finally,

$$R_{C2} = \frac{V_{CC} - V_{CE2}}{I_{C2}} = 200 \; \Omega.$$

To verify the β-dependence, I_{C2} is evaluated at several points:

β	50	75	100	125	150
I_{C2}, mA	7.5	9.38	10	9.38	7.5

Although active biasing offers a number of performance advantages over passive networks, certain disadvantages also arise: specifically, additional circuit board space, possible layout complications, and added power requirements.

Another active biasing network for a BJT in common-emitter configuration is shown in Figure 8-34. Here, diodes D_1 and D_2 provide a fixed reference for the voltage drop across the base-emitter junctions of both transistors. Resistor R_1 is used to adjust the biasing current to the base of transistor Q_1 and R_2 limits the range of this adjustment. Ideally, for temperature compensation, transistor Q_1 and one of the diodes should remain at ambient temperature, whereas the second diode should be placed on the same heat sink as RF transistor Q_2.

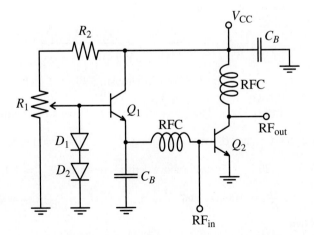

Figure 8-34 Active biasing network containing low-frequency transistor and two diodes.

Amplifier Classes of Operation and Biasing Networks

As a final remark, it is important to point out that in all biasing networks, the operating conditions (common-base, common-emitter, or common-collector) of the transistor at RF frequencies are entirely independent of the DC configuration. For instance, we can take an active biasing network, shown in Figure 8-33, and modify it for common-base RF operation, as seen in Figure 8-35.

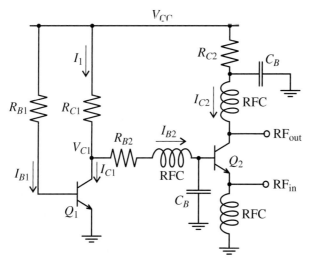

Figure 8-35 Modification of the active biasing network shown in Figure 8-33 for common-base RF operation.

At DC, all blocking capacitors represent an open circuit and all RFCs behave like short circuits. Therefore, this biasing network can be redrawn as shown in Figure 8-36(a), indicating the common-emitter configuration. However, at RF frequency all blocking capacitors become short circuits and all RFCs behave like open circuits. This transforms the biasing network into common-base configuration, as depicted in Figure 8-36(b).

8.3.3 Field Effect Transistor Biasing Networks

The biasing networks for field effect transistors are in many ways very similar to the BJT networks covered in the previous section. One key distinction is that MESFETs usually require a negative gate-source voltage as part of the bias conditions.

The most basic passive dual supply biasing network for FETs is shown in Figure 8-37. The main disadvantage of such a network is the need of a bipolar power supply for $V_G < 0$ and $V_D > 0$. If such a bipolar power supply is unavailable, one can resort to a strategy where

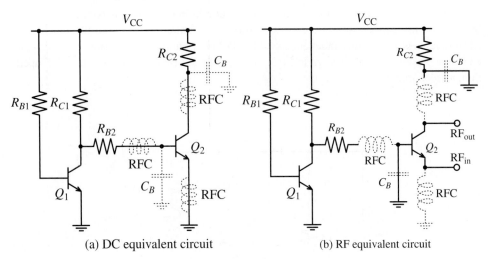

Figure 8-36 DC and RF equivalent circuits for the active biasing network in Figure 8-35.

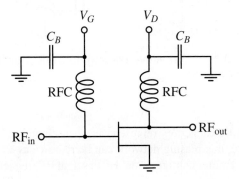

Figure 8-37 Bipolar passive biasing network for FETs.

instead of the gate, the source terminal of the transistor is biased. The gate in this case is grounded. Two examples of such networks are shown in Figure 8-38.

Temperature compensation of FET biasing networks is typically accomplished through the use of thermistors.

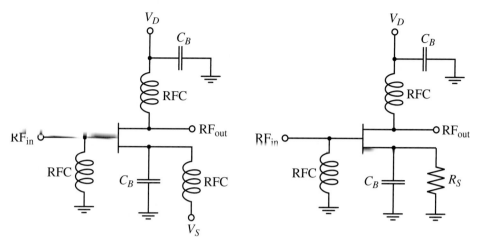

Figure 8-38 Unipolar passive biasing networks for FETs.

  RF&MW→

Practically Speaking. Design of a 7 GHz HEMT amplifier matching and biasing networks

In this section, we will design and simulate biasing and matching networks for a HEMT amplifier operated at 7 GHz. The goal is to obtain a simultaneous power match on both ports of this amplifier, achieving the maximum power gain.

The device selected for this amplifier is the Agilent ATF551M4, an Enhancement-Mode Pseudomorphic HEMT (E-pHEMT). This transistor offers good noise figure, high linearity, and significant gain up to 18 GHz, making it useful for a variety of applications from low-noise amplifiers to small transmitter power amplifiers. This device is also designed for a positive gate-source voltage V_{GS} (hence its classification as enhancement mode), allowing easy biasing under the conditions of a grounded source and a single-rail power supply. To achieve high gain, we will operate the device at a drain current of $I_D = 20$ mA and a drain-source voltage of $V_{DS} = 3$ V, which are at the higher end of the operating range. Higher drain voltage and current also increase the maximum output power and improve linearity, but we will not evaluate these parameters here.

We start by analyzing the biasing network used with the E-pHEMT, as detailed in Figure 8-39(a). This active-bias circuit has the advantageous ability to accurately bias an enhancement-mode FET or a BJT with a grounded source or emitter. If we had to bias a BJT, the resistor R_2 could be omitted. The primary BJT in the biasing network, Q_2, creates a feedback loop that controls the E-pHEMT's gate voltage to achieve the specified drain voltage V_{DS} and current I_D. The transistor Q_1, together with resistors R_1 and R_3, create a reference for the drain voltage. Furthermore, Q_1 provides additional temperature stability, a fact that is not essential to the functionality of the circuit. For the DC bias analysis, we can substitute the BJTs with their DC models (known as constant V_{BE} active region models), as seen in Figure 8-39(b). For the highest bias stability over temperature, we would like the two BJTs to be matched, and the collector currents to be equal, resulting in $V_{BE1} = V_{BE2} = V_{BE}$. Starting from these assumptions and neglecting the base current (assuming very large β), it is straightforward to determine the resistor values. We notice that the same current, labeled I_{R3}, flows through resistors R_1, R_2 and R_3, while the current through R_4 is $I_{R4} = I_D + I_{R3}$, where I_D is the desired drain current. The voltages across R_3 and R_4 are equal, and can be expressed as $V_{R3} = V_{R4} = V_{DD} - V_D$, where $V_D = V_{DS}$ is the desired drain voltage, and V_{DD} is the supply voltage. Thus,

$$R_3 = \frac{V_{DD} - V_D}{I_{R3}}$$

$$R_4 = \frac{V_{DD} - V_D}{I_D + I_{R3}}.$$

Since the voltage across R_1 is computed to be $V_D - |V_{BE}|$, we can approximately state:

$$R_1 \approx \frac{V_D - |V_{BE}|}{I_{R3}}.$$

The voltage across R_2 is the desired gate voltage $V_G = V_{GS}$, for which an estimate is sufficient (the datasheet suggest $V_{GS} \approx 500$ mV). Then,

$$R_2 \approx \frac{V_G}{I_{R3}}.$$

A more rigorous analysis of the circuit in Figure 8-39(b) yields slightly modified expressions for R_1 and R_2 (see also the problem section at the end of this chapter):

Amplifier Classes of Operation and Biasing Networks

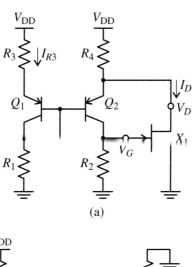

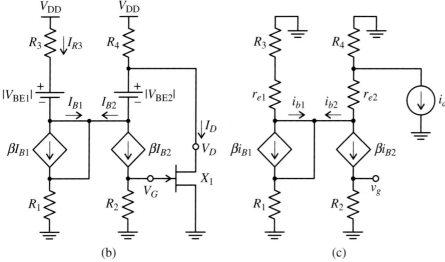

Figure 8-39 E-pHEMT biasing network: (a) DC bias circuit, (b) DC model, (c) low-frequency small-signal model.

$$R_1 = \frac{V_D - |V_{BE}|}{I_{R3}}\left(\frac{\beta+1}{\beta+2}\right)$$

$$R_2 = \frac{V_G}{I_{R3}}\left(\frac{\beta+1}{\beta}\right).$$

We observe that the β-dependence is nearly insignificant, and only R_1 depends on V_{BE} (which is temperature-dependent). In all these expressions, I_{R3} is an adjustable current. To help determine this parameter, we examine the feedback loop functionality of this circuit using the low-frequency small-signal model shown in Figure 8-39(c). Here, we use the small-signal T-model with the dynamic emitter resistance $r_e = V_T/I_E$, where V_T is the thermal voltage $V_T = kT/q \approx 26$ mV at room temperature and I_E is the DC emitter current. The circuit in Figure 8-39(c) is configured for determining the open-loop transresistance (a type of gain) $r = \partial v_G/\partial i_D = v_g/i_d$, which specifies to what degree the feedback circuit would move the gate voltage in response to a small change in drain current. The transresistance should be negative to achieve negative feedback. Furthermore, its magnitude should be sufficiently large to guarantee bias stability with temperature and device-to-device variations. Analyzing Figure 8-39(c), the transresistance can be expressed as

$$ r = \frac{v_g}{i_d} = -R_2 \frac{R_4}{R_4 + r_e + \frac{R_1 \| (R_3 + r_e)}{\beta + 1}} \left(\frac{\beta}{\beta + 1} \right). $$

Substituting the previously derived resistor values and making some simplifying approximations (such as $\beta + 1 \approx \beta$, $V_{DD} + V_T \approx V_{DD}$, $I_D + I_{R3} \approx I_D$), the following result is obtained:

$$ r \approx -\frac{V_G}{I_{R3} + \left(\frac{V_T}{V_{DD} - V_D} + \frac{V_D - |V_{BE}|}{\beta(V_{DD} - |V_{BE}|)} \right) I_D}. $$

We observe that I_{R3} should be minimized in order to optimize the feedback gain. However, decreasing the BJT collector current reduces its β and increases the influence of leakage currents. As a result, it is prudent to keep I_{R3} above some minimum. Besides I_{R3}, the denominator includes a second term, which is proportional to I_D and only dependent on the bias parameters. This term ultimately limits the transresistance; therefore it makes sense to pick the magnitude of the current I_{R3} to be only a few times smaller than the term proportional to I_D. Upon further reflection, it is also apparent that $V_{DD} - V_D$ (voltage drop across R_4) should not be too small to prevent the feedback loop from becoming ineffective.

We can next determine the resistor values, based on the bias parameters $I_D = 20$ mA, $V_D = 3$ V, $V_G \approx 500$ mV, and a supply voltage of $V_{DD} = 5$ V. The BJTs, which are assumed to possess parameter values similar to type 2N3906, are set to $\beta \approx 100$, $|V_{BE}| \approx 0.6$ V, and $V_T = 26$ mV. From these, the second term in the denominator of the transresistance expression ($\propto I_D$) is 369 µA, limiting r to -1.36 kΩ. We choose $I_{R3} = 100$ µA to supply the BJTs with sufficient bias current for proper operation, while keeping the transresistance nearly optimal. The resulting $r = -1.07$ kΩ implies that for a 0.1 mA increase in drain current, the gate voltage would decrease by 0.107 V. Alternatively, if the required gate voltage varies by ± 0.2 V (the maximum range of variation), the drain current would only vary by ± 0.187 mA or 0.94%. The resistors are calculated to be $R_1 = 23.77$ kΩ, $R_2 = 5.05$ kΩ, $R_3 = 20$ kΩ, and $R_4 = 99.5$ Ω, and then rounded to the nearest 1% tolerance values (or 23.7 kΩ, 5.11 kΩ, 20 kΩ, and 100 Ω respectively). Conducting a DC analysis in the ADS circuit simulator produces $I_D = 19.9$ mA and $V_D = 3.00$ V, which are very close to our original goal.

A basic microstrip layout, which serves as a starting point for the 7 GHz amplifier, is shown in Figure 8-40. We need to assure that the biasing network does not interfere with the RF operation of the E-pHEMT; the DC biases are supplied through quarter-wavelength sections of microstrips terminated by large bypass capacitors (C_B) to ground on the DC side. The length of these microstrips is determined from the knowledge of the wavelength which in turn depends on the effective dielectric constant ε_{eff}. For this design, we intend to deploy for simplicity microstrips with $Z_0 = 50$ Ω. Using equation (2.46b) with a substrate of $\varepsilon_r = 3.48$, we obtain for the microstrip a width-to-height ratio of $w/h = 2.266$, corresponding to a strip width of 45.3 mil on a 20 mil substrate. Compensating for 1.4 mil strip thickness (1 oz copper) using equation (2.47), the final strip width is 43.4 mil. The effective dielectric constant is found to be $\varepsilon_{eff} = 2.734$ according to equation (2.44). This determines the wavelength at $f = 7$ GHz (the center frequency) to be $\lambda = c/(f\sqrt{\varepsilon_{eff}}) = 1019.7$ mil. The $\lambda/4$ bias stubs are thus 254.9 mil long.

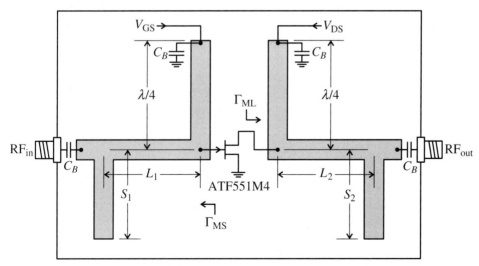

Figure 8-40 Conceptual PCB layout of the 7 GHz amplifier, showing the biasing filters and matching networks implemented with microstrips.

At this point, we simulate the biased E-pHEMT without any matching networks to determine the S-parameters at 7 GHz:

$$S_{11} = 0.550\angle 147.25°, \quad S_{12} = 0.092\angle 14.69°$$
$$S_{21} = 3.538\angle 29.13°, \quad S_{22} = 0.200\angle -83.87°.$$

The simulation also indicates that the amplifier is unconditionally stable at 7 GHz with $\mu = 1.09 > 1$ (see Chapter 9 for a discussion of stability), implying that both ports can be simultaneously matched for maximum power transfer, achieving the maximum power gain. Calculations according to formulas in Chapter 9 reveal that for simultaneous power match at both ports, the source reflection coefficient seen by the transistor must be $\Gamma_{MS} = 0.791\angle -145.13°$, and the load reflection coefficient must be $\Gamma_{ML} = 0.650\angle 94.88°$.

One way to approach the source-side matching is to transform the desired port impedance of 50 Ω to the corresponding source impedance specified by Γ_{MS}. In general, this can be accomplished by using a shunt microstrip stub followed by a series length of microstrip, as in Figure 8-40. The open-circuited stub (which looks capacitive when shorter than $\lambda/4$) moves the source impedance downwards away from the origin of the Smith Chart along the $g = 1$ circle, and can be sized to achieve the desired reflection coefficient magnitude. A series length of 50 Ω microstrip can then be inserted to rotate the source reflection coefficient to any desired angle. Plotting Γ_{MS} on the Smith Chart reveals that the desired source admittance $y_{MS} = 1.14 + 2.76j$ is already very close to the $g = 1$ circle. Accepting a small error, the shunt stub by itself is sufficient to achieve source-side matching; we can therefore set $L_1 = 0$. Using the Smith Chart, we determine the length of the shunt open-circuited stub that would replicate the imaginary part of y_{MS} to be 0.195λ. In terms of actual trace length, this translates into $S_1 = 199$ mil.

Plotting Γ_{ML} in the Smith Chart, we find that the normalized admittances becomes $y_{ML} = 0.440 - 0.988j$. For this case, both a shunt stub and a series transmission line are required. The stub length is determined by quantifying the intersection of the $|\Gamma| = |\Gamma_{ML}|$ circle with the $g = 1$ circle in the lower half of the Smith Chart. This occurs at $\Gamma = 0.650\angle -130.5°$, or $y = 1 + 1.71j$. The length of the shunt open-circuited stub that would replicate the imaginary part of this admittance is $S_2 = 0.166\lambda = 169$ mil. Finally, the rotation angle of the series transmission line is determined by subtracting the phase of Γ_{ML} from the phase of Γ at the intersection point: $-130.5° - 94.88° + 360° = 134.6°$. Since the rotation of the reflection coefficient occurs by the signal traveling the length of line twice (forward and back), the required electrical length of the transmission line is $134.6°/2 = 67.3°$, translating to $L_2 = 0.187\lambda = 191$ mil. Alternatively, we could use the Wavelengths

Toward Generator scale of the Smith Chart to graphically determine L_2 in wavelengths.

Figure 8-41 presents the schematic of the amplifier design using the previous outlined steps. Simulating the design in ADS produces the S-parameter results depicted in Figure 8-42. We observe that the amplifier comes close to achieving the goal of simultaneous power match at the input and output ports ($S_{11} = S_{22} = 0$) at 7 GHz. The error in S_{11} is the greatest due to the omission of the series transmission line section in the input matching network. However, this error is insignificant in practical terms because the tolerances of a practical realization are expected to be worse. The gain $|S_{21}|^2$ of 13.8 dB is very close to the maximum available from this device at 7 GHz, calculated from the S-parameters to be 14.0 dB. It is worth noting that this simulation contains significant inaccuracies due to the omission of the microstrip interconnecting patches (junctions), component solder pads (and related microstrip width discontinuities) and edge effects on the open-circuited stubs. All these features, supported by ADS as regular circuit

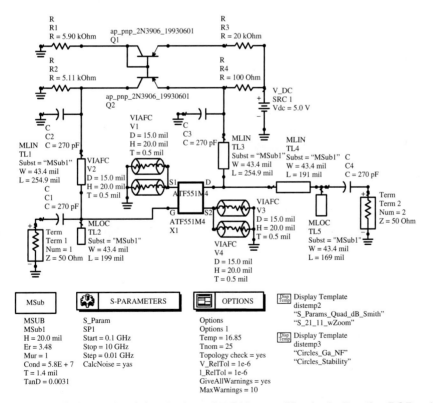

Figure 8-41 Schematic of the designed 7 GHz amplifier including the PCB substrate description and the S-parameter simulation setup.

components, have significant effects at 7 GHz due to their relatively large dimensions in comparison to the wavelength.

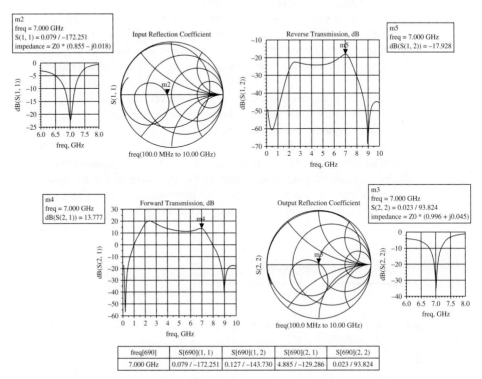

Figure 8-42 Simulated S-parameters of 7 GHz amplifier.

8.4 Summary

The material covered in this chapter is geared toward providing an understanding of two key issues encountered in any RF/MW system: interfacing various components of different impedance values, and suitably biasing the active devices depending on their class of operation.

To ensure optimal power transfer between systems of different impedances, we investigate at first two-element L-type matching configurations. In the context of two-port network analysis, the conjugate complex matching requirement at the input and output ports results in optimal power transfer at a particular target frequency. The technique is simple and can be compared with the design of a bandpass or bandstop filter. Care must be exercised in selecting a suitable L-type network to avoid the forbidden regions for which a given load impedance cannot be matched to the desired input impedance. From the knowledge of the network transfer function, the loaded quality factor

$$Q_L = \frac{f_0}{\text{BW}}$$

and the simpler to determine nodal quality factor

$$Q_n = \frac{|X_S|}{R_S} = \frac{|B_P|}{G_P}$$

can be utilized to assess the frequency behavior of the matching networks. Unfortunately, L-type networks do not allow any flexibility in conditioning the frequency response and are, therefore, mostly used for narrow-band RF designs. To affect the frequency behavior, a third element must be added, resulting in T- and Pi-type networks. With these configurations, a certain nodal quality factor, and indirectly a desired bandwidth, can be implemented.

While lumped element design is appropriate at low frequencies, distributed transmission line elements must be employed when the frequency extends into the GHz range. Hybrid configurations using series-connected transmission line elements and shunt-connected capacitors are very attractive for prototyping since the location and value of the capacitors can easily be varied. If the capacitors are replaced by open- and short-circuited transmission lines, one arrives at the single- and double-stub matching networks.

Depending on the application (for instance, linear small signal or nonlinear large signal amplification), various classes of transistor amplifiers are identified. The classification is done by computing the RF to supply power ratio, known as efficiency:

$$\eta = \frac{P_{RF}}{P_S} 100\%$$

which can be expressed in terms of the conduction angle Θ_0 quantifying the amount of load current flow through the relation

$$\eta = -\frac{\Theta_0 - \sin\Theta_0}{2[\Theta_0 \cos(\Theta_0/2) - 2\sin(\Theta_0/2)]}.$$

For instance, Class A offers the highest linearity at the expense of the lowest efficiency of 50%, whereas Class B compromises linearity but improves efficiency to 78.5%.

Once the class of operation is identified, a biasing network is chosen to set the appropriate quiescent point of the transistor. Passive biasing networks are normally easy to implement. However, they are not as flexible as biasing networks involving active devices. The biasing not only sets the DC operating conditions but must also ensure isolation of the RF signal through the use of RFCs and blocking capacitors.

Further Reading

W. A. Davis, *Microwave Semiconductor Circuit Design,* Van Nostrand Reinhold Company, New York, 1984.

N. Dye and H. Granberg, *Radio Frequency Transistors: Principles and Practical Applications,* Butterworth-Heinemann, 1993.

G. Gonzalez, *Microwave Transistor Amplifiers: Analysis and Design,* Prentice Hall, Upper Saddle River, NJ, 1997.

P. Horowitz and W. Hill, *The Art of Electronics,* Cambridge University Press, Cambridge, UK, 1993.

D. Pozar, *Microwave Engineering,* John Wiley & Sons, New York, 1998.

P. Rizzi, *Microwave Engineering: Passive Circuits,* Prentice Hall, Englewood Cliffs, NJ, 1988.

Problems

8.1 Obtain the "forbidden" regions for the two-element matching networks shown in Figures 8-1(c)–(f). Assume that the load is matched to the normalized input impedance (i.e., $z_{in} = 1$).

8.2 Use the analytical approach and design a two-component matching network that matches the $Z_L = (100 + j20)\Omega$ load impedance to a given $Z_S = (10 + j25)\Omega$ source, at the frequency of $f_0 = 960$ MHz.

8.3 Develop a two-component matching network for a $Z_L = (30 - j40)\Omega$ load and a 50 Ω source. How many network topologies exist that can be used? Find the values of the components if a perfect match is desired at $f_0 = 450$ MHz.

8.4 Repeat Problem 8.3 for a $Z_L = (40 + j10)\Omega$ load and a matching frequency of $f_0 = 1.2$ GHz.

8.5 Measurements indicate that the source impedance in Problem 8.3 is not purely resistive, but has a parasitic inductance of $L_S = 2$ nH. Recompute the values for the matching network components that take into account the presence of L_S.

8.6 A load $Z_L = (20 + j10)\Omega$ consisting of a series RL combination is to be matched to a 50 Ω microstrip line at $f_0 = 800$ MHz. Design two two-element matching networks and specify the values of their components. Plot the frequency response at the load resistance for both networks, and find the corresponding bandwidths.

8.7 In Example 8-5, a T-network is discussed that matches a load impedance of $Z_L = (60 - j30)\Omega$ to an input impedance of $Z_{in} = (10 + j20)\Omega$ at 1 GHz under the constraint that Q_n does not exceed 3. Go through this design step by step and identify each point in the Smith Chart in terms of its impedance or admittance value. Verify the final results shown in Figure 8-16.

8.8 Go through Example 8-6 and find each point in the Smith Chart shown in Figure 8-17, and verify the final network components depicted in Figure 8-18.

8.9 Repeat the Pi-type matching network design in Example 8-6 for a nodal quality factor of $Q_n = 2.5$. Plot $Z_{in}(f)$ for this Q_n value and compare it against the $Q_n = 2$ design in Example 8-6. As frequency range, choose 1 GHz $< f <$ 4 GHz.

8.10 Design two T-type matching networks that transform a $Z_L = 100$ Ω load to a $Z_{in} = (20 - j40)\Omega$ input impedance at a nodal quality factor of $Q_n = 4$. The matching should be achieved at $f_0 = 600$ MHz.

8.11 Design two Pi-type matching networks for the same conditions as in Problem 8.10.

Problems

8.12 To achieve matching conditions for a specified Q_n, the circuit designer often has to use more than two or three elements in the matching network. Using a graphical approach, design a multisection matching network that transforms $Z_L = 10\ \Omega$ into $Z_S = 250\ \Omega$ at $f_0 = 500$ MHz while maintaining a nodal quality factor of $Q_n = 1$. The multisection matching network should consist of a series of two-element sections each of which is a series inductor, shunt capacitor combination (see Figure 8-1(h)).

8.13 For an increased frequency of $f_0 = 1$ GHz, it was decided that the network designed in Problem 8.12 should be replaced by a combined matching network shown in Figure 8-19. Determine the total number of capacitors and transmission line sections necessary to achieve matching and find the values of all components in the network.

8.14 Using the design from Example 8-7, find the length and width of each transmission line if an FR4 substrate with dielectric constant of $\varepsilon_r = 4.6$ and height of $h = 25$ mil is used. Find the maximum deviation of the input impedance of the matching network if the capacitor that is used in the circuit has a $\pm 10\%$ tolerance and the automatic component placement equipment has a ± 2 mil precision (i.e., the capacitor can be placed within ± 2 mil of the intended position).

8.15 In Example 8-7, it is argued that open-circuited stubs can be replaced by short-circuited ones if the length is increased by a quarter wavelength. Matching is achieved only at a single frequency, and over a broader frequency range the network response can significantly differ from the target impedance values. Design a single-stub matching network that transforms a $Z_L = (80 + j20)\Omega$ load impedance into a $Z_{in} = (30 - j10)\Omega$ input impedance. Compare the frequency behavior of the input impedance over the $\pm 0.8 f_0$ frequency range for two different realizations of the matching network: open-circuited stub, and using an equivalent short-circuited stub. Assume that the matching frequency is $f_0 = 1$ GHz and the load is a series combination of resistance and inductance.

8.16 Using the matching network shown in Figure 8-23(b), find the stub length l_S, the characteristic line impedance Z_{0L}, and the transmission line length l_L such that the $Z_L = (80 - j40)\Omega$ load impedance is matched to 50 Ω source. Assume that the characteristic impedance of the stub is $Z_{0S} = 50\ \Omega$.

8.17 For a double-stub tuner shown in Figure 8-26 with parameters $l_1 = \lambda/(8)$, $l_2 = 5\lambda/8$, and $l_3 = 3\lambda/8$, determine to which end of the tuner a $Z_L = (20 - j20)\Omega$ load has to be connected and find the length of the short-circuited stubs such that the load is matched to a 50 Ω line. Assume that all stubs and transmission lines in the tuner have a 50 Ω characteristic impedance.

8.18 Discuss a circuit configuration that replaces the stubs in the previous problem with varactor diodes in series with inductors. Choose the appropriate inductances if the varactor diodes can change their capacitances in the range from 1 pF to 6 pF. For a frequency of 1.5 GHz, discuss the tuning capabilities in terms of possible load impedance variations.

8.19 An ideal amplifier has a transfer characteristic given by the equation

$$V_{out} = \begin{cases} 30(V_{in} - V^*), & V_{in} \geq V^* \\ 0, & V_{in} < V^* \end{cases}$$

where $V^* = 60$ mV. Find the quiescent point V_{out}^Q and the corresponding maximum efficiency such that the amplifier is operated in the AB class and has conduction angle of $\Theta_0 = 270°$. Assume that the input signal is a sinusoidal voltage wave of 100 mV amplitude.

8.20 Find the component values for a low-GHz range biasing network for a BJT with bypassed emitter resistor R_3, as shown below:

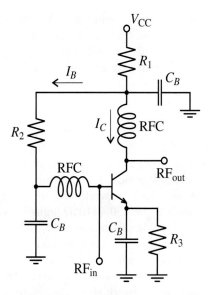

Assume that the power supply voltage is $V_{CC} = 12$ V and the transistor has the following parameters: $I_C = 20$ mA, $V_{CE} = 5$ V, $\beta = 125$, and $V_{BE} = 0.75$ V.

8.21 For stability purposes, a feedback resistor $R_F = 1$ kΩ has been added between base and collector of the transistor in the biasing network shown in Figure 8-32(b). Compute the values of all resistors in the biasing network if the following biasing conditions must be satisfied: supply voltage of $V_{CC} = 5$ V, collector current of $I_C = 10$ mA, and collector-emitter voltage of $V_{CE} = 3$ V. Assume that the transistor has a $\beta = 100$ and a $V_{BE} = 0.8$ V.

8.22 Design a biasing network (shown in the following figure) for $I_{C2} = 10$ mA, $V_{CE2} = 3$ V, and $V_{CC} = 5$ V. Assume that $\beta_1 = 150$, $\beta_2 = 80$, and both transistors have $V_{BE} = 0.7$ V.

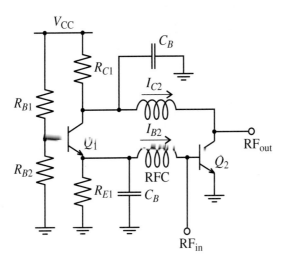

8.23 Redraw the active biasing network shown in Figure 8-34 for a common-base and a common-collector operating mode, respectively.

8.24 For the passive FET biasing network shown in Figure 8-38, find the value of the source resistance R_S if $V_{GS} = -4$ V, $V_{DS} = 10$ V, and the drain current is given as $I_D = 50$ mA.

8.25 In the Practically Speaking section of this chapter, we derived expressions for R1 and R2, namely

$$R_1 = \frac{V_D - |V_{BE}|}{I_{R3}}\left(\frac{\beta+1}{\beta+2}\right)$$

$$R_2 = \frac{V_G}{I_{R3}}\left(\frac{\beta+1}{\beta}\right).$$

Prove the validity of these design equations, which are based on the simplified transistor circuit models shown.

9

RF Transistor Amplifier Design

9.1 Characteristics of Amplifiers 486
9.2 Amplifier Power Relations 487
    9.2.1   RF Source 487
    9.2.2   Transducer Power Gain 488
    9.2.3   Additional Power Relations 489
9.3 Stability Considerations 492
    9.3.1   Stability Circles 492
    9.3.2   Unconditional Stability 494
    9.3.3   Stabilization Methods 501
9.4 Constant Gain 504
    9.4.1   Unilateral Design 504
    9.4.2   Unilateral Figure of Merit 510
    9.4.3   Bilateral Design 512
    9.4.4   Operating and Available Power Gain Circles 515
9.5 Noise Figure Circles 521
9.6 Constant VSWR Circles 525
9.7 Broadband, High-Power, and Multistage Amplifiers 529
    9.7.1   Broadband Amplifiers 529
    9.7.2   High-Power Amplifiers 540
    9.7.3   Multistage Amplifiers 543
9.8 Summary 550

Amplifier designs at RF differ significantly from conventional low-frequency circuit approaches, and consequently require special considerations. In particular, the fact that *voltage and current waves* impinge on the active device necessitates appropriate matching to reduce the VSWR. In addition, we will find that most amplifiers can oscillate when terminated with certain source and load impedances. Matching networks can help stabilize the amplifier by keeping the source and load impedances in the appropriate range. For this reason, a stability analysis is usually the first step in the design process and, in conjunction with gain and noise figure circles, is a basic ingredient needed to develop amplifier circuits that meet the often competing requirements of gain, gain flatness, output power, bandwidth, and bias conditions.

This chapter expands upon the material covered in Chapters 2 and 3, where power relations of terminated transmission lines are investigated. However, unlike the passive circuit presentations, Chapter 9 deals with active devices, where gain and feedback considerations assume central importance. Issues such as power gain, unilateral and bilateral circuit designs and their graphical display in the Smith Chart constitute the starting point of an extensive analysis into quantifying high-frequency transistor amplifier performance. The reader will note the flexibility of the Smith Chart, which allows constant gain, VSWR, and stability circle displays to be superimposed over the reflection coefficient and impedance representation discussed in Chapter 3. Moreover, even noise analysis can be conducted by converting the noise figure of an amplifier into circles that are displayed in the Smith Chart.

After covering the basic design tools, Chapter 9 also investigates various types of power amplifiers and their characteristics such as gain flatness, bandwidth, and intermodulation distortion. We will also examine the differences between single- and multistage amplifiers.

9.1 Characteristics of Amplifiers

Perhaps the most important and complex task in analog circuit theory is the amplification of an input signal through either a single or multistage transistor circuit. A generic single-stage amplifier configuration embedded between input and output matching networks is shown in Figure 9-1.

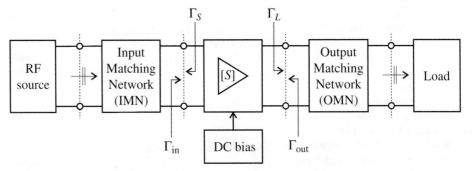

Figure 9-1 Generic amplifier system.

Input and output matching networks, discussed in Chapter 8, are needed to reduce undesired reflections and thus improve the power flow capabilities. In Figure 9-1, the amplifier is characterized through its S-parameter matrix at a particular DC bias point. In terms of performance specifications, the following list constitutes a set of key amplifier parameters:

- Gain and gain flatness (in dB)
- Operating frequency and bandwidth (in Hz)
- Output power (in dBm)
- Power supply requirements (in V and A)
- Input and output reflection coefficients (VSWR)
- Noise figure (in dB)

Amplifier Power Relations

In addition, one often must consider such parameters as intermodulation distortion (IMD) products, harmonics, feedback, and heating effects, all of which can seriously affect the amplifier performance.

To approach the amplifier design process systematically, we need to first establish a number of definitions regarding various power relations. This is followed by several important analysis tools required to define stability, gain, noise, and VSWR performance. The common denominator of all four tools is that they can be expressed as circle equations and displayed in the Smith Chart.

9.2 Amplifier Power Relations

9.2.1 RF Source

There are various power gain definitions that are critical to the understanding of how an RF amplifier functions. For this reason, let us examine Figure 9-1 in terms of its power flow relations under the assumption that the two matching networks are included in the source and load impedances. This simplifies our system to the configuration shown in Figure 9-2(a). The starting point of the analysis is the RF source connected to the amplifier network. For the convention depicted in Figure 9-2, we recall our signal flow discussion in Section 4.4.5 (see (4.82) and (4.83)) and write for the source:

$$b_S = \frac{\sqrt{Z_0}}{Z_S + Z_0} V_S = b_1' - a_1' \Gamma_S = b_1'(1 - \Gamma_{in}\Gamma_S). \tag{9.1}$$

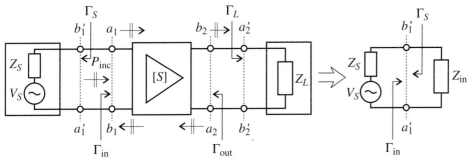

(a) Simplified schematics of a single-stage amplifier

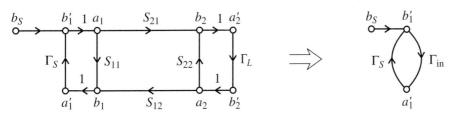

(b) Signal flow graph

Figure 9-2 Source and load connected to a single-stage amplifier network.

The incident power wave associated with b'_1 is given as

$$P_{\text{inc}} = \frac{|b'_1|^2}{2} = \frac{1}{2}\frac{|b_S|^2}{|1 - \Gamma_{\text{in}}\Gamma_S|^2} \quad (9.2)$$

which is the power launched toward the amplifier. The actual input power P_{in} observed at the input terminal of the amplifier is composed of the incident and reflected power waves. With the aid of the input reflection coefficient Γ_{in}, we can therefore write:

$$P_{\text{in}} = P_{\text{inc}}(1 - |\Gamma_{\text{in}}|^2) = \frac{1}{2}\frac{|b_S|^2}{|1 - \Gamma_{\text{in}}\Gamma_S|^2}(1 - |\Gamma_{\text{in}}|^2). \quad (9.3)$$

> **MAXIMUM AVAILABLE POWER FROM THE SOURCE**
>
> If the source is matched to the transmission line $(Z_0 = Z_s)$, we observe
>
> $$P_A = \frac{|b_s|^2}{2} = \frac{|V_s|^2}{8Z_s}$$
>
> which is the maximum available power from the source.

The **maximum power transfer** from the source to the amplifier is achieved if the input impedance is complex conjugate matched $(Z_{\text{in}} = Z_S^*)$ or, in terms of the reflection coefficients, if $\Gamma_{\text{in}} = \Gamma_S^*$. Under maximum power transfer condition, we define the **available power** P_A as

$$P_A = P_{\text{in}}\Big|_{\Gamma_{\text{in}} = \Gamma_S^*} = \frac{1}{2}\frac{|b_S|^2}{|1 - \Gamma_{\text{in}}\Gamma_S|^2}(1 - |\Gamma_{\text{in}}|^2)\Big|_{\Gamma_{\text{in}} = \Gamma_S^*} = \frac{1}{2}\frac{|b_S|^2}{1 - |\Gamma_S|^2}. \quad (9.4)$$

This expression makes clear the dependence on Γ_S. If $\Gamma_{\text{in}} = 0$ and $\Gamma_S \neq 0$, it is seen from (9.2) that $P_{\text{inc}} = |b_S|^2/2$.

9.2.2 Transducer Power Gain

We can next investigate the **transducer power gain** G_T, which quantifies the gain of the amplifier placed between source and load.

$$G_T = \frac{\text{power delivered to the load}}{\text{available power from the source}} = \frac{P_L}{P_A}$$

or with $P_L = \frac{1}{2}|b_2|^2 \cdot (1 - |\Gamma_L|^2)$ we obtain

$$G_T = \frac{P_L}{P_A} = \frac{|b_2|^2}{|b_S|^2}(1 - |\Gamma_L|^2)(1 - |\Gamma_S|^2). \quad (9.5)$$

In this expression, the ratio b_2/b_S has to be determined. With the help of our signal flow discussion in Section 4.4.5, and based on Figure 9-2, we establish

$$b_2 = \frac{S_{21}a_1}{1 - S_{22}\Gamma_L} \quad (9.6\text{a})$$

$$b_S = \left[1 - \left(S_{11} + \frac{S_{21}S_{12}\Gamma_L}{1 - S_{22}\Gamma_L}\right)\Gamma_S\right]a_1. \quad (9.6\text{b})$$

The required ratio is given by

$$\frac{b_2}{b_S} = \frac{S_{21}}{(1 - S_{11}\Gamma_S)(1 - S_{22}\Gamma_L) - S_{21}S_{12}\Gamma_L\Gamma_S}. \quad (9.7)$$

Inserting (9.7) into (9.5) results in

$$G_T = \frac{(1 - |\Gamma_L|^2)|S_{21}|^2(1 - |\Gamma_S|^2)}{|(1 - S_{11}\Gamma_S)(1 - S_{22}\Gamma_L) - S_{21}S_{12}\Gamma_L\Gamma_S|^2} \quad (9.8)$$

which can be rearranged by defining the input and output reflection coefficients (see Problem 9.2)

$$\Gamma_{in} = S_{11} + \frac{S_{21}S_{12}\Gamma_L}{1 - S_{22}\Gamma_L} \quad (9.9a)$$

$$\Gamma_{out} = S_{22} + \frac{S_{12}S_{21}\Gamma_S}{1 - S_{11}\Gamma_S}. \quad (9.9b)$$

With these two definitions, two more transducer power gain expressions can be derived. First, by incorporating (9.9a) into (9.8), it is seen that

$$G_T = \frac{(1 - |\Gamma_L|^2)|S_{21}|^2(1 - |\Gamma_S|^2)}{|1 - \Gamma_S\Gamma_{in}|^2|1 - S_{22}\Gamma_L|^2}. \quad (9.10)$$

Second, using (9.9b) in (9.8) results in the expression

$$G_T = \frac{(1 - |\Gamma_L|^2)|S_{21}|^2(1 - |\Gamma_S|^2)}{|1 - \Gamma_L\Gamma_{out}|^2|1 - S_{11}\Gamma_S|^2}. \quad (9.11)$$

An often employed approximation for the transducer power gain is the so-called **unilateral power gain** G_{TU}, which neglects the feedback effect of the amplifier ($S_{12} = 0$). This simplifies (9.11) to

$$G_{TU} = \frac{(1 - |\Gamma_L|^2)|S_{21}|^2(1 - |\Gamma_S|^2)}{|1 - \Gamma_L S_{22}|^2|1 - S_{11}\Gamma_S|^2}. \quad (9.12)$$

As discussed in Section 9.4.1, equation (9.12) is often used as a basis for developing approximate designs of an amplifier and its input and output matching networks.

9.2.3 Additional Power Relations

The transducer power gain is a fundamental expression from which additional important power relations can be derived. For instance, the **available power gain** for load side matching ($\Gamma_L = \Gamma_{out}^*$) is defined as

$$G_A = G_T\big|_{\Gamma_L = \Gamma_{out}^*} = \frac{\text{power available from the network}}{\text{power available from the source}} = \frac{P_N}{P_A}$$

or, with the aid of (9.11),

$$G_A = \frac{|S_{21}|^2(1-|\Gamma_S|^2)}{(1-|\Gamma_{out}|^2)|1-S_{11}\Gamma_S|^2}. \qquad (9.13)$$

Further, the **power gain (operating power gain)** is defined as the ratio of the power delivered to the load to the power supplied to the amplifier.

$$G = \frac{\text{power delivered to the load}}{\text{power supplied to the amplifier}} = \frac{P_L}{P_{in}} = \frac{P_L}{P_A} \cdot \frac{P_A}{P_{in}} = G_T \frac{P_A}{P_{in}}.$$

Combining (9.3), (9.4), and (9.10), we find

$$G = \frac{(1-|\Gamma_L|^2)|S_{21}|^2}{(1-|\Gamma_{in}|^2)|1-S_{22}\Gamma_L|^2}. \qquad (9.14)$$

It is worth noting that (9.14) can be obtained from (9.10) by setting $\Gamma_S = \Gamma_{in}^*$, since in this case $P_{in} = P_A$. The following example goes through the computation of some of these expressions for an amplifier with given S-parameters.

Example 9-1. Power relations for an RF amplifier

An RF amplifier has the following S-parameters: $S_{11} = 0.3\angle-70°$, $S_{21} = 3.5\angle 85°$, $S_{12} = 0.2\angle-10°$, and $S_{22} = 0.4\angle-45°$. Furthermore, the input side of the amplifier is connected to a voltage source with $V_S = 5V\angle 0°$ and source impedance $Z_S = 40\ \Omega$. The output is utilized to drive an antenna, which has an impedance of $Z_L = 73\ \Omega$. Assuming that the S-parameters of the amplifier are measured with reference to a $Z_0 = 50\ \Omega$ characteristic impedance, find the following quantities:

(a) transducer gain G_T, unilateral transducer gain G_{TU}, available gain G_A, operating power gain G, and
(b) power delivered to the load P_L, available power from the source P_A, and incident power to the amplifier P_{inc}.

Solution. First, we find the source and load reflection coefficients assuming a $Z_0 = 50\ \Omega$ characteristic impedance:

$$\Gamma_S = \frac{Z_S - Z_0}{Z_S + Z_0} = -0.111 \quad \text{and} \quad \Gamma_L = \frac{Z_L - Z_0}{Z_L + Z_0} = 0.187.$$

Next, the input and output impedances, as given in (9.9a) and (9.9b), are determined:

$$\Gamma_{in} = S_{11} + \frac{S_{21}S_{12}\Gamma_L}{1-S_{22}\Gamma_L} = 0.146 - j0.151$$

$$\Gamma_{out} = S_{22} + \frac{S_{12}S_{21}\Gamma_S}{1-S_{11}\Gamma_S} = 0.265 - j0.358.$$

Substituting the obtained values along with the S-parameters into (9.11), (9.12), (9.13), and (9.14), the transducer gain G_T, unilateral transducer gain G_{TU}, available gain G_A, and operating power gain G are computed as follows:

$$G_T = \frac{(1-|\Gamma_L|^2)|S_{21}|^2(1-|\Gamma_S|^2)}{|1-\Gamma_L\Gamma_{out}|^2|1-S_{11}\Gamma_S|^2} = 12.56 \text{ or } 10.99 \text{ dB}$$

$$G_{TU} = \frac{(1-|\Gamma_L|^2)|S_{21}|^2(1-|\Gamma_S|^2)}{|1-\Gamma_L S_{22}|^2|1-S_{11}\Gamma_S|^2} = 12.67 \text{ or } 11.03 \text{ dB}$$

$$G_A = \frac{|S_{21}|^2(1-|\Gamma_S|^2)}{|1-|\Gamma_{out}|^2||1-S_{11}\Gamma_S|^2} = 14.74 \text{ or } 11.68 \text{ dB}$$

$$G = \frac{(1-|\Gamma_L|^2)|S_{21}|^2}{|1-|\Gamma_{in}|^2||1-S_{22}\Gamma_L|^2} = 13.74 \text{ or } 11.38 \text{ dB}.$$

Using (9.2) in conjunction with (9.1) allows us to find the incident power flow into the amplifier:

$$P_{inc} = \frac{1}{2}\frac{|b_S|^2}{|1-\Gamma_{in}\Gamma_S|^2} = \frac{1}{2}\frac{Z_0}{(Z_S+Z_0)^2}\frac{|V_S|^2}{|1-\Gamma_{in}\Gamma_S|^2} = 74.7 \text{ mW}.$$

Often P_{inc} is expressed in dBm as

$$P_{inc}[\text{dBm}] = 10\log(P_{inc}/(1 \text{ mW})) = 18.73 \text{ dBm}.$$

Similarly, from (9.4) we find the available power from the source to be $P_A = 78.1$ mW or 18.93 dBm. Finally, the power delivered to the load is the available power from the source multiplied by the transducer gain. This results in $P_L = P_A G_T = 981.4$ mW, or

$$P_L[\text{dBm}] = P_A[\text{dBm}] + G_T[\text{dB}] = 29.92 \text{ dBm}.$$

The unilateral power gain often matches the actual transducer power gain very closely. As discussed further, the use of the unilateral gain significantly simplifies the amplifier design task.

9.3 Stability Considerations

9.3.1 Stability Circles

One of the first requirements that an amplifier circuit must meet is stable performance over the entire frequency range. This is a particular concern when dealing with RF circuits, which tend to oscillate depending on operating frequency and termination. The phenomenon of oscillations can be understood in the context of a voltage wave along a transmission line. If $|\Gamma| > 1$, then the return voltage increases in magnitude (positive feedback), possibly causing instability. Conversely, $|\Gamma| < 1$ causes a diminished return voltage wave (negative feedback).

Let us regard the amplifier as a two-port network characterized through its S-parameters, with external terminations described by Γ_L and Γ_S. Stability then implies that the magnitudes of the reflection coefficients are less than unity. Namely,

$$|\Gamma_L| < 1, |\Gamma_S| < 1 \tag{9.15a}$$

$$|\Gamma_{in}| = \left|\frac{S_{11} - \Gamma_L \Delta}{1 - S_{22}\Gamma_L}\right| < 1 \tag{9.15b}$$

$$|\Gamma_{out}| = \left|\frac{S_{22} - \Gamma_S \Delta}{1 - S_{11}\Gamma_S}\right| < 1 \tag{9.15c}$$

where $\Delta = S_{11}S_{22} - S_{12}S_{21}$ has been used to re-express (9.9a) and (9.9b). Since the S-parameters are fixed for a particular frequency, the only factors that have a parametric effect on stability are Γ_L and Γ_S.

In terms of the amplifier's *output port,* we need to establish the condition for which (9.15b) is satisfied. To this end, the complex quantities

$$S_{11} = S_{11}^R + jS_{11}^I, S_{22} = S_{22}^R + jS_{22}^I, \Delta = \Delta^R + j\Delta^I, \Gamma_L = \Gamma_L^R + j\Gamma_L^I \tag{9.16}$$

are substituted into (9.15b), resulting after some algebra in the **output stability circle** equation

$$(\Gamma_L^R - C_{out}^R)^2 + (\Gamma_L^I - C_{out}^I)^2 = r_{out}^2 \tag{9.17}$$

where the circle radius is given by

$$r_{out} = \frac{|S_{12}S_{21}|}{\left||S_{22}|^2 - |\Delta|^2\right|} \tag{9.18}$$

Stability Considerations

and the center of this circle is located at

$$C_{out} = C_{out}^R + jC_{out}^I = \frac{(S_{22} - S_{11}^*\Delta)^*}{|S_{22}|^2 - |\Delta|^2} \qquad (9.19)$$

as depicted in Figure 9-3(a). In terms of the *input port*, substituting (9.16) into (9.15c) yields the **input stability circle** equation

$$(\Gamma_S^R - C_{in}^R)^2 + (\Gamma_S^I - C_{in}^I)^2 = r_{in}^2 \qquad (9.20)$$

where

$$r_{in} = \frac{|S_{12}S_{21}|}{\left||S_{11}|^2 - |\Delta|^2\right|} \qquad (9.21)$$

and

$$C_{in} = C_{in}^R + jC_{in}^I = \frac{(S_{11} - S_{22}^*\Delta)^*}{|S_{11}|^2 - |\Delta|^2}. \qquad (9.22)$$

When plotted in the Γ_S-plane, we obtain a response as schematically shown in Figure 9-3(b).

To interpret the meaning of Figure 9-3 correctly, an important issue arises that is investigated for the output circle (Figure 9-3(a)), although the same argument holds for the

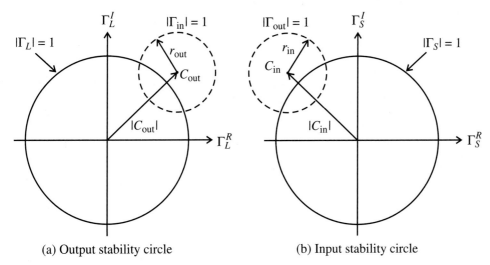

(a) Output stability circle   (b) Input stability circle

Figure 9-3 Stability circle $|\Gamma_{in}| = 1$ in the complex Γ_L-plane and stability circle $|\Gamma_{out}| = 1$ in the complex Γ_S-plane.

input circle. If $\Gamma_L = 0$, then $|\Gamma_{in}| = |S_{11}|$, and two cases have to be differentiated depending on $|S_{11}| < 1$ or $|S_{11}| > 1$. For $|S_{11}| < 1$, the origin (the point $\Gamma_L = 0$) is part of the stable region, as in Figure 9-4(a). However, for $|S_{11}| > 1$ the condition $\Gamma_L = 0$ results in $|\Gamma_{in}| = |S_{11}| > 1$, i.e. the origin is part of the unstable region. In this case, the only stable region is the shaded domain between the output stability circle $|\Gamma_{in}| = 1$ and the $|\Gamma_L| = 1$ circle, as shown in Figure 9-4(b).

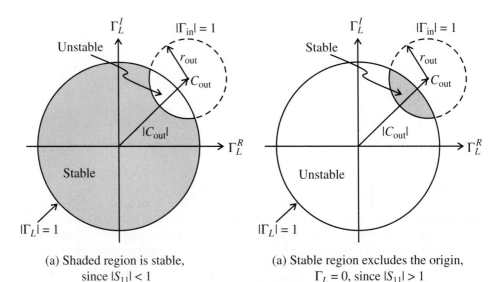

(a) Shaded region is stable, since $|S_{11}| < 1$

(a) Stable region excludes the origin, $\Gamma_L = 0$, since $|S_{11}| > 1$

Figure 9-4 Output stability circles denoting stable and unstable regions.

For completeness, Figure 9-5 shows the two stability domains for the input stability circle. The rule is that if $|S_{22}| < 1$, the center ($\Gamma_S = 0$) must be stable; otherwise the center becomes unstable for $|S_{22}| > 1$.

Care has to be exercised in correctly interpreting the stability circles if the circle radius is larger than $|C_{in}|$ or $|C_{out}|$. Figure 9-6 depicts the input stability circles for $|S_{22}| < 1$ and the two possible stability domains depending on $r_{in} < |C_{in}|$ or $r_{in} > |C_{in}|$.

9.3.2 Unconditional Stability

As the name implies, unconditional stability refers to the situation where the amplifier remains stable for any passive source and load at the selected frequency and bias conditions. For $|S_{11}| < 1$ and $|S_{22}| < 1$, it is stated as

$$||C_{in}| - r_{in}| > 1 \tag{9.23a}$$

$$||C_{out}| - r_{out}| > 1. \tag{9.23b}$$

In other words, the stability circles have to reside completely outside the $|\Gamma_S| = 1$ and $|\Gamma_L| = 1$ circles. In the following discussion, we concentrate on the $|\Gamma_S| = 1$ circle shown

Stability Considerations

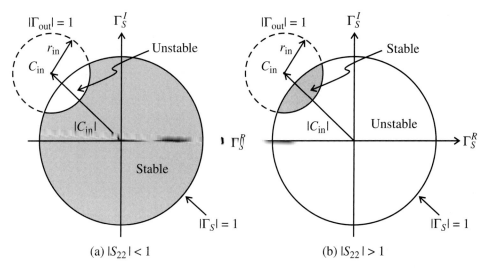

Figure 9-5 Input stability circles denoting stable and unstable regions.

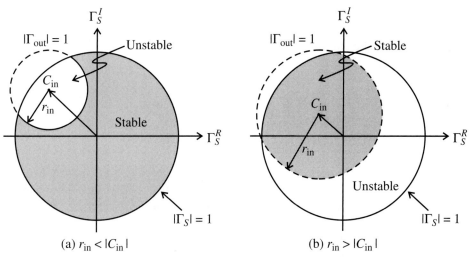

Figure 9-6 Different input stability regions for $|S_{22}| < 1$ depending on ratio between r_S and $|C_{in}|$.

in Figure 9-7(a). It is shown in Example 9-2 that the condition (9.23a) can be re-expressed in terms of the stability or Rollett factor k:

$$k = \frac{1 - |S_{11}|^2 - |S_{22}|^2 + |\Delta|^2}{2|S_{12}||S_{21}|} > 1. \tag{9.24}$$

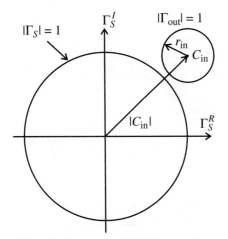 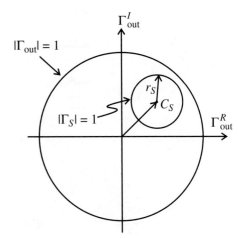

(a) $|\Gamma_{out}| = 1$ circle must reside outside (b) $|\Gamma_S| = 1$ circle must reside inside

Figure 9-7 Unconditional stability in the Γ_S and Γ_{out} planes for $|S_{11}| < 1$.

Alternatively, unconditional stability can also be viewed in terms of the Γ_S behavior in the complex $\Gamma_{out} = \Gamma_{out}^R + j\Gamma_{out}^I$ plane. Here, the $|\Gamma_S| \leq 1$ domain must reside completely within the $|\Gamma_{out}| = 1$ circle, as depicted in Figure 9-7(b). Plotting $|\Gamma_S| = 1$ in the Γ_{out}-plane produces a circle whose center is located at

$$C_S = S_{22} + \frac{S_{12}S_{21}S_{11}^*}{1 - |S_{11}|^2} \tag{9.25}$$

and which possesses a radius of

$$r_S = \frac{|S_{12}S_{21}|}{1 - |S_{11}|^2} \tag{9.26}$$

where the condition $|C_S| + r_S < 1$ must hold. We note that (9.25) can be rewritten as $C_S = (S_{22} - \Delta S_{11}^*)/(1 - |S_{11}|^2)$. Employing $|C_S| + r_S < 1$ and (9.26), it is seen that

$$|S_{22} - \Delta S_{11}^*| + |S_{12}S_{21}| < 1 - |S_{11}|^2 \tag{9.27a}$$

and since $|S_{12}S_{21}| \leq |S_{22} - \Delta S_{11}^*| + |S_{12}S_{21}|$, we conclude

$$|S_{12}S_{21}| < 1 - |S_{11}|^2. \tag{9.27b}$$

A similar analysis can be established for Γ_L in the complex Γ_{in}-plane. From the corresponding circle center C_L and radius r_L, we set $|C_L| = 0$ and $r_S < 1$. Thus,

$$|S_{12}S_{21}| < 1 - |S_{22}|^2. \tag{9.28}$$

Stability Considerations

However, as long as $|\Delta| < 1$, (9.24) remains the sufficient requirement to ensure unconditional stability. This follows from the fact that when (9.27b) and (9.28) are added, it is seen that

$$2|S_{12}S_{21}| < 2 - |S_{11}|^2 - |S_{22}|^2.$$

Introducing the inequality $|\Delta| = |S_{11}S_{22} - S_{12}S_{21}| \leq |S_{11}S_{22}| + |S_{12}S_{21}|$ results in

$$|\Delta| < 1 - \frac{1}{2}(|S_{11}|^2 + |S_{22}|^2 - 2|S_{11}||S_{22}|) = 1 - \frac{1}{2}(|S_{11}| - |S_{22}|)^2.$$

Since $(1/2)(|S_{11}| - |S_{22}|)^2 < 1$, it is seen that (9.27b) and (9.28) are equivalent to

$$|\Delta| < 1. \tag{9.29}$$

Example 9-2. Stability factor derivation

Derive the stability factor k (Rollett factor) from (9.23a).

Solution. Substituting (9.21) and (9.22) into (9.23a) gives

$$\left|\frac{|S_{11} - S_{22}^*\Delta| - |S_{12}S_{12}|}{|S_{11}|^2 - |\Delta|^2}\right| > 1. \tag{9.30a}$$

Squaring and rearranging (9.30a) results in

$$2|S_{11} - S_{22}^*\Delta||S_{12}S_{21}| <$$
$$|S_{11} - S_{22}^*\Delta|^2 + |S_{12}S_{21}|^2 - ||S_{11}|^2 - |\Delta|^2|^2. \tag{9.30b}$$

The term $|S_{11} - S_{22}^*\Delta|^2$ in (9.30b) can be re-expressed as

$$|S_{11} - S_{22}^*\Delta|^2$$
$$= |S_{12}S_{21}|^2 + (1 - |S_{22}|^2)(|S_{11}|^2 - |\Delta|^2). \tag{9.30c}$$

Squaring (9.30b) again and rearranging terms finally gives

$$(|S_{11}|^2 - |\Delta|^2)^2 \times$$
$$\{[(1 - |S_{22}|^2) - (|S_{11}|^2 - |\Delta|^2)]^2$$
$$- 4|S_{12}S_{21}|^2\} > 0 \tag{9.30d}$$

The terms inside the curly brackets are recognized as the desired stability factor:

$$k = \frac{1 - |S_{11}|^2 - |S_{22}|^2 + |\Delta|^2}{2|S_{12}||S_{21}|} > 1. \tag{9.30e}$$

> **μ STABILITY FACTOR**
>
> An alternative stability factor can be derived from geometric considerations:
>
> $$\mu = \frac{1 - |S_{11}|^2}{|S_{22} - S_{11}^*\Delta| + |S_{21}S_{12}|}$$
>
> The condition $\mu > 1$ is necessary and sufficient for unconditional stability. The physical meaning of μ is the minimum distance from the origin of the Smith Chart to the stability circle in the Γ_L-plane; a larger μ implies greater stability. Negative values indicate that the unstable region contains the origin. A dual of this stability factor for the Γ_S-plane can be defined by interchanging S_{11} and S_{22}.

A stability analysis starting from (9.23b) would have resulted in exactly the same inequality. Thus, the stability factor k applies for both input and output ports.

It is always prudent to determine that both the $|\Delta| < 1$ and $k > 1$ conditions are fulfilled to ensure an unconditionally stable design. The next example investigates a transistor in common-emitter configuration in terms of its input and output stability behavior.

Example 9-3. Stability circles for a BJT at different operating frequencies

Determine the stability regions of the bipolar junction transistor BFG505W (NXP) biased at $V_{CE} = 6$ V and $I_C = 4$ mA. The corresponding S-parameters as functions of frequency are given in Table 9-1.

Table 9-1 BFG505W S-parameters as functions of frequency

Frequency	S_{11}	S_{12}	S_{21}	S_{22}
500 MHz	0.70∠–57°	0.04∠47°	10.5∠136°	0.79∠–33°
750 MHz	0.56∠–78°	0.05∠33°	8.6∠122°	0.66∠–42°
1000 MHz	0.46∠–97°	0.06∠22°	7.1∠112°	0.57∠–48°
1250 MHz	0.38∠–115°	0.06∠14°	6.0∠104°	0.50∠–52°

Solution. Based on the definitions for k, $|\Delta|$, C_{in}, r_{in}, C_{out}, and r_{out}, we compute the values via a Matlab routine (see m-file ex9_3.m). A summary of the results is given in Table 9-2 for the four frequencies listed in Table 9-1.

The example input and output stability circles for the frequencies of $f = 750$ MHz and $f = 1.25$ GHz are shown in Figure 9-8. We notice that $|S_{11}| < 1$ and $|S_{22}| < 1$ in all cases. This implies that the $\Gamma_L = 0$ and $\Gamma_S = 0$ points are stable, indicating that the interior domain of the Smith Chart up to the stability circles denotes the stable region.

Stability Considerations

Table 9-2 Stability parameters for BFG505W

f, MHz	k	$|\Delta|$	C_{in}	r_{in}	C_{out}	r_{out}
500	0.41	0.69	39.04∠108°	38.62	3.56∠70°	3.03
750	0.60	0.56	62.21∠119°	61.60	4.12∠70°	3.44
1000	0.81	0.45	206.23∠131°	205.42	4.39∠69°	3.54
1250	1.02	0.37	42.42∠143°	41.40	4.24∠68°	3.22

As can be seen from Figure 9-8 and Table 9-2, the transistor is unconditionally stable at $f = 1.25$ GHz—both input and output stability circles are located completely outside of the $|\Gamma| = 1$ circle. At other frequencies, the transistor is potentially unstable.

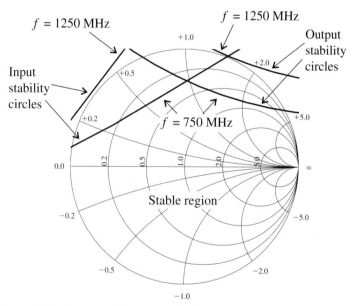

Figure 9-8 Input and output stability circles for BFG505W computed at $f = 750$ MHz and $f = 1.25$ GHz.

The stability circles are not only affected by frequency, but also by the bias conditions. We recall that the S-parameters are given for particular bias conditions. The entire stability analysis must be repeated if biasing, or even temperature, changes.

Even though k can vary widely, most unstable practical designs fall into the range $0 \leq k \leq 1$. Oscillators, discussed in Chapter 10, target the entire Smith Chart as the unstable domain, resulting in negative values of k. It is also interesting to observe that in the absence of any output to input feedback ($S_{12} = 0$) the transistor is inherently stable, since the stability factor yields $k \rightarrow \infty$. In practice, one often examines k alone without paying attention to the $|\Delta| < 1$ condition. This can cause potential problems, as the following example highlights.

&RF&MW&

Example 9-4. Stable versus unstable region of a transistor

Investigate the stability regions of a transistor whose S-parameters are recorded as follows: $S_{11} = 0.7\angle-70°$, $S_{12} = 0.2\angle-10°$, $S_{21} = 5.5\angle 85°$, and $S_{22} = 0.7\angle-45°$.

Solution. We compute the values k, $|\Delta|$, C_{in}, r_{in}, C_{out}, and r_{out}. The results are $k = 1.15$, $|\Delta| = 1.58$, $C_{in} = 0.21\angle 52°$, $r_{in} = 0.54$, $C_{out} = 0.21\angle 27°$, and $r_{out} = 0.54$ (see Figure 9-9). It is seen that even though $k > 1$, the transistor is still potentially unstable because $|\Delta| > 1$. This results in input and output stability circles being located inside of the Smith Chart. Since both $|S_{11}|$ and $|S_{22}|$ are less than unity, the center of the Smith Chart is a stable point. Therefore, since $|C_{in}| < r_{in}$ and $|C_{out}| < r_{out}$, the area inside of the stability circles represents the stable region, as shown in Figure 9-9.

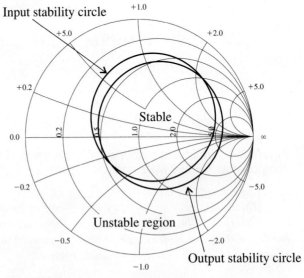

Figure 9-9 Stability circles for $k > 1$ and $|\Delta| > 1$.

> Usually, manufacturers avoid producing transistors with both $k > 1$ and $|\Delta| > 1$ by incorporating matching networks housed inside the transistor casing. We note that the μ-stability factor ($\mu = 0.335$) correctly identifies the conditional stability.

9.3.3 Stabilization Methods

If the operation of a FET or BJT is found to be unstable, an attempt can be made to stabilize the transistor. We recall that the conditions $|\Gamma_{in}| > 1$ and $|\Gamma_{out}| > 1$ can be written in terms of the input and output impedances:

$$|\Gamma_{in}| = \left|\frac{Z_{in} - Z_0}{Z_{in} + Z_0}\right| > 1 \text{ and } |\Gamma_{out}| = \left|\frac{Z_{out} - Z_0}{Z_{out} + Z_0}\right| > 1$$

which imply $\text{Re}(Z_{in}) < 0$ and $\text{Re}(Z_{out}) < 0$. One way to stabilize the active device is to add a series resistance or a shunt conductance to the port. Figure 9-10 shows the configuration for the input port. This loading in conjunction with $\text{Re}(Z_S)$ must compensate the negative contribution of $\text{Re}(Z_{in})$. Thus, we require

$$\text{Re}(Z_{in} + R'_{in} + Z_S) > 0 \text{ or } \text{Re}(Y_{in} + G'_{in} + Y_S) > 0. \quad (9.31a)$$

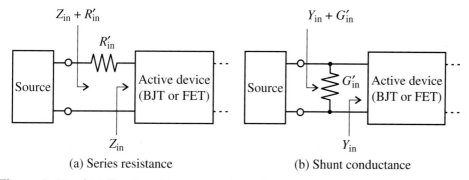

(a) Series resistance (b) Shunt conductance

Figure 9-10 Stabilization of input port through series resistance or shunt conductance.

Following an identical argument, Figure 9-11 shows the stabilization of the output port. The corresponding condition is

$$\text{Re}(Z_{out} + R'_{out} + Z_L) > 0 \text{ or } \text{Re}(Y_{out} + G'_{out} + Y_L) > 0. \quad (9.31a)$$

The next example explains the various stabilization procedures for a transistor.

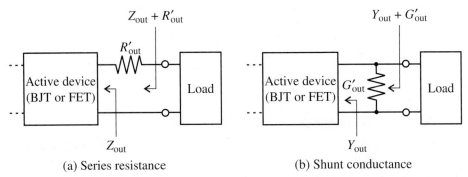

(a) Series resistance (b) Shunt conductance

Figure 9-11 Stabilization of output port through series resistance or shunt conductance.

Example 9-5. Stabilization of a BJT

Using the transistor BFG505W from Example 9-3 operated at $f = 750$ MHz, with the S-parameters given as $S_{11} = 0.56\angle{-78°}$, $S_{21} = 0.05\angle 33°$, $S_{12} = 8.64\angle 122°$, and $S_{22} = 0.66\angle{-42°}$, attempt to stabilize the transistor by finding a series resistor or shunt conductance for the input and output ports.

Solution. With the given S-parameters, we can identify the input and output stability circles by computing their radii and center positions: $C_{in} = 62.21\angle 119°$, $r_{in} = 61.60$, and $C_{out} = 4.12\angle 70°$, $r_{out} = 3.44$. The corresponding stability circles are shown in Figure 9-12. A constant resistance circle $r' = 0.33$ in the Z-Chart indicates the minimal series resistance that has to be connected to the input of the transistor to make this port stable. If a passive network is connected in series to the resistor with the value of $R'_{in} = r'Z_0 = 16.5$ Ω, then the combined impedance will be located inside of the $r' = 0.33$ circle, and therefore in the stable region. Similarly, by tracing a constant conductance circle $g' = 2.8$ we find the shunt admittance $G'_{in} = g'/Z_0 = 56$ mS that stabilizes the input of the transistor. Here, any passive network connected to G'_{in} will have the combined admittance residing inside of the $g' = 2.8$ circle in the Y-Chart, which is inside the stable region for the input port of the transistor.

ATTENUATOR FOR STABILIZATION

An attenuator can also be used to restrict Γ_L or Γ_S to the stable region, provided the origin of the Smith Chart is in the stable region. If the attenuator is attached between the output and the load, the minimum attenuation for unconditional stability is

$$\alpha_{min}[dB] = -10 \log \mu$$

As mentioned, the μ stability factor is the minimum distance from the origin to the stability circle in the Γ_L-plane. This approach makes it easy to stabilize the network over a wide frequency range: simply find the minimum μ.

Stability Considerations

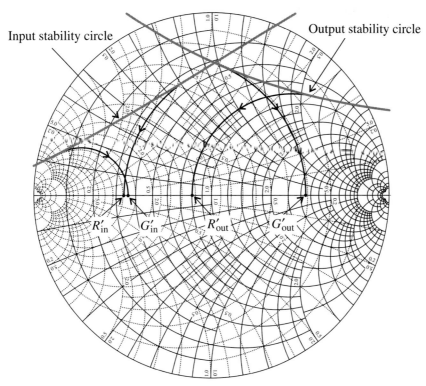

Figure 9-12 Input and output stability circles and circles for finding stabilizing series resistance and shunt conductances.

Following an identical procedure, we can find a series resistance of $R'_{out} = 40\ \Omega$ and a shunt conductance of $G'_{out} = 6.2$ mS, which stabilize the output port of the transistor.

Due to the coupling between input and output ports of the transistor, it is usually sufficient to stabilize one port. The choice of which port is generally up to the circuit designer. However, one attempts to avoid resistive elements at the input port since they cause additional noise to be amplified.

Stabilization through the addition of resistors comes at a price: the impedance matching can suffer, there may be a loss in power flow, and the noise figure typically worsens due to the additional thermal noise sources that the resistors present.

9.4 Constant Gain

9.4.1 Unilateral Design

Besides ensuring stability, the need to obtain a desired gain performance is another important consideration in the amplifier design task. If, as sometimes done in practice, the influence of the transistor's feedback is neglected ($S_{12} \approx 0$), we can employ the unilateral power gain G_{TU} described by (9.12). This equation is rewritten such that the individual contributions of the matching networks become identifiable. With reference to Figure 9-13, we write

$$G_{TU} = \frac{1-|\Gamma_S|^2}{|1-S_{11}\Gamma_S|^2} \times |S_{21}|^2 \times \frac{1-|\Gamma_L|^2}{|1-\Gamma_L S_{22}|^2} = G_S \times G_0 \times G_L \quad (9.32)$$

where the individual blocks are

$$G_S = \frac{1-|\Gamma_S|^2}{|1-S_{11}\Gamma_S|^2}, \quad G_0 = |S_{21}|^2, \quad G_L = \frac{1-|\Gamma_L|^2}{|1-\Gamma_L S_{22}|^2}. \quad (9.33)$$

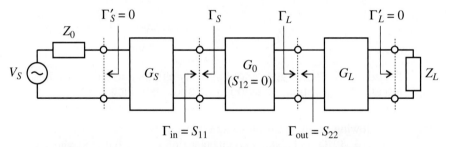

Figure 9-13 Unilateral power gain system arrangement.

Because most gain calculations are done in dB, (9.32) is frequently expressed as

$$G_{TU}[\text{dB}] = G_S[\text{dB}] + G_0[\text{dB}] + G_L[\text{dB}] \quad (9.34)$$

where G_S and G_L are gains associated with input and output matching networks, and G_0 is the insertion gain of the transistor. As seen from (9.33), the network gains can be greater than unity, which at first glance might appear surprising since they do not contain any active devices. The reason for this seemingly contradictory behavior is that without any matching, a significant power loss can occur at the input and output sides of the amplifier. The use of G_S and G_L attempts to reduce these inherent losses, which is considered a gain.

Constant Gain

If $|S_{11}|$ and $|S_{22}|$ are less than unity, the maximum unilateral power gain $G_{TU\max}$ results when both input and output are matched (i.e., $\Gamma_S = S_{11}^*$ and $\Gamma_L = S_{22}^*$). For this case, it is seen that

$$G_{S\max} = \frac{1}{1 - |S_{11}|^2} \tag{9.35}$$

$$G_{L\max} = \frac{1}{1 - |S_{22}|^2}. \tag{9.36}$$

The contributions from G_S and G_L can be normalized with respect to their maximum values such that

$$g_S = \frac{G_S}{G_{S\max}} = \frac{1 - |\Gamma_S|^2}{|1 - S_{11}\Gamma_S|^2}(1 - |S_{11}|^2) \tag{9.37a}$$

$$g_L = \frac{G_L}{G_{L\max}} = \frac{1 - |\Gamma_L|^2}{|1 - S_{22}\Gamma_L|^2}(1 - |S_{22}|^2) \tag{9.37b}$$

where the normalized gain is given in both cases as $0 \le g_i \le 1$, with $i = S, L$.

Even though we have explicit gain equations for the input and output matching networks, they are not directly usable in terms of providing parametric curves of constant gain. The key question that must be answered is formulated as follows: for a given S_{11} (or S_{22}) and a desired normalized gain g_S (or g_L), what is the range of values for Γ_S (or Γ_L) that achieves a particular gain? The solution requires the inversion of (9.37)

$$g_i = \frac{1 - |\Gamma_i|^2}{|1 - S_{ii}\Gamma_i|^2}(1 - |S_{ii}|^2) \tag{9.38}$$

for the reflection coefficient Γ_i. Here, $ii = 11, 22$ depending on $i = S, L$. The result is a set of circles with center locations

$$d_{g_i} = \frac{g_i S_{ii}^*}{1 - |S_{ii}|^2(1 - g_i)} \tag{9.39}$$

and radii

$$r_{g_i} = \frac{\sqrt{1 - g_i}(1 - |S_{ii}|^2)}{1 - |S_{ii}|^2(1 - g_i)}. \tag{9.40}$$

Example 9-6 details the necessary steps to derive the unilateral constant gain circle equations (9.39) and (9.40).

Example 9-6. Derivation of the constant gain circles

Find the expressions for d_{g_i} and r_{g_i} as given in (9.39) and (9.40).

Solution. The derivation begins with (9.38), which is rewritten as

$$g_i(1 + |S_{ii}\Gamma_i|^2 - S_{ii}^*\Gamma_i^* - S_{ii}\Gamma_i)$$
$$= 1 - |S_{ii}|^2 - |\Gamma_i|^2 + |S_{ii}|^2|\Gamma_i|^2. \qquad (9.41a)$$

The reflection coefficient Γ_i can be factored out such that

$$|\Gamma_i|^2 - \frac{g_i S_{ii}}{1 - |S_{ii}|^2(1 - g_i)}\Gamma_i - \frac{g_i S_{ii}^*}{1 - |S_{ii}|^2(1 - g_i)}\Gamma_i^* \qquad (9.41b)$$
$$+ \frac{g_i^2 |S_{ii}|^2}{(1 - |S_{ii}|^2(1 - g_i))^2} = \frac{(1 - g_i)(1 - |S_{ii}|^2)^2}{(1 - |S_{ii}|^2(1 - g_i))^2}.$$

This equation is the complex form of a circle expression

$$(\Gamma_i - d_{g_i})(\Gamma_i^* - d_{g_i}^*) = r_{g_i}^2 \qquad (9.41c)$$

with

$$d_{g_i} = \frac{g_i S_{ii}^*}{1 - |S_{ii}|^2(1 - g_i)} \quad \text{and} \quad r_{g_i} = \frac{\sqrt{1 - g_i}(1 - |S_{ii}|^2)}{1 - |S_{ii}|^2(1 - g_i)}.$$

Multiplying out (9.41c) results in the more familiar from

$$(\Gamma_i^R - d_{g_i}^R)^2 + (\Gamma_i^I - d_{g_i}^I)^2 = r_{g_i}^2 \qquad (9.41d)$$

where superscripts R and I denote real and imaginary parts of Γ_i and d_{g_i}.

Because of the unilateral assumption, we are able to derive separate gain circle equations for input and output ports.

The following observations can be made from the constant gain circle equations (9.39) and (9.40):

- The maximum gain $G_{i\max} = 1/(1 - |S_{ii}|^2)$ is obtained at $\Gamma_i = S_{ii}^*$, which coincides with the gain circle whose center is at $d_{g_i} = S_{ii}^*$ and of radius $r_{g_i} = 0$.

Constant Gain

- The constant gain circles all have their centers on a line connecting the origin to S_{ii}^*. The smaller the gain values, the closer the center d_{g_i} moves to the origin and the larger the radius r_{g_i}.
- For the special case $\Gamma_i = 0$, the normalized gain becomes $g_i = 1 - |S_{ii}|^2$ and both $|d_{g_i}|$ and r_{g_i} have the same value $|d_{g_i}| = r_{g_i} = |S_{ii}|/(1+|S_{ii}|^2)$. This implies that the $G_i = 1$ (or 0 dB) circle always passes through the origin of the Γ_i-plane.

Example 9-7 demonstrates the source gain circles for an amplifier design under unilateral approximation.

Example 9-7. Computation of the source gain circles for a unilateral design

A FET is operated at $f = 4$ GHz and is biased such that $S_{11} = 0.7 \angle 125°$. It is assumed that the transistor is unconditionally stable. Find the maximum source gain $G_{S\max}$ and plot the constant source gain circles for several values of G_S.

Solution. First, we find the maximum source gain $G_{S\max}$ using (9.35). The result is

$$G_{S\max} = \frac{1}{1-|S_{11}|^2} = \frac{1}{1-0.7^2} = 1.96 \text{ or } G_{S\max} = 2.92 \text{ dB}$$

We can now plot the constant gain circles by using (9.39) and (9.40) for the computation of circle centers d_{g_S} and radii r_{g_S}. A summary of several source gains G_S is presented in Table 9-3.

Table 9-3 Parameters for constant source gain circles in Example 9-7

G_S	g_S	d_{g_S}	r_{g_S}
2.6 dB	0.93	0.67∠–125°	0.14
2 dB	0.81	0.62∠–125°	0.25
1 dB	0.64	0.54∠–125°	0.37
0 dB	0.51	0.47∠–125°	0.47
–1 dB	0.41	0.40∠–125°	0.56

As seen from Table 9-3, the radius r_{g_S} of the $G_S = 0$ dB circle is equal to the magnitude of its center position d_{g_S} and the circle indeed passes through the center of the Smith Chart. We also observe that the centers for all G_S circles are located on the $\Theta = \angle S_{11}^* = -125°$ line, and as G_S approaches $G_{S\max}$, the radius of the corresponding circle reduces to zero and its center position becomes $S_{11}^* = 0.7\angle -125°$.

Figure 9-14 illustrates the source gain circles based on the computed numerical values given in Table 9-3. The figure points out clearly that, despite the input matching network being passive, the gain can be greater than 0 dB, indicating amplification. The physical meaning for such a behavior lies in the fact that the matching network reduces the input reflection coefficient of the overall system, thus effectively creating an "additional" gain.

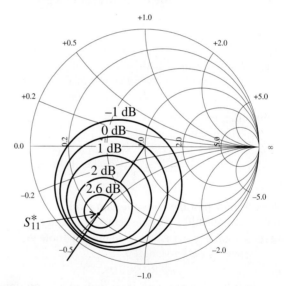

Figure 9-14 Constant source gain circles in the Smith Chart.

The underlying assumption of this example is that the gain associated with the matched input port is not affected by the output, since the unilateral approximation neglects the reverse gain.

We next discuss a typical application that requires the use of the constant gain circle approach. Specifically, let us develop a unilateral amplifier for a predetermined fixed gain value.

Constant Gain

Example 9-8. Design of a 18 dB single-stage MESFET amplifier operated at 5.7 GHz

A MESFET operated at 5.7 GHz has the following S-parameters:
$S_{11} = 0.5\angle-60°$, $S_{12} = 0.02\angle 0°$, $S_{21} = 6.5\angle 115°$, and $S_{22} = 0.6\angle-35°$.

(a) Determine if the circuit is unconditionally stable.
(b) Find the maximum power gain under optimal choice of reflection coefficients, assuming unilateral design ($S_{12} = 0$).
(c) Adjust the load reflection coefficient such that the desired gain is realized using the concept of constant gain circles.

Solution. (a) The stability of the device is tested via (9.24) and (9.29), with the results

$$k = \frac{1 - |S_{11}|^2 - |S_{22}|^2 + |\Delta|^2}{2|S_{12}||S_{21}|} = 2.17$$

and

$$|\Delta| = |S_{11}S_{22} - S_{12}S_{21}| = 0.42.$$

Because $k > 1$ and $|\Delta| < 1$, the transistor is unconditionally stable.
(b) We next compute the maximum gain for the optimal choice of the reflection coefficients (i.e., $\Gamma_L = S_{22}^*$ and $\Gamma_S = S_{11}^*$)

$$G_{S\max} = \frac{1}{1 - |S_{11}|^2} = 1.33 \text{ or } 1.25 \text{ dB}$$

$$G_{L\max} = \frac{1}{1 - |S_{22}|^2} = 1.56 \text{ or } 1.94 \text{ dB}$$

$$G_0 = |S_{21}|^2 = 42.25 \text{ or } 16.26 \text{ dB}.$$

Therefore, the maximum unilateral transducer gain is given by

$$G_{TU\max} = G_{S\max}G_0G_{L\max} = 88.02 \text{ or } 19.45 \text{ dB}.$$

(c) Since the source matching network ($\Gamma_S = S_{11}^*$) and the transistor combined already provide a gain of 17.51 dB, we have to choose Γ_L in such a way that $G_L = 0.49$ dB. This means that Γ_L has to reside on the $r_{g_L} = 0.38$, $d_{g_L} = 0.48\angle 35°$ circle, as shown in Figure 9-15. If we choose $\Gamma_L = 0.03 + j0.17$, the output matching network reduces to a single element (i.e., a series inductor with a value of $L = 0.49$ nH), provided the load is equal to the characteristic impedance ($Z_L = Z_0$).

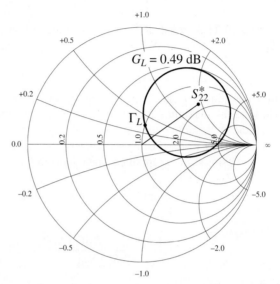

Figure 9-15 Constant load gain circle in the Smith Chart.

If the amplifier is operated over a range of frequencies, the gain has to be determined for a corresponding number of discrete frequency points due to the changing S-parameters.

For the case where $|S_{ii}| > 1$ ($ii = 11$ for the input port and $ii = 22$ for the output port) it is possible for a passive network to produce an infinite value of G_i ($i = S$ or L, respectively). This situation occurs when $\Gamma_i = S_{ii}^{-1}$, meaning that the real component of the impedance associated with Γ_i is equal in magnitude to the negative resistance related to S_{ii}. Thus, the two resistances cancel each other and oscillations will result: the amplifier is unstable. To avoid this problem, we plot the constant gain circles for $|S_{ii}| > 1$ and the corresponding stability circle and choose Γ_i in such a way that it is located on the desired gain circle, but also resides inside the stable region.

9.4.2 Unilateral Figure of Merit

The unilateral design approach discussed in Example 9-8 involves the approximation that the feedback effect, or the reverse gain, of the amplifier is negligible ($S_{12} = 0$). To estimate the error due to this assumption, the ratio between the transducer gain G_T, which takes into

Constant Gain

account S_{12}, and the unilateral transducer gain G_{TU} can be formulated. Using definitions (9.8) and (9.12), we find

$$\frac{G_T}{G_{TU}} = \frac{1}{\left|1 - \frac{S_{12}S_{21}\Gamma_L\Gamma_S}{(1-S_{11}\Gamma_S)(1-S_{22}\Gamma_L)}\right|^2} \quad (9.42)$$

where $G_T \leq G_{TU}$.

The maximum value of G_{TU}, and therefore the maximum error, is obtained for the input and output matching conditions ($\Gamma_S = S_{11}^*$ and $\Gamma_L = S_{22}^*$). Therefore, (9.42) becomes

$$\frac{G_T}{G_{TU\max}} = \frac{1}{\left|1 - \frac{S_{12}S_{21}S_{22}^*S_{11}^*}{(1-|S_{11}|^2)(1-|S_{22}|^2)}\right|^2}. \quad (9.43)$$

This can be used to set bounds on the error fluctuation

$$(1+U)^{-2} \leq \frac{G_T}{G_{TU}} \leq (1-U)^{-2} \quad (9.44)$$

where U is known as the frequency-dependent **unilateral figure of merit**:

$$U = \frac{|S_{12}||S_{21}||S_{22}||S_{11}|}{(1-|S_{11}|^2)(1-|S_{22}|^2)}. \quad (9.45)$$

To justify a unilateral amplifier design approach, this figure of merit should be as small as possible. In the limit, as G_T approaches G_{TU} for the ideal case of $S_{12} = 0$, we see that the error does indeed vanish (i.e., $U = 0$).

─────────────────────────────────RF&MW→

Example 9-9. Unilateral design applicability test

For the amplifier discussed in Example 9-8, estimate the error that is introduced by making the unilateral design approximation.

Solution. Substituting the S-parameter values into (9.45), the unilateral figure of merit is found to be

$$U = \frac{|S_{12}||S_{21}||S_{22}||S_{11}|}{(1-|S_{11}|^2)(1-|S_{22}|^2)} = 0.0812.$$

The maximum error can then be estimated from (9.44):

$$0.86 \leq \frac{G_T}{G_{TU}} \leq 1.18.$$

This implies that the theoretical value for the transducer gain can deviate from its unilateral approximation by as much as 18%. Practically, however, the actual difference often is much smaller. This becomes apparent if we substitute the values obtained in Example 9-8 into the transducer power gain definition (9.8). It is found that $G_T = 62.86$ or 17.98 dB, which compares favorably with $G_{TU} = 63.10$ or 18 dB. In other words, we introduced an error of less than 1%.

The unilateral figure of merit computation constitutes a very conservative, worst case error estimate.

9.4.3 Bilateral Design

For many practical situations, the unilateral approach may not be appropriate because the error committed by setting $S_{12} = 0$ could result in an intolerably imprecise design. The **bilateral design** takes into account this feedback. Instead of the unilateral matching $\Gamma_S^* = S_{11}$ and $\Gamma_L^* = S_{22}$, it deals with the complete equations (see (9.15b) and (9.15c)) for the input and output reflection coefficients

$$\Gamma_S^* = S_{11} + \frac{S_{12}S_{21}\Gamma_L}{1 - S_{22}\Gamma_L} = \frac{S_{11} - \Gamma_L \Delta}{1 - S_{22}\Gamma_L} \quad (9.46a)$$

$$\Gamma_L^* = S_{22} + \frac{S_{12}S_{21}\Gamma_S}{1 - S_{11}\Gamma_S} = \frac{S_{22} - \Gamma_S \Delta}{1 - S_{11}\Gamma_L} \quad (9.46b)$$

which require a **simultaneous conjugate match**. The meaning of *simultaneous* implies that matched source and load reflection coefficients Γ_{MS} and Γ_{ML} have to be found that satisfy both coupled equations. If the device is potentially unstable, then a simultaneous complex conjugate match does not exist. The solution approach to obtain these optimal coefficients is outlined in Example 9-10. The final result for the **matched source reflection coefficient** Γ_{MS} is

$$\Gamma_{MS} = \frac{B_1}{2C_1} - \frac{1}{2}\sqrt{\left(\frac{B_1}{C_1}\right)^2 - 4\frac{C_1^*}{C_1}} \quad (9.47)$$

where

$$C_1 = S_{11} - S_{22}^* \Delta \text{ and } B_1 = 1 - |S_{22}|^2 - |\Delta|^2 + |S_{11}|^2. \quad (9.48)$$

Similarly, the **matched load reflection coefficient** Γ_{ML} is

$$\Gamma_{ML} = \frac{B_2}{2C_2} - \frac{1}{2}\sqrt{\left(\frac{B_2}{C_2}\right)^2 - 4\frac{C_2^*}{C_2}} \quad (9.49)$$

Constant Gain

where

$$C_2 = S_{22} - S_{11}^* \Delta \text{ and } B_2 = 1 - |S_{11}|^2 - |\Delta|^2 + |S_{22}|^2. \quad (9.50)$$

The solutions (9.47) and (9.49) are derived under the assumption of unconditional stability.

With Γ_{ML} and Γ_{MS} given by (9.47) and (9.49), the optimal matching can be rewritten as

$$\Gamma_{MS}^* = S_{11} + \frac{S_{12}S_{21}\Gamma_{ML}}{1 - S_{22}\Gamma_{ML}} \quad (9.51a)$$

and

$$\Gamma_{ML}^* = S_{22} + \frac{S_{12}S_{21}\Gamma_{MS}}{1 - S_{11}\Gamma_{MS}}. \quad (9.51b)$$

It is noted that the unilateral approach, which decouples input and output ports, is a subset of the bilateral design approach.

Example 9-10. Derivation of simultaneous conjugate matched reflection coefficients

Derive the reflection coefficient expression (9.47).

Solution. Starting from (9.46a) and (9.46b), we see that

$$(1 - S_{22}\Gamma_L)(\Gamma_S^* - S_{11}) = \Gamma_L S_{12} S_{21} \quad (9.52a)$$

$$(1 - S_{11}\Gamma_S)(\Gamma_L^* - S_{22}) = \Gamma_S S_{12} S_{21}. \quad (9.52b)$$

Solving (9.52a) for Γ_L yields

$$\Gamma_L = \frac{S_{11} - \Gamma_S^*}{\Delta - S_{22}\Gamma_S^*}. \quad (9.52c)$$

Substituting (9.52c) into (9.52b) results, after some algebra, in

$$\Gamma_S^2(S_{11} - S_{22}^*\Delta) - \Gamma_S(1 + |S_{11}|^2 - |S_{22}|^2 - |\Delta|^2)$$

$$= -S_{11}^* + S_{22}\Delta^*. \quad (9.52d)$$

Identifying $C_1 = (S_{11} - S_{22}^*\Delta)$ and $B_1 = (1 + |S_{11}|^2 - |S_{22}|^2 - |\Delta|^2)$ leads to the standard quadratic equation

$$\Gamma_S^2 - \frac{B_1}{C_1}\Gamma_S = -C_1^* \quad (9.52e)$$

whose solution is

$$\Gamma_{MS} = \frac{B_1}{2C_1} - \frac{1}{2}\sqrt{\left(\frac{B_1}{C_1}\right)^2 - 4\frac{C_1^*}{C_1}}. \quad (9.52f)$$

The negative sign in front of the square root is selected to ensure stability ($k > 1$).

An identical analysis approach for the load side leads to a similar quadratic equation for Γ_L, whose solution yields Γ_{ML}.

Example 9-11 demonstrates the use of simultaneous complex conjugate reflection coefficients for the design of an amplifier with maximum gain.

Example 9-11. Amplifier design for maximum gain

MAXIMUM POWER GAIN

When the network is unconditionally stable ($k > 1, |\Delta| < 1$), a simultaneous conjugate match results in the maximum power gain, known as the Maximum Available Gain (MAG):

$$G_{A\max} = G_{T\max}$$
$$= \frac{|S_{21}|}{|S_{12}|}(k - \sqrt{k^2 - 1})$$

When $k < 1$, the gain can reach infinity, implying an oscillating condition. Here, we define the Maximum Stable Gain (MSG)

$$G_{MSG} = \frac{|S_{21}|}{|S_{12}|}$$

which results from operation on the verge of instability ($k = 1$). Any reasonable design should target a gain significantly lower than the MSG.

A BJT with $I_C = 10$ mA and $V_{CE} = 6$ V is operated at a frequency of $f = 2.4$ GHz. The corresponding S-parameters are: $S_{11} = 0.3\angle 30°$, $S_{12} = 0.2\angle -60°$, $S_{21} = 2.5\angle -80°$, and $S_{22} = 0.2\angle -15°$. Determine whether the transistor is unconditionally stable and find the values for source and load reflection coefficients that provide maximum gain.

Solution. The stability of the transistor is determined by computing k and $|\Delta|$ based on (9.24) and (9.29), with the explicit result of $k = 1.18$, $|\Delta| = 0.56$. Since $k > 1$ and $|\Delta| < 1$, the transistor is unconditionally stable.

As we see from the S-parameters of the transistor, S_{12} has a relatively large magnitude, suggesting the unilateral design method for the amplifier is not appropriate. Therefore, we choose the bilateral approach.

Using (9.48) and (9.50), we find the coefficients $C_1 = 0.19 + j0.06$, $B_1 = 0.74$, and $C_2 = 0.03 + j0.07$, $B_2 = 0.64$, which allow us to compute the simultaneously matched source and load reflection coefficients $\Gamma_{MS} = 0.30\angle -18°$ and $\Gamma_{ML} = 0.12\angle 69°$, respectively. It should be noted that these values differ significantly from S_{11}^* and S_{22}^*, which are the basis for the unilateral design.

Applying (9.8), with Γ_L and Γ_S replaced by Γ_{ML} and Γ_{MS}, we find the transducer gain to be $G_T = 8.42$ dB. This also happens to be the maximum transducer gain $G_{T\max}$.

The discrepancy between the unilateral and bilateral approaches is best seen in the large differences in phase between S_{11}^* and Γ_{MS}, as well as S_{22}^* and Γ_{ML}.

9.4.4 Operating and Available Power Gain Circles

For the situation where the reverse gain of S_{12} cannot be neglected, the input impedance is dependent on the load reflection coefficient. Conversely, the output impedance becomes a function of the source reflection coefficient. Because of this mutual coupling, the unilateral approach described in Section 9.4.1 is not appropriate to design an amplifier for a predetermined gain.

In the bilateral case, which takes into account the mutual coupling between input and output ports, there are two alternative design methods to develop amplifiers with a specified gain.

The first method is based on the use of the operating power gain G given by (9.14). Here, we attempt to find the load reflection coefficient Γ_L, assuming that the source is complex conjugate matched to the input reflection coefficient (i.e., $\Gamma_S = \Gamma_{in}^*$, where Γ_{in} is computed based on (9.9a)). This method yields an input voltage standing wave ratio of $VSWR_{in} = 1$.

The second method uses the available power gain G_A definition of (9.13). In this case, we assume perfect match on the output side of the amplifier ($\Gamma_L = \Gamma_{out}^*$), and the source impedance is chosen in such a way as to satisfy the gain requirement. This method is preferable if the output standing wave ratio should be unity (i.e., $VSWR_{out} = 1$).

Operating Power Gain

To develop the design procedure based on the operating power gain (and thus ensuring $VSWR_{in} = 1$), we rewrite (9.14) in the form

$$G = \frac{(1-|\Gamma_L|^2)|S_{21}|^2}{(1-|\Gamma_{in}|^2)|1-S_{22}\Gamma_L|^2}$$

$$= \frac{(1-|\Gamma_L|^2)|S_{21}|^2}{\left(1-\left|S_{11}+\frac{S_{21}S_{12}\Gamma_L}{1-S_{22}\Gamma_L}\right|^2\right)|1-S_{22}\Gamma_L|^2} = g_o|S_{21}|^2 \quad (9.53)$$

where we use (9.9a) for Γ_{in}. The factor g_o defines a proportionality factor given by

$$g_o = \frac{1-|\Gamma_L|^2}{\left(1-\left|S_{11}+\frac{S_{21}S_{12}\Gamma_L}{1-S_{22}\Gamma_L}\right|^2\right)|1-S_{22}\Gamma_L|^2} = \frac{1-|\Gamma_L|^2}{|1-S_{22}\Gamma_L|^2-|S_{11}-\Delta\Gamma_L|^2}. \quad (9.54)$$

As shown in Example 9-12, (9.54) can be rewritten in terms of a circle equation for the load reflection coefficient Γ_L; that is,

$$|\Gamma_L - d_{g_o}|^2 = r_{g_o}^2 \qquad (9.55)$$

where the center position d_{g_o} is

$$d_{g_o} = \frac{g_o(S_{22} - \Delta S_{11}^*)^*}{1 + g_o(|S_{22}|^2 - |\Delta|^2)} \qquad (9.56)$$

and the radius r_{g_o} is defined as

$$r_{g_o} = \frac{\sqrt{1 - 2kg_o|S_{12}S_{21}| + g_o^2|S_{12}S_{21}|^2}}{|1 + g_o(|S_{22}|^2 - |\Delta|^2)|}. \qquad (9.57)$$

Here, k denotes the Rollett stability factor, as defined in (9.24).

Example 9-12. Operating power gain circle derivation

Starting from (9.54), derive the circle equation (9.55) in the complex Γ_L-plane.

Solution. First we rewrite (9.54) in the form

$$g_o = \frac{1 - |\Gamma_L|^2}{1 - |S_{11}|^2 + |\Gamma_L|^2(|S_{22}|^2 - |\Delta|^2) - 2\mathrm{Re}[\Gamma_L(S_{22} - \Delta S_{11}^*)]}. \qquad (9.58)$$

After multiplying both sides by the denominator and rearranging terms, we see that

$$|\Gamma_L|^2[1 + g_o(|S_{22}|^2 - |\Delta|^2)] - 2g_o\mathrm{Re}[\Gamma_L(S_{22} - \Delta S_{11}^*)] \qquad (9.59)$$
$$= 1 - g_o(1 - |S_{11}|^2).$$

Dividing (9.59) by $1 + g_o(|S_{22}|^2 - |\Delta|^2)$, we find

$$|\Gamma_L|^2 - \frac{2g_o\mathrm{Re}[\Gamma_L(S_{22} - \Delta S_{11}^*)]}{1 + g_o(|S_{22}|^2 - |\Delta|^2)} = \frac{1 - g_o(1 - |S_{11}|^2)}{1 + g_o(|S_{22}|^2 - |\Delta|^2)}.$$

This equation can already be recognized as a circle equation of the form $|\Gamma_L - d_{g_o}|^2 = r_{g_o}^2$, where the circle center d_{g_o} is given by (9.56) and the radius r_{g_o} is computed from

Constant Gain

$$r_{g_o}^2 = \frac{1-g_o(1-|S_{11}|^2)}{1+g_o(|S_{22}|^2-|\Delta|^2)} + \left|\frac{g_o(S_{22}-\Delta S_{11}^*)^*}{1+g_o(|S_{22}|^2-|\Delta|^2)}\right|^2$$

$$= \frac{[1-g_o(1-|S_{11}|^2)][1+g_o(|S_{22}|^2-|\Delta|^2)]+(g_o|S_{22}-\Delta S_{11}^*|)^2}{[1+g_o(|S_{22}|^2-|\Delta|^2)]^2}$$

$$= \frac{1-g_o(1-|S_{11}|^2-|S_{22}|^2+|\Delta|^2)-g_o^2 M}{[1+g_o(|S_{22}|^2-|\Delta|^2)]^2}$$

$$= \frac{1-2g_o|S_{12}S_{21}|k-g_o^2 M}{[1+g_o(|S_{22}|^2-|\Delta|^2)]^2}$$

where k is the stability factor defined in (9.24) and M is a constant given by

$$M = (1-|S_{11}|^2)(|S_{22}|^2-|\Delta|^2) - |S_{22}-\Delta S_{11}^*|^2 = -|S_{12}S_{21}|^2$$

Thus, for the square of the circle radius we obtain

$$r_{g_o}^2 = \frac{1-2g_o|S_{12}S_{21}|k+g_o^2|S_{12}S_{21}|^2}{[1+g_o(|S_{22}|^2-|\Delta|^2)]^2}.$$

which agrees with (9.57).

The following example demonstrates the design of an amplifier based on the bilateral method. It targets a specified gain using the constant operating gain approach.

Example 9-13. Amplifier design using the constant operating gain circles

Use the same BJT as described in Example 9-11, but instead of $G_{T\max} = 8.42$ dB, design an amplifier with 8 dB power gain. In addition, ensure a perfect match on the input port of the amplifier.

Solution. As shown in Example 9-11, the transistor is unconditionally stable. Because a perfect match on the input port must be maintained, we employ the operating power gain circles in our design.

First, we compute the value of factor g_o; that is,

$$g_o = \frac{G}{|S_{21}|^2} = 1.0095$$

where $G = 6.31$ is the required 8 dB operating gain. Substituting g_o into (9.56) and (9.57), we find the center and radius of the constant operating gain circle in the Γ_L-plane. The corresponding values are $d_{g_o} = 0.11\angle 69°$ and $r_{g_o} = 0.35$. The constant gain circle is shown in Figure 9-16.

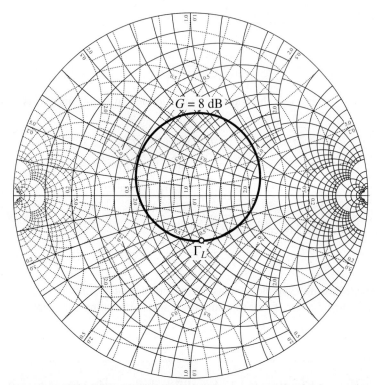

Figure 9-16 Constant operating power circle in the Γ_L-plane.

There is a great variety of possible choices for the load reflection coefficient that ensures a $G = 8$ dB operating gain. To simplify the output matching network, we choose Γ_L at the intersection of the constant gain circle with the constant resistance circle $r = 1$ (see Figure 9-16). The value obtained at that point is $\Gamma_L = 0.26\angle -75°$. With Γ_L known, we can next find the source reflection coefficient that must be the conjugate to the input refection coefficient as given in (9.9a):

$$\Gamma_S = \left(\frac{S_{11} - \Delta\Gamma_L}{1 - S_{22}\Gamma_L}\right)^* = 0.28\angle -55°$$

Constant Gain

Based on the previously computed values, we check the correctness of our approach. Substituting Γ_{in} and Γ_L into (9.10), we find that the transducer power gain is indeed 8 dB.

The complexity of the input matching network is directly affected by the appropriate choice of Γ_L because of the requirement $\Gamma_S = \Gamma_{in}^*$, where Γ_{in} is a function of Γ_L.

In Example 9-13, we pick the value of Γ_L arbitrarily, residing on the desired gain circle, and compute a corresponding input impedance such that $\Gamma_S = \Gamma_{in}^*$, assuming that there are no restrictions imposed on the value of Γ_S. Unfortunately, in many practical applications, Γ_S has to satisfy certain constraints (for example, to stay within a desired noise performance). Such additional conditions may therefore restrict our freedom in selecting Γ_S and, as a consequence, limit the possible choices for Γ_L. One way to satisfy both requirements (Γ_L residing within an appropriate gain circle, and Γ_S satisfy a particular noise requirement) is via trial and error, whereby we arbitrarily pick Γ_L and see whether the corresponding Γ_S meets design specifications. This method is simple but very tedious and time consuming.

A more systematic approach relies upon mapping the constant gain circle (9.55) in the Γ_L-plane into a circle in the Γ_S-plane, i.e.,

$$\left|\Gamma_S - d_{gs}\right| = r_{gs} \tag{9.60}$$

where the equations for the circle radius r_{gs} and its center d_{gs} are obtained from the requirement that $\Gamma_S = \Gamma_{in}^*$. This can be written as

$$\Gamma_S^* = \frac{S_{11} - \Delta\Gamma_L}{1 - S_{22}\Gamma_L} \tag{9.61}$$

or

$$\Gamma_L = \frac{S_{11} - \Gamma_S^*}{\Delta - S_{22}\Gamma_S^*}. \tag{9.62}$$

Substituting (9.62) into (9.55) gives us

$$\left|\frac{S_{11} - \Gamma_S^*}{\Delta - S_{22}\Gamma_S^*} - d_{g_o}\right|^2 = r_{g_o}^2 \tag{9.63}$$

which can be rewritten in the form of (9.60), where the circle radius is

$$r_{gs} = \frac{r_{g_o}|S_{12}S_{21}|}{\left||1 - S_{22}d_{g_o}|^2 - r_{g_o}^2|S_{22}|^2\right|} \tag{9.64}$$

and the center is given by

$$d_{g_s} = \frac{(1 - S_{22}d_{g_o})(S_{11} - \Delta d_{g_0})^* - r_{g_o}^2 \Delta^* S_{22}}{|1 - S_{22}d_{g_o}|^2 - r_{g_o}^2 |S_{22}|^2}. \quad (9.65)$$

The derivation of (9.64) and (9.65) is left as a problem at the end of this chapter. The example of constant gain circle mapping is discussed further in Section 9.5, Example 9-14.

Available Power Gain

In those cases where perfect matching at the output side of the amplifier is required ($VSWR_{out} = 1$), the available power gain approach should be used instead of the previously presented operating gain method. For this situation, a constant available gain circle equation can be derived in the same fashion as (9.55) is obtained. The result of such a derivation is a circle equation that relates the source reflection coefficient to the desired gain:

$$|\Gamma_S - d_{g_a}| = r_{g_a} \quad (9.66)$$

where the center position d_{g_a} is

$$d_{g_a} = \frac{g_a(S_{11} - \Delta S_{22}^*)^*}{1 + g_a(|S_{11}|^2 - |\Delta|^2)} \quad (9.67)$$

and the radius r_{g_a} is defined as

$$r_{g_a} = \frac{\sqrt{1 - 2kg_a|S_{12}S_{21}| + g_a^2|S_{12}S_{21}|^2}}{|1 + g_a(|S_{11}|^2 - |\Delta|^2)|}. \quad (9.68)$$

The proportionality factor g_a is computed as

$$g_a = \frac{G_A}{|S_{21}|^2} \quad (9.69)$$

where G_A is the desired gain.

Similarly to the constant operating power circles, a constant available power gain circle can be mapped into the Γ_L-plane using

$$|\Gamma_L - d_{g_l}| = r_{g_l} \quad (9.70)$$

with the circle radius given by

$$r_{g_l} = \frac{r_{g_a}|S_{12}S_{21}|}{||1 - S_{11}d_{g_a}|^2 - r_{g_a}^2|S_{11}|^2|} \quad (9.71)$$

and the center location defined by

$$d_{g_l} = \frac{(1 - S_{11}d_{g_a})(S_{22} - \Delta d_{g_a})^* - r_{g_a}^2 \Delta^* S_{11}}{|1 - S_{11}d_{g_a}|^2 - r_{g_a}^2|S_{11}|^2}. \quad (9.72)$$

We see that r_{g_l} and d_{g_l} for $\text{VSWR}_{\text{out}} = 1$ are analogous to r_{g_S} and d_{g_S} for $\text{VSWR}_{\text{in}} = 1$, with S_{11} in (9.71) and (9.72) replaced by S_{22}.

9.5 Noise Figure Circles

In many RF amplifiers, the need for signal amplification at low noise level becomes an essential system requirement. Unfortunately, designing a low-noise amplifier competes with such factors as stability and gain. For instance, minimum noise performance at maximum gain cannot be obtained. It is therefore important to develop a method that allows us to display the influence of noise as part of the Smith Chart to conduct comparisons and observe trade-offs between gain and stability.

The generated noise of a two-port network can be quantified by investigating the decrease in the signal-to-noise ratio (SNR) from the input to the output. The noise figure F (≥ 1) is defined as the ratio of the input SNR to the output SNR. For a practical two-port amplifier, the noise figure can be cast in the admittance form:

$$F = F_{\min} + \frac{R_n}{G_S}|Y_S - Y_{\text{opt}}|^2 \qquad (9.73)$$

or in the equivalent impedance representation

$$F = F_{\min} + \frac{G_n}{R_S}|Z_S - Z_{\text{opt}}|^2 \qquad (9.74)$$

where $Z_S = 1/Y_S$ is the source impedance.

Both expressions are derived in Appendix H. As the following will show, (9.73) can conveniently be displayed in the Smith Chart as circles. When using transistors, typically four noise parameters are known either through datasheets from the FET or BJT manufacturers or through direct measurements. They are:

- The **minimum** (also called optimum) **noise figure** $F_{\min}$, whose behavior depends on biasing condition and operating frequency. If the device were noise free, we would obtain $F_{\min} = 1$.
- The **equivalent noise resistance** $R_n = 1/G_n$ of the device.
- The **optimum source admittance** $Y_{\text{opt}} = G_{\text{opt}} + jB_{\text{opt}} = 1/Z_{\text{opt}}$. Instead of the impedance or admittance, the **optimum reflection coefficient** Γ_{opt} is often listed. The relationship between Y_{opt} and Γ_{opt} is given by

$$Y_{\text{opt}} = Y_0 \frac{1 - \Gamma_{\text{opt}}}{1 + \Gamma_{\text{opt}}}. \qquad (9.75)$$

> **NOISE FACTOR AND NOISE FIGURE**
>
> To be precise, the noise figure (often denoted NF) is recorded in dB and related to the noise factor via $\text{NF} = 10 \log(F)$. However, in practice no distinction is made, and F is reported in dB, as done in our text.

Since the S-parameter representation is a more suitable choice for high-frequency designs, we convert (9.73) into a form that replaces the admittances by reflection coefficients. Besides (9.75), we use

$$Y_S = Y_0 \frac{1 - \Gamma_S}{1 + \Gamma_S} \qquad (9.76)$$

in (9.73). Recognizing that G_S can be written as $G_S = Y_0(1-|\Gamma_S|^2)/|1+\Gamma_S|^2$, the final result becomes

$$F = F_{\min} + \frac{4R_n}{Z_0}\frac{|\Gamma_S - \Gamma_{opt}|^2}{(1-|\Gamma_S|^2)|1+\Gamma_{opt}|^2}. \tag{9.77}$$

In (9.77), the quantities $F_{\min}$, R_n, and Γ_{opt} are known. In general, the design engineer has the freedom to adjust Γ_S to affect the noise figure. For $\Gamma_S = \Gamma_{opt}$, we see that the lowest possible noise figure is achieved, $F = F_{\min}$. To answer the question of how a particular noise figure, let us say F_k, relates to Γ_S, (9.77) is re-stated as

$$|\Gamma_S - \Gamma_{opt}|^2 = (1-|\Gamma_S|^2)|1+\Gamma_{opt}|^2\left(\frac{F_k - F_{\min}}{4R_n/Z_0}\right) \tag{9.78}$$

which on the right-hand side already suggests the form of a circle equation. Introducing a constant Q_k such that

$$Q_k = |1+\Gamma_{opt}|^2\frac{F_k - F_{\min}}{4R_n/Z_0} \tag{9.79}$$

and rearranging terms gives

$$(1+Q_k)|\Gamma_S|^2 - 2\text{Re}(\Gamma_S\Gamma_{opt}^*) + |\Gamma_{opt}|^2 = Q_k. \tag{9.80}$$

Division by $1+Q_k$ and completing the square yields, after some algebra,

$$\left|\Gamma_S - \frac{\Gamma_{opt}}{1+Q_k}\right|^2 = Q_k\left[\frac{1}{1+Q_k} - \frac{|\Gamma_{opt}|^2}{(1+Q_k)^2}\right] = \frac{Q_k^2 + Q_k(1-|\Gamma_{opt}|^2)}{(1+Q_k)^2}. \tag{9.81}$$

This is the required circle equation in standard form that can be displayed as part of the Smith Chart:

$$|\Gamma_S - d_{F_k}|^2 = (\Gamma_S^R - d_{F_k}^R)^2 + (\Gamma_S^I - d_{F_k}^I)^2 = r_{F_k}^2 \tag{9.82}$$

with the circle center location d_{F_k} denoted by the complex number

$$d_{F_k} = d_{F_k}^R + jd_{F_k}^I = \frac{\Gamma_{opt}}{1+Q_k} \tag{9.83}$$

and the associated radius

$$r_{F_k} = \frac{\sqrt{(1-|\Gamma_{opt}|^2)Q_k + Q_k^2}}{1+Q_k}. \tag{9.84}$$

Noise Figure Circles

Two noteworthy conclusions can be drawn from (9.83) and (9.84):

- The minimum noise figure is obtained for $F_k = F_{min}$, which coincides with the location $d_{F_k} = \Gamma_{opt}$ and radius $r_{F_k} = 0$.
- All constant noise circles have their centers located along a line drawn from the origin to the point Γ_{opt}. The larger the noise figure, the closer the center d_{F_k} moves to the origin and the larger the radius r_{F_k}.

The following example points out the trade-offs between gain and noise figure for a small-signal amplifier.

RF&MW→

Example 9-14. Design of a small-signal amplifier for minimum noise figure and specified gain

Using the same transistor as in Example 9-13, design a low-noise amplifier with 8 dB gain and a noise figure that is less than 1.6 dB. Assume that the transistor has the following noise parameters: $F_{min} = 1.5$ dB, $R_n = 4\ \Omega$, and $\Gamma_{opt} = 0.5\angle 45°$.

Solution. The noise figure is independent of the load reflection coefficient. However, it is a function of the source impedance. It is therefore convenient to map the constant gain circle obtained in Example 9-13 into the Γ_S-plane. Applying equations (9.64) and (9.65) and values from Example 9-13, we find the center and radius of the mapped constant gain circle: $d_{g_S} = 0.29\angle -18°$ and $r_{g_S} = 0.18$. A Γ_S residing anywhere on this circle will satisfy our gain requirement. However, for the noise figure specifications to be met, we have to ensure that Γ_S resides inside the $F_k = 1.6$ dB constant noise circle.

The noise circle center and its radius are computed using (9.83) and (9.84), respectively. They are listed below together with the coefficient Q_k, see (9.79):

$$Q_k = 0.2,\quad d_{F_k} = 0.42\angle 45°,\quad r_{F_k} = 0.36$$

The obtained $G = 8$ dB and $F_k = 1.6$ dB circles are shown in Figure 9-17.

Notice that the maximum power gain is obtained at the point $\Gamma_S = \Gamma_{MS} = 0.30\angle -18°$ (see Example 9-11 for the detailed computations). However, the minimum noise figure is obtained at $\Gamma_S = \Gamma_{opt} = 0.5\angle 45°$, which shows for this example that it is

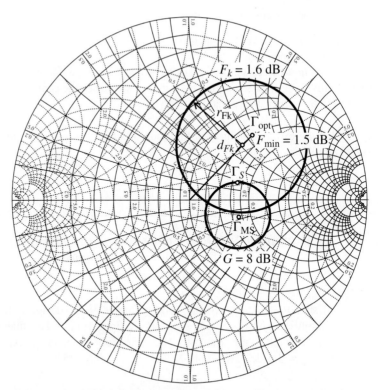

Figure 9-17 Constant noise figure circle and constant operating gain circle mapped into the Γ_S-plane.

impossible to achieve maximum gain and minimum noise figure simultaneously. Clearly, some compromise has to be made.

To minimize the noise figure for a given gain, we should pick the source reflection coefficient as close as possible to the location of Γ_{opt} while still residing on the constant gain circle. Approximately choosing $\Gamma_S = 0.29 \angle 19°$, the corresponding load reflection coefficient is found to be $\Gamma_L = 0.45 \angle 50°$ by applying (9.62). The obtained amplifier noise figure is then computed using (9.77):

$$F = F_{min} + \frac{4R_n}{Z_0} \frac{|\Gamma_S - \Gamma_{opt}|^2}{(1 - |\Gamma_S|^2)|1 + \Gamma_{opt}|^2} = 1.54 \text{ dB}$$

The requirements of maximum gain and minimum noise figure will always be design trade-offs and cannot be met simultaneously.

9.6 Constant VSWR Circles

In many cases, the amplifier has to stay below a specified VSWR as measured at the input or output port of the amplifier. Typical values range between $1.5 \leq \text{VSWR} \leq 2.5$. As we know from our discussion in Chapter 8, the purpose of matching networks is primarily to reduce the VSWR at the ports. The complication arises from the fact that the input VSWR (or VSWR_{IMN}) is determined at the input matching network (IMN), which in turn is affected by the active device, and, through feedback, by the output matching network (OMN). Conversely, the output VSWR (or VSWR_{OMN}) is determined by the OMN and, again through feedback, by the IMN. This calls for a bilateral design approach, as discussed in Section 9.4.3.

To set the stage, let us consider the arrangement depicted in Figure 9-18. The two VSWRs that are part of an RF amplifier specification are

$$\text{VSWR}_{\text{IMN}} = \frac{1 + |\Gamma_{\text{IMN}}|}{1 - |\Gamma_{\text{IMN}}|} \text{ and } \text{VSWR}_{\text{OMN}} = \frac{1 + |\Gamma_{\text{OMN}}|}{1 - |\Gamma_{\text{OMN}}|}. \tag{9.85}$$

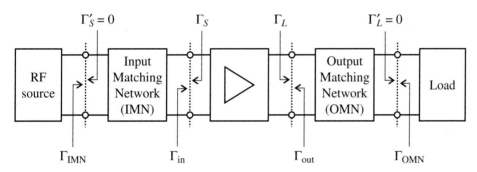

Figure 9-18 System configuration for input and output VSWR.

The reflection coefficients Γ_{IMN}, Γ_{OMN} require further clarification. If we concentrate on Γ_{IMN}, it is apparent from Section 9.2.1 that the input power P_{in} (under the assumption $\Gamma'_S = 0$) can be expressed as a function of the available power P_A:

$$P_{\text{in}} = P_A(1 - |\Gamma_{\text{IMN}}|^2). \tag{9.86}$$

Postulating that the matching network is lossless, the same power is also present at the input terminal of the active device

$$P_{\text{in}} = P_A \frac{(1 - |\Gamma_S|^2)(1 - |\Gamma_{\text{in}}|^2)}{|1 - \Gamma_S \Gamma_{\text{in}}|}. \tag{9.87}$$

Equating (9.86) to (9.87) and solving for $|\Gamma_{\text{IMN}}|$ yields

$$|\Gamma_{\text{IMN}}| = \sqrt{1 - \frac{(1-|\Gamma_S|^2)(1-|\Gamma_{\text{in}}|^2)}{|1-\Gamma_S\Gamma_{\text{in}}|}} = \left|\frac{\Gamma_{\text{in}} - \Gamma_S^*}{1-\Gamma_S\Gamma_{\text{in}}}\right| = \left|\frac{\Gamma_{\text{in}}^* - \Gamma_S}{1-\Gamma_S\Gamma_{\text{in}}}\right|. \tag{9.88}$$

Equation (9.88) can be converted into a circle equation for Γ_S that is centered at location $d_{V_{\text{IMN}}}$ with radius $r_{V_{\text{IMN}}}$ such that

$$(\Gamma_S^R - d_{V_{\text{IMN}}}^R)^2 + (\Gamma_S^I - d_{V_{\text{IMN}}}^I)^2 = r_{V_{\text{IMN}}}^2 \tag{9.89}$$

where

$$d_{V_{\text{IMN}}} = d_{V_{\text{IMN}}}^R + j d_{V_{\text{IMN}}}^I = \frac{(1-|\Gamma_{\text{IMN}}|^2)\Gamma_{\text{in}}^*}{1-|\Gamma_{\text{IMN}}\Gamma_S|^2} \tag{9.90}$$

and

$$r_{V_{\text{IMN}}} = \frac{(1-|\Gamma_{\text{in}}|^2)|\Gamma_{\text{IMN}}|}{1-|\Gamma_{\text{IMN}}\Gamma_S|^2}. \tag{9.91}$$

Here, the subscript V_{IMN} in $d_{V_{\text{IMN}}}$ and $r_{V_{\text{IMN}}}$ is used to denote the VSWR at the IMN location.

In an identical procedure, the circle equation for the output VSWR is found.

$$|\Gamma_{\text{OMN}}| = \sqrt{1 - \frac{(1-|\Gamma_L|^2)(1-|\Gamma_{\text{out}}|^2)}{|1-\Gamma_L\Gamma_{\text{out}}|}} = \left|\frac{\Gamma_{\text{out}} - \Gamma_L^*}{1-\Gamma_L\Gamma_{\text{out}}}\right| = \left|\frac{\Gamma_{\text{out}}^* - \Gamma_L}{1-\Gamma_L\Gamma_{\text{out}}}\right|. \tag{9.92}$$

We convert (9.92) into a circle equation for Γ_L that is centered at the location $d_{V_{\text{OMN}}}$ with the radius $r_{V_{\text{OMN}}}$ such that

$$(\Gamma_L^R - d_{V_{\text{OMN}}}^R)^2 + (\Gamma_L^I - d_{V_{\text{OMN}}}^I)^2 = r_{V_{\text{OMN}}}^2 \tag{9.93}$$

where

$$d_{V_{\text{OMN}}} = d_{V_{\text{OMN}}}^R + j d_{V_{\text{OMN}}}^I = \frac{(1-|\Gamma_{\text{OMN}}|^2)\Gamma_{\text{out}}^*}{1-|\Gamma_{\text{OMN}}\Gamma_L|^2} \tag{9.94}$$

and

$$r_{V_{\text{OMN}}} = \frac{(1-|\Gamma_{\text{out}}|^2)|\Gamma_{\text{OMN}}|}{1-|\Gamma_{\text{OMN}}\Gamma_L|^2}. \tag{9.95}$$

Constant VSWR Circles

The previous derivations allow us to draw the following conclusions regarding the constant VSWR circles:

- For minimum VSWR (on the input side: $VSWR_{IMN} = 1$, $|\Gamma_{IMN}| = 0$; on the output side: $VSWR_{OMN} = 1$, $|\Gamma_{OMN}| = 0$), the circles are located at $d_{V_{IMN}}|_{|\Gamma_{IMN}|=0} = \Gamma_{in}^*$ (for the input) and $d_{V_{OMN}}|_{|\Gamma_{OMN}|=0} = \Gamma_{out}^*$ (for the output) with both radii equal to zero.
- All VSWR circles reside on the line extending from the origin to Γ_{in}^* (input) or Γ_{out}^* (output).

We should be aware of the fact that under *bilateral matching*, the input and output reflection coefficients are functions of source and load reflection coefficients (Γ_S, Γ_L). Therefore, the input and output VSWR circles cannot be plotted simultaneously, but rather have to be considered one at a time in the iterative process of adjusting Γ_S and Γ_L.

────RF&MW→

Example 9-15. Constant VSWR design for given gain and noise figure

Using the results of Example 9-14, plot the $VSWR_{IMN} = 1.5$ circle in the Γ_S-plane as part of the Smith Chart. Plot the graph of $VSWR_{OMN}$ as a function of the Γ_S position for a $VSWR_{IMN} = 1.5$. Find Γ_S that gives a minimum reflection on the output port of the amplifier and compute its corresponding gain.

Solution. In Example 9-14, we have found $\Gamma_S = 0.29\angle 19°$ and $\Gamma_L = 0.45\angle 50°$ as source and load reflection coefficients that meet specifications in terms of power gain and noise figure. Since the design is based on constant operating gain circles, we obtain a perfect match at the input port of the amplifier. However, the output port is mismatched and the $VSWR_{OMN}$ can be computed from $|\Gamma_{OMN}|$, which is found from (9.92) in conjunction with (9.9b):

$$|\Gamma_{OMN}| = \left|\frac{\Gamma_{out}^* - \Gamma_L}{1 - \Gamma_L \Gamma_{out}}\right| = 0.26$$

The result is

$$VSWR_{OMN} = \frac{1 + |\Gamma_{OMN}|}{1 - |\Gamma_{OMN}|} = 1.69.$$

To improve the VSWR_{OMN}, we can relax the requirements on VSWR_{IMN} and introduce some mismatch at the input. If we set $\text{VSWR}_{\text{IMN}} = 1.5$, the corresponding input VSWR circle can be plotted in the Smith Chart, as shown in Figure 9-19.

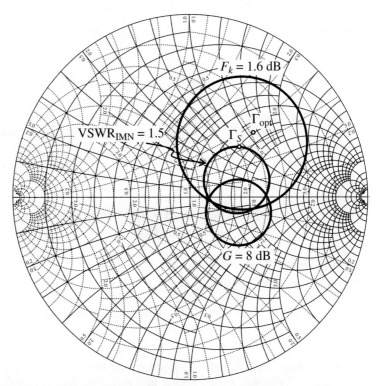

Figure 9-19 Constant operating power gain, noise figure, and input VSWR circle in Γ_S-plane.

The center of the $\text{VSWR}_{\text{IMN}} = 1.5$ circle and its radius are found from (9.90) and (9.91), respectively: $d_{V_{\text{IMN}}} = 0.28 \angle 19°$ and $r_{V_{\text{IMN}}} = 0.18$.

The points on the $\text{VSWR}_{\text{IMN}} = 1.5$ circle can be expressed in polar form

$$\Gamma_S = d_{V_{\text{IMN}}} + r_{V_{\text{IMN}}} \exp(j\alpha)$$

where the angle α ranges from 0 to 360°. As α varies, we obtain a changing Γ_S, which in turn results in a corresponding Γ_{out} and VSWR_{OMN}. The graph of such a dependence is shown in Figure 9-20.

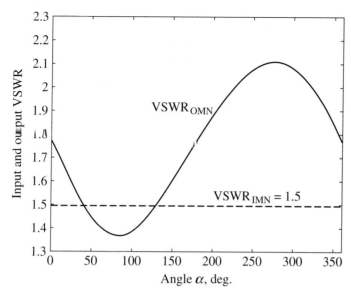

Figure 9-20 Input and output VSWR as a function of angle α.

As can be observed in Figure 9-20, the VSWR_{OMN} reaches its minimum value of 1.37 at approximately $\alpha = 85°$. The corresponding source and output reflection coefficients, transducer gain, and noise figure are as follows:

$$\Gamma_S = 0.39\angle 45°, \quad \Gamma_{\text{out}} = 0.32\angle -52°$$

$$G_T = 7.82 \text{ dB}, \quad F = 1.51 \text{ dB}.$$

An improvement in VSWR_{OMN} has been achieved at the expense of reduced gain. If the gain reduction becomes unacceptable, then both source and load reflection coefficients have to be adjusted simultaneously.

Design specifications often explicitly prescribe a maximum tolerable VSWR that the amplifier must meet. This becomes particularly important when dealing with system integration issues where several blocks are cascaded.

9.7 Broadband, High-Power, and Multistage Amplifiers

9.7.1 Broadband Amplifiers

Many modulation and coding circuits require amplifiers with a wide frequency band of operation. From the RF point of view, one of the major problems in broadband amplifier design is the limitation imposed by the gain-bandwidth product of the active device. As pointed out in Chapter 7, any active device has a gain rolloff at higher frequencies due to the parasitic capacitances. Eventually, as the frequency reaches the transition frequency f_T, the transistor stops functioning as an amplifier and turns attenuative.

Unfortunately, the forward gain $|S_{21}|$ seldom remains constant over the wide frequency band of operation, necessitating compensation measures. Besides gain degradation, other complications that arise in the design of broadband amplifiers include:

- Increase in the reverse gain $|S_{12}|$, which degrades the overall gain even further and increases the possibility for a device to fall into oscillation
- Frequency variation of S_{11} and S_{22}
- Noise figure degradation at high frequencies.

To account for these effects, two different amplifier design approaches are used: frequency-compensated matching networks and negative feedback. In the subsequent sections, we investigate both design techniques.

> **f_T vs. f_{max}**
>
> Strictly speaking, the maximum frequency of oscillation, f_{max}, refers to the point where the maximum available power gain reaches unity, and the transistor no longer functions as an amplifier. Even though f_{max} is typically close to f_T, it can significantly deviate in either direction.

Frequency-Compensated Matching Networks

Frequency-compensated matching networks introduce a mismatch on either the input or output port of the device to compensate for the frequency variation introduced by the S-parameters. The difficulty with these types of matching networks is that they are rather difficult to design, and the procedures involved are more an art than a well-defined engineering approach that guarantees success. Frequency-compensated matching networks have to be custom tailored for each particular case.

The example below demonstrates some of the key steps required to design a frequency-compensated matching network.

Example 9-16. Design of a broadband amplifier using a frequency-compensated matching network

Design a broadband amplifier with 7.5 dB nominal gain and ±0.2 dB gain flatness in the frequency range from 2 GHz to 4 GHz. For the design use Avago Technologies' AT41410 BJT, which is biased with $I_C = 10$ mA collector current and $V_{CE} = 8$ V collector-emitter voltage. The corresponding S-parameters, measured at frequencies of 2, 3, and 4 GHz under the unilateral assumption, are summarized in Table 9-4.

Table 9-4 S-parameters of AT41410 BJT (I_C = 10 mA, V_{CE} = 8 V)

f, GHz	$|S_{21}|$	S_{11}	S_{22}
2	3.72	0.61∠165°	0.45∠–48°
3	2.56	0.62∠149°	0.44∠–58°
4	1.96	0.62∠130°	0.48∠–78°

Solution. According to the data provided in Table 9-4, the insertion gain of the transistor is $|S_{21}|^2$ = 11.41 dB at f = 2 GHz, 8.16 dB at 3 GHz, and 5.85 dB at 4 GHz. To realize an amplifier with a nominal gain of 7.5 dB, the source and load matching networks must be designed so as to decrease the gain by 3.91 dB at 2 GHz, 0.66 dB at 3 GHz, and increase the gain by 1.65 dB at 4 GHz.

The maximum gain provided by the source and load are found from (9.35) and (9.36), and are as follows:

f = 2 GHz: $G_{S\max}$ = 2.02 dB, $G_{L\max}$ = 0.98 dB

f = 3 GHz: $G_{S\max}$ = 2.11 dB, $G_{L\max}$ = 0.93 dB

f = 4 GHz: $G_{S\max}$ = 2.11 dB, $G_{L\max}$ = 1.14 dB.

Although for the general case, source and load matching networks would have to be designed, in this example the additional gain G_S that can be produced by the source matching is already sufficient to meet the amplifier specifications. Therefore, we concentrate on the development of the source matching network and leave the output port of the transistor without any matching network.

Since the output of the transistor is directly connected to the load, we have G_L = 0 dB. The input matching network should produce an additional gain of (-3.9 ± 0.2) dB at f = 2 GHz, (-0.7 ± 0.2) dB at 3 GHz, and (1.7 ± 0.2) dB at 4 GHz. The corresponding constant gain circles are shown in Figure 9-21.

The required input matching network must be capable of transforming points on the constant gain circles in Figure 9-21 to the center of the Smith Chart. There are a number of networks that can accomplish this task. One solution involves a combination of two capacitors, one in shunt with the transistor and one in series with the input port of the amplifier, as shown in Figure 9-22. From a known Γ_S, we can compute the transducer gain by setting Γ_L = 0 in (9.10). We can next

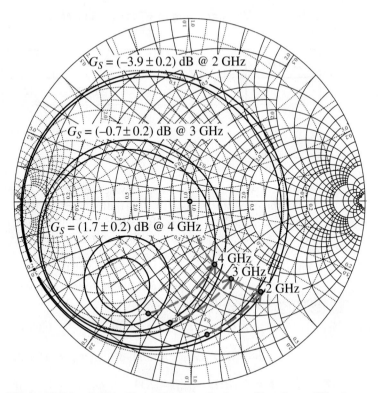

Figure 9-21 Smith Chart design of a broadband amplifier in Example 9-16.

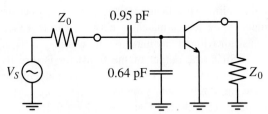

Figure 9-22 Broadband amplifier with 7.5 dB gain and ±0.2 dB gain flatness over a frequency range from 2 to 4 GHz.

find the input and output VSWR. Since $\Gamma_L = 0$, the values for VSWR_{OMN} is equal to VSWR_{out} and is found as

$$\text{VSWR}_{\text{out}} = \frac{1 + |S_{22}|}{1 - |S_{22}|}.$$

For the computation of the VSWR at the input port, we use

$$\text{VSWR}_{\text{IMN}} = \frac{1+|\Gamma_{\text{IMN}}|}{1-|\Gamma_{\text{IMN}}|}$$

where $|\Gamma_{\text{IMN}}|$ is computed based on (9.88):

$$|\Gamma_{\text{IMN}}| = \left|\frac{\Gamma_{\text{in}}^* - \Gamma_S}{1-\Gamma_S\Gamma_{\text{in}}}\right| = \left|\frac{S_{11}^* - \Gamma_S}{1-\Gamma_S S_{11}}\right|$$

The obtained values are summarized in Table 9-5.

Table 9-5 Parameters of a broadband amplifier

f, GHz	Γ_S	G_T, dB	VSWR$_{\text{IMN}}$	VSWR$_{\text{OMN}}$
2	0.74∠−83°	7.65	13.1	2.6
3	0.68∠−101°	7.57	5.3	2.6
4	0.66∠−112°	7.43	2.0	2.8

As seen from the values provided in Table 9-5, gain linearity is achieved at the expense of significantly higher VSWR.

As demonstrated in Example 9-16, the addition of a frequency-compensated matching network to improved gain flatness may result in significant impedance mismatch, degrading the amplifier performance. To circumvent this problem, a balanced amplifier can be employed.

Balanced Amplifier Design

The typical balanced amplifier block diagram using a 3 dB Lange, or hybrid, coupler and a 3 dB Wilkinson power divider and combiner are shown in Figures 9-23(a) and (b), respectively. The input signal power is split into two, amplified, and combined at the output. A complete discussion of the theory behind the operation of couplers and power dividers is given in Appendix G.

Let us first discuss the operation of the balanced amplifier in Figure 9-23(a). Here, the input power launched into port 1 of the input coupler is equally divided in magnitude, but with a 90° phase shift between ports 2 and 3. The output coupler combines the output signals of amplifiers A and B by introducing an additional 90° phase shift, thus bringing them in phase again. We denote the S-parameters of amplifier A as S_{11}^A, S_{12}^A, S_{21}^A, S_{22}^A, and the corresponding S-parameters of amplifier B with superscript B. The equations that

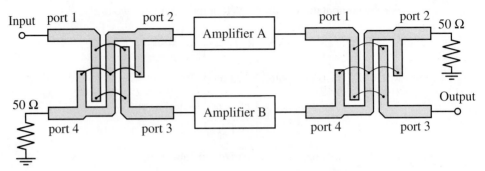

(a) Balanced amplifier using 3 dB coupler

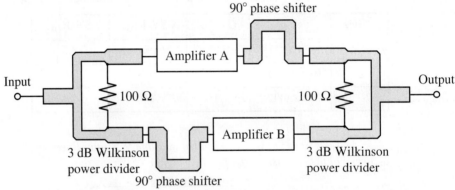

(b) Balanced amplifier using 3 dB Wilkinson power divider and combiner

Figure 9-23 Block diagram of a balanced broadband amplifier.

relate the *S*-parameters of the entire amplifier to the *S*-parameters of the individual branches are

$$|S_{11}| = \frac{1}{2}|S_{11}^A - S_{11}^B|$$
$$|S_{21}| = \frac{1}{2}|S_{21}^A + S_{21}^B|$$
$$|S_{12}| = \frac{1}{2}|S_{12}^A + S_{12}^B|$$
$$|S_{22}| = \frac{1}{2}|S_{22}^A - S_{22}^B|$$
(9.96)

where coefficients 1/2 take into account the 3 dB attenuation, and the minus sign is due to the 90° phase shift at port 3 that is encountered in both directions, adding up to 180°.

If the amplifiers in the two branches are identical, then $|S_{11}| = |S_{22}| = 0$, and the forward and reverse gains of the balanced amplifier are equal to the corresponding gains of each branch.

Broadband, High-Power, and Multistage Amplifiers

The operation of the balanced amplifier with Wilkinson power dividers (see Figure 9-23(b)) is identical. The only difference compared to the hybrid coupler is that the signals are in phase, and we need to add 90° phase shifters to achieve the desired effect.

The main advantages of balanced amplifiers are that they possess very good impedance match at the input and output ports (provided that the amplifiers in both branches have similar characteristics), and one of the two amplifiers can continue operating even if the other branch should fail completely. The chief disadvantages of balanced amplifiers include increased circuit size and a reduction in frequency response introduced by the bandwidth of the couplers.

Negative Feedback Circuits

The alternative to frequency-compensating networks is the use of negative feedback. This allows a flat gain response and reduces the input and output VSWR over a wide frequency range. An additional advantage of negative feedback is that it makes the circuit less sensitive to transistor-to-transistor parameter variations. However, such circuits tend to limit the maximum power gain of the transistor and increase its noise figure.

The term *negative feedback* implies that part of the signal from the output of the transistor is coupled back to the input with opposite phase so that it subtracts from the input signal, thereby reducing it. If the signals add in phase, the resulting response will grow, and positive feedback is obtained. The most general resistive feedback circuits for BJTs and FETs are shown in Figure 9-24, where resistor R_1 constitutes shunt feedback, and resistor R_2 series feedback.

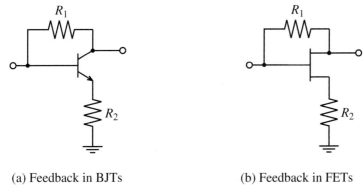

(a) Feedback in BJTs (b) Feedback in FETs

Figure 9-24 Negative resistive feedback circuits.

As discussed in Chapter 7, both circuits in Figure 9-24 at low frequencies can be replaced by the equivalent π-models, as shown in Figure 9-25, where the input resistance r_π is infinite for FETs.

If we assume for the BJT that

$$r_\pi(1 + g_m R_2) \gg R_1 \qquad (9.97)$$

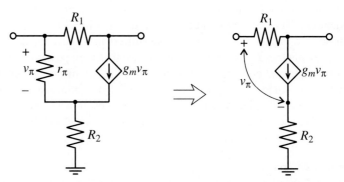

Figure 9-25 Low-frequency model of negative feedback circuit.

then r_π in Figure 9-25 can be replaced by an open circuit, and the h-parameter representation written as

$$[\mathbf{h}] = \begin{bmatrix} R_1 & 1 \\ \dfrac{g_m R_1}{1 + g_m R_2} - 1 & \dfrac{g_m}{1 + g_m R_2} \end{bmatrix}. \tag{9.98}$$

Using the matrix conversion formula from Appendix D, we find the corresponding S-parameter representation

$$[\mathbf{S}] = \dfrac{1}{\Delta} \begin{bmatrix} \dfrac{R_1}{Z_0} - \dfrac{g_m Z_0}{1 + g_m R_2} & 2 \\ 2\left(1 - \dfrac{g_m R_1}{1 + g_m R_2}\right) & \dfrac{R_1}{Z_0} - \dfrac{g_m Z_0}{1 + g_m R_2} \end{bmatrix} \tag{9.99}$$

where

$$\Delta = 2 + \dfrac{R_1}{Z_0} + \dfrac{g_m Z_0}{1 + g_m R_2}. \tag{9.100}$$

Assuming ideal matching conditions $S_{11} = S_{22} = 0$ (i.e., the input and output VSWRs equal unity) yields an equation relating the value of the shunt feedback resistor to the series feedback resistor R_1:

$$R_2 = \dfrac{Z_0^2}{R_1} - \dfrac{1}{g_m} \tag{9.101}$$

where the characteristic impedance Z_0 and transconductance g_m are used.

Substituting (9.101) into (9.100) and (9.99) gives

$$[S] = \begin{bmatrix} 0 & \dfrac{Z_0}{R_1 + Z_0} \\ 1 - \dfrac{R_1}{Z_0} & 0 \end{bmatrix}. \qquad (9.102)$$

As seen from (9.99) and (9.102), a perfect match can be achieved by choosing appropriate values for the feedback resistors R_1 and R_2. The only limitation arises from the requirement that R_2 in (9.101) must be nonnegative; that is, there exists a minimum value $g_{m_{\min}}$ that limits the range of g_m to

$$g_m \geq g_{m_{\min}} = \frac{R_1}{Z_0^2} = \frac{1 - S_{21}}{Z_0}. \qquad (9.103)$$

Any transistor with g_m satisfying condition (9.103) can be used in the negative feedback configuration shown in Figure 9-24.

The analysis of the feedback circuit is applicable only for ideal devices operated in the low-frequency range where all reactances are neglected. In practical applications, the presence of parasitic resistances in the transistor must be taken into account, resulting in modified values of the feedback resistors. In addition, at RF and MW frequencies the influence of internal capacitances and inductances cannot be neglected, and reactive components in the feedback loops enter the analysis. The most common practice is to add an inductance in series with the feedback resistor R_1. This is done to reduce the feedback from higher frequencies and thus compensate for S_{21}-related roll-off.

The following example demonstrates the use of negative feedback for a broadband amplifier design, where the feedback resistors are first computed theoretically and then adjusted using a CAD software package.

Example 9-17. Design of a negative feedback loop broadband amplifier

The BJT BFG403W is biased with $V_{CE} = 3$ V and $I_C = 3.3$ mA ($\beta = 125$). The corresponding S-parameters in the common-emitter configuration are listed in Table 9-6. Design a broadband amplifier with $G_T = 10$ dB over a frequency range from 10 MHz to 2 GHz by using a negative feedback loop.

Table 9-6 S-parameters for the transistor in Example 9-17

f, MHz	$|S_{11}|$	$\angle S_{11}$	$|S_{21}|$	$\angle S_{21}$	$|S_{12}|$	$\angle S_{12}$	$|S_{22}|$	$\angle S_{22}$
10	0.877	−0.3°	7.035	179.6°	1×10^{-4}	66.8°	0.805	−0.1°
100	0.876	−2.4°	7.027	176.1°	7×10^{-4}	85.9°	0.805	−1.4°
250	0.870	−5.9°	6.983	170.2°	0.002	84.3°	0.803	−3.4°
500	0.850	−11.5°	6.834	160.6°	0.003	80.5°	0.797	−6.6°
750	0.820	−16.9°	6.607	151.4°	0.004	76.0°	0.789	−9.8°
1000	0.783	−21.7°	6.327	142.8°	0.005	68.2°	0.777	−12.7°
1500	0.700	−29.6°	5.711	127.2°	0.007	74.1°	0.755	−18.1°
2000	0.619	−35.7°	5.119	113.8°	0.007	74.1°	0.735	−23.0°

Solution. As seen from Table 9-6, the minimum gain of 14.2 dB is attained at $f = 2$ GHz, which is well above the required transducer power gain of $G_T = 10$ dB.

Before continuing our approximate analysis, we have to ensure that condition (9.103) is satisfied. The value of r_π is found to be $r_\pi = \beta/g_m = 984$ Ω, where the transconductance g_m is computed as $g_m = I_C/V_T = 0.127$ S. Thus, the negative feedback analysis is applicable since condition (9.103) is satisfied even for $R_2 = 0$.

The next step involves the estimation of the resistances R_1 and R_2. Because the desired gain is $G = 10$ dB, the low-frequency S_{21} coefficient should be equal to −3.16. Here, the minus sign is due to the 180° phase shift of the common-emitter configuration. Substituting this value into (9.103) yields

$$R_1 = Z_0(1 - S_{21}) = 208 \ \Omega$$

Applying (9.101), we compute the value for the series feedback resistor R_2:

$$R_2 = \frac{Z_0^2}{R_1} - \frac{1}{g_m} = 4.1 \ \Omega.$$

The resulting insertion gain of the feedback network is listed in the second column of Table 9-7. It is observed that the negative feedback makes the gain response of the amplifier more uniform at the lower frequencies, unfortunately at too low a level. The discrepancy between the expected gain of 10 dB and the obtained value of $|S_{21}|^2 = 7.5$ dB is largely due to the fact that we neglected all parasitic resistances in the transistor. Such parasitics include the base resistance that is connected in series with r_π and thus reduces the effective transconductance g_m. Furthermore, the emitter resistance, which is in series with R_2, has to be subtracted from the obtained value of R_2.

Table 9-7 Insertion gain of the feedback amplifier

f, MHz	$|S_{21}|^2$, dB		
	$R_1 = 208\,\Omega$, $R_2 = 4.1\,\Omega$	$R_1 = 276\,\Omega$, $R_2 = 1.4\,\Omega$	$R_1 = 276\,\Omega$, $R_2 = 1.4\,\Omega$, $L_1 = 4.5\,\text{nH}$
10	7.50	10.01	10.01
100	7.50	10.01	10.01
250	7.50	10.00	10.01
500	7.50	9.97	10.00
750	7.50	9.93	10.00
1000	7.50	9.88	10.00
1500	7.51	9.75	9.99
2000	7.54	9.59	9.99

Optimization of the circuit for frequencies up to 500 MHz using CAD tools results in the following modified values of the feedback resistances: $R_1 = 276\,\Omega$ and $R_2 = 1.43\,\Omega$. The corresponding insertion gain is listed in the third column of Table 9-7.

As observed from Table 9-7, these new values for the feedback resistances bring the transistor gain closer to the 10 dB specification at lower frequencies, but the gain degrades as the frequency increases. This indicates that a $R_1 = 276\,\Omega$ feedback resistor is too small at those frequencies and has to be increased. This can be effectively accomplished by connecting an additional $L_1 = 4.5$ nH inductor in series with the resistor R_1 (the value of L_1 is predicted by a separate CAD optimization procedure).

The resulting gain is listed in the last column in Table 9-7. As seen from the values presented, the addition of an inductor improves the gain flatness to better than 0.1% over the entire bandwidth.

As the frequency increases, the negative feedback design approach becomes increasingly prone to parasitic influences. Above approximately 5 GHz, this lumped element method begins to lose effectiveness.

LOAD-PULL

In its simplest form, load-pull involves the design of matching circuits for power amplifiers. It starts from a complex conjugate match at the input and a tunable load. The load is varied over many discrete points covering a subset of the Smith Chart. At each point, the parameters of interest, like linearity, efficiency, gain, stability, VSWR, are evaluated. The designer can thus select the optimal load, given the various design tradeoffs.

9.7.2 High-Power Amplifiers

Thus far we have discussed the design of amplifiers based on linear, small-signal S-parameters. When dealing with high-power amplifiers, however, a small-signal approximation is usually not valid because the amplifier operates in a nonlinear region, and **large-signal** S-parameters or impedances have to be obtained to conduct the appropriate design. Small-signal S-parameters can still be used when designing a Class A amplifier. Here, the signal amplification is largely restricted to the linear region of the transistor. However, the small-signal S-parameters become progressively unsuitable for Class AB, B, or C amplifiers, which operate in the cutoff region part of the time.

One of the important characteristics of a high-power amplifier is the so-called **gain compression**. As the input signal to the amplifier reaches a sufficiently high level, the gain begins to fall off, or compress. The typical relationship between input and output power can be plotted on a log-log scale, as shown in Figure 9-26.

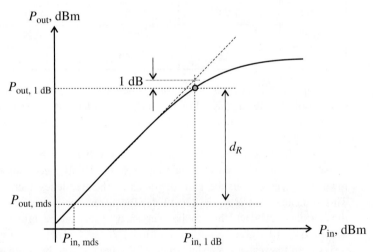

Figure 9-26 Output power of the amplifier as a function of input power.

At low drive levels, the output is proportional to the input power. However, as the power increases beyond a certain point, the gain of the transistor decreases, and eventually the output power saturates. The point where the gain of the amplifier deviates from the linear, or small-signal, gain by 1 dB is called the **1 dB compression point** and is used to characterize the power handling capabilities of the amplifier. The gain corresponding to the 1 dB compression point is referred to as $G_{1\text{dB}} = G_0 - 1$ dB, where G_0 is the small-signal gain.

If the output power $P_{out, 1dB}$ at the 1 dB compression point is expressed in dBm, it can be related to the corresponding input power $P_{in, 1dB}$ as

$$P_{out, 1dB}[\text{dBm}] = G_{1dB}[\text{dB}] + P_{in, 1dB}[\text{dBm}]$$
$$= G_0[\text{dB}] - 1 \text{ dB} + P_{in, 1dB}[\text{dBm}]. \quad (9.104)$$

Another important characteristic of an amplifier is its **dynamic range** labeled d_R. The dynamic range signifies the region where the amplifier has a linear power gain expressed as the difference between $P_{out, 1dB}$ and the output power of the **minimum detectable signal** $P_{out, mds}$. The quantity $P_{out, mds}$ is defined as a level X dB above the output noise power $P_{n, out}$. In most specifications, X is chosen to be 3 dB. The output noise power of an amplifier with noise figure F is given as

$$P_{n, out} = kTBG_0F \quad (9.105)$$

which, if expressed in dBm, can be cast in the form

$$P_{n, out}[\text{dBm}] = 10 \log(kT) + 10 \log B + G_0[\text{dB}] + F[\text{dB}] \quad (9.106)$$

where $10 \log(kT) = -173.8$ dBm/Hz at $T = 300°$K and B is the bandwidth, in Hertz.

Being nonlinear circuits, high-power amplifiers create **harmonic distortion** (the output contains multiples of the fundamental frequency). The harmonics appear as a power loss in the fundamental frequency. In general, Class A operation produces the lowest distortion. For higher-power applications, where Class A operation is not feasible, due to low efficiency, Class AB push-pull amplifiers are employed to achieve nearly comparable distortion levels. Harmonic distortion is specified as the harmonic content of the overall output expressed in dB below the output power at the fundamental frequency.

An undesirable property of power amplifiers is the occurrence of so-called **intermodulation distortion** (IMD). Although present in any amplifier (like harmonic distortion) it is most prominent in the high-power region of an active device, where the nonlinear behavior has to be taken into account. Unlike harmonic distortion, IMD is the result of applying two unmodulated harmonic signals of slightly different frequencies to the input of an amplifier and observing the output, as shown in Figure 9-27.

Due to third-order nonlinearities of the amplifier, the input signals $P_{in}(f_1)$ and $P_{in}(f_2)$ create, besides the expected output signals $P_{out}(f_1)$ and $P_{out}(f_2)$, additional tones $P_{out}(2f_1 - f_2)$ and $P_{out}(2f_2 - f_1)$, known as third-order intermodulation products. The additional frequency components can serve a desirable purpose when dealing with mixer circuits (see Chapter 10). However, for an amplifier one would like to see these contributions be as small as possible. The difference between the desired and the undesired power level (in dBm) at the output port is typically defined as IMD in dB; that is,

$$\text{IMD}[\text{dB}] = P_{out}(f_2)[\text{dBm}] - P_{out}(2f_2 - f_1)[\text{dBm}] \quad (9.107)$$

In Figure 9-28, the output powers $P_{out}(f_2)$ and $P_{out}(2f_2 - f_1)$ are plotted versus the input power $P_{in}(f_2)$ on a log-log scale. In the region of linear amplification, the output

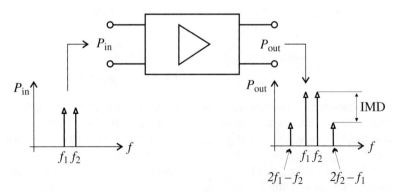

Figure 9-27 Observing the intermodulation distortion of an amplifier.

power $P_{\text{out}}(f_2)$ increases in proportion to the input power $P_{\text{in}}(f_2)$; however, the third-order product $P_{\text{out}}(2f_2 - f_1)$ increases in proportion to the third power, i.e., $P_{\text{out}}(2f_2 - f_1) \propto P_{\text{in}}^3(f_2)$. Thus, consistent with (9.107), the IMD increases in proportion to the square of the input power. Projecting the linear region of $P_{\text{out}}(f_2)$ and $P_{\text{out}}(2f_2 - f_1)$ results in a fictitious point called the third order **intercept point** (IP3).

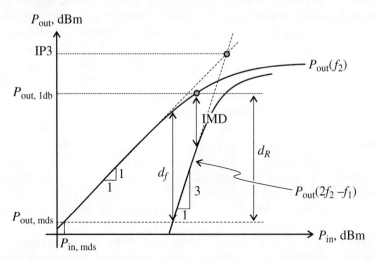

Figure 9-28 Recording of IMD based on input-output power relation.

Also shown in Figure 9-28 is a quantity called **spurious free dynamic range**, d_f, which is defined as

$$d_f[\text{dB}] = \frac{2}{3}(\text{IP3}[\text{dBm}] - G_0[\text{dB}] - P_{\text{in, mds}}[\text{dBm}]) \qquad (9.108)$$

Typical values for a MESFET are $P_{\text{in, mds}} = -100$ dBm, IP3 = 40 dBm, and $d_f = 85$ dB.

9.7.3 Multistage Amplifiers

A multistage amplifier circuit should be considered if the power gain requirement of the amplifier is so high that a single stage may not be able to achieve it. A typical example of a dual-stage BJT amplifier is shown in Figure 9-29.

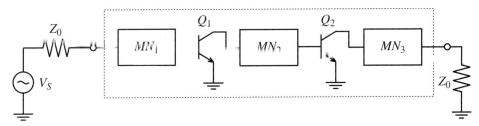

Figure 9-29 Dual-stage transistor amplifier.

Besides the typical input and output matching networks (MN_1 and MN_3), this configuration features an additional so-called **interstage matching network** (MN_2) for matching the output of stage 1 with the input of stage 2. In addition to providing appropriate matching, MN_2 can also be used to condition the gain flatness.

Under the assumption of optimally matched and lossless networks, let us summarize the most important dual-stage performance parameters. The total power gain G_{tot} of a dual-stage amplifier under linear operating conditions results in a multiplication of the individual gains G_1 and G_2, or in dB

$$G_{\text{tot}}[\text{dB}] = G_1[\text{dB}] + G_2[\text{dB}]. \tag{9.109}$$

An increase in gain performance is, unfortunately, accompanied by an increase in the noise figure, as discussed in Appendix H. Specifically, if F_1 and F_2 denote the noise figures associated with stages 1 and 2, we obtain a total noise figure

$$F_{\text{tot}} = F_1 + \frac{F_2 - 1}{G_1}. \tag{9.110}$$

In addition, if the minimum detectable signal $P_{\text{in, mds}}$ at 3 dB above thermal noise at the input is given by $P_{\text{in, mds}}[\text{dBm}] = k\text{TB}[\text{dBm}] + 3\text{ dB} + F_1[\text{dB}]$, the minimum detectable output power $P_{\text{out, mds}}$ becomes

$$P_{\text{out, mds}}[\text{dBm}] = k\text{TB}[\text{dBm}] + 3\text{dB} + F_{\text{tot}}[\text{dB}] + G_{\text{tot}}[\text{dB}]. \tag{9.111}$$

The dynamic properties are also affected. For instance, Rhode and Bucher (see Further Reading) have shown that the previously mentioned third-order intercept point changes to

$$\text{IP3}_{\text{tot}} = \frac{1}{1/\text{IP3}_2 + 1/(G_2 \text{IP3}_1)} \tag{9.112}$$

where $IP3_1$ and $IP3_2$ are the third-order intercept points associated with stages 1 and 2. Finally, the total spurious-free dynamic range d_{ftot} is approximately

$$d_{ftot}[\text{dBm}] = \frac{2}{3}(IP3_{tot}[\text{dBm}] - P_{out,\,mds}[\text{dBm}]). \qquad (9.113)$$

Equation (9.113) also reveals that the addition of a second stage reduces the total dynamic range.

RF&MW→

Example 9-18. Transistor choices for multistage amplifier design

Design an amplifier with $P_{out,\,1\,dB} = 18$ dBm and a power gain not less than 20 dB. Using the transistor choices listed in Table 9-8, which shows pertinent characteristics at the operating frequency of $f = 2$ GHz, determine the number of stages for the amplifier and discuss the choice of an appropriate transistor for each stage. In addition, estimate the noise figure F_{tot} and the third-order intercept point $IP3_{tot}$ of the amplifier.

Table 9-8 Transistor characteristics for Example 9-18

Transistor	F[dB]	G_{max}[dB]	$P_{out,\,1dB}$[dBm]	IP3[dBm]
BFG505	1.9	10	4	10
BFG520	1.9	9	17	26
BFG540	2	7	21	34

Solution. Since the output power should be 18 dBm, the only transistor choice for the output stage of the amplifier is BFG540.

Because the output power of the amplifier $P_{out,\,1dB} = 18$ dBm is much lower than $P_{out,\,1dB}$ of the BFG540, it can operate at maximum gain of $G = 7$ dB. This means that the remaining stages of the amplifier must be able to provide at least 20 dB − 7 dB = 13 dB of gain. Thus, our amplifier should have at least three stages.

For the last stage to have 18 dBm output power, the second-stage transistor should be able to produce a power level of $P_{out_2,\,1dB} = 18$ dBm − 7 dBm = 11 dBm, which eliminates BFG505 from the list of possible candidates. Since the BFG540 has a much higher power handling capability than necessary for the second stage, we choose BFG520.

Due to the fact that $P_{out,1\,dB} = 11$ dBm is much lower than the 1 dB compression power of the BFG520, the second-stage transistor will also operate well below the compression point and the maximum gain will be equal to $G_{max} = 9$ dB. Therefore, the transistor in the first stage has to have a minimum gain of $G = 13\text{ dB} - 9\text{ dB} = 4$ dB and be able to provide $P_{out_1} = 11\text{ dBm} - 9\text{ dB} = 2$ dBm. Thus, the BFG505 is more than adequate for the task with $P_{out_1} = 2$ dBm and $G_1 = 4$ dB. The input power to the amplifier is then $P_{in} = -2$ dBm.

As shown in Appendix H, the noise figure of the entire amplifier is computed as

$$F_{tot} = F_1 + \frac{F_2 - 1}{G_1} + \frac{F_3 - 1}{G_1 G_2}$$

and is minimized if the gain of the first stage is high. The BFG505 cannot provide a gain higher than 6 dB, because in this case (for a given P_{in}) it reaches the compression point. This difficulty is avoided if BFG520 is used as the first stage. We can design the first stage for maximum gain, and the second stage for necessary power to drive the output transistor. We can also adjust the gains of the individual stages so that none of the transistors reaches the compression point.

The block diagram of the resulting amplifier is shown in Figure 9-30, where the gain of each stage is chosen according to the preceding discussion. The noise figure of this amplifier is predicted as

$$F_{tot} = F_1 + \frac{F_2 - 1}{G_1} + \frac{F_3 - 1}{G_1 G_2} = 2.13 \text{ dB}.$$

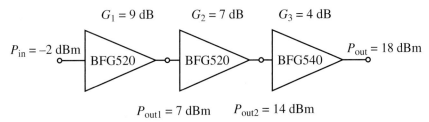

Figure 9-30 Block diagram of a three-stage amplifier.

The output power at the third-order intercept point is calculated using (9.112) and modified for a three-stage amplifier

$$IP3_{tot} = \frac{1}{1/IP3_3 + 1/(G_3 IP3_2) + 1/(G_3 G_2 IP3_1)} = 28 \text{ dBm}$$

where the preceding formula was obtained from (9.112) by first computing the IP3 of the first two stages and then resubstituting it into (9.112).

> **IP3 AND $P_{out, 1dB}$**
> It can be shown that IP3 is linked to $P_{out, 1dB}$ via IP3[dBm] $\approx P_{out,1dB}$[dBm] + 10dB. However, depending on the design, this offset can vary by as much as ± 3 dB in practice.

The above analysis is actually one of the first steps required in an amplifier design process. Here, the crucial steps of selecting suitable transistor types and deciding on the number of stages are made. They then become the starting point of a detailed performance analysis.

Practically Speaking. **Performance measurements of a medium-power amplifier module**

In this section, we perform measurements on a commercial single-chip RF power amplifier. Using a network analyzer, we collect S-parameter data over a large frequency range and determine the stability of this amplifier. Using a power sweep, we are also able to determine the 1 dB gain compression point.

The device, BGA6589 MMIC (Monolithic Microwave Integrated Circuit) medium power amplifier, is available from NXP. A key advantage of this MMIC is that it is extremely easy to use, requiring no sophisticated matching, biasing, or stabilization networks. This BGA6589 device is internally matched to 50 Ω over a wide frequency range from 200 MHz to 3.5 GHz, and is unconditionally stable over these frequencies. Measurements indicate that the amplifier is capable of up to 20 dBm output power, making it useful in such applications as RF and IF buffer amplifiers, oscillator amplifiers, high-linearity small-signal amplifiers, as well as power amplifiers for small wireless transmitters. Since the device has approximately a 3 dB noise figure, it is not well-suited for high-performance preamplifier applications.

For the test, the MMIC device was installed on a manufacturer-supplied demonstration board, shown in Figure 9-31 with its schematic. We observe that power is supplied to the output of the amplifier with an RF choke $L1$, bypass capacitor combination $C3$, $C4$, $C5$, and one series resistor $R1$. According to the datasheet, the resistor assures DC bias current stability with temperature. Aside from the bias supply, DC blocking capacitors are needed at the input and output. The values of $L1$, $C1$, $C2$ and $C4$ can be adjusted to fine-tune the matching in a narrower frequency range. The datasheet includes a table of optimal values for several center frequencies.

Broadband, High-Power, and Multistage Amplifiers

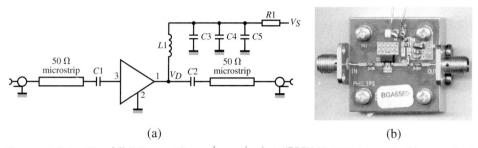

Figure 9-31 The MMIC amplifier schematic (courtesy of NXP) installed in a typical application circuit (a), and the corresponding layout (b).

Before performing measurements, the network analyzer was calibrated over its entire frequency range using the manufacturer's two-port calibration procedure. This calibration allows removing the effects of the interconnecting cables from the measurements, but it does not remove the effects of the microstrip transmission lines and other circuit components present on the amplifier PCB. As a result, the measured S-parameters are those of the entire amplifier block rather than the MMIC device itself. Small deviations are expected between the measurements and the device S-parameters listed in the datasheet.

The amplifier was connected to a 9 V DC power supply and drew 82 mA of current, consistent with the datasheet specification of 73 to 89 mA, with 81 mA regarded as typical. The network analyzer was configured for -30 dBm output power, coinciding with the power level at which the datasheet S-parameters are recorded. Figure 9-32 presents the S-parameter measurements over the frequency range from 300 kHz to 3 GHz. Examining the magnitude plots, we observe fairly good matching on the input (S_{11}) and output (S_{22}) ports over the frequency range from 300 MHz to 3 GHz. The VSWR on the output port reaches slightly above 2:1 at 2 GHz, indicating that the matching is not ideal. However, consistent with the manufacturer's claims, no additional matching is needed in many applications. The insertion power gain ($|S_{21}|^2$) rolls off smoothly, crossing 20 dB at around 1.2 GHz, and 15 dB at around 2.3 GHz. The reverse gain ($|S_{12}|^2$) stays below -22 dB, which is acceptable for most applications. Such reverse isolation is typical of single-stage amplifiers or amplifiers utilizing feedback.

Using the measured S-parameters, we can evaluate many useful amplifier characteristics, the most important being stability. Figure 9-33 presents two popular stability metrics: the Rollett stability factor (requiring $k > 1$ and $|\Delta| < 1$ for unconditional stability, see Section 9.3) and the μ-test ($\mu > 1$ for unconditional stability). As the manufacturer claims, the device is unconditionally stable over the full measurement

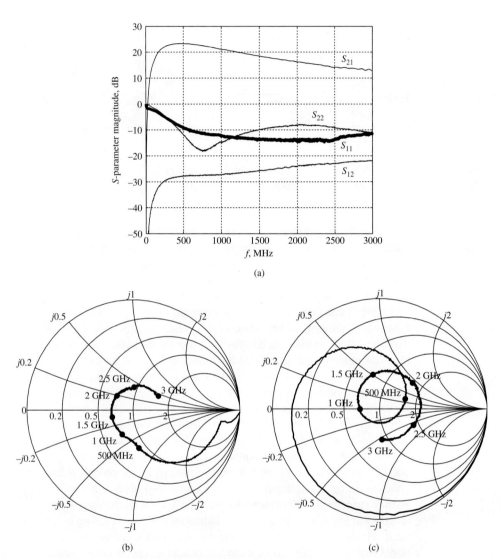

Figure 9-32 Measured MMIC amplifier S-parameters: (a) magnitude, (b) S_{11} values displayed in Smith Chart, and (c) S_{22} values displayed in Smith Chart.

frequency range. The Rollett factor k and the μ-test present two different viewpoints of stability. Although k is more often quoted (such as in the datasheet), μ is a better indicator of relative stability. According to μ, the stability is best near 500 MHz and drops off to bordering on conditional stability at lower frequencies. Such conclusions cannot be directly reached from the k and $|\Delta|$ plots.

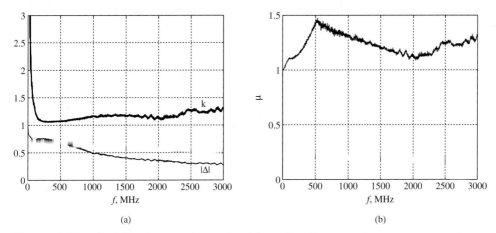

Figure 9-33 Stability factors determined from the S-parameter measurements: (a) Rollett stability factor, (b) μ-test.

The network analyzer allows limited evaluation of the nonlinear characteristics of the device under test by performing a power sweep at a fixed frequency. Figure 9-34 presents output power as a function of input power at 850 MHz and 1.95 GHz, the two frequencies with nonlinear characteristics quoted in the datasheet. We clearly observe gain compression and output power leveling off as the input power increases. The measured output power at 1 dB gain compression is 20.3 dBm at 850 MHz and 19.9 dBm at 1.95 GHz. More sophisticated

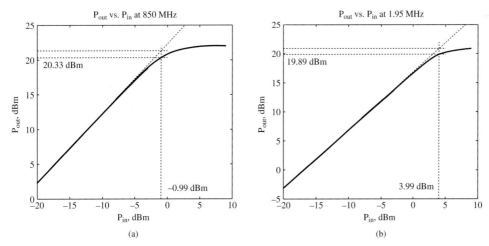

Figure 9-34 Gain compression observed in power sweeps at (a) 850 MHz and (b) 1.95 GHz.

analysis, such as determining the third-order intercept point, would require a spectrum analyzer. However, the simple relationship $IP3_{out}[dB] - P_{out,1dB}[dB] \approx 10$ dB, can be used to obtain an estimate (and other more sophisticated methods are available in the literature).

Finally, Table 9-9 compares measured and datasheet-derived amplifier performance parameters. Since the datasheet applies to the device only, and the measurements are performed on a complete amplifier with an included biasing network, there are some discrepancies in the return loss results. Nonetheless, the measured gain and stability factor k closely follow datasheet predictions. The output power at 1 dB gain compression is slightly lower than the datasheet value at 850 MHz, possibly a side effect of the biasing network being tuned for a higher operating frequency (it uses a relatively small inductor).

Table 9-9 Comparison between the datasheet and network analyzer measurements

Parameter	850 MHz		1.95 GHz	
	Datasheet	Measured	Datasheet	Measured
Insertion power gain ($|S_{21}|^2$)	22 dB	22.1 dB	17 dB	16.6 dB
Input return loss	9 dB	11.7 dB	11 dB	13.8 dB
Output return loss	10 dB	17.2 dB	13 dB	8.4 dB
Stability factor k	1.1	1.13	1.1	1.15
Output power at 1 dB gain compression	21 dBm	20.3 dBm	20 dBm	19.9 dBm

9.8 Summary

This chapter deals with a multitude of amplifier design concepts. First, the various power relations are defined. Specifically, the transducer power gain

$$G_T = \frac{(1-|\Gamma_L|^2)|S_{21}|^2(1-|\Gamma_S|^2)}{|1-\Gamma_S\Gamma_{in}|^2|1-S_{22}\Gamma_L|^2}$$

as well as the available and operating power gains are of key importance. We next establish the various input and output stability circle equations and examine the meaning of unconditional stability. Specifically, the factors

$$k = \frac{1-|S_{11}|^2-|S_{22}|^2+|\Delta|^2}{2|S_{12}||S_{21}|} > 1 \text{ and } |\Delta| < 1$$

Summary

are employed to assess the unconditional stability of an active device. If the transistor turns out to be unstable, additional series or shunt resistances can be used to stabilize the device. Next, the constant unilateral gain circles are established and displayed in the Smith Chart. The location and radius equations

$$d_{g_i} = \frac{g_i S_{ii}^*}{1 - |S_{ii}|^2(1-g_i)} \quad \text{and} \quad r_{g_i} = \frac{\sqrt{1-g_i}(1-|S_{ii}|^2)}{1-|S_{ii}|^2(1-g_i)}$$

provide insight in to where certain constant gain values are located under unilateral design conditions (inverse power gain is assumed negligible). The error committed by using the unilateral design approach over the bilateral method is quantified through the unilateral figure of merit. If the unilateral approach turns out to be too imprecise, a bilateral design has to be pursued, leading to the simultaneous conjugate matched reflection coefficients (Γ_{MS}, Γ_{ML}), at the input and output ports. The optimal matching

$$\Gamma_{MS}^* = S_{11} + \frac{S_{12}S_{21}\Gamma_{ML}}{1-S_{22}\Gamma_{ML}} \quad \text{and} \quad \Gamma_{ML}^* = S_{22} + \frac{S_{12}S_{21}\Gamma_{MS}}{1-S_{11}\Gamma_{MS}}$$

results in amplifier designs with maximum gain. Based on the operating power gain expression, circles of constant gain under optimal source matching are derived. Alternatively, starting with the available power gain expression, circles of constant gain under optimal load matching are derived.

We then investigate the influence of noise generated by an amplifier. Using the noise figure of a generic two-port network

$$F = F_{min} + \frac{R_n}{G_S}|Y_S - Y_{opt}|^2$$

circle equations for the Smith Chart are computed. The noise figure circles can be used by the circuit designer to make trade-offs with the previously conducted constant gain analysis.

An investigation into reducing the VSWR as part of various input and output matching network strategies results in an additional set of circle equations that quantify the VSWR at the matching network ports:

$$VSWR_{IMN} = \frac{1+|\Gamma_{IMN}|}{1-|\Gamma_{IMN}|} \quad \text{and} \quad VSWR_{OMN} = \frac{1+|\Gamma_{OMN}|}{1-|\Gamma_{OMN}|}$$

Combining the various circle representations permits small-signal amplifier design based on constant operating gain, noise figure, and VSWR circles, jointly displayed in the Smith Chart.

For broadband design, we discuss the need to develop frequency-compensated matching networks. The use of negative feedback loops is introduced as a method of flattening the power gain over a wide frequency range.

In high-power amplifier applications, issues related to the output power compression are of major concern, since they limit the dynamic range of amplification. An important figure of merit is the 1-dB compression point:

$$P_{\text{out, 1dB}}[\text{dBm}] = G_0[\text{dB}] - 1 \text{ dB} + P_{\text{in, 1dB}}[\text{dBm}].$$

Furthermore, an additional undesirable property is the occurrence of intermodulation distortion due to the presence of nonlinearities. Finally, the influences of power compression, noise figure, and gain are investigated in the context of a multistage amplifier design.

Further Reading

P. L.D. Abrie, *Design of RF and Microwave Amplifiers and Oscillators*, Artech House, 1999, ISBN 0-89006-797-X.

S. A. Maas, *The RF and Microwave Circuit Design Cookbook*, Artech House, 1998, ISBN 0-89006-973-5.

S. Akamatsu, C. Baylis, and L. Dunleavy, "Accurate simulation models yield high-efficiency power amplifier design," *IEEE Microwave Magazine*, pp. 114–124, 2005.

I. Bahil and P. Bhartia, *Microwave Solid State Circuit Design*, John Wiley, New York, 1988.

G. Gonzalez, *Microwave Transistor Amplifiers, Analysis and Design*, Prentice Hall, Upper Saddle River, NJ, 1997.

K. C. Gupta, R. Garg, and R. Chada, *Computer-Aided Design of Microwave Circuits*, Artech, Dedham, MA, 1981.

Hewlett-Packard, RF Design and Measurement Seminar, Seminar Notes, Burlington, MA, 1999.

Hewlett-Packard, *S*-Parameter Techniques for Faster and more Accurate Network Design, Application Notes 95-1, 1968.

H. Krauss, C. Bostian, and F. Raab, *Solid Radio Engineering*, John Wiley, New York, 1980.

M. L. Edwards and J. H. Sinsky, "A new criterion for linear 2-port stability using a single geometrically derived parameter," *IEEE Transactions on Microwave Theory and Techniques*, vol. 40, pp. 2303–2311, 1992.

S. Y. Liao, *Microwave Circuit Analysis and Amplifier Design*, Prentice Hall, Englewood Cliffs, NJ, 1987.

S. J. Mason, "Power Gain in Feedback Amplifiers, *IRE Trans.*, Vol. 1, pp. 20–25, 1954.

D. Pozar, *Microwave Engineering*, John Wiley, New York, 1998.

B. Razavi, *RF Microelectronics*, Prentice Hall, Upper Saddle River, NJ, 1998.

U. L. Rohde and T. T. N. Bucher, *Communication Receivers, Principle and Design,* McGraw-Hill, New York, 1988.

J. M. Rollett, "Stability and Power-Gain Invariants of Linear Two-Ports," *IRE Trans.*, Vol. 9, pp. 29–32, 1962.

G. D. Vendelin, *Design of Amplifiers and Oscillators by the S-Parameter Method,* John Wiley, New York, 1982.

Problems

9.1 The available power of an RF source driving an amplifier connected to a load $Z_L = 80\ \Omega$ can be represented as

$$P_A = \frac{1}{2}\frac{|b_S|^2}{1-|\Gamma_S|^2}$$

Based on the signal flow graph shown on the right side of Figure 9-2(b), substituting Γ_L for Γ_{in},
(a) Find the power at the load P_L in terms of Γ_L, Γ_S, and b_S.
(b) For $Z_S = 40\ \Omega$, $Z_0 = 50\ \Omega$, $V_S = 5V\angle 0°$, find the available power P_A and the power at the load P_L.

9.2 Based on the signal flow graph in Figure 9-2(b), establish the validity of equation (9.8) in Section 9.2.2.

9.3 An amplifier is characterized by the following S-parameters: $S_{11} = 0.78\angle -65°$, $S_{21} = 2.2\angle 78°$, $S_{12} = 0.11\angle -21°$, $S_{22} = 0.9\angle -29°$. The input side of the amplifier is connected to a voltage source with $V_S = 4V\angle 0°$, and impedance $Z_S = 65\ \Omega$. The output is utilized to drive an antenna that has an impedance of $Z_L = 85\ \Omega$. Assuming that the S-parameters of the amplifier are measured with reference to a $Z_0 = 75\ \Omega$ characteristic impedance, find the following quantities:
(a) transducer gain G_T, unilateral transducer gain G_{TU}, available gain G_A, operating power gain G
(b) power delivered to the load P_L, available power P_A, and incident power to the amplifier P_{inc}

9.4 A FET is operated at $f = 5.5$ GHz and under bias conditions $V_{DS} = 3.2$ V and $I_D = 24$ mA. The S-parameters are $S_{11} = 0.73\angle 176°$, $S_{21} = 3.32\angle 75°$, $S_{12} = 0.05\angle 34°$, $S_{22} = 0.26\angle -107°$. In the absence of matching networks, a load of $Z_L = 75\ \Omega$ and a source of $Z_S = 30\ \Omega$ are attached. Assume $Z_0 = 50\ \Omega$.
(a) Find G_{TU}, G_T, G_A, and plot the magnitude of G_{TU} for $10\ \Omega \le Z_L \le 100\ \Omega$.
(b) Match the input side for the unilateral case and find G_{TU}.
(c) Match both input and output for the unilateral case and compute $G_{TU} = G_{TU\max}$.

9.5 Unconditional stability in the complex Γ_{out}-plane requires that the $|\Gamma_S| < 1$ domain resides completely within the $|\Gamma_{out}| = 1$ circle, or $||C_s| - r_s| < 1$, where

$$C_S = S_{22} + \frac{S_{12}S_{21}S_{11}^*}{1 - |S_{11}|^2} \text{ and } r_S = \frac{|S_{12}S_{21}|}{1 - |S_{11}|^2}.$$

(a) Derive these two equations.
(b) Find the circle equations for C_L and r_L and show that $|S_{12}S_{21}| < 1 - |S_{22}|^2$ is a necessary condition for stability.

9.6 Prove that $|S_{11} - S_{22}^*\Delta|^2 = |S_{12}S_{21}|^2 + (1 - |S_{11}|^2)(|S_{22}|^2 - |\Delta|^2)$. This is a key identity in the stability factor derivation of Example 9.2.

9.7 A BJT has the following S-parameters (see the table below) given at four frequencies. Determine the stability regions and sketch them in the Smith Chart.

Frequency	S_{11}	S_{12}	S_{21}	S_{22}
500 MHz	0.70∠–57°	0.04∠47°	10.5∠136°	0.79∠–33°
750 MHz	0.56∠–78°	0.05∠33°	8.6∠122°	0.66∠–42°
1000 MHz	0.46∠–97°	0.06∠22°	7.1∠112°	0.57∠–48°
1250 MHz	0.38∠–115°	0.06∠14°	6.0∠104°	0.50∠–52°

9.8 The S-parameters for a BJT at a particular bias point and operating frequency are as follows: $S_{11} = 0.60\angle 157°$, $S_{21} = 2.18\angle 61°$, $S_{12} = 0.09\angle 77°$, $S_{22} = 0.47\angle -29°$. Check the transistor stability, stabilize it if necessary, and design an amplifier for maximum gain.

9.9 In this chapter, we have derived the circle equations for constant operating power gain. It can be concluded that the maximum gain is obtained when the radius of the constant gain circle is equal to zero. Using this condition, prove that the maximum achievable power gain in the unconditionally stable case is

$$G_{T\max} = \frac{|S_{21}|}{|S_{12}|}(k - \sqrt{k^2 - 1})$$

where k is the stability factor ($k > 1$).

9.10 A BJT is operated at $f = 750$ MHz with the S-parameters given as follows: $S_{11} = 0.56\angle -78°$, $S_{21} = 0.05\angle 33°$, $S_{12} = 8.64\angle 122°$, and $S_{22} = 0.66\angle -42°$. Attempt to stabilize the transistor by finding a series resistor or shunt conductance for the input and output ports.

9.11 In Example 9-2, the stability factor k is derived based on the input stability circle equation. Start with the output stability circle equation and show that the same result (9.24) is obtained.

9.12 A BJT is operated at $f = 7.5$ GHz, and is biased such that the S-parameter is given as $S_{11} = 0.85\angle 105°$. It is assumed that the transistor is unconditionally stable, and the unilateral approximation can be applied. Find the maximum source gain and plot the constant source gain circles for several appropriately chosen values of g_S.

9.13 A MESFET is used as a single-stage amplifier at 2.25 GHz. The S-parameters at that frequency and under given bias conditions are reported as $S_{11} = 0.83\angle -132°$, $S_{12} = 0.03\angle 22°$, $S_{21} = 4.9\angle 71°$, $S_{22} = 0.36\angle -82°$. For a required 18 dB gain, use the unilateral assumption by setting $S_{12} = 0$, and
(a) Determine if the circuit is unconditionally stable.
(b) Find the maximum power gain under the optimal choice of the reflection coefficients.
(c) Adjust the load reflection coefficient such that the desired gain is realized using the concept of constant gain circles.

9.14 A BJT is used in an amplifier at 7.5 GHz. The S-parameters at that frequency and under given bias conditions are reported as $S_{11} = 0.63\angle -140°$, $S_{12} = 0.08\angle 35°$, $S_{21} = 5.7\angle 98°$, $S_{22} = 0.47\angle -57°$. The design requires a 19 dB gain. Use the unilateral assumption and
(a) Find the maximum power gain under the optimal choice of the reflection coefficients.
(b) Adjust the load reflection coefficient such that the desired gain under stable operating conditions is realized.

9.15 A small-signal BJT amplifier operated at 4 GHz is appropriately biased and has the following S-parameters: $S_{11} = 0.57\angle -150°$, $S_{12} = 0.12\angle 45°$, $S_{21} = 2.0\angle 56°$, $S_{22} = 0.35\angle -85°$. If a unilateral design approach is pursued, estimate the transducer gain error involved.

9.16 A BJT with $I_C = 10$ mA and $V_{CE} = 6$ V is operated at a frequency of $f = 2.4$ GHz. The corresponding S-parameters are $S_{11} = 0.54\angle -70°$, $S_{12} = 0.017\angle 176°$, $S_{21} = 1.53\angle 91°$, and $S_{22} = 0.93\angle -15°$. Determine whether the transistor is unconditionally stable and find the values for source and load reflection coefficients that provide maximum gain.

9.17 Using the same BJT discussed in the Problem 9.16, design an amplifier whose transducer power gain is 60% of $G_{T\max}$. In addition, ensure a perfect match on the input port of the amplifier.

9.18 A MESFET operated at 9 GHz under appropriate bias conditions has the following S-parameters: $S_{11} = 1.2\angle -60°$, $S_{12} = 0.02\angle 0°$, $S_{21} = 6.5\angle 115°$, and $S_{22} = 0.6\angle -35°$. Design an amplifier that stays within 80% of $G_{TU\max}$. Moreover, ensure that $\text{VSWR}_{\text{out}} = 1$.

9.19 In Section 9.4.4, it is mentioned that the constant-gain design for a matched input results in the circle equation

$$\left| \frac{S_{11} - \Gamma_S^*}{\Delta - S_{22}\Gamma_S^*} - d_{g_0} \right|^2 = r_{g_0}^2.$$

Show that the center d_{g_s} and radius r_{g_s} are given by

$$r_{g_s} = \frac{r_{g_0}|S_{12}S_{21}|}{\left||1-S_{22}d_{g_0}|^2 - r_{g_0}^2|S_{22}|^2\right|}$$

and

$$d_{g_s} = \frac{(1-S_{22}d_{g_0})(S_{11}-\Delta d_{g_0})^* - r_{g_0}^2\Delta^*S_{22}}{|1-S_{22}d_{g_0}|^2 - r_{g_0}^2|S_{22}|^2}.$$

9.20 For the constant available gain circle $|\Gamma_S - d_{g_a}| = r_{g_a}$ (see (9.66)), show that

$$d_{g_a} = \frac{g_a(S_{11}-\Delta S_{22}^*)^*}{1+g_a(|S_{11}|^2-|\Delta|^2)} \text{ and } r_{g_a} = \frac{\sqrt{1-2kg_a|S_{12}S_{21}|+g_a^2|S_{12}S_{21}|^2}}{|1+g_a(|S_{11}|^2-|\Delta|^2)|}$$

9.21 A BFG197X transistor is biased at $V_{CE} = 8$ V and $I_C = 10$ mA and has the following S-parameters measured at $f = 1$ GHz: $S_{11} = 0.73\angle176°$, $S_{12} = 0.07\angle35°$, $S_{21} = 3.32\angle75°$, and $S_{22} = 0.26\angle107°$. Determine the unilateral figure of merit and compare the transducer gain of the amplifier designed for maximum gain under the unilateral and bilateral assumptions.

9.22 The BFG33 BJT is biased under $V_{CE} = 5$ V and $I_C = 5$ mA and has the following noise and S-parameters:

	S_{11}	S_{12}	S_{21}	S_{22}	F_{min}, dB	Γ_{opt}	R_n, Ω
500 MHz	0.72∠−39°	0.05∠63°	6.22∠135°	0.78∠−32°	2.3	0.64∠5°	58.5
1000 MHz	0.45∠−70°	0.08∠56°	5.13∠109°	0.61∠−43°	2.5	0.56∠13°	67.5
2000 MHz	0.18∠−115°	0.12∠54°	3.24∠82°	0.49∠−54°	3.0	0.52∠39°	49.7

Design a broadband, low-noise amplifier with minimum gain of 10 dB and a noise figure not exceeding 3.5 dB.

9.23 Design a microwave amplifier using a GaAs FET whose S-parameters at $f = 10$ GHz are $S_{11} = 0.79\angle100°$, $S_{12} = 0.20\angle-21°$, $S_{21} = 6.5\angle-73°$, $S_{22} = 0.74\angle152°$. For a constant VSWR$_{in}$ = 6.5, find the transducer gain that results in the lowest VSWR$_{out}$.

9.24 A broadband amplifier under unilateral assumption with nominal characteristics of VSWR$_{in}$ = 4, VSWR$_{out}$ = 2.8, and G_T = 10 dB is used as part of a balanced amplifier design. Compute the worst input and output VSWR and the insertion gain of the balanced amplifier, if the S-parameters can vary by as much as 10%.

9.25 In Section 9.7.3, we have listed equation (9.112) for the IP3 definition of a two-stage amplifier.
(a) Derive a generalized formula for the IP3 computation of an N-stage amplifier.

(b) Compute the total IP3 and the noise figure of the N-stage amplifier assuming that all stages are identical and have IP3 = 35 dBm, $F = 2$ dB, and $G = 8$ dB.

9.26 Design a 15 dB broadband amplifier using a BJT with a feedback loop. Calculate the values of the feedback resistors and find the minimum collector current of the transistor. Assume that the amplifier is operated at $T = 300$ K, and $Z_0 = 50\ \Omega$.

9.27 A transistor has the following S-parameters: $S_{11} = 0.61\angle 152°$, $S_{12} = 0.1\angle 79°$, $S_{21} = 1.89\angle 55°$, and $S_{22} = 0.47\angle -30°$. Design an amplifier for minimum noise figure if $F_{min} = 3$ dB, $\Gamma_{opt} = 0.52\angle 170°$, and $R_n = 9\ \Omega$.

9.28 Prove equation (9.113), which states the total spurious-free dynamic range.

9.29 An amplifier has a transducer gain of $G_T = 25$ dB, and a 200 MHz bandwidth. The noise figure is given as $F = 2.5$ dB and the 1 dB gain compression point is measured as $P_{out,\ 1dB} = 20$ dBm. Calculate the dynamic range and the spurious-free dynamic range of the amplifier if IP3 = 32 dBm. Assume that the amplifier is operated at room temperature.

9.30 An amplifier has a gain of $G = 8$ dB at 1 GHz and lists a 1 dB compression point of $P_{out,\ 1\ dB} = 12$ dBm and the third-order intercept point at IP3$_{tot}$ = 25 dBm. Find the third-order intercept points for the cascaded amplifiers consisting of two and three of these stages. What value of IP3$_{tot}$ is obtained in the limit of an infinite number of stages?

9.31 Derive a formula for the noise figure of a balanced amplifier. Make the assumption that the power gains and noise figures of the amplifiers in the individual branches are G_A, G_B, and F_A, F_B, respectively. Assume that the balanced amplifier uses 3 dB hybrid couplers at the input and output ports.

10

Oscillators and Mixers

10.1	Basic Oscillator Models	560
	10.1.1 Feedback Oscillator	560
	10.1.2 Negative Resistance Oscillator	562
	10.1.3 Oscillator Phase Noise	574
	10.1.4 Feedback Oscillator Design	578
	10.1.5 Design Steps	581
	10.1.6 Quartz Oscillators	585
10.2	High-Frequency Oscillator Configuration	587
	10.2.1 Fixed-Frequency Oscillators	591
	10.2.2 Dielectric Resonator Oscillators	598
	10.2.3 YIG-Tuned Oscillator	603
	10.2.4 Voltage-Controlled Oscillator	604
	10.2.5 Gunn Element Oscillator	608
10.3	Basic Characteristics of Mixers	609
	10.3.1 Basic Concepts	610
	10.3.2 Frequency Domain Considerations	612
	10.3.3 Single-Ended Mixer Design	614
	10.3.4 Single-Balanced Mixer	622
	10.3.5 Double-Balanced Mixer	623
	10.3.6 Integrated Active Mixers	624
	10.3.7 Image Reject Mixer	628
10.4	Summary	641

With the advent of modern radio and radar systems came the need to provide stable harmonic oscillations at particular carrier frequencies to establish the required modulation and mixing conditions. While the carrier frequencies in the early days mostly reached into the low to mid MHz range, today's RF systems easily surpass the 1 GHz point. This has resulted in the need for specialized oscillator circuits capable of providing stable and pure sinusoidal signals. What makes the design of oscillators such a difficult task is that we exploit an inherently nonlinear circuit behavior that can only be described incompletely with linear system tools. Specifically, the small-signal linear circuit models utilized to represent the active device afford limited capabilities to handle the complicated feedback mechanism. Moreover, since an oscillator has to provide power to subsequent circuits, frequency-dependent output loading often plays an important role. For these reasons, the design process for oscillators remains more of an art than an exact engineering design task. This holds particularly true for the high-frequency regime, where parasitic component influences can significantly impact the overall system performance. Affected in part by the additional resonance effects of the passive circuit elements, it is possible that the oscillator operates not only at the intended frequency, but also at lower or higher harmonics. Certain system realizations may even cease to oscillate completely.

In the first part of this chapter, we concentrate on the negative resistance and feedback harmonic oscillators. Once the fundamental idea of how to generate oscillations is mastered, we investigate the basic Colpitts and Hartley oscillators before moving to the modern RF circuit design approaches involving the S-parameters of the active device in conjunction with the various network configurations.

In the second part of this chapter, we turn our attention to the basic frequency translation tasks performed by mixers. Of the many different mixer implementations for a wide range of applications, the main emphasis in this chapter is placed on downconverters. A typical application of a mixer in a receiver system is to convert the RF input signal into a lower intermediate frequency signal that is generally more suitable for subsequent signal conditioning and processing. This conversion is accomplished by combining the RF input with a local oscillator signal as part of a multiplication operation that requires a nonlinear, at least quadratic, transfer function. Nowadays, transistors and diodes are in use, with present FET technology permitting the construction of mixer circuits up to 100 GHz, and diode mixers well exceeding the 100 GHz mark.

10.1 Basic Oscillator Models

10.1.1 Feedback Oscillator

At the core of any oscillator circuit is a loop that causes positive feedback at a selected frequency. Figure 10-1(a) illustrates the generic closed-loop system representation, while Figure 10-1(b) provides a two-port network description.

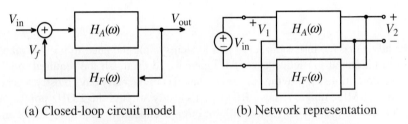

(a) Closed-loop circuit model   (b) Network representation

Figure 10-1 Basic oscillator configuration.

Basic Oscillator Models

The mathematical condition for a circuit to oscillate can be established by combining the transfer functions of the amplification stage $H_A(\omega)$ with the feedback stage $H_F(\omega)$ to obtain the closed-loop transfer function:

$$\frac{V_{out}}{V_{in}} = H_{CL}(\omega) = \frac{H_A(\omega)}{1 - H_F(\omega)H_A(\omega)}. \tag{10.1}$$

Since there is no input to an oscillator, $V_{in} = 0$, to yield a nonzero output voltage V_{out}, the denominator in (10.1) has to be zero. This requirement leads to the **Barkhausen criterion**, which is also known as the loop gain equation:

$$H_F(\omega)H_A(\omega) = 1. \tag{10.2}$$

If the feedback transfer function $H_F(\omega)$ is written as a complex quantity (that is, $H_F(\omega) = H_{Fr}(\omega) + jH_{Fi}(\omega)$) and the amplifier transfer function possesses a real valued gain $H_A(\omega) = H_{A0}$, we can reexpress (10.2) as

$$H_{A0} = \frac{1}{H_{Fr}(\omega)} \tag{10.3a}$$

$$H_{Fi}(\omega) = 0. \tag{10.3b}$$

The conditions (10.2), (10.3a), and (10.3b) apply only for a steady-state situation. Initially, we have to require that $H_{A0}H_{Fr}(\omega) > 1$. In other words, the loop gain has to be larger than unity to obtain an increasing output voltage. However, the voltage must reach a steady state (i.e., the amplitude eventually must stabilize). This nonlinear behavior of the oscillator is shown in Figure 10-2.

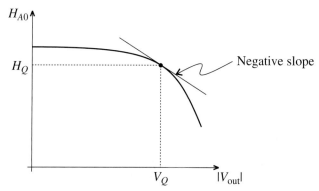

Figure 10-2 Output voltage versus gain characteristic.

A negative slope of the curve is needed to ensure a decrease in gain for increasing voltage. At the point $|V_{out}| = V_Q$ for $H_{A0} = H_Q = H_{Fr}(\omega)$, a stable operating point is reached. A similar curve can be established for the frequency versus loop gain with a stable resonance frequency f_Q.

10.1.2 Negative Resistance Oscillator

An oscillator can also be understood as a resonator coupled to an active circuit that creates "negative resistance". This negative resistance exactly cancels the resonator's internal resistance, allowing oscillations to continue ad infinitum at the resonance frequency. Figure 10-3 illustrates a generic oscillator circuit, where a passive resonator having an impedance $Z = R + jX$ is connected to an active circuit having an input impedance $Z_{in} = R_{in} + jX_{in}$. We would like a finite current I to flow at the oscillation frequency. According to Kirchhoff's voltage law,

$$(Z + Z_{in})I = 0. \tag{10.4}$$

Therefore, to achieve nonzero current, we require $Z + Z_{in} = 0$. Separating this equation into real and imaginary parts, we obtain the oscillation criterion:

$$R_{in} = -R \tag{10.5a}$$

$$X_{in} = -X. \tag{10.5b}$$

Since the external resonator is passive, $R > 0$, implying $R_{in} < 0$. This negative (AC) resistance is only necessary around the oscillation frequency. The second condition (10.5b) determines the oscillation frequency.

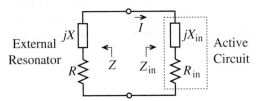

Figure 10-3 Generic negative resistance oscillator circuit.

How can we create and maintain the required negative input resistance? Using two popular negative resistance circuits in Figure 10-4, we will demonstrate that it is possible to predict both the small-signal and the asymptotic large-signal negative resistance of a transistor circuit. With the help of simple approximations and circuit simulations, we will observe that the negative resistance predictably increases or decreases as a function of the signal amplitude. If we build an oscillator from one of these circuits, the oscillation amplitude would stabilize at the point where the negative resistance magnitude equals the resistance of the resonator. Consequently, we can predict and control the magnitude of the oscillations.

We need to keep in mind that the negative input resistance can decrease or increase with increasing oscillation amplitude. A common rule of thumb is that under small-signal conditions, the negative resistance magnitude should be about 3 times the external resonator series resistance. This rule only holds for active circuits whose negative resistance magnitude

Basic Oscillator Models

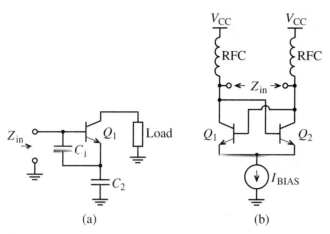

Figure 10-4 Common negative resistance circuits for oscillators: (a) emitter/source feedback (no biasing shown) and (b) cross-coupled.

decreases with rising oscillation amplitude. Such a circuit is best used with a series resonator, as shown in Figure 10-5(a). As we will discover, there are negative resistance circuits with the exact opposite behavior that will fail to oscillate if this rule of thumb is followed. The circuits whose negative resistance magnitude increases with rising oscillation amplitude can be classified as "negative conductance" circuits. This type of circuit is best coupled to a parallel resonator, as in Figure 10-5(b), and the small-signal negative conductance magnitude should be 3 times the resonator shunt conductance.

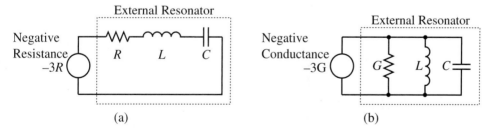

Figure 10-5 Negative resistance oscillator (a) and negative conductance oscillator (b).

Emitter/Source Feedback Circuit

Figure 10-6(a) presents a version of the circuit in Figure 10-4(a) with an idealized bias arrangement. The boxed capacitors and inductors denote, respectively, idealized DC blocking and feed (RFC) conditions. Sources SRC2-SRC4 provide the base voltage, emitter current, and collector voltage, respectively. In order to analyze this circuit under small-signal conditions, we will use the simplest model for the transistor: an ideal transconductance. The small-signal model for this circuit is shown in Figure 10-6(b). Analyzing this circuit for the input impedance yields

$$Z_{in} = -\frac{g_m}{\omega^2 C_1 C_2} - j\frac{1}{\omega}\left(\frac{1}{C_1} + \frac{1}{C_2}\right) \tag{10.6}$$

where $g_m = I_E/V_T$ is the transistor transconductance, I_E is the DC emitter bias current and $V_T \approx 26$ mV is the thermal voltage at room temperature. The negative real part of Z_{in} corresponds to the negative resistance. This resistance is in series with a reactance that is the series combination of the two capacitors. To create an oscillator, we need to externally connect a (lossy) inductive impedance at the input terminals.

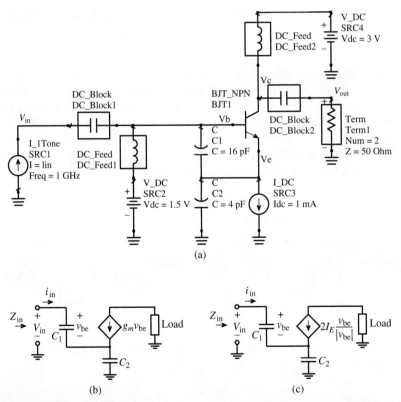

Figure 10-6 Negative resistance circuit using emitter feedback: (a) ADS schematic including biasing, (b) small-signal model, and (c) large-signal model at the first harmonic frequency. For the ADS model legend, see next page.

Basic Oscillator Models

As the oscillations increase in amplitude, the transistor leaves the linear Class A regime, and spends more and more time in cutoff. The conduction angle decreases until the transistor is operating deep in the Class C regime. In Class C, the transistor conducts in short bursts when v_{BE} is near its positive peaks. As the duration of the current bursts decreases, the emitter current waveform approaches an impulse train with a DC value still equal to the DC emitter bias current. We will represent the emitter current as a Fourier series and analyze the circuit only at the frequency of interest, which corresponds to the first harmonic. The phasor representation of the first harmonic of the emitter current $i_{e(1)}$ can be determined as:

$$i_{e(1)} = \frac{2}{T}\int_{-T/2}^{T/2} i_E e^{-j\omega t} dt \approx \frac{2}{T}\int_{-T/2}^{T/2} I_E \delta\left(t + \frac{\theta}{\omega}\right) e^{-j\omega t} dt$$

$$= 2I_E e^{j\theta} = 2I_E \angle \theta \qquad (10.7)$$

where θ is the phase shift of the i_E pulses. We use the Dirac δ-function to approximate the single emitter current pulse occurring during the waveform cycle. As we can see, in the deep Class C regime, the first harmonic of the emitter current becomes independent of the input signal magnitude, and is simply twice the DC bias current in amplitude. The input signal only controls the phase θ that is determined later.

Figure 10-6(c) shows the large-signal (deep class C) model of the negative resistance circuit at the first harmonic frequency ω. To determine the input impedance at this frequency, we inject a sinusoidal test current i_{in} into the input terminal. The test current flows through the capacitor C_1, giving rise to the voltage v_{be}:

$$v_{be} = -j\frac{1}{\omega C_1} i_{in}. \qquad (10.8)$$

> **MODEL LEGEND**
> ADS circuit elements used in this chapter:
> **V_1Tone, I_1Tone** – single-frequency sinusoidal voltage and current sources, respectively; normally used in frequency-domain analysis.
> **V_DC, I_DC** – DC voltage and current sources, respectively.
> **DC_Block** – open circuit at DC, short circuit at other frequencies.
> **DC_Feed** – short circuit at DC, open circuit at other frequencies.
> **Term** – port termination in Z_0, usually used to compute S-parameters; also serves as a signal source in S-parameter simulations.
> **I_Probe** – current measurement point; behaves as a short circuit.
> **OscPort** – specialized simulator component needed for oscillator analysis; behaves as a short circuit.

The i_E pulses occur at the positive peaks of v_{be}; thus, the first harmonic $i_{e(1)}$ is in phase with v_{be}. Therefore, $\theta = -90°$ with respect to i_{in}, and the controlled current source is

$$i_{e(1)} = -j2I_E \frac{i_{in}}{|i_{in}|}. \qquad (10.9)$$

Solving for v_{in} and dividing by i_{in}, we obtain the input impedance at the first harmonic frequency:

$$Z_{in(1)} = -\frac{2I_E}{|i_{in}|}\frac{1}{\omega C_2} - j\frac{1}{\omega}\left(\frac{1}{C_1} + \frac{1}{C_2}\right). \qquad (10.10)$$

HARMONIC BALANCE ANALYSIS

The harmonic balance (HB) analysis is an iterative method of simulating nonlinear systems. Within the ADS simulator, a circuit is separated into linear and nonlinear portions. The nonlinear part is then iteratively solved in the time domain before being converted back into the frequency domain via the Fourier transform and combined with the linear frequency-domain circuit. All signals can consist of only a finite number of frequency tones.

It is noticed that the circuit still produces negative resistance, but its magnitude is now inversely proportional to the input current amplitude. The reactance of this circuit remains unchanged from the small-signal case, so the resonance frequency does not shift with increasing oscillation amplitude.

Performing a so-called Harmonic Balance (large-signal) simulation of the circuit in Figure 10-6(a) in ADS at 1 GHz, we plot the input impedance $Z_{in} = R_{in} + jX_{in}$ as a function of input current amplitude in Figure 10-7 together with the approximations from (10.6) and (10.10). The approximated and simulated input resistances R_{in} match reasonably well in the small-signal and large-signal regime, with a transition region in the middle, as must be anticipated. We observe that for the largest input currents, the circuit deviates from normal Class C behavior as the base-collector junction starts forward biasing. The input reactance X_{in} stays close to the anticipated -49.7 Ω in (10.6) and (10.10) for almost the entire range except the very largest $|i_{in}|$, where the deviation is for the same reason as observed in R_{in}.

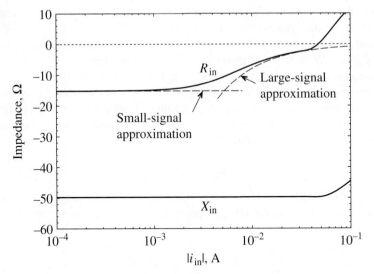

Figure 10-7 Harmonic Balance simulation results for the example emitter feedback circuit at 1 GHz.

Attaching an external resonator with series resistance R, we can estimate the magnitude of the oscillations:

$$|i_{in}| = \frac{2I_E}{\omega C_2 R} \tag{10.11a}$$

Basic Oscillator Models

$$|v_{in}| = |Z_{in}||i_{in}| \approx \frac{2I_E}{\omega^2 C_2 R}\left(\frac{1}{C_1} + \frac{1}{C_2}\right). \tag{10.11b}$$

The power dissipated in the external resistance is:

$$P = \frac{1}{2}|i_{in}|^2 R = \frac{2I_E^2}{\omega^2 C_2^2 R} \tag{10.11c}$$

As we will find out, maximizing resonator power is important for reducing oscillator phase noise. This can be accomplished by increasing the bias current, as well as reducing C_2 or R. If we follow the rule of thumb $R = -R_{in}/3$ (small-signal), then the oscillation magnitude becomes:

$$|i_{in}| = 6\omega C_1 \frac{I_E}{g_m} = 6\omega C_1 V_T \tag{10.12a}$$

$$|v_{in}| \approx 6\frac{I_E}{g_m}\left(\frac{C_1}{C_2} + 1\right) = 6V_T\left(\frac{C_1}{C_2} + 1\right) \tag{10.12b}$$

$$P = 6\frac{I_E^2 C_1}{g_m C_2} = 6I_E V_T \frac{C_1}{C_2} \tag{10.12c}$$

where the formula that uses V_T is only valid for BJTs, while the version that uses g_m is valid for any transistor.

We can still increase the resonator power by increasing the bias current or by increasing the C_1/C_2 ratio. We also observe that higher C_1/C_2 ratio necessitates lower resonator series resistance. Practical experience with Colpitts oscillators (see Section 10.1.5) has shown that a suitable C_1/C_2 ratio is 4, which we used in the circuit of Figure 10-6(a). Using this ratio, the following simplifications can be made:

$$|v_{in}| \approx 30\frac{I_E}{g_m} = 30V_T \approx 0.78 \text{ V} \tag{10.13a}$$

$$P = 24\frac{I_E^2}{g_m} = 24I_E V_T. \tag{10.13b}$$

At this point, we are ready to simulate an oscillator based on this circuit. Following the resistance rule of thumb, a 5 Ω resistance in series with an 8 nH inductance ($X = 50.3$ Ω at 1 GHz) is attached at the input of the circuit in Figure 10-6(a). According to our approximations, the oscillator should stabilize at the frequency of 0.995 GHz with oscillation voltage amplitude of $|v_{in}| = 0.78\ V_{pk} = 1.56\ V_{p\text{-}p}$. Figure 10-8 shows the base, emitter, and collector voltage waveforms for the transistor, simulated using Harmonic Balance. The simulation indicates oscillation at 0.9937 GHz. The base voltage v_B, which is v_{in} plus a DC bias, is not perfectly sinusoidal. Its actual swing is 1.457 $V_{p\text{-}p}$, while the first harmonic is 1.438 $V_{p\text{-}p}$, which is reasonably close to our estimate. It should be noticed that the collector voltage v_C is far from sinusoidal. This is the result of the pulsing Class C collector current flowing into the load resistor.

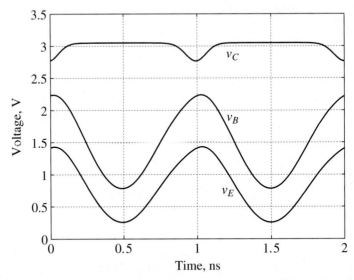

Figure 10-8 Harmonic Balance simulation of a 1 GHz oscillator with the emitter feedback negative resistance circuit.

It is important to observe that the bias point of the transistor slightly changes as it progresses deeper into Class C regime. As the input voltage amplitude increases, the v_E peaks move with an approximate 0.7 V offset from the v_B peaks. Since the v_E swing is less than that of v_B due to the voltage divider effect in the capacitors, the DC average V_E rises, and thereby reducing V_{CE}. Thus, if we used an emitter resistor instead of a current source to bias this circuit (a common practice), the emitter bias current during oscillation would be slightly higher than the DC analysis would indicate.

Since the circuit operates in deep Class C regime, the output power of this oscillator at the fundamental frequency depends only on the transistor bias current:

$$P_{out(1)} = \frac{1}{2}|i_{load(1)}|^2 R_{load} = 2I_E^2 R_{load} \qquad (10.14)$$

where $|i_{load(1)}| = 2I_E$, which is the same as the emitter current. The output power with $I_E = 1$ mA and $R_{load} = 50\ \Omega$ is predicted to be -10 dBm, while the simulation result is -10.95 dBm.

A compelling feature of this circuit is that the load is isolated from the resonator. This isolation diminishes as the operating frequency approaches the transition frequency of the transistor and the device parasitics gain more influence. Another consequence of device parasitics is that the internal base-emitter capacitance is often sufficiently large such that the external capacitor C_1 is not required.

Basic Oscillator Models

An additional item to note about the load is that its impedance should be relatively low. If that were not the case, the collector voltage pulses (see Figure 10-8) would grow too large and cause the transistor to exit the Class C regime. Also, since the device parasitics couple the load to the resonator, a low load impedance minimizes the influence of load variations on the oscillation frequency.

Cross-Coupled Circuit

Another popular negative resistance circuit is the cross-coupled transistor pair in Figure 10-9. Analyzing this circuit for the small-signal differential input impedance (using the ideal transconductance model for the transistors), we obtain

$$Z_{in} = -\frac{2}{g_m} = -\frac{4V_T}{I_{BIAS}} \qquad (10.15)$$

where $g_m = I_{BIAS}/(2V_T)$ since the common bias current branches evenly between the two transistors. We observe that this circuit generates pure negative resistance, with no reactive term (obviously, this is only true if the device parasitics are negligible). As the input voltage amplitude increases, the transistor collector current approaches a square wave, switching between zero current for half of the cycle and full I_{BIAS} during the other half-cycle. In this limiting case, the input impedance at the first harmonic frequency can be written as:

$$Z_{in(1)} = -\frac{\pi |v_{in}|}{2I_{BIAS}}. \qquad (10.16)$$

Here, the negative resistance magnitude increases with increasing input signal. Thus, this can be considered a "negative conductance" circuit, and should be used with a parallel resonator, as shown in Figure 10-5(b). The parallel resonator reaches its highest shunt resistance (or lowest shunt conductance) at resonance, which is the oscillation target for our circuit.

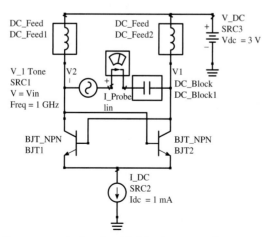

Figure 10-9 Cross-coupled pair with biasing, creating negative resistance.

Figure 10-10 presents the Harmonic Balance simulation results for the input impedance at the first harmonic frequency (1 GHz), plotted together with the approximations from (10.15) and (10.16). Once again, we obtain a good match in the small-signal and large-signal regions, with a transition region in between and a breakdown of the model at the largest signal levels.

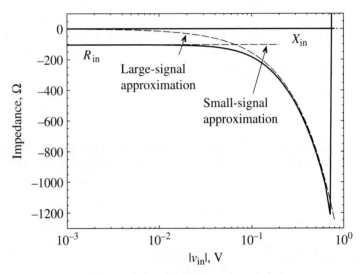

Figure 10-10 Harmonic balance simulation results for the example cross-coupled negative resistance circuit at 1 GHz.

Knowing the resonator shunt conductance G, we can estimate the oscillation voltage and current amplitudes

$$|v_{in}| = \frac{2I_{BIAS}}{\pi G} \tag{10.17a}$$

$$|i_{in}| = \frac{2}{\pi}I_{BIAS} \tag{10.17b}$$

and the associated power dissipation

$$P = \frac{2I_{BIAS}^2}{\pi^2 G}. \tag{10.17c}$$

If we use the negative conductance rule of thumb $G = -G_{in}/3$ (small-signal), we can make the following simplifications:

$$|v_{in}| = \frac{12 I_{BIAS}}{\pi g_m} = \frac{24}{\pi}V_T \approx 0.20 \text{ V} \tag{10.18a}$$

$$P = \frac{12 I_{BIAS}^2}{\pi^2 g_m} = \frac{24}{\pi^2}I_{BIAS}V_T. \tag{10.18b}$$

Basic Oscillator Models

Figure 10-11 presents an oscillator built from the cross-coupled negative resistance circuit. The parallel resonator consists of the capacitor C_1 in parallel with the series combination of the inductors L_1 and L_2. Here, we used the (small signal) negative conductance rule of thumb $G = -G_{in}/3 = 0.0032$ S $= (1/312)$ Ω. However, there is no explicit 312 Ω shunt resistor; rather, the inductive impedance consists of a total of 5 nH inductance in series with 3.1 Ω of resistance. At 1 GHz, this is equivalent to 5.05 nH inductance in shunt with a 321 Ω resistance. The approximate formulas predict an oscillation frequency of 0.992 GHz with a voltage amplitude of $|v_{in}| = 0.20\ V_{pk} = 0.40\ V_{p-p}$. Simulating the circuit in Figure 10-11 produces the waveforms shown in Figure 10-12. The simulated oscillation frequency is 0.9913 GHz. The collector voltage waveforms v_{c1} and v_{c2} are almost perfectly sinusoidal, with $|v_{in}| = |v_{c1(1)} - v_{c2(1)}| = 0.382\ V_{p-p}$ being a reasonable match to our estimate. The collector current waveform i_{C1} approximates the expected square wave.

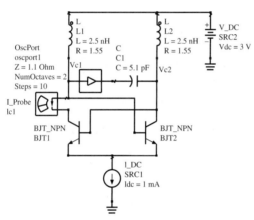

Figure 10-11 Cross-coupled oscillator. OscPort and I_Probe are simulator features that behave as short circuits.

Unlike the previous circuit, the cross-coupled pair does not provide a natural way of extracting oscillator power into a load, since loading would directly affect the resonator. Therefore, in most applications, either differential or single-ended buffer stages are installed between the cross-coupled pair's collectors and the load.

Origin of the 1/3 Negative Resistance Rule of Thumb

The $R = -R_{in}/3$ (small-signal) rule of thumb is based on the maximization of the power dissipated in the resonator and a simple linear approximation of the dependence of the negative resistance on the signal amplitude:

$$R_{in(1)} = R_{in}\left(1 - \frac{|i_{in}|}{I_{max}}\right). \tag{10.19}$$

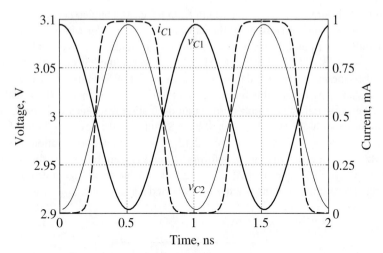

Figure 10-12 Harmonic balance simulation of a 1 GHz cross-coupled oscillator.

This expression simply states that the negative resistance starts out at the value R_{in} in the small-signal case, and linearly shrinks with increasing input current amplitude until it reaches zero at some input current amplitude I_{max}, at which point oscillation is no longer possible. If we couple this negative resistance with a resonator having a series resistance $R = -R_{in(1)} < -R_{in}$ (small-signal), we can solve (10.19) for $|i_{in}|$

$$|i_{in}| = I_{max}\left(1 + \frac{R}{R_{in}}\right) \qquad (10.20)$$

and express the resonator power as

$$P = \frac{1}{2}|i_{in}|^2 R = \frac{1}{2}I_{max}^2\left(\frac{R^3}{R_{in}^2} + \frac{2R^2}{R_{in}} + R\right). \qquad (10.21)$$

When examining these expressions, we need to remember that R_{in} is negative. In order to maximize the resonator power, we set to zero the derivative of P with respect to R:

$$0 = \frac{dP}{dR} = \frac{1}{2}I_{max}^2\left(\frac{3R^2}{R_{in}^2} + \frac{4R}{R_{in}} + 1\right). \qquad (10.22)$$

Solving for R, two solutions are obtained:

$$R = -R_{in}, \text{ and } R = -\frac{R_{in}}{3}. \qquad (10.23)$$

Since $|i_{in}|$ drops to zero at $R = -R_{in}$, this solution is a local minimum. The remaining solution turns out to be the point which optimizes the resonator power. Hence the rule of thumb is justified.

Basic Oscillator Models

Previously, we found the dependence of the negative resistance on the input current to be closer to the inverse proportionality of (10.10), rather than the linear decrease of (10.19). The resonator power expression in (10.11c) does not have a local maximum with respect to R. In fact, series resistance R should be minimized to optimize resonator power. However, reviewing the input impedance graph in Figure 10-7, we notice that the relationship (10.10) only holds down to $R \approx -R_{in}/5$, at which point the oscillation amplitude becomes too high, breaking the normal Class C regime and rapidly reducing the negative resistance to zero. With a different choice of bias and reactive components, the relationship (10.10) may hold for much lower R/R_{in} ratios. However, there is an additional caveat. As the oscillation amplitude increases, the negative resistance diminishes, in turn reducing the oscillation amplitude, and so on. As with any feedback loop, this adjustment is not instantaneous, and instability is possible. An unstable amplitude feedback loop causes the signal amplitude to vary, usually quasi-sinusoidally with a much lower frequency than the fundamental. This difficult-to-predict phenomenon is called "squegging". It is most prominent at low R/R_{in} and high C_1/C_2 ratios (referring to the circuit in Figure 10-6). To avoid this pitfall, it is recommended to stay away from the extremes and take the compromise of $R \approx -R_{in}/3$ and $C_1/C_2 \approx 4$, although it may result in poorer phase noise performance.

Negative Resistance versus Negative Conductance

We have seen two types of negative resistance circuits where the negative resistance increases or decreases with increasing signal amplitude. Without discussing its derivation, both of these cases (and a few practically less relevant possibilities) are covered by Kurokawa's condition for stable oscillation:

$$\left.\frac{\partial R_{in}(I_{in}, \omega_0)}{\partial I_{in}}\right|_{I_{in}=I_0} \left.\frac{dX(\omega)}{d\omega}\right|_{\omega=\omega_0} - \left.\frac{\partial X_{in}(I_{in}, \omega_0)}{\partial I_{in}}\right|_{I_{in}=I_0} \left.\frac{dR(\omega)}{d\omega}\right|_{\omega=\omega_0} > 0 \quad (10.24)$$

where R_{in} and X_{in} are the resistance and reactance of the active negative resistance circuit, R and X are the resistance and reactance of the resonator, and $I_{in} = |i_{in}|$ is the amplitude of the first harmonic of the input current to the active circuit. The resistance in most resonators is nearly constant with frequency (at least in a narrow range around the center frequency); thus the second term in (10.24) can be neglected. This results in the simplified stable oscillation condition:

$$\left.\frac{\partial R_{in}(I_{in}, \omega_0)}{\partial I_{in}}\right|_{I_{in}=I_0} \left.\frac{dX(\omega)}{d\omega}\right|_{\omega=\omega_0} > 0 \quad (10.25)$$

As we can see, only the sign of each derivative matters. If the negative resistance R_{in} increases (becomes less negative) with increasing input current, then the oscillator reactance must increase with frequency (transition from capacitive to inductive). This condition is satisfied by a series resonator, where capacitance dominates at low frequencies, and inductance at high frequencies. In the case of R_{in} decreasing (becoming more negative) with increasing signal, the resonator reactance must decrease with frequency (transition from inductive to capacitive). Obviously, a parallel resonator satisfies this condition.

10.1.3 Oscillator Phase Noise

Of all oscillator performance characteristics, phase noise is perhaps the most important. Phase noise and its time domain equivalent, jitter, refer to the random variation of the phase (or frequency) of the output signal. Phase noise is often measured in the frequency domain using a spectrum analyzer that does not differentiate between phase and amplitude noise. This measurement can still be called "phase noise" because amplitude variation is attenuated by the amplitude feedback mechanism of the oscillator, while phase variation remains unaltered, resulting in phase noise dominating the overall noise close to the carrier. However, farther away from the carrier, a significant amount of amplitude noise remains.

Oscillator phase noise originates due to the various noise sources of the oscillator circuit, including thermal, shot, and $1/f$ noise. To gain an intuitive understanding of phase noise, consider an idealized oscillator consisting of a parallel resonator with a normal, noisy conductance, in shunt with an active negative conductance circuit that is completely noise-free. The resonator conductance generates only white thermal (Johnson) noise, which can be represented by a Norton equivalent current source with a mean-square spectral density of

$$\frac{I_n^2}{\Delta f} = 2kTG \tag{10.26}$$

where k is Boltzmann's constant, T is the temperature, G is the resonator conductance, and Δf is the noise bandwidth. Close to the oscillation frequency, the negative conductance of the active circuit effectively cancels the resonator conductance G. Assuming the output of the oscillator is the voltage across the resonator (or the power delivered to the conductance G), the output noise is simply the result of the above noise current flowing through the *purely reactive* parallel resonator

$$\frac{V_n^2}{\Delta f} = \frac{I_n^2}{\Delta f}|Z(\omega)|^2 \tag{10.27}$$

where the impedance of the lossless parallel resonator is written as

$$Z(\omega) = \frac{1}{j\omega C + \frac{1}{j\omega L}} = j\frac{1}{GQ\left(\frac{\omega_0}{\omega} - \frac{\omega}{\omega_0}\right)}. \tag{10.28}$$

In (10.28), Q is the quality factor of the original, lossy parallel resonator. If we are concerned with only small frequency offsets from the center frequency, $\Delta\omega \ll \omega_0$, then we can approximate the resonator impedance as:

$$Z(\omega_0 + \Delta\omega) \approx -j\frac{\omega_0}{2GQ\Delta\omega}. \tag{10.29}$$

Substituting (10.29) into (10.27), we obtain

$$\frac{V_n^2}{\Delta f} = 4kTG\left(\frac{\omega_0}{2GQ\Delta\omega}\right)^2 = \frac{4kT}{G}\left(\frac{\omega_0}{2Q\Delta\omega}\right)^2 \tag{10.30}$$

where Δf and $\Delta\omega$ refer to different quantities: noise bandwidth and frequency offset from the carrier, respectively. The above expression is close to the phase noise of this oscillator. We only need to account for the amplitude feedback in the oscillator. According to the *equipartition theorem* of thermodynamics, the noisy conductance generates equal amounts of amplitude and phase noise power. However, the amplitude feedback in the oscillator strongly attenuates any amplitude fluctuations. As a result, the oscillator output contains only half the mean-square noise indicated by (10.30).

Finally, the standard phase noise specification of an oscillator compares the phase noise power to the carrier power. The typical specification is the ratio of noise power in a 1 Hz bandwidth at a given offset from the carrier frequency to the carrier power:

$$L(\Delta\omega) = \frac{P_n/(\Delta f)}{P_c} = \frac{V_n^2 G}{\Delta f}\frac{1}{P_c} = \frac{2kT}{P_c}\left(\frac{\omega_0}{2Q\Delta\omega}\right)^2 \quad (10.31)$$

where P_n is the noise power and P_c is the carrier power (both dissipated in the conductance G). The above ratio is often given in decibels, written as dBc/Hz at a given frequency offset from the carrier. Although the unit dB/Hz is technically incorrect, since the 1/Hz is inside the argument of a logarithm, it is widely accepted in the literature.

The oscillator phase noise spectral density strongly depends on frequency, unlike its white thermal noise source. The phase noise spectral density approaches infinity at the carrier and falls off as a square of the frequency offset from the carrier. On a log-log plot, this rolloff is stated as –20 dB/decade. We notice that at any given frequency offset, the phase noise is inversely proportional to the square of the resonator quality factor. Thus, increasing the resonator Q is the most effective method of reducing oscillator phase noise. Furthermore, the phase noise to carrier ratio depends inversely on the carrier power. This particular type of noise is independent of carrier power. Thus, maximizing the power dissipated in the resonator helps reduce phase noise. Finally, we notice that the phase noise power grows as a square of the oscillation frequency.

Even though the above phase noise derivation is extremely simplified, it nevertheless establishes a relatively accurate qualitative model of phase noise in real oscillators. The shape of the phase noise spectral density is generally correct, but real oscillators will produce significantly more noise than (10.31) implies. The active part of a real oscillator is a nonlinear, time-varying circuit that acts as a mixer, up-converting and down-converting noise from near carrier harmonic frequencies to frequencies near the carrier. Thus, the amount of thermal noise contributing to phase noise is increased. In addition, the active device contributes some of its own white noise, both of the thermal and shot variety. To make matters worse, transistor 1/f noise, which is normally significant only at very low frequencies (kHz to a few MHz), is up-converted to near the carrier. Finally, at relatively large frequency offsets from the carrier, the noise reaches a noise floor.

> **EQUIPARTITION THEOREM**
>
> In classical statistical mechanics, the equipartition theorem states that at thermal equilibrium, the thermal energy is equally shared among its various forms on average. For example, if gas molecules can move and rotate, the kinetic energies in the translational and rotational motions are equal on average. Oscillator noise obeys the same principle; its orthogonal components, magnitude and phase noise, share noise power equally.

Leeson has developed a formula for oscillator phase noise that includes all of the above-described effects:

$$L(\Delta\omega) = \frac{2FkT}{P_c}\left[1+\left(\frac{\omega_0}{2Q\Delta\omega}\right)^2\right]\left(1+\frac{\Delta\omega_{1/f^3}}{|\Delta\omega|}\right). \quad (10.32)$$

We recognize that this is simply a modified version of (10.31). The noise factor F describes the increase in noise compared to the ideal case of a noiseless negative resistance. The additive term of unity to the $1/\Delta\omega^2$ part of the noise spectrum indicates the noise floor, which is reached at approximately the 3 dB bandwidth point of the resonator. An additional multiplicative factor on the right causes the noise to vary as $1/\Delta\omega^3$ (−30 dB/decade) close to the carrier. This is the contribution of the low-frequency $1/f$ noise, up-converted and modified into phase noise by the oscillator. The frequency offset of the transition from −30 dB/decade falloff to −20 dB/decade falloff is $\Delta\omega_{1/f^3}$, which is close to the $1/f$ noise corner frequency for the transistor, but not exactly the same. Figure 10-13 graphically depicts the shape of Leeson's phase noise spectrum according to (10.32), compared to the phase noise of the ideal oscillator in (10.31).

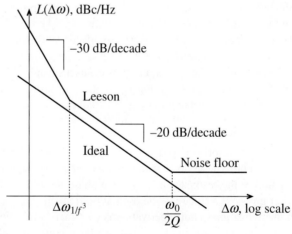

Figure 10-13 Oscillator phase noise according to Leeson's model and the ideal oscillator model.

Leeson's phase noise model of (10.32) is a good approximation of the phase noise spectral shape, but it does not provide a method of deriving its parameters from the circuit. The noise factor F has no relationship to the noise figure of an amplifier built from this transistor, because the circuit operates in the large-signal, strongly nonlinear regime. Although the formula states that the noise floor is reached at the 3 dB point of the resonator, the actual noise floor depends on the buffer stages after the oscillator. Finally, the $\Delta\omega_{1/f^3}$

Basic Oscillator Models

corner frequency offset is not exactly equal to the low-frequency $1/f$ noise corner for the transistor. In fact, in a well-optimized oscillator, the $\Delta\omega_{1/f^3}$ corner can be significantly lower than the $1/f$ corner.

There have been many attempts to derive the values of the Leeson's formula parameters from oscillator circuits. One interesting development is Hajimiri's time-varying noise model. It attempts to intuitively, yet accurately, describe the up- and down-conversion of noise and the special nature of the transistor noise, such as shot noise or channel noise. Shot noise is time varying because it is proportional to the collector current, which varies considerably within one oscillation cycle. We encountered this strong variation in all the oscillator circuits simulated so far.

Hajimiri's model is based on the concept of the Impulse Sensitivity Function (ISF), which describes the effect of noise impulses on the phase of the oscillations in the time domain. Because resonator energy is periodically moved among different energy storage elements of the resonator, a unit impulse (noise) current source would shift the waveform phase by a different amount depending on the time instant of its occurrence. Let us consider the ideal parallel resonator based oscillator. If at the moment of the noise current impulse, all the energy is stored in the capacitor, the result will be a change in voltage amplitude, but no change in phase. The amplitude feedback mechanism would soon correct the amplitude of the oscillation. Therefore, the ISF at this point in the oscillation cycle is zero. If, on the other hand, all the energy is stored in the inductor, an impulse current source would introduce a phase shift for which the oscillator cannot compensate. The ISF reaches an extremum (positive or negative peak) at this point. It turns out that the noise up- and down-conversion is simply described by the Fourier series coefficients of the ISF. Noise from near DC is up-converted to near the carrier in proportion to the DC value of the ISF. Noise from near the carrier is multiplied by the first harmonic of the ISF. Noise from around twice the carrier frequency is down-converted to near the carrier according to the second harmonic of the ISF, and so on.

The ISF still cannot be easily derived from the circuit, but Hajimiri presents three methods of estimating it. In one of the methods, the ISF is simply proportional to the derivative of the oscillator voltage waveform (more precisely the resonator capacitor voltage). The best method of determining the ISF is by using a circuit simulator.

One conclusion from the ISF theory indicates that the overall ISF RMS value should be minimized to reduce phase noise, which can be interpreted as minimizing the waveform rise and fall times. Maximizing P_c and Q is still predicted to be a reliable method of reducing phase noise, because this maximizes the energy stored in the resonator, such that noise impulses perturb it less. An additional conclusion is that the $1/\Delta\omega^3$ noise can be minimized by reducing the DC value of the ISF. Waveforms with even symmetry or half-wave symmetry theoretically have near-zero ISF DC values. Half-wave symmetry is when half of the waveform period is an inverted version of the other half-period. Even symmetry is difficult to realize, but half-wave symmetry can be improved by eliminating even harmonics from the waveform (for example by installing a short-circuited $\lambda/4$ transmission line stub that appears as a short circuit at all even harmonic frequencies), or by using balanced complementary transistor pairs in CMOS technology.

To properly treat shot noise, the shape of the collector current waveform must be taken into account. For this noise source, an effective ISF is developed, which is, within a constant factor, the usual ISF multiplied by the instantaneous collector current. An additional conclusion from this is that in a good oscillator, the transistor current should occur in short bursts, timed to coincide with the zeros of the ISF (when all the energy is in the capacitance). If we examine the waveforms of the first circuit we considered in Figure 10-8, we will notice that this is exactly how this circuit operates. The (noisy) emitter current occurs in short bursts right around the waveform voltage peak, where most of the resonator energy is stored in capacitors. Thus, we expect a circuit like this to exhibit lower phase noise compared to other oscillator types with the same Q and P_c. This has been verified in practice. The cross-coupled circuit waveform of Figure 10-12 indicates a different behavior. Here, the bias current flows through at least one of the transistors all the time, continuously generating shot noise. Therefore, we expect this oscillator to have comparatively poorer phase noise.

A practically relevant oscillator that produces perhaps the worst phase noise is the ring oscillator. It consists of logic inverters connected in a ring. During operation, a logic transition chases itself around the loop. The oscillation frequency is determined by the propagation delay of the inverters. The inverters may or may not have load capacitors, but there are no inductors in this circuit. Thus, the quality factor Q of this circuit is, unfortunately, about unity. Moreover, the inverter transistors conduct (noisy) current during the waveform voltage transitions, where the phase of the oscillation is most sensitive to noise impulses (peak of the ISF). This is perhaps exactly the wrong way to design an oscillator, but it is still useful in integrated circuits because it does not require sizeable inductors. The output frequency of this type of oscillator can be controlled by the DC supply voltage to the inverters. A ring oscillator is almost always used in a phased-locked loop (PLL). The feedback in a PLL can significantly reduce phase noise close to the carrier.

> **PLL**
>
> A phase-locked loop (PLL) produces an output signal whose frequency is a constant multiple of that of the input (reference) signal, and which is also phase locked to the input signal. The heart of the PLL is a phase/frequency comparator that controls a tunable oscillator, often a VCO. It works somewhat analogously to an operational amplifier, except it compares the frequencies/phases of the two input signals, and increases or decreases the output frequency depending on the result of the comparison. Negative feedback is established by passing the output signal through a frequency divider (programmable counter) on the way to the inverting input of the comparator, which is analogous to the voltage divider in a non-inverting amplifier.

10.1.4 Feedback Oscillator Design

Because of their fundamental importance in the development of low-frequency as well as RF oscillators, let us next focus on the two-port feedback networks, shown generically in Figure 10-14.

It is straightforward to find the transfer function of the feedback loop. For instance, for the Pi-network we obtain under high-impedance input and output assumptions

$$H_F(\omega) = \frac{V_1}{V_{out}} = \frac{Z_1}{Z_1 + Z_3}. \tag{10.33}$$

Basic Oscillator Models

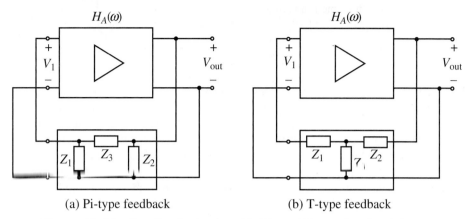

(a) Pi-type feedback (b) T-type feedback

Figure 10-14 Feedback circuits with Pi- and T-type feedback loops.

More complicated is the computation of the transfer function $H_A(\omega)$ of the amplifier. This depends on the chosen active element and its electric equivalent circuit model. To demonstrate the concept, we use a simple low-frequency FET model with voltage gain μ_V and output resistance R_B. The corresponding loop equation for the circuit depicted in Figure 10-15 is

$$\mu_V V_1 + I_B R_B + I_B Z_C = 0 \qquad (10.34)$$

where $1/Z_C = Y_C = 1/Z_2 + 1/(Z_1 + Z_3)$.

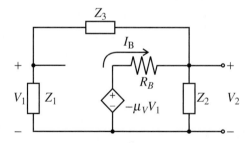

Figure 10-15 Feedback oscillator with FET electric circuit model.

Solving (10.34) for I_B and multiplying by Z_C gives us the output voltage V_{out} from which the voltage gain is found to be

$$H_A(\omega) = \frac{V_{out}}{V_1} = \frac{-\mu_V}{Y_C R_B + 1}. \qquad (10.35)$$

The closed-loop transfer function is thus

$$H_F(\omega)H_A(\omega) = \frac{-\mu_V Z_1 Z_2}{Z_2 Z_1 + Z_2 Z_3 + R_B(Z_1 + Z_2 + Z_3)} \equiv 1. \qquad (10.36)$$

This equation allows us to design various oscillator types depending on the choice of the three impedances in the feedback loop. To eliminate resistive losses, we choose purely reactive components $Z_i = jX_i$ ($i = 1,2,3$). This ensures that the numerator is real. Further, to make the denominator real, it is necessary that $X_1 + X_2 + X_3 = 0$, which implies that one of the reactances has to be the negative sum of the others. It is understood that negative-valued reactances correspond to capacitors and positive-valued reactances identify inductors. For instance, if we decide to use $X_3 = -(X_1 + X_2)$, then, upon substitution into (10.36), the result is

$$\frac{\mu_V X_1 X_2}{-X_2 X_1 + X_2(X_1 + X_2)} = \frac{\mu_V}{X_2}X_1 = 1. \qquad (10.37)$$

It is apparent that X_1 and X_2 must have the same sign but different values according to (10.37). In Table 10-1, a few possible configurations of the feedback loop are summarized.

Table 10-1 Various feedback configurations for oscillator designs based on Figure 10-14(a)

X_1, X_2	inductor	capacitor
X_3	capacitor	inductor
Z_1, Z_3, Z_2	Hartley (L_1, C_3, L_2)	Colpitts (C_1, L_3, C_2) / Clapp (C_1, C_3, L_3, C_2)

Basic Oscillator Models

Two often used realizations are the **Hartley oscillator**, where $X_1 = \omega L_1$, $X_2 = \omega L_2$, $X_3 = 1/(\omega C_3)$, and the **Colpitts oscillator**, where $X_1 = 1/(\omega C_1)$, $X_2 = 1/(\omega C_2)$, and $X_3 = \omega L_3$, as depicted in Figure 10-16, where a FET is employed as the active device. Here, the resistors R_A, R_B, R_D, and R_S set the DC bias point. C_S is an RF bypass capacitor, and C_B denotes DC blocking capacitors.

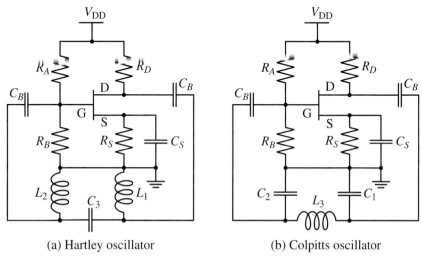

(a) Hartley oscillator (b) Colpitts oscillator

Figure 10-16 Hartley and Colpitts oscillators.

The various choices of L and C element combinations are in practice limited by the range of realizable values for a given frequency. Often, hybrid configurations are used; if, for instance, the inductance becomes very small, a capacitor connected in series can yield a smaller effective inductive reactance (**Clapp oscillator**).

Besides the standard common-source (or common-emitter for a BJT) configuration, common-gate (common-base) and common-drain (common-collector) type oscillators can be constructed, as shown in Figure 10-17, where all DC biasing elements are omitted.

10.1.5 Design Steps

What often makes oscillator design so complicated is that the nonlinear electric equivalent circuit describing the active device (BJT, FET) becomes increasingly complex as the frequency increases. Moreover, the oscillator has to drive additional circuits, and must therefore provide a certain amount of power. This output loading affects the oscillator in terms of frequency stability and waveform purity.

To provide the reader with a glimpse of the essential steps involved, we will at first examine the design of a low-frequency Colpitts oscillator. The h-parameter configuration with the appropriate feedback loop is depicted in Figure 10-18. The corresponding

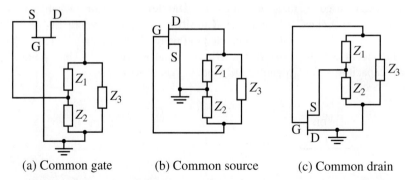

(a) Common gate (b) Common source (c) Common drain

Figure 10-17 Common gate, source, and drain configurations.

Kirchhoff voltage mesh equations involving input, output, and feedback loops are established by utilizing the output voltage $V_2 = V_{out} = I_2/h_{22} - I_1(h_{21}/h_{22})$.

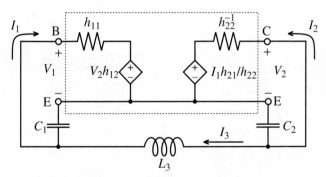

Figure 10-18 Colpitts oscillator design.

For the unknown currents, we obtain in matrix form

$$\begin{bmatrix} \left(h_{11}-jX_{C1}-\dfrac{h_{12}h_{21}}{h_{22}}\right) & \dfrac{h_{12}}{h_{22}} & jX_{C1} \\ -\dfrac{h_{21}}{h_{22}} & \left(\dfrac{1}{h_{22}}-jX_{C2}\right) & -jX_{C2} \\ jX_{C1} & -jX_{C2} & j(X_{L3}-X_{C1}-X_{C2}) \end{bmatrix} \begin{Bmatrix} I_1 \\ I_2 \\ I_3 \end{Bmatrix} = \begin{Bmatrix} 0 \\ 0 \\ 0 \end{Bmatrix}. \quad (10.38)$$

Computing the determinant and setting its imaginary portion to zero results, after lengthy algebra, in the form

$$f = \frac{1}{2\pi}\frac{1}{\sqrt{C_1 C_2}}\sqrt{\frac{h_{22}}{h_{11}}+\frac{C_1+C_2}{L_3}}. \quad (10.39)$$

Basic Oscillator Models

Furthermore, setting the real part of the determinant of (10.38) to zero, and assuming that $h_{12} \ll 1$, yields a quadratic equation in terms of the capacitor ratio C_1/C_2:

$$\frac{C_1^2}{C_2^2}(h_{11}h_{22} - h_{12}h_{21}) - \frac{C_1}{C_2}h_{21} + 1 = 0 \quad (10.40)$$

which, under the assumption that $h_{21}^2 \gg 4(h_{11}h_{22} - h_{12}h_{21})$, can be simplified to

$$C_1 \approx \frac{h_{21}}{h_{11}h_{22} - h_{12}h_{21}} C_2. \quad (10.41)$$

The preceding treatment deals with the *h*-parameters as real quantities, an assumption that generally may not be applicable. In fact, even for moderately high frequencies, the *h*-parameters can attain a significant phase angle. To incorporate the actual frequency-dependent behavior, we need to resort to the equations given in Section 4.3.2. For these situations, explicit formulas as (10.39) and (10.41) are impossible to derive, and we must rely on a mathematical spreadsheet to find numerical results.

Example 10-1. Design of a Colpitts oscillator

For a 200 MHz oscillation frequency, a Colpitts BJT oscillator in common-emitter configuration has to be designed. For the bias point of $V_{CE} = 3$ V and $I_C = 3$ mA, the following circuit parameters are given at room temperature of 25°C: $C_{BC} = 0.1$ fF, $r_{BE} = 2$ kΩ, $r_{CE} = 10$ kΩ, $C_{BE} = 100$ fF. If the inductance should not exceed $L_3 = L = 50$ nH, find values for the capacitances in the feedback loop.

Solution. The first step involves the determination of the *h*-parameters. We compute the values for DC (i.e., $f \to 0$).

$$h_{11} = h_{ie} = \frac{r_{BE}}{1 + j\omega(C_{BE} + C_{BC})r_{BE}} = 2000 \, \Omega$$

$$h_{12} = h_{re} = \frac{j\omega C_{BC} r_{BE}}{1 + j\omega(C_{BE} + C_{BC})r_{BE}} = 0$$

$$h_{21} = h_{fe} = \frac{r_{BE}(g_m - j\omega C_{BC})}{1 + j\omega(C_{BE} + C_{BC})r_{BE}} = 233.32$$

$$h_{22} = h_{oe} = \frac{1}{r_{CE}} + \frac{j\omega C_{BC}(1 + g_m r_{BE} + j\omega C_{BE} r_{BE})}{1 + j\omega(C_{BE} + C_{BC})r_{BE}} = 0.1 \text{ mS}$$

At DC, the h-parameters are real, and we can find from (10.41) the ratio between the capacitances C_1 and C_2:

$$C_1 = \frac{h_{21}}{h_{11}h_{22} - h_{12}h_{21}} C_2 = 1166.6 C_2.$$

Introducing a proportionality factor K such that $C_1 = KC_2$, equation (10.39) is rewritten as

$$f = \frac{1}{2\pi C_2 \sqrt{K}} \sqrt{\frac{h_{22}}{h_{11}} + (1+K)\frac{C_2}{L}}. \tag{10.42}$$

Solving the resonance condition (10.39) for C_2, we obtain

$$C_2 = \frac{\frac{1+K}{L} + \sqrt{\left(\frac{1+K}{L}\right)^2 + 16K\pi^2 f^2 \frac{h_{22}}{h_{11}}}}{8K\pi^2 f^2} = 12.68 \text{ pF}$$

where the inductance $L = 50$ nH has been used.

From the known C_2, we next find $C_1 = 1166.6 C_2$, or $C_1 = 14.79$ nF. In the preceding design, the transistor's h-parameters are given under DC conditions. In reality, however, the oscillator is operated at the resonance frequency of 200 MHz. Here, the h-parameters have the following values:

$$h_{11} = h_{ie} = \frac{r_{BE}}{1 + j\omega(C_{BE} + C_{BC})r_{BE}} = (1881 - j473)\Omega$$

$$h_{12} = h_{re} = \frac{j\omega C_{BC} r_{BE}}{1 + j\omega(C_{BE} + C_{BC})r_{BE}} = 5.9 \times 10^{-5} + j2.4 \times 10^{-4}$$

$$h_{21} = h_{fe} = \frac{r_{BE}(g_m - j\omega C_{BC})}{1 + j\omega(C_{BE} + C_{BC})r_{BE}} = 219 - j55$$

$$h_{22} = h_{oe} = \frac{1}{r_{CE}} + \frac{j\omega C_{BC}(1 + g_m r_{BE} + j\omega C_{BE} r_{BE})}{1 + j\omega(C_{BE} + C_{BC})r_{BE}}$$
$$= (0.11 + j0.03) \text{ mS}.$$

As seen, the h-parameters at this frequency differ only slightly from the DC conditions. Therefore, the analysis should equally apply for this frequency setting and the oscillator will require only a minimal amount of tuning.

Basic Oscillator Models

In practice, the situation often arises where the h-parameters at a given oscillation frequency differ significantly from their DC values, necessitating substantial tuning. The difference becomes more significant as the frequency increases. Also, the assumption of lossless reactive components makes the ratio C_1/C_2 unrealistically large. As discussed before, a ratio of 4 is more typical.

10.1.6 Quartz Oscillators

Unlike electric resonance circuits, quartz resonators can offer a number of advantages. A much higher quality factor (in the range 10^5 to 10^6), improved frequency stability, and near immunity to temperature fluctuations are among the chief benefits. Unfortunately, because quartz crystals are mechanical systems, they cannot be constructed to exceed a resonance frequency approximately 250 MHz.

A quartz crystal exploits the piezoelectric effect, whereby an applied electric field causes a mechanical deformation of the crystal. Depending on the geometric configuration and crystal cut, the crystal performs either longitudinal or shear vibrations at distinct resonance frequencies.

A typical electric circuit representation for a quartz crystal is shown in Figure 10-19. The circuit approximates the electric behavior at one of the resonance points for which the quartz is designed.

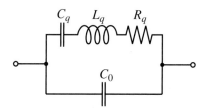

Figure 10-19 Quartz resonator equivalent electric circuit representation.

The capacitor C_q along with R_q and L_q describes the mechanical resonance behavior while C_0 denotes the capacitance due to the external contacting of the crystal through electrodes. Normally, the ratio between C_q and C_0 can reach values as high as 1000. Moreover, the inductance L_q is typically in the range from 0.1 mH to 100 H.

The admittance of this model can be stated as

$$Y = j\omega C_0 + \frac{1}{R_q + j[\omega L_q - 1/(\omega C_q)]} = G + jB. \tag{10.43}$$

The angular resonance frequency ω_0 is found by setting the imaginary component B to zero, or

$$\omega_0 C_0 - \frac{\omega_0 L_q - 1/(\omega_0 C_q)}{R_q^2 + [\omega_0 L_q - 1/(\omega_0 C_q)]^2} = 0. \tag{10.44}$$

Solution of this equation (see Problem 10.3) using a Taylor series expansion (and retaining the first two terms) leads to approximate expressions for the series and parallel resonance frequencies:

$$\omega_0 = \omega_S \approx \omega_{S0}\left[1 + \frac{R_q^2}{2}\left(\frac{C_0}{L_q}\right)\right] \tag{10.45a}$$

$$\omega_0 = \omega_P \approx \omega_{P0}\left[1 - \frac{R_q^2}{2}\left(\frac{C_0}{L_q}\right)\right] \tag{10.45b}$$

where $\omega_{S0} = 1/\sqrt{L_q C_q}$ and $\omega_{P0} = \sqrt{(C_q + C_0)/(L_q C_q C_0)}$. A representative model is discussed next.

Example 10-2. Prediction of resonance frequencies of quartz crystal

A crystal is characterized by the parameters $L_q = 0.1$ H, $R_q = 25\ \Omega$, $C_q = 0.3$ pF, and $C_0 = 1$ pF. Determine the series and parallel resonance frequencies and compare them against the imaginary component of the admittance given by (10.43).

Solution. As a first approach, to compute series and parallel resonance frequencies of the quartz crystal, we use (10.45a) and (10.45b), respectively:

$$f_S = f_{S0}\left[1 + \frac{R_q^2}{2}\left(\frac{C_0}{L_q}\right)\right] = \frac{1}{2\pi\sqrt{L_q C_q}}\left[1 + \frac{R_q^2}{2}\left(\frac{C_0}{L_q}\right)\right]$$

$$= 0.919 \text{ MHz}$$

$$f_P = f_{P0}\left[1 - \frac{R_q^2}{2}\left(\frac{C_0}{L_q}\right)\right] = \frac{1}{2\pi}\sqrt{\frac{C_q + C_0}{L_q C_q C_0}}\left[1 - \frac{R_q^2}{2}\left(\frac{C_0}{L_q}\right)\right]$$

$$= 1.048 \text{ MHz}.$$

The second approach is graphical. At resonance, reactance and susceptance of the circuit equal zero; thus, we can plot the imaginary portion of the admittance given by (10.43). Such a plot is shown in Figure 10-20, where the absolute value of the suceptance is plotted versus frequency.

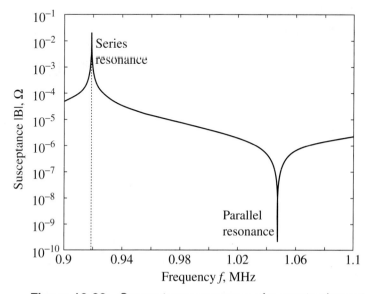

Figure 10-20 Susceptance response of a quartz element.

Comparing the graphical results with the analytical approach (10.45a) and (10.45b), we see that they are virtually the same.

Care has to be exercised in selecting quartz crystals due to their multiple resonances. Depending on the crystal, these responses can be closely spaced and may result in an undesired oscillation frequency.

10.2 High-Frequency Oscillator Configuration

As the operating frequency approaches the GHz range, the wave nature of voltages and currents cannot be neglected. As outlined in previous chapters, reflection coefficients and the associated S-parameter representation are required to represent the circuit's functionality. This requires us to reexamine (10.1) from a transmission line point of view. The Barkhausen criterion has to be reformulated in the context of the reflection coefficients.

To reexpress the loop gain in terms of transmission line principles, we recall our signal flow graph representation in Section 4.4.5, see Figure 10-21.

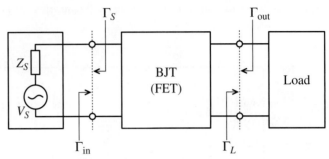

(a) Sourced and loaded transistor

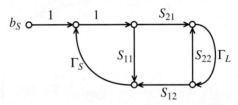

(b) Equivalent signal flow graph

Figure 10-21 Sourced and loaded transistor and its flow graph model.

The input reflection coefficient for matched source impedance ($Z_S = Z_0$) is

$$\Gamma_{in} = \frac{b_1}{a_1} = S_{11} + \frac{S_{12}S_{21}}{1 - S_{22}\Gamma_L}\Gamma_L = \frac{S_{11} - \Delta\Gamma_L}{1 - S_{22}\Gamma_L} \tag{10.46}$$

where $\Delta = S_{11}S_{22} - S_{12}S_{21}$. This is consistent with definitions given in Example 4-8. Conducting the computation with respect to the source term, $b_S = V_G\sqrt{Z_0}/(Z_G + Z_0)$, we can define the loop gain:

$$\frac{b_1}{b_S} = \frac{\Gamma_{in}}{1 - \Gamma_S\Gamma_{in}}. \tag{10.47}$$

The equation implies that if

$$\Gamma_{in}\Gamma_S = 1 \tag{10.48}$$

at a particular frequency, the circuit is unstable and begins to oscillate.
The identical circuit situation applies if the output side is considered, implying the condition

$$\Gamma_{out}\Gamma_L = 1 \tag{10.49}$$

for oscillations to occur.

High-Frequency Oscillator Configuration

When the stability factor $k = (1 - |S_{11}|^2 - |S_{22}|^2 + |\Delta|^2)/(2|S_{12}||S_{21}|)$ is included, see Chapter 9, the preceding conditions for oscillation can be summarized as follows:

$$k < 1 \tag{10.50a}$$

$$\Gamma_{in}\Gamma_S = 1 \tag{10.50b}$$

$$\Gamma_{out}\Gamma_L = 1. \tag{10.50c}$$

Since the stability factor is dependent on the S-parameters of the active device, we have to ensure that condition (10.50a) is satisfied. If the S-parameters at the desired frequency do not ensure this requirement, we can switch to a common-base or common-collector configuration or add positive feedback to increase instability, as the following example illustrates.

—RF&MW→

Example 10-3. Adding a positive feedback element to initiate oscillations

A BJT is operated at 2 GHz and has the following S-parameters specified in common-base configuration: $S_{11} = 0.94\angle 174°$, $S_{12} = 0.013\angle -98°$, $S_{21} = 1.9\angle -28°$, and $S_{22} = 1.01\angle -17°$. Determine how the Rollett stability factor is affected by adding an inductance to the base of the transistor ranging from 0 to 2 nH.

Solution. Using the definition for k gives us, without the inductance, the value

$$k = (1 - |S_{11}|^2 - |S_{22}|^2 + |\Delta|^2)/(2|S_{12}||S_{21}|) = -0.25.$$

Accounting for the inductance can be accomplished by redrawing the circuit in terms of two networks, as depicted in Figure 10-22.

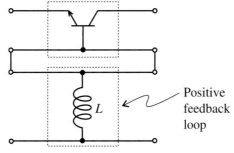

Figure 10-22 Network representation of the BJT with base inductance.

In this case, the overall S-parameter representation can be found by first converting the transistor's S-parameters into impedance representation, followed by adding the Z-parameters of the inductor, and finally converting the result back into S-parameter form.

Using the conversion formulas described in Chapter 4, we find the Z-representation of the transistor in common-base configuration:

$$[\mathbf{Z}]_{tr} = \begin{bmatrix} -0.42 + j3.43 & -2.17 - j0.097 \\ -95.23 - j303.06 & -6.88 - j321.03 \end{bmatrix}.$$

For the inductor, the **Z**-matrix is given by

$$[\mathbf{Z}]_{ind} = j\omega L \begin{bmatrix} 1 & 1 \\ 1 & 1 \end{bmatrix} = \begin{bmatrix} j\omega L & j\omega L \\ j\omega L & j\omega L \end{bmatrix}.$$

Adding $[\mathbf{Z}]_{tr}$ and $[\mathbf{Z}]_{ind}$ results in the Z-parameters of the entire circuit, which can then be converted into S-parameters.

To obtain the dependence of the Rollett stability factor as a function of feedback inductance, we have to repeat the preceding computations for each value of L. The result of such calculations is shown in Figure 10-23.

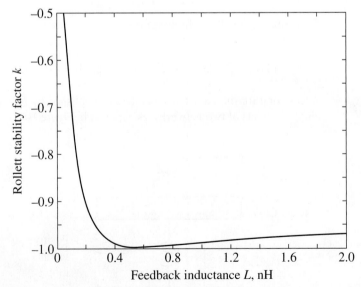

Figure 10-23 Rollett stability factor (k) as a function of feedback inductance in common-base configuration.

As seen in Figure 10-23, a maximum instability (minimum value of k) is obtained by adding a 0.6-nH inductor to the base.

At frequencies in the GHz range, even the lengths of the leads can be sufficient to create the desired inductance value at the base of the transistor.

It is interesting to note that if the oscillation condition is met either at the input or output port, the circuit is oscillating at both ports. This is directly seen by comparing the reflection coefficients at the input and output ports. We know that

$$\frac{1}{\Gamma_{in}} = \frac{1 - S_{22}\Gamma_L}{S_{11} - \Delta\Gamma_L} \equiv \Gamma_S \qquad (10.51)$$

and solving for Γ_L yields

$$\Gamma_L = \frac{1 - S_{11}\Gamma_S}{S_{22} - \Delta\Gamma_S}. \qquad (10.52)$$

However, Γ_{out} can also be written as

$$\Gamma_{out} = \frac{S_{22} - \Delta\Gamma_S}{1 - S_{11}\Gamma_S}. \qquad (10.53)$$

Therefore, we conclude that (10.52) is the inverse of (10.53), and thus

$$\Gamma_L = 1/\Gamma_{out} \qquad (10.54)$$

as required by (10.50c).

10.2.1 Fixed-Frequency Oscillators

An approximate oscillator design approach involves two-port design where the transistor configuration is first chosen such that it meets the requirement of $k < 1$ (inductive feedback may have to be added). Next, we select Γ_L such that $|\Gamma_{in}| > 1$ or Γ_S such that $|\Gamma_{out}| > 1$. When the circuit reaches a stable oscillating point, either case implies the other condition. For instance, if $|\Gamma_{out}| > 1$ we conclude that $|\Gamma_{in}| > 1$ and vice versa. A proof is left as an exercise. Once the termination is chosen at one port, we can determine the termination of the other port based on the Barkhausen criterion. However, during stable oscillations one has to recognize that the S-parameters change due to gain compression.

A rule-of-thumb in conjunction with a high-Q resonator allow us to base the oscillator design on linear S-parameters. The following example details these steps.

Example 10-4. Design of a fixed-frequency lumped element oscillator

A BFQ65 BJT manufactured by NXP is used in the common-base configuration. For this case, the transistor has the following S-parameters measured at 1.5 GHz: $S_{11} = 1.47\angle 125°$, $S_{12} = 0.327\angle 130°$, $S_{21} = 2.2\angle -63°$, and $S_{22} = 1.23\angle -45°$. Design a series feedback oscillator that satisfies conditions (10.50a)-(10.50c) at $f = 1.5$ GHz.

Solution. As the first step in the design process, we have to ensure that the transistor is at least potentially unstable. This can be tested by computing the Rollett stability factor:

$$k = (1 - |S_{11}|^2 - |S_{22}|^2 + |\Delta|^2)/(2|S_{12}||S_{21}|) = -0.975.$$

Since k is less than unity, the transistor is indeed potentially unstable.

Next, we plot the input stability circle to choose a reflection coefficient for the input matching network. The center and radius of the input stability circle are computed based on the formulas provided in Chapter 9:

$$r_{in} = \left|\frac{S_{12}S_{21}}{|S_{11}|^2 - |\Delta|^2}\right| = 0.82$$

$$C_{in} = \frac{(S_{11} - \Delta S_{22}^*)^*}{|S_{11}|^2 - |\Delta|^2} = 0.27\angle -57°.$$

Since $|C_{in}| < r_{in}$ and $|S_{22}| > 1$, the stable region is outside the shaded circle, as illustrated in Figure 10-24.

According to Figure 10-24, we have a considerable flexibility in choosing the reflection coefficient for the input matching network. Theoretically, any Γ_S residing inside the stability circle would satisfy our requirements. In practice, however, we would like to choose Γ_S so that it maximizes the output reflection coefficient

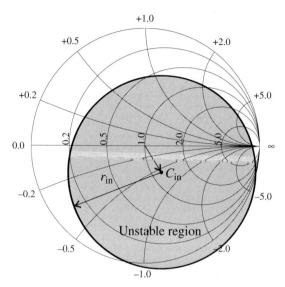

Figure 10-24 Input stability circle for the oscillator design.

$$\Gamma_{out} = S_{22} + \frac{S_{12}S_{21}}{1 - S_{11}\Gamma_S}\Gamma_S. \qquad (10.55)$$

From (10.55), it is obvious that Γ_{out} achieves its maximum value when $\Gamma_S = S_{11}^{-1}$. In this case, we obtain an infinite output reflection coefficient, which from (10.50c) results in $\Gamma_L = 0$ (i.e., $Z_L = Z_0 = 50\ \Omega$). The problem with such an approach is that in practical realizations it is almost impossible to achieve perfect 50 Ω matching. Moreover, as we approach $\Gamma_S = S_{11}^{-1}$, the oscillator becomes increasingly sensitive to changes in the load impedance. At $\Gamma_S = S_{11}^{-1}$, the slightest deviation from the 50 Ω value results in ceasing all oscillations. Because of this phenomenon, we choose Γ_S away from S_{11}^{-1}, while simultaneously achieving a sufficiently large $|\Gamma_{out}|$.

After attempting several values for the source reflection coefficient, we finally select $\Gamma_S = 0.65\angle-125°$. From the knowledge of Γ_S, the source impedance is computed as $Z_S = (13-j25)\Omega$, which is realized by a series combination of a 13 Ω resistor and a 4.3 pF capacitor, as shown in Figure 10-25.

Next, the output reflection coefficient is computed using (10.55) with the result $\Gamma_{out} = 14.67\angle-36.85°$. To determine the output matching network, we utilize (10.50c) and obtain the value $\Gamma_L = \Gamma_{out}^{-1} = 0.068\angle 36.85°$. This corresponds to the impedance

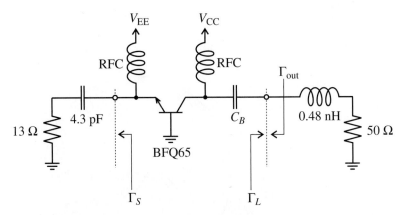

Figure 10-25 Series-feedback BJT oscillator circuit.

$Z_L = (55.6 + j4.57)\Omega = -Z_{\text{out}}$, and can be realized as a series combination of a 55.6 Ω resistor with a 0.48 nH inductor.

The final point that has to be taken into account in our design is the fact that as the output power of the oscillator begins to build up, the transistor's small-signal S-parameters become invalid. Usually, the power dependence of the transistor's S-parameters results in a less negative $R_{\text{out}} = \text{Re}(Z_{\text{out}})$ for increasing output power. Thus, it is necessary to choose $R_L = \text{Re}(Z_L)$ such that $R_L + R_{\text{out}} < 0$. In practice, a value of $R_L = -R_{\text{out}}/3$ is often used. However, we have to be careful with such a choice because it is only applicable if Γ_S is sufficiently far away from S_{11}^{-1}, as discussed previously. Another implication of $R_L \neq -R_{\text{out}}$ is a shift in the oscillation frequency. To minimize this shift, a high-Q resonator is almost universally employed. More comprehensive examples, including the design of resonators, are presented in the Practically Speaking section at the end of this chapter.

Although the component values assure that this oscillator meets design specifications and the electric behavior is successfully modeled, the final circuit implementation will pose additional problems. This is apparent when considering, for instance, the 0.48 nH inductor, which is comparable with the inductance of PCB through-hole connections (vias) and parasitics of the individual components.

High-Frequency Oscillator Configuration

For high-frequency applications, a more realistic design requires the use of distributed elements. A typical oscillator example involving a FET with connection to a 50 Ω load is seen in Figure 10-26. Here, TLi ($i = 1, \ldots, 6$) represent microstrip lines.

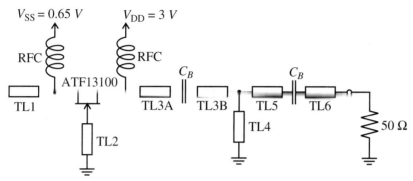

Figure 10-26 GaAs FET oscillator implementation with microstrip lines.

The design approach is presented in the following example, which provides details of how to increase the instability through a microstrip line attached to the common gate and how to select appropriate microstrip lines to match the load impedance.

Example 10-5. Microstrip design of a GaAs FET oscillator

The S-parameters of the GaAs FET (Avago Technologies ATF13100) in common-gate configuration are measured at 10 GHz and have the following values: $S_{11} = 0.37\angle-176°$, $S_{12} = 0.17\angle 19.8°$, $S_{21} = 1.37\angle-20.7°$, and $S_{22} = 0.90\angle-25.6°$. Design an oscillator with 10 GHz fundamental frequency. Furthermore, match the oscillator to a 50 Ω load impedance.

Solution. Similar to Example 10-4, we first check the stability of the transistor by computing the Rollett stability factor:

$$k = (1 - |S_{11}|^2 - |S_{22}|^2 + |\Delta|^2)/(2|S_{12}||S_{21}|) = 0.776.$$

Even though $k < 1$ indicates that the transistor is potentially unstable, we can attempt to increase the instability by connecting a feedback inductor to the gate of the transistor. Following the same approach as discussed in Example 10-3, we plot the stability factor versus the feedback inductance in Figure 10-27.

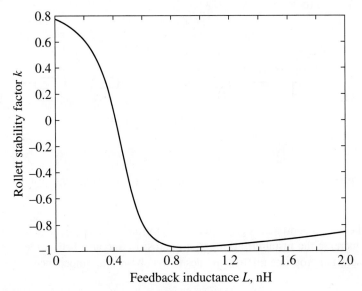

Figure 10-27 Stability factor for FET in common-gate configuration as a function of gate inductance.

It is noted that maximum instability is achieved for $L = 0.9$ nH. Due to the high operating frequency of the oscillator, the use of lumped elements is undesirable, and we have to replace the inductor by its distributed equivalent. One way to realize an inductance is to replace it with a short-circuited transmission line stub. Referring to Chapter 2, we can calculate the electric length of the transmission line assuming 50 Ω characteristic line impedance:

$$\Theta = \beta l = \tan^{-1}\left(\frac{\omega L}{Z_0}\right) = 48.5°.$$

The resulting S-parameters for the FET with a short-circuited stub connected to the gate contact are as follows:

$$[S] = \begin{bmatrix} 1.01\angle 169° & 0.29\angle 148° \\ 2.04\angle -33° & 1.36\angle -34° \end{bmatrix}.$$

The next step in the design procedure is the development of an input matching network. As mentioned in Example 10-4, for a realizable oscillator we should choose a source reflection coefficient close to the inverse of the S_{11} parameter of the transistor. In our design, we have selected $\Gamma_S = 1\angle-160°$, which corresponds to a source impedance of $Z_S = -j8.8\ \Omega$, and which can be realized as an open-circuited stub with a 50 Ω characteristic impedance and 80° electric length.

The output reflection coefficient is computed as

$$\Gamma_{out} = S_{22} + \frac{S_{12}S_{21}}{1 - S_{11}\Gamma_S}\Gamma_S = 4.18\angle 26.7°$$

which is equivalent to $Z_{out} = (-74.8 + j17.1)\Omega$. To satisfy (10.50c), we would have to choose a load impedance of $Z_L = -Z_{out}$, but due to the power dependence of the transistor's S-parameters (see Example 10-4) we choose the real portion of the load impedance to be slightly smaller than $-R_{out}$:

$$Z_L = (70 - j17.1)\Omega.$$

The transformation of the 50 Ω load impedance to Z_L is done through a matching network consisting of a 50 Ω transmission line with an electric length of 67°, and a short-circuited stub of 66° length.

The conversion of the electric parameters of the transmission lines into physical dimensions is done using the same approach as described in Example 2-5 in Chapter 2. The dimensions of the lines computed for a FR4 substrate of 40 mil thickness are summarized in Table 10-2.

Table 10-2 Dimensions of the transmission lines in the FET oscillator

Transmission line	Electrical length, deg.	Width, mil	Length, mil
TL1	80	74	141
TL2	48.5	74	86
TL3	67	74	118
TL4	66	74	116

Based on the oscillator circuit diagram shown in Figure 10-26, the TL3 line is cut into two halves, TL3A and TL3B, to accommodate the blocking capacitor. The lines TL5 and TL6 can have arbitrary length, because they are connected to a 50 Ω load.

The microstrip line design allows for an extremely small circuit board implementation as seen by the individual line lengths.

10.2.2 Dielectric Resonator Oscillators

When dealing with microstrip line realizations, a **dielectric resonator** (DR) can be added to provide a very high-Q oscillator design (up to 10^5) with extraordinary temperature stability of better than ±10 ppm/°C. This resonator, simply called a **puck**, can either be placed on top or next to the microstrip line in a metallic enclosure. The electromagnetic field coupling between the microstrip and the cylindrical resonator (see Figure 10-28), can be modeled near resonance as a parallel RLC circuit. The tuning screw permits geometric adjustment, which translates into a change of the resonance frequency.

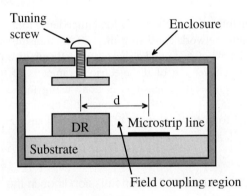

Figure 10-28 Dielectric resonator (DR) placed in proximity to a microstrip line.

We will not investigate the various waveguide modes (TE and TM modes) that are established inside the resonator, but rather concentrate on the use of DRs under common transmission line conditions.

In general, the circuit model of a DR (see Figure 10-29) is specified in terms of the **unloaded Q**, or Q_U,

$$Q_U = \frac{R}{\omega_0 L} = \omega_0 RC \qquad (10.56)$$

and the **coupling coefficient**, β,

$$\beta = \frac{R}{R_{ext}} = \frac{R}{2Z_0} = \frac{\omega_0 Q_U L}{2Z_0} \qquad (10.57)$$

at the desired angular resonance frequency $\omega_0 = 1/(\sqrt{LC})$. The value of the external resistance R_{ext} is equal to twice the line impedance because of the symmetric termination into Z_0. Similar to a transformer, the coupling coefficient quantifies the electromagnetic linkage between the resonator and the microstrip line, with typical values in the range of 2 to 20. Additionally, β is also employed to describe the relationship between the unloaded (Q_U), loaded (Q_L), and external (Q_E) quality factors:

$$Q_U = \beta Q_E = (1+\beta) Q_L. \qquad (10.58)$$

For oscillator design, it is required to specify the DR behavior in terms of the S-parameters. The modified transmission line configuration is illustrated in Figure 10-29(b).

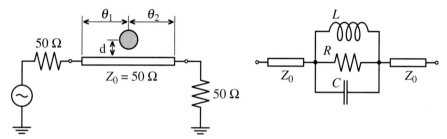

(a) Terminated microstrip line with DR (b) Transmission line model

Figure 10-29 Placement of DR along a transmission line and equivalent circuit representation for S-parameter computation.

Recalling our discussion of parallel resonance circuits in Section 5.1.4, we can compute the impedance Z_{DR} as

$$Z_{DR} = \frac{R}{1 + j\omega RC - jR/(\omega L)} = \frac{R}{1 + jQ_U(\omega/\omega_0) - jQ_U(\omega_0/\omega)} \qquad (10.59)$$

which simplifies to

$$Z_{DR} = \frac{R}{1 + jQ_U\left(\frac{\omega^2 - \omega_0^2}{\omega\omega_0}\right)} \approx \frac{R}{1 + j2Q_U \Delta f/f_0} \qquad (10.60)$$

where $\Delta f = f - f_0$ is the deviation from the center frequency. The last equation is only valid around the resonance point, where $\omega + \omega_0 \approx 2\omega_0$. Normalized with respect to Z_0 near resonance, it is seen that

$$z_{DR} \approx \frac{R/Z_0}{1 + j2Q_U(\Delta f/f_0)} = 2\beta. \qquad (10.61)$$

The transmission line segments on either side can now be included, leading to

$$[S]_{DR} = \begin{bmatrix} 0 & e^{-j\theta_1} \\ e^{-j\theta_1} & 0 \end{bmatrix} \begin{bmatrix} \dfrac{\beta}{\beta+1} & \dfrac{1}{\beta+1} \\ \dfrac{1}{\beta+1} & \dfrac{\beta}{\beta+1} \end{bmatrix} \begin{bmatrix} 0 & e^{-j\theta_2} \\ e^{-j\theta_2} & 0 \end{bmatrix} = \begin{bmatrix} \dfrac{\beta e^{-j2\theta_1}}{\beta+1} & \dfrac{e^{-j(\theta_1+\theta_2)}}{\beta+1} \\ \dfrac{e^{-j(\theta_1+\theta_2)}}{\beta+1} & \dfrac{\beta e^{-j2\theta_2}}{\beta+1} \end{bmatrix}. \quad (10.62)$$

Depending on the direction, we can determine the reflection coefficient as either S_{11}^{DR} or S_{22}^{DR}. If the electric line length is equal on both sides of the DR, we obtain $\theta_1 = \theta_2 = \theta = (2\pi/\lambda)(l/2)$, and therefore

$$\Gamma_{in}(\omega_0) = \frac{\beta}{\beta+1} e^{-j2\theta} = \Gamma_{out}(\omega_0). \quad (10.63)$$

The selection and purchase of a DR can be carried out quickly and efficiently, often over the manufacturers' websites. The design engineer specifies a particular resonance frequency and board material (thickness, dielectric constants) and the manufacturer will provide a particular DR in terms of diameter, length, tuning screw extension, distance d from the microstrip line, and cavity material. In addition, the coupling parameter and the unloaded Q are given, as well as the lumped parallel resonant circuit elements needed in the CAD simulation programs.

Example 10-6. Dielectric resonator oscillator design

Design an 8 GHz dielectric resonator oscillator (DRO) using a GaAs FET whose S-parameters at $f_0 = 8$ GHz are $S_{11} = 1.1\angle 170°$, $S_{12} = 0.4\angle -98°$, $S_{21} = 1.5\angle -163°$, and $S_{22} = 0.9\angle -170°$. A dielectric resonator that is used in the design has the following parameters at the resonant frequency $f_{res} = f_0$: $\beta = 7$, $Q_U = 5000$. Find the length of the 50 Ω microstrip line at the input port side of the FET, if the DR is located in the middle; assume the DR is terminated with a 50 Ω resistor. Examine the sensitivity in the DRO response to frequency fluctuations as compared to the conventional designs discussed previously.

Solution. The input stability circle of the FET at $f_0 = 8$ GHz is shown in Figure 10-30. To satisfy the oscillation conditions, we have to choose a source reflection coefficient somewhere in the nonshaded area of Figure 10-30. Since the termination resistance for the dielectric resonator is equal to the characteristic line impedance, the output reflection coefficient of the DR is computed according to (10.63):

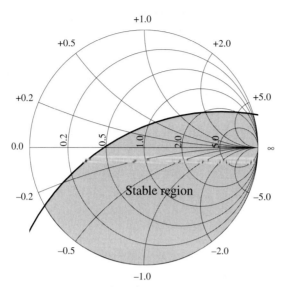

Figure 10-30 Input stability circle of the FET in the DRO design example.

$$\Gamma_S = \frac{\beta}{\beta+1} e^{-j2\Theta} = 0.875 e^{-j2\Theta}.$$

As discussed in the previous examples, to maximize the output reflection of the transistor, we have to select Γ_S close to the inverse of the S_{11} parameter. Since the absolute value of Γ_S is fixed, the best we can do is to select Θ such that the phase angle of Γ_S is equal to the phase angle of S_{11}^{-1}, or $-2\Theta = \angle S_{11}^{-1} = -\angle S_{11}$, leading to $\Theta = 85°$. The resulting electric circuit for the input matching network of the oscillator is shown in Figure 10-31.

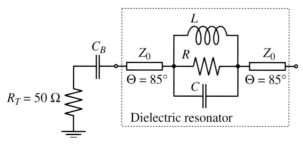

Figure 10-31 DR-based input matching network of the FET oscillator.

If the DR is not used in the input matching network of the transistor, then the simplest network that yields the same $\Gamma_S = 0.875\angle-170°$ at the oscillation frequency f_0 would be a series combination of a 3.35 Ω resistor and a 4.57 pF capacitor. A comparison of $|\Gamma_{out}|$ for the DR versus no DR realization as a function of frequency is shown in Figure 10-32. Here, the FET S-parameters are

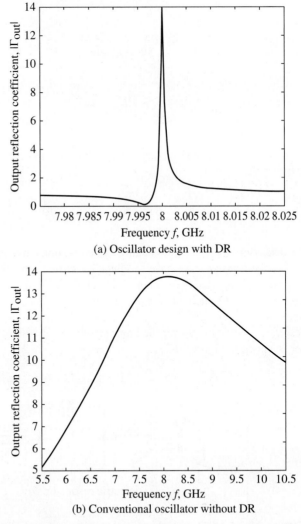

Figure 10-32 Frequency response of the output reflection coefficient for an oscillator design with and without DR.

assumed to be frequency-independent and the DR is approximated by its equivalent circuit shown in Figure 10-29 with parameters computed using (10.56) and (10.57):

$$R = 2\beta Z_0 = 700 \text{ } \Omega$$

$$L = R/(\omega_0 Q_U) = 2.79 \text{ pH}$$

$$C = \omega_0^{-2} L^{-1} = 14.2 \text{ nF}.$$

As clearly seen in Figure 10-32, the DRO design has a $|\Gamma_{out}| > 1$ in a much narrower frequency band than the conventional oscillator without the DR. This approach generally results in high selectivity and reduced drifts of the oscillation frequency. With the tuning screw, small frequency adjustments can be accomplished, typically in the range $\pm 0.01 f_0$ around the target frequency.

The dielectric resonator is an inexpensive and easy way to improve the quality factor of an oscillator. Unfortunately, its geometric size depends on the resonance frequency and typically becomes too large at low frequencies.

10.2.3 YIG-Tuned Oscillator

The dielectric resonator allows tuning over a very narrow band around the resonance frequency, typically between 0.01 and 1%. As an alternative, a magnetic element offers a wide-band tunable oscillator design with a tuning range of more than a decade. Such a tunable element, often of spherical shape, derives its name from **yttrium iron garnet** (YIG), a ferrimagnetic material whose effective permeability can be externally controlled through a static magnetic bias field H_0. This applied field directly influences the resonance frequency of the equivalent parallel resonant circuit consisting of conductance G_0, inductance L_0, and capacitance C_0. Figure 10-33 depicts a typical YIG element oscillator circuit.

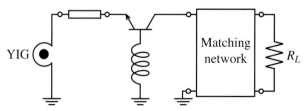

Figure 10-33 Oscillator design based on a YIG tuning element.

YIG RESONATORS

With physical principles similar to magnetic resonance imaging (MRI) in biomedicine, a YIG resonator exploits the magnetic resonance phenomenon that occurs when a static magnetic field is applied to a medium (here a single crystal of gallium-doped YIG medium), which is perpendicularly oriented to the RF resonance field. The resonance frequencies can range between 0.5 to 50 GHz, depending on size and material composition. Linear tuning over a broad frequency band is possible by changing the electric current that establishes the static field.

The unloaded quality factor is given as

$$Q_U = \frac{-4\pi(M_s/3) + H_0}{H_L} \qquad (10.64)$$

where M_s is the saturation magnetization in the sphere, and H_L is the resonance line width of 0.2 Oe. The saturation magnetization can be linked to the precessional motion of the magnetic moments at the angular frequency ω_m via

$$\omega_m = 2\pi\gamma(4\pi M_S) = 8\pi^2 \gamma M_s \qquad (10.65)$$

where γ is the gyromagnetic ratio recorded as 2.8 MHz/Oe. The resonance frequency follows from the bias field:

$$\omega_0 = 2\pi\gamma H_0. \qquad (10.66)$$

From these equations, the circuit elements of the parallel resonance circuit can be quantified. Specifically, the inductance is found to be

$$L_0 = \frac{\mu_0 \omega_m}{\omega_0 d^2}\left(\frac{4}{3}\pi a^3\right) \qquad (10.66a)$$

with a being the radius of the YIG sphere. This also determines C_0 from the resonance condition $\omega_0^2 = 1/(L_0 C_0)$; that is,

$$C_0 = 1/(L_0 \omega_0^2). \qquad (10.66b)$$

Finally, the conductance is

$$G_0 = \frac{d^2}{\mu_0 \omega_m Q_U \left(\frac{4}{3}\pi a^3\right)}. \qquad (10.66c)$$

In equations (10.66a)–(10.66c), d is the diameter of the coupling loop.

10.2.4 Voltage-Controlled Oscillator

It is mentioned in Chapter 6 that certain diodes exhibit a large change in capacitance in response to an applied bias voltage. A typical example is the varactor diode, with its variable capacitance $C_V = C_{V0}(1 - V_Q/V_{\text{diff}})^{-1/2}$ that can be affected by the reverse bias V_Q. Figure 10-34 illustrates how the feedback loop for the Clapp oscillator can be modified, by replacing C_3 in Figure 10-34(a) with a varactor diode. The modified circuit is shown in Figure 10-34(b). This circuit can readily be analyzed if a simplified BJT model ($R_L \ll h_{22}$) is employed.

High-Frequency Oscillator Configuration

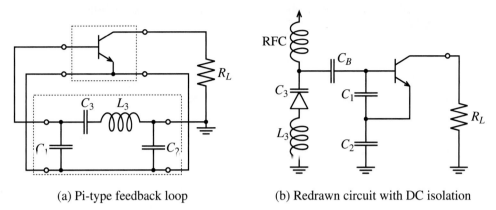

(a) Pi-type feedback loop (b) Redrawn circuit with DC isolation

Figure 10-34 Varactor diode oscillator.

In Figure 10-35, the varactor diode and a transmission line element, whose length is adjusted to be inductive, form the termination circuit connected to the input of the oscillator. If the varactor diode and the transmission line segment is disconnected, the input impedance Z_{IN} can be computed from two loop equations:

$$v_{IN} - i_{IN}X_{C1} - i_{IN}X_{C2} + i_B X_{C1} - \beta i_B X_{C2} = 0 \qquad (10.67a)$$

$$h_{11} i_B + i_B X_{C1} - i_{IN} X_{C1} = 0 \qquad (10.67b)$$

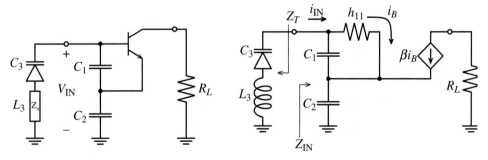

Figure 10-35 Circuit analysis of varactor diode oscillator.

Rearranging leads to

$$Z_{IN} = \frac{1}{h_{11} + X_{C1}}[h_{11}(X_{C1} + X_{C2}) + X_{C1}X_{C2}(1 + \beta)]. \qquad (10.68)$$

The equation can be simplified by noting that $(1 + \beta) \approx \beta$ and assuming that $h_{11} \gg X_{C1}$, which results in

$$Z_{IN} = \frac{1}{j\omega}\left[\frac{1}{C_1} + \frac{1}{C_2}\right] - \frac{\beta}{h_{11}}\left(\frac{1}{\omega^2 C_1 C_2}\right). \tag{10.69}$$

As expected from our previous discussion, the input resistance is negative. Therefore, with $g_m = \beta/h_{11}$,

$$R_{IN} = -\frac{g_m}{\omega^2 C_1 C_2} \tag{10.70a}$$

and

$$X_{IN} = \frac{1}{j\omega C_{IN}} \tag{10.70b}$$

where $C_{IN} = C_1 C_2/(C_1 + C_2)$. The resonance frequency follows from the previously established condition $X_1 + X_2 + X_3 = 0$ (see Section 10.1.4), or

$$j\left(\omega_0 L_3 - \frac{1}{\omega_0 C_3}\right) - \frac{1}{j\omega_0}\left[\frac{1}{C_1} + \frac{1}{C_2}\right] = 0 \tag{10.71}$$

with the result

$$f_0 = \frac{1}{2\pi}\sqrt{\frac{1}{L_3}\left(\frac{1}{C_3} + \frac{1}{C_2} + \frac{1}{C_1}\right)}. \tag{10.72}$$

It can be concluded from (10.70a) that the combined resistance of the varactor diode must be equal to or less than $|R_{IN}|$ in order to create sustained oscillations.

Example 10-7. Design of a varactor-controlled oscillator

A typical varactor diode has an equivalent series resistance of 15 Ω and a capacitance ranging from 10 pF to 30 pF for reverse voltages between 30 V and 2 V. Design a voltage controlled Clapp-type oscillator with center frequency of 300 MHz and ±10% tuning capability. Assume that the transconductance of the transistor is constant and equal to $g_m = 115$ mS.

Solution. To create sustained oscillations, we have to ensure that the series resistance of the varactor diode is smaller or equal to $|R_{IN}|$ over the entire frequency range as computed in (10.70a). From (10.70a) we can conclude that $|R_{IN}|$ achieves its minimum value at the maximum frequency of operation. Substituting $\omega_{max} = 2\pi f_{max}$ (with $f_{max} = 1.1 f_0 = 330$ MHz being the maximum oscillation frequency) into (10.70a), it is found that the capacitances C_1 and C_2 are related as

$$C_1 = -\frac{g_m}{\omega_{max}^2 R_{IN} C_2} = \frac{1}{kC_2} = \frac{1}{1.68 \times 10^{21} \text{pF}^{-2} C_2} \tag{10.73}$$

where $R_{IN} = -3R_S = -45\ \Omega$ according to our rule of thumb, and R_S being the varactor's equivalent series resistance.

Since the maximum oscillation frequency is obtained when the varactor capacitance C_3 has its minimum value, and the minimum frequency corresponds to the maximum C_3, we can rewrite (10.72) as

$$f_{min} = \frac{1}{2\pi}\sqrt{\frac{1}{L_3}\left(\frac{1}{C_{3max}} + \frac{1}{C_2} + kC_2\right)} \tag{10.74}$$

$$f_{max} = \frac{1}{2\pi}\sqrt{\frac{1}{L_3}\left(\frac{1}{C_{3min}} + \frac{1}{C_2} + kC_2\right)} \tag{10.75}$$

where the relation (10.73) is used to eliminate C_1. Dividing (10.74) by (10.75) and squaring the result, the following quadratic equation is obtained for C_2:

$$k(1-\alpha^2)C_2^2 + \left(\frac{1}{C_{3max}} - \frac{\alpha^2}{C_{3min}}\right)C_2 + (1-\alpha^2) = 0 \tag{10.76}$$

where $\alpha = f_{min}/f_{max}$. Solving (10.76) and substituting the result in (10.73) and (10.74) or (10.75), we find $C_1 = 12.4$ pF, $C_2 = 48$ pF, and $L_3 = 46.9$ nH as our desired values.

Unlike a mechanically adjustable dielectric resonator, the varactor diode permits dynamic tuning over a substantial frequency range.

10.2.5 Gunn Element Oscillator

The **Gunn element** can be employed to create oscillators from 1 to 100 GHz at low power outputs of roughly up to 1 W. It exploits a unique negative resistance phenomenon first discovered by Gunn in 1963. When certain semiconductor structures are subjected to an increasing electric field, they begin to shift, or transfer, electrons from the main valley to side valleys in the energy band structure. The accumulation of up to 90–95% of the electron concentration into these valleys results in a substantial decrease in effective carrier mobility and produces a technologically interesting I-V characteristic. Semiconductors with these band structures are primarily GaAs and InP. Figure 10-36 depicts a Gunn element and its current versus applied voltage response.

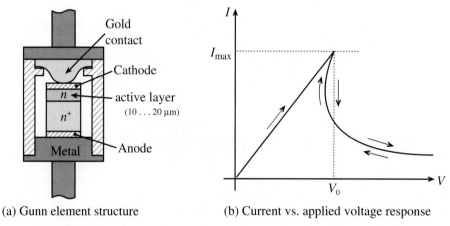

(a) Gunn element structure   (b) Current vs. applied voltage response

Figure 10-36 Gunn element and current versus voltage response.

We notice that in the presence of an applied DC voltage to the Gunn element, it behaves like a normal ohmic contact resistor for low field strength. However, if a certain threshold voltage V_0 is exceeded, **dipole domains** begin to be created below the cathode triggered by doping fluctuations. The formation of these domains lowers the current, as indicated in Figure 10-36(b). The current then remains constant while the domains travel from cathode to anode. After collection, the process repeats itself. The frequency can be estimated from the drift velocity of the domain motion $v_d \approx 10^5$ m/s and the travel length L of the active zone of the Gunn element. For a length of 10 μm, we obtain

$$f = \frac{v_d}{L} = \frac{10^5 \text{ m/s}}{10 \times 10^{-6} \text{ m}} = 10 \text{ GHz}. \tag{10.77}$$

If an external DC voltage is applied, the domain motion can be influenced and thus the resonance frequency is varied. The tuning range is approximately within 1% of the resonance frequency.

Figure 10-37 shows a microstrip line implementation of a Gunn element oscillator. Here, the Gunn element is connected to a $\lambda/4$ microstrip line, which in turn is coupled to a dielectric resonator. The bias voltage for the Gunn element is fed through an RFC onto the microstrip line.

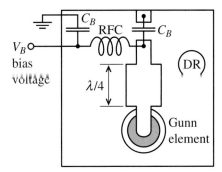

Figure 10-37 Gunn element oscillator circuit with dielectric resonator (DR).

10.3 Basic Characteristics of Mixers

Mixers are commonly used to multiply signals of different frequencies in an effort to achieve frequency translation. The motivation for this translation stems from the fact that filtering out a particular RF signal channel centered among many densely populated, narrowly spaced neighboring channels would require extremely high-Q filters. The task, however, becomes more manageable if the RF signal carrier frequency is reduced, or downconverted, within the communication system. Perhaps one of the best known systems is the downconversion in a **heterodyne receiver**, schematically depicted in Figure 10-38.

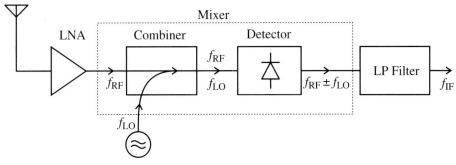

Figure 10-38 Heterodyne receiver system incorporating a mixer.

Here, the received RF signal is, after preamplification in a low-noise amplifier (LNA), supplied to a mixer whose task is to multiply the input signal of center frequency f_{RF} with a **local oscillator** (LO) frequency f_{LO}. The signal obtained after the **mixer** contains the frequencies $f_{RF} \pm f_{LO}$. After low-pass (LP) filtering, the lower frequency component $f_{RF} - f_{LO}$, known as the **intermediate frequency** (IF), is then selected for further processing.

The two key ingredients constituting a mixer are the **combiner** and **detector**, although we will learn of other implementations in Section 10.3.6. The combiner can be implemented through the use of a 90° (or 180°) directional coupler. A discussion of couplers and hybrids is found in Appendix G. The detector traditionally employs a single diode as a nonlinear device. However, antiparallel dual diode and double-balanced quadruple diode configurations are also utilized, as discussed later. In addition to diodes, BJT and MESFET mixers with low noise figure and high conversion gain, have been designed up to the X-band.

10.3.1 Basic Concepts

Before delving into details of the circuit design, let us briefly review how a mixer is capable of taking two frequencies at its input and producing multiple frequency components at the output. Clearly, a linear system cannot perform such a task, and we need to select a nonlinear device such as a diode, FET, or BJT that can generate multiple harmonics. Figure 10-39 depicts the basic system arrangement of a mixer connected to an RF signal $V_{RF}(t)$ and local oscillator signal $V_{LO}(t)$ known as the **pump signal**.

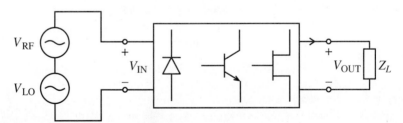

Figure 10-39 Basic mixer concept: two input frequencies are used to create new frequencies at the output of the system.

It is seen that the RF input voltage signal is combined with the LO signal and supplied to a semiconductor device with a nonlinear transfer characteristic. Both diode and BJT have an exponential transfer characteristic, as expressed by the Shockley diode equation discussed in Chapter 6:

$$I = I_0(e^{V/V_T} - 1). \tag{10.78a}$$

Basic Characteristics of Mixers

Alternatively, for a MESFET we have approximately a square behavior

$$I(V) = I_{DSS}(1 - V/V_{T0})^2 \qquad (10.78b)$$

where the subscripts denoting drain current and gate-source voltage are omitted for simplicity. The input voltage is represented as the sum of the RF signal $v_{RF} = V_{RF}\cos(\omega_{RF}t)$, the LO signal $v_{LO} = V_{LO}\cos(\omega_{LO}t)$, and a bias V_Q; that is,

$$V = V_Q + V_{RF}\cos(\omega_{RF}t) + V_{LO}\cos(\omega_{LO}t). \qquad (10.79)$$

This voltage is applied to the nonlinear device whose current output characteristic can be found via a Taylor series expansion around the Q-point:

$$I(V) = I_Q + V\left(\frac{dI}{dV}\right)\bigg|_{V_Q} + \frac{1}{2}V^2\left(\frac{d^2I}{dV^2}\right)\bigg|_{V_Q} + \ldots = I_Q + VA + V^2B + \ldots \qquad (10.80)$$

where the constants A and B refer to $(dI/dV)|_{V_Q}$ and $(1/2)(d^2I/dV^2)|_{V_Q}$, respectively. Neglecting the constant bias V_Q and I_Q, the substitution of (10.79) into (10.80) yields

$$\begin{aligned} I = {} & A(V_{RF}\cos(\omega_{RF}t) + V_{LO}\cos(\omega_{LO}t)) \\ & + B(V_{RF}^2\cos^2(\omega_{RF}t) + V_{LO}^2\cos^2(\omega_{LO}t)) \\ & + 2BV_{RF}V_{LO}\cos(\omega_{RF}t)\cos(\omega_{LO}t) + \ldots \end{aligned} \qquad (10.81)$$

The factors containing the cosine square terms can be rewritten, via the trigonometric identity $\cos^2(\omega t) = (1/2)(1 + \cos(2\omega t))$, into DC terms and terms involving $2\omega_{RF}t$ and $2\omega_{LO}t$. The key lies in the last term of (10.81), which becomes

$$I = \ldots + BV_{RF}V_{LO}(\cos[(\omega_{RF} + \omega_{LO})t] + \cos[(\omega_{RF} - \omega_{LO})t]). \qquad (10.82)$$

This expression makes clear that the nonlinear action of a diode, or transistor, can generate new frequency components of the form $\omega_{RF} \pm \omega_{LO}$. It is also noted that the amplitudes are multiplied and scaled by B, a device-dependent factor.

Equation (10.82) is the Taylor series representation up to the third term, and thus up to the **second-order intermodulation product** V^2B. Any higher-order products, such as the **third-order intermodulation product** V^3C, are neglected. For diodes and BJTs, these higher-order harmonic terms can significantly affect the performance of a mixer. However, the second-order intermodulation product is the only surviving term if a FET with a quadratic transfer characteristic is utilized. Thus, a FET is less prone to generate undesired higher-order intermodulation products.

The following example discusses the downconversion process from a given RF signal frequency to a desired intermediate frequency.

Example 10-8. Local oscillator frequency selection

An RF channel with a center frequency of 1.89 GHz and bandwidth of 20 MHz is to be downconverted to an IF of 200 MHz. Select an appropriate f_{LO}. Find the quality factor Q of a bandpass filter to select this channel if no downconversion is involved, and determine the Q of the bandpass filter after downconversion.

Solution. As seen in (10.82), by mixing RF and LO frequencies through a nonlinear device, we produce an IF frequency that is equal to either $f_{IF} = f_{RF} - f_{LO}$ or $f_{IF} = f_{LO} - f_{RF}$, depending on whether f_{RF} or f_{LO} is higher. Thus, to produce an $f_{IF} = 200$ MHz from $f_{RF} = 1.89$ GHz, we can use either

$$f_{LO} = f_{RF} - f_{IF} = 1.69 \text{ GHz or}$$
$$f_{LO} = f_{RF} + f_{IF} = 2.09 \text{ GHz}.$$

These two choices are equally valid and are both used in practice. When $f_{RF} > f_{LO}$ is chosen, the mixer is said to have **low-side injection,** whereas when $f_{RF} < f_{LO}$ the design is called **high-side injection**. The second approach is generally preferred since the relative tuning range of the LO is reduced.

Prior to downconversion, the signal has a bandwidth of BW = 20 MHz at a center frequency of $f_{RF} = 1.89$ GHz. Therefore, if we attempted to filter out the desired signal, we would have to use a filter with $Q = f_{RF}/\text{BW} = 94.5$. However, after downconversion, the bandwidth of the signal does not change, but the center frequency shifts to $f_{IF} = 200$ MHz, thus requiring a bandpass filter with a quality factor of only $Q = f_{IF}/\text{BW} = 10$.

This generic example highlights that less selective filtering is required once the mixer has downconverted the RF signal. Also, the IF is fixed, while the RF depends on the channel.

10.3.2 Frequency Domain Considerations

It is important to place the previous section into a frequency-domain perspective. To this end, it is assumed that the angular RF signal is centered at ω_{RF} with two extra frequency components situated ω_W above and below ω_{RF}. The LO signal contains one single component at ω_{LO}. After performing mixing, according to (10.82), the resulting spectral representation contains both **upconverted** and **downconverted** frequency components. Figure 10-40 graphically displays this process.

Basic Characteristics of Mixers

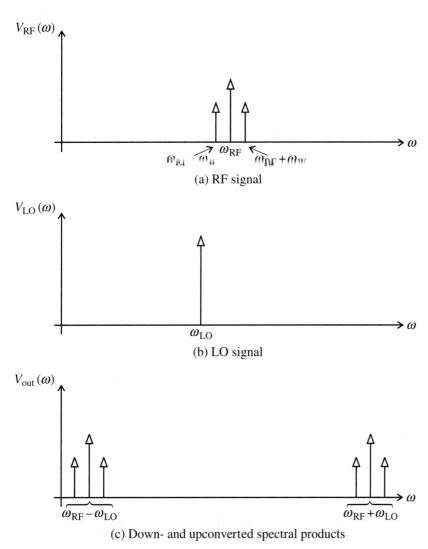

Figure 10-40 Spectral representation of mixing process.

Typically, the upconversion process is associated with the modulation in a **transmitter**, whereas the downconversion is encountered in a **receiver**. Specifically, when dealing with modulation, the following terminology is common:

- Lower sideband, or LSB ($\omega_{RF} - \omega_{LO}$)
- Upper sideband, or USB ($\omega_{RF} + \omega_{LO}$)
- Double sideband, or DSB ($\omega_{RF} + \omega_{LO}$, $\omega_{RF} - \omega_{LO}$).

A critical question to answer is the choice of an LO frequency that shifts the RF frequency to a suitable IF level.

An interrelated issue is the problem of **image** frequencies mapping into the same downconverted frequency range. To understand this problem, assume an RF signal is downconverted with a given LO frequency. In addition to the desired signal, we have placed an interferer symmetrically about the LO (see Figure 10-41). The desired RF signal transforms as expected:

$$\omega_{RF} - \omega_{LO} = \omega_{IF}. \qquad (10.83a)$$

However, the image frequency ω_{IM} transforms as

$$\omega_{IM} - \omega_{LO} = (\omega_{LO} - \omega_{IF}) - \omega_{LO} = -\omega_{IF}. \qquad (10.83b)$$

Since $\cos(-\omega_{IF}t) = \cos(\omega_{IF}t)$, we see that both frequency spectra are shifted to the same frequency location, as Figure 10-41 illustrates.

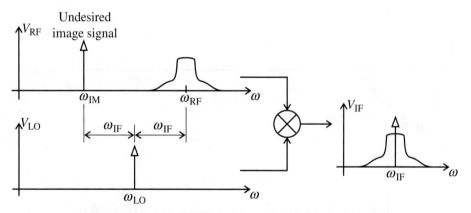

Figure 10-41 Problem of image frequency mapping.

To avoid the presence of undesired image signals that can be greater in magnitude than the RF signal, a so-called **image filter** is placed in front of the mixer circuit to suppress this influence, provided sufficient spectral separation is assured. More sophisticated measures involve an image rejection mixer.

10.3.3 Single-Ended Mixer Design

The simplest mixer is the single-ended design involving a Schottky diode, as shown in Figure 10-42(a). The RF and LO sources are supplied to an appropriately biased diode followed by a resonator circuit tuned to the desired IF. In contrast, Figure 10-42(b) shows an improved design involving a FET, which, unlike a diode, is able to provide a gain to the incoming RF and LO signals.

Basic Characteristics of Mixers

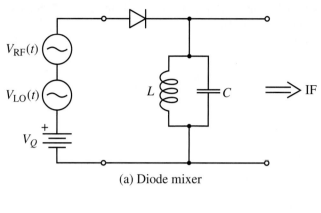

(a) Diode mixer

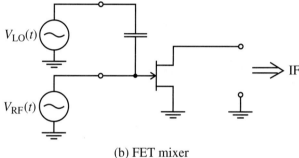

(b) FET mixer

Figure 10-42 Two single-ended mixer types.

In both cases, the combined RF and LO signal is subjected to a nonlinear device with exponential (diode) or nearly quadratic (FET) transfer characteristic followed by a bandpass filter whose task is to isolate the IF signal. The two very different mixer realizations allow us to contrast a number of parameters important when developing suitable designs:

- Conversion loss or gain between the RF and IF signal powers
- Noise figure
- Isolation between LO and RF signal ports
- Nonlinearity.

Since LO and RF signals are not electrically separated in Figure 10-42(a), there is a potential danger that the LO signal can interfere with the RF reception, possibly even reradiating portions of the LO energy through the receiving antenna. Unlike the diode mixer, the FET realization in Figure 10-42(b) provides signal gain. The **conversion loss** (CL) of a mixer is generally defined in dB as the ratio of supplied input power P_{RF} to the obtained IF power P_{IF}:

$$\text{CL [dB]} = 10\log\left(\frac{P_{RF}}{P_{IF}}\right) \qquad (10.84)$$

When dealing with BJTs and FETs, it is preferable to specify a **conversion gain** (CG) defined as the inverse of the power ratio.

Additionally, the noise figure of a mixer is generically defined as

$$F = \frac{P_{n_\text{out}}}{\text{CG}\, P_{n_\text{in}}} \qquad (10.85)$$

with CG being the conversion gain. P_{n_out}, P_{n_in} are the noise power at the output (at IF) and the noise power at the input (at RF). The FET generally has a lower noise figure than a BJT, and because of a nearly quadratic transfer characteristic (see Section 7.2), the influence of higher-order nonlinear terms is minimized. Instead of the FET design, a BJT finds application when high conversion gain and low voltage bias conditions are needed (for instance, battery powered systems).

Nonlinearities are customarily quantified in terms of **conversion compression** and **intermodulation distortion** (IMD). Conversion compression relates to the fact that the IF output power as a function of RF input power begins to deviate from the linear curve at a certain point. The point where the deviation reaches 1 dB is a typical mixer performance specification. As already encountered in the amplifier discussion, the intermodulation distortion is related to the influence of a second frequency component in the RF input signal, giving rise to distortion. To quantify this influence, a two-tone test is typically employed. If f_RF is the desired signal and f_2 is a second input frequency, then the mixing process produces a frequency component at $2f_2 - f_\text{RF} \pm f_\text{LO}$, where the +/− sign denotes up- or downconversion. The influence of this intermodulation product can be plotted in the same graph as the conversion compression (see Figure 10-43).

The intercept point between the desired linear output response and the undesirable third-order IMD response is a common figure of merit, indicating the ability of a mixer to suppress this influence.

Additional mixer characteristics include distortion generated inside the mixer, defined as **harmonic IMD,** as well as **isolation** between RF and IF, LO and IF, LO and RF ports. Specifically, RF and LO isolation is directly linked to the influence of the combiner (hybrid coupler; see Appendix G). Another characteristic is the **dynamic range,** which specifies the amplitude range over which no performance degradation occurs.

The circuit design of an RF mixer follows a similar approach as discussed when dealing with an RF amplifier. The RF and LO signals are supplied to the input of an appropriately biased transistor or diode. The matching techniques of the input and output side are presented in Chapter 8 and directly apply for mixers as well. However, one has to pay special attention to the fact that there is a large difference in frequencies between RF, LO on the input side, and IF on the output side.

RECEIVER CHARACTERISTICS

The design of RF front-end receivers for cellphones deals with different performance parameters depending on the required mobile phone standards. For instance, typical LNA parameters for the Global System Mobile (**GSM**) standard in the 925-960 MHz band entail a gain of 13 dB, a noise figure of 3 dB, and a 1 dB (output referred) gain compression point of −16 dBm, whereas the Universal Mobile Telecommunications System (**UMTS**) standard in the 1930-1990 MHz band, involve a gain of 16 dB, a 2 dB noise figure, and a 11 dBm 1 dB gain compression. For mixers, the GSM and the UMTS standards are similar with respect to conversion gain of 5 dB and single sideband noise figure of 12 dB, but differ in the 1dB (input referred) gain compression of −4 dB for GSM versus −12 dB for UMTS.

Basic Characteristics of Mixers

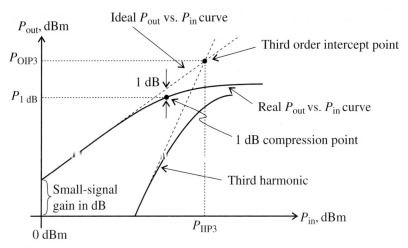

Figure 10-43 Conversion compression and intermodulation product of a mixer.

Since both sides have to be matched to the typical 50 Ω line impedance, the transistor port impedances (or S-parameter representation) at these two different frequencies have to be specified. Furthermore, to minimize interference at the output side of the device, it is important to short circuit the input to IF, and conversely short circuit the output to RF (see Figure 10-44). Including these requirements as part of the matching networks can be difficult.

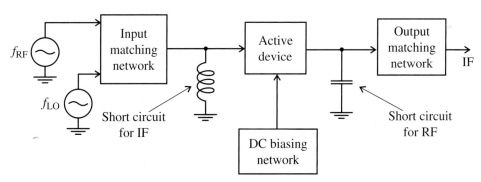

Figure 10-44 General single-ended mixer design approach.

These short-circuit conditions affect the transistor's behavior through an internal feedback mechanism. Ideally, $\Gamma_{in}(\omega_{RF})$ should be known based on the short-circuit output condition. Similarly, $\Gamma_{out}(\omega_{IF})$ requires a short-circuit input condition. Typically, an additional load resistance is added to the output port to adjust the conversion gain. In the following example, the salient design steps are explained.

Example 10-9. Design of a single-ended BJT mixer

For the topology shown in Figure 10-45, compute the values of the resistors R_1 and R_2 such that biasing conditions are satisfied. Using this network as a starting point, design a low-side injection mixer for $f_{RF} = 1900$ MHz and $f_{IF} = 200$ MHz. The BJT is measured at IF to have an output impedance of $Z_{out} = (677.7 - j2324)\Omega$ for short-circuited input, and an input impedance of $Z_{in} = (77.9 - j130.6)\Omega$ for short-circuited output at RF. Attempt to minimize the component count in this design.

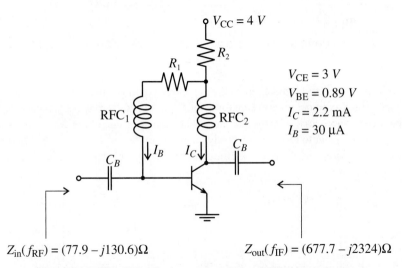

$Z_{in}(f_{RF}) = (77.9 - j130.6)\Omega$ $\qquad Z_{out}(f_{IF}) = (677.7 - j2324)\Omega$

Figure 10-45 DC-biasing network for BJT mixer design.

Solution. Since the voltage drop across the resistor R_2 is equal to the difference between V_{CC} and V_{CE}, and the current is the sum of the base and collector currents, R_2 is computed as

$$R_2 = \frac{V_{CC} - V_{CE}}{I_C + I_B} = 448 \; \Omega.$$

Similarly, the base resistor R_1 is computed as a ratio of $V_{CE} - V_{BE}$ to the base current:

$$R_1 = \frac{V_{CE} - V_{BE}}{I_B} = 70.3 \; k\Omega.$$

Before beginning the design of an input matching network, we have to decide on how to supply the LO signal. The simplest arrangement is to connect the LO source directly to the base of the transistor via a decoupling capacitor, as shown in Figure 10-46.

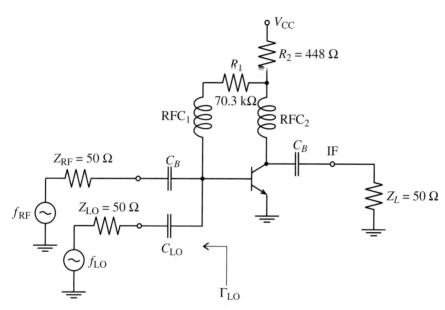

Figure 10-46 Connection of RF and LO sources to the BJT.

The value of this capacitor C_{LO} has to be chosen sufficiently small so as to prevent RF signal coupling into the LO source. We select $C_{LO} = 0.2$ pF. In this case, the series combination of C_{LO} and Z_{LO} creates a return loss RL_{RF} of only 0.24 dB, since

$$RL_{RF} = -20\log|\Gamma_{LO}(f_{RF})| = -20\log(0.9727) = 0.24 \text{ dB}.$$

Unfortunately, the LO frequency is very close to f_{RF} so that the same capacitance will attenuate not only the RF signal but the LO as well. We can compute the insertion loss IL_{RF} due to this capacitor at $f_{LO} = f_{RF} - f_{IF}$,

$$IL_{RF} = -10\log(1 - |\Gamma_{LO}(f_{LO})|^2) = 13.6 \text{ dB}.$$

Thus, if the LO source pumps at −20 dBm, only −33.6 dBm reaches the transistor. This seemingly high power loss is still tolerable since we can adjust the power provided by the local oscillator. However, we should be aware of significant LO power coupling into the RF port; additional filtering may be required.

The presence of C_{LO} and Z_{LO} modifies the value of the input impedance. A new total input impedance Z'_{in} can be computed as a parallel combination of C_{LO} in series with Z_{LO} and the input impedance of the transistor

$$Z'_{in} = \left(Z_{LO} + \frac{1}{j\omega_{RF}C_{LO}}\right) \| Z_{in} = (47.2 - j103.5)\Omega.$$

The output impedance does not change since the input is shorted during the measurement of Z_{out}.

Knowing Z'_{in}, we can next design an input matching network using any of the methods described in Chapter 8. One of the possible topologies consists of a shunt inductor followed by a series capacitor, as shown in Figure 10-47, where $C_1 = 0.8$ pF, $L_1 = 5.2$ nH, and we added the blocking capacitor C_{B1} to prevent a DC short circuit to ground.

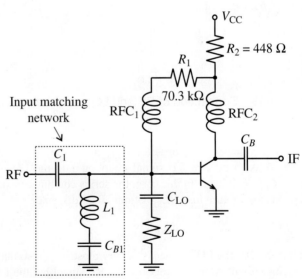

Figure 10-47 Input matching network for a single-ended BJT mixer.

There are several modifications that can be made to the circuit in Figure 10-47. First, we notice that instead of biasing the base of the transistor through an RFC, we can connect R_1 directly to the contact between L_1 and C_{B1}. In this case, we still bias the base of the transistor through L_1 and maintain isolation of the RF signal from the DC supply by grounding the RF through C_{B1}. One more objective of this matching network is to provide a short-circuit condition for the IF signal. Even though the impedance of the inductor L_1 is rather small at f_{IF}, we can still lower it by choosing the value of C_{B1} such that L_1 and C_{B1} exhibit a series resonance at IF. For example, if we choose $C_{B1} = 120$ pF, we create a solid short circuit for the IF signal, and maintain the path to ground for the RF signal. The modified input matching network is shown in Figure 10-48.

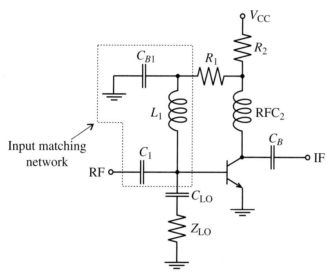

Figure 10-48 Modified input matching network.

The output matching network is developed in Figure 10-49 using a similar approach. The first version of the matching network consists of a shunt inductance L_2 followed by a series capacitance C_2. The values are $L_2 = 416$ nH and $C_2 = 1.21$ pF. This topology allows us to eliminate the RFC at the collector terminal of the transistor. However, the problem with this topology is that it does not provide a short circuit to ground for the RF signal that may interfere with the output. To remedy this drawback we replace L_2 with an equivalent LC combination. The additional capacitance $C_3 = 120$ pF is chosen to provide solid ground condition for the RF signal, and L_2 is adjusted to $L_2 = 5.2$ nH. The complete circuit of the designed single-ended BJT mixer is shown in Figure 10-49.

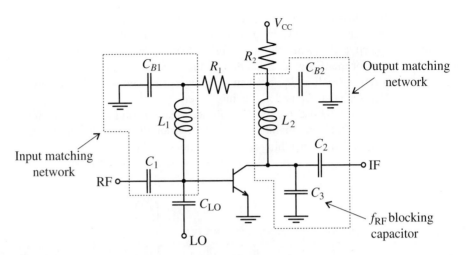

Figure 10-49 Complete electrical circuit of the low-side injection, single-ended BJT mixer with f_{RF} = 1900 MHz and f_{IF} = 200 MHz.

> *This design shows the multiple purposes that a matching network can serve. At first glance, they are often difficult to understand. Specifically, the dual purpose of matching and isolation provides challenges for the circuit designer.*

10.3.4 Single-Balanced Mixer

From the previous section, it is seen that the single-ended mixers are conceptually simple, but difficult to implement circuits. The main disadvantage of these designs is the difficulty associated with providing LO energy while maintaining isolation between LO, RF, and IF signals for broadband applications. The balanced dual-diode or dual-transistor mixer in conjunction with a hybrid coupler can operate over a wide bandwidth. Moreover, it provides further advantages related to noise suppression and spurious product rejection. Spurious products arise from the mixing of higher harmonics of the RF and LO, and potentially occur within the IF band. Thermal noise can critically raise the noise floor in the receiver. Figure 10-50 shows the basic mixer design featuring a quadrature coupler and a dual-diode detector followed by a capacitor acting as summation point.

Besides an excellent VSWR (see Appendix G), it can be shown that this design is capable of suppressing a considerable amount of noise because of the opposite diode arrangement in conjunction with the 90° phase shift. The proof is left as an exercise; see Problem 10.21.

A more sophisticated design, involving two MESFETs and 90° and 180° hybrid couplers, is shown in Figure 10-51. The 180° phase shift is needed since the second MESFET cannot easily be reversed as done in the anti-parallel diode configuration. It is also

Basic Characteristics of Mixers

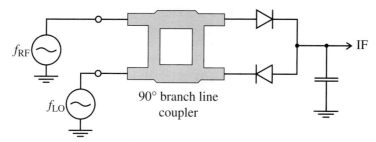

Figure 10-50 Balanced mixer involving a hybrid coupler.

important to point out that this circuit exhibits no RF-to-IF signal isolation. For this reason, a low-pass filter is typically incorporated into the output matching networks of each of the transistors in Figure 10-51. To achieve good LO-to-RF isolation, this mixer may be built with a 180° coupler instead of the 90° coupler. However, this is done at the expense of poorer input VSWR.

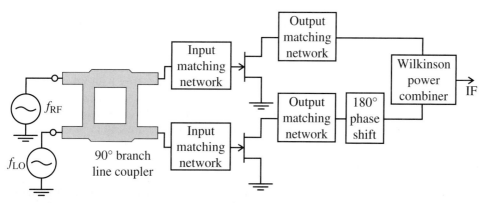

Figure 10-51 Single-balanced MESFET mixer with coupler and power combiner.

10.3.5 Double-Balanced Mixer

The double-balanced mixer can be constructed by using four diodes arranged in a ring configuration. The additional diodes provided better isolation and an improved suppression of spurious modes. Unlike the single-balanced approach, the double-balanced design eliminates all even harmonics of both the LO and RF signals. However, the disadvantages are a considerably higher LO drive power and increased conversion loss. Figure 10-52 depicts a typical circuit of the double-balanced design. All three signal paths are decoupled, and the input and output transformers enable a symmetric mixing with the LO signal.

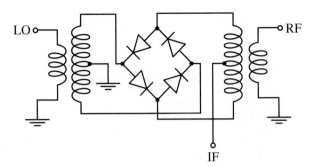

Figure 10-52 Double-balanced mixer design.

For design details of double-balanced mixers the reader is referred to the books by Vendelin and Mass listed at the end of this chapter.

10.3.6 Integrated Active Mixers

Integrated active mixers have gained considerable popularity due to the proliferation of consumer wireless communication devices (mobile phones, wireless LAN, Bluetooth). The main advantages of such mixers are conversion gain and ease of implementation on low-cost CMOS or bipolar-CMOS (BiCMOS) processes that allow integration of almost the entire receiver on one chip. Table 10-3 presents the advantages and disadvantages of active versus passive mixers, as well as of several active mixer topologies. Besides contributing conversion gain, active mixers can typically achieve better linearity with lower LO drive than passive mixers. Although active mixers can theoretically achieve better noise figure than diode mixers, practical implementations, particularly in CMOS, usually exhibit inferior noise figure. In addition, active mixer performance is more difficult to model than that of diode mixers due to the complexity of the circuits.

We have already examined one possible unbalanced active mixer topology in Example 10-9. This topology can achieve the best noise figure due to the simplicity of the mixer core. Although this mixer circuit is seemingly simple, it is difficult to implement properly; the design is complicated by the poor inherent port-to-port isolation. Strong RF-to-IF and LO-to-IF feed-through is typically filtered by a shunt capacitance at the IF output. The poor to nonexistent LO-to-RF isolation requires good filtering at both the RF and LO ports, as Example 10-9 points out. In addition to the port isolation issues, the mixing of LO phase noise at the IF frequency with baseband noise (which can be large due to $1/f$ noise), as well as RF port noise at the IF frequency with baseband noise add to the output noise at IF. To achieve a good noise figure, the LO must have low phase noise at the IF frequency, possibly requiring additional filtering of the LO signal. Furthermore, capacitive degeneration may be used with the transistor driven by the RF port to reduce the gain for the RF port noise at the IF frequency, which is lower than the RF frequency. However, the capacitance in series with the emitter can also destabilize the circuit and cause oscillations, so caution is needed. Moreover, the achievable linearity of an unbalanced mixer is low compared with

Basic Characteristics of Mixers

Table 10-3 Comparison of different active mixer topologies.

Mixer type	Advantages	Disadvantages
Active (vs. passive)	• Conversion gain • Better linearity • Lower LO power • Simpler to implement	• Typically higher noise figure • Less-predictable performance • Limited to lower frequencies
Unbalanced (active)	• Lowest noise figure • All ports single-ended	• Poor port-to-port isolation • Poor linearity • Difficult to implement
Single-balanced (active)	• LO-to-RF isolation • RF-to-IF isolation • Best linearity • Good noise figure	• Differential IF output • LO-to-IF feed-through
Double-balanced (active)	• LO-to-RF, LO-to-IF and RF-to-IF isolation • Good spurious product rejection • Good linearity • Simple to implement	• High noise figure • High power consumption

other topologies due to the fact that the RF and LO signals feed the same transistor. More complex mixer topologies have separate transistor blocks optimized for linear RF amplification and nonlinear LO switching.

Single-Balanced Active Mixer

Figure 10-53 presents the single-balanced active mixer topology, excluding some biasing components. The first stage of this mixer is the RF driver stage (Q_1) that contributes gain and converts the RF signal into a current-only output. For this reason, this stage is sometimes referred to as the transconductor. The current output of the driver stage biases a differential pair (Q_2 and Q_3), whose input is the differential LO. The LO signal is usually large enough to completely switch Q_2 and Q_3 on and off, shifting the bias current from one transistor to the other in a near square-wave fashion. The resulting differential IF output is the RF signal (plus a DC bias) multiplied by a square wave at the LO frequency. Thus, this active mixer behaves like a multiplier.

The primary advantages of the single-balanced mixer are good noise figure and linearity. The noise figure is worse than can be achieved with an unbalanced mixer, but better than in more complex mixers. Linearity can be very good because the RF driver stage mostly controls this parameter, provided the LO voltage is large enough to fully switch the differential pair. The driver stage can be optimized for linearity by carefully choosing the

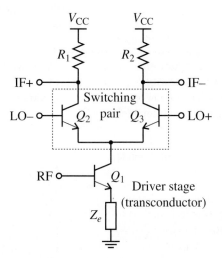

Figure 10-53 Single-balanced active mixer with drive stage.

device parameters, bias point, and installing emitter degeneration impedance Z_e if necessary, as seen in Figure 10-53. In general, inductive degeneration proves the most advantageous. The unbalanced mixer provides far less control over linearity, while more complex mixers require extra power to achieve the same linearity.

The single-balanced mixer also provides good RF-to-IF isolation if the IF output is taken differentially, and good LO-to-RF isolation if the LO input is driven differentially. The LO-to-RF isolation, a function of the symmetry of the circuit and the LO drive, is a welcome feature that eliminates costly filters. Driving the LO port differentially does not constitute a serious problem, because a single-ended LO can be converted to differential form using an additional buffer stage. The differential IF output is, however, a liability. Converting it back to single-ended form is difficult with passive components due to the low IF frequency. Taking the IF output as single-ended directly from the mixer carries a heavy (>6 dB) noise figure penalty owing to many common-mode noise contributions. Because the main advantage of an active single-balanced mixer is an improved noise figure, the IF output of this mixer is always taken differentially. Differential IF filters have been designed to avoid converting the IF signal to single-ended (single-ended signals also suffer from poor isolation from one another on the chip).

It is evident from the circuit in Figure 10-53 that a large LO current is present at the IF output (strong LO-to-IF feedthrough). This LO current must be shunted by a capacitor, otherwise it would saturate the IF output. Finally, just as was the case in the unbalanced mixer, the LO must have low noise power at the IF frequency. If this is not the case, the LO noise at the IF frequency can mix with the baseband noise and increase noise power at the IF output. However, the same restriction no longer applies to the RF signal due to good RF-to-IF isolation.

Double-Balanced Active Mixer

Figure 10-54 presents the double-balanced active mixer topology (without bias circuits), widely known as the Gilbert cell mixer. This, and its many variants, is by far the most popular mixer topology, and for good reasons. This mixer has excellent LO-to-RF, LO-to-IF, and RF-to-IF isolation. It is a very good approximation of a multiplier, with low spurious products. Its differential driver stage cancels second-order nonlinearity. Overall, this is an easy-to-implement and forgiving mixer, despite its seemingly complex circuit.

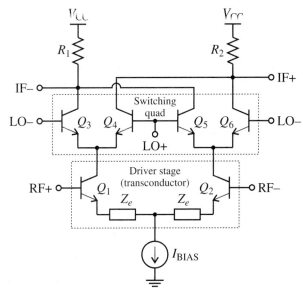

Figure 10-54 Double-balanced (Gilbert cell) active mixer.

The double-balanced (Gilbert cell) mixer is a natural extension of the single-balanced topology. It replaces the common emitter RF driver stage with a differential pair (Q_1 and Q_2), requiring approximately twice the bias current to replicate the linearity and conversion gain of the single-balanced mixer. The differential output current of the RF driver stage (transconductor) is then commutated by a four-transistor (quad) switch driven by the LO. Thus, the IF output is simply the RF input multiplied by a near-square wave at the LO frequency. Symmetry cancels both the RF and the LO signals at the IF output. The LO signal is also prevented from penetrating to the RF port by circuit symmetry.

For best results, all ports of the Gilbert cell mixer need to be differential. However, this mixer can also tolerate single-ended signals. For instance, the RF input can be single-ended, with the other input AC grounded. The cost of this is slightly degraded linearity. The LO should be differential to assure good LO-to-RF isolation, but it is easily converted from the single-ended form by an additional buffer stage; this is usually required anyway to achieve the desired LO drive. The IF output can be directly taken as single-ended at a relatively modest cost of reduced conversion gain and increased noise figure. The noise figure increase is due to the common-mode noise of the bias current source, which is relatively low at the IF frequency.

The main disadvantages of the Gilbert cell mixer are its higher noise figure and higher power requirements. The higher noise figure is due to the large number of active components contributing noise, although it is not extraordinarily worse than in other topologies. Noise figure and linearity generally improve with increasing LO power. With its four transistors to switch, the Gilbert cell mixer requires higher LO power than other mixer topologies, increasing the power consumption of the LO driver. Additionally, the RF driver stage requires more bias current to achieve the same conversion gain and linearity as a single balanced design.

Degeneration in the RF driver stage (Z_e in Figure 10-54) helps linearity and impedance matching, but reduces conversion gain. Inductive degeneration can improve noise figure by not contributing much additional noise and attenuating noise at higher RF frequencies. Since inductive degeneration reduces the RF gain at higher frequencies, it works best with high-side injection, such that the image frequency experiences less gain than the desired RF frequency. Resistive degeneration is also common, although it increases the noise figure. Its advantages are simpler implementation and wider useable RF frequency range. FET mixers can often omit degeneration due to their naturally more linear response.

10.3.7 Image Reject Mixer

We already know that a simple multiplying mixer converts both the $(\omega_{LO} - \omega_{IF})$ and the $(\omega_{LO} + \omega_{IF})$ frequencies to the same ω_{IF}. In a typical high-side injection scheme, $(\omega_{LO} - \omega_{IF})$ is the desired RF signal, while $(\omega_{LO} + \omega_{IF})$ is the image. After mixing, the image is indistinguishable from the desired signal. Thus, the image must be carefully filtered out before it reaches the mixer. It is difficult, however, to integrate the high-Q image filter onto the receiver chip, and an external filter is costly. One solution that can eliminate the need for an image filter is the image-reject mixer, shown in Figure 10-55. This mixer can separate the desired signal and the image into two independent outputs at the IF frequency. An alternative implementation uses a 90° splitter (hybrid) in the RF path and a 0° splitter for the LO. When used for upconversion (in transmitters), this mixer is also known as the single-sideband modulator.

Shown here with idealized components, it is easy to derive how the image-reject mixer works. Let the RF voltage be a combination of the desired signal and the image signal (high-side injection):

$$v_{RF}(t) = V_{Sig}\cos((\omega_{LO} - \omega_{IF})t) + V_{Img}\cos((\omega_{LO} + \omega_{IF})t). \quad (10.86)$$

After mixing with two LO signals of assumed unity amplitude, but differing by 90° in phase, the upper branch, known as the in-phase branch, contains the voltage

$$v_I(t) = v_{RF}(t)\cos(\omega_{LO}t) = \frac{1}{2}[(V_{Sig} + V_{Img})\cos(\omega_{IF}t) \quad (10.87)$$
$$+ V_{Sig}\cos((2\omega_{LO} - \omega_{IF})t) + V_{Img}\cos((2\omega_{LO} + \omega_{IF})t)].$$

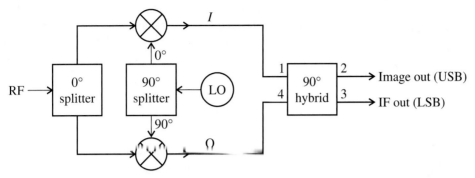

Figure 10-55 Image reject mixer.

We can drop the $(2\omega_{LO} \pm \omega_{IF})$ terms, since the IF filter would strongly attenuate these out-of-band signals, leaving only the ω_{IF} term where the signal and the image are superimposed. Similarly, the lower branch, known as the quadrature branch, contains

$$v_Q(t) = v_{RF}(t)\sin(\omega_{LO}t) = \frac{1}{2}[(V_{Sig} - V_{Img})\sin(\omega_{IF}t) + \ldots]. \qquad (10.88)$$

Passing through the quadrature hybrid, the two output signals are

$$v_{USB}(t) = \frac{1}{2\sqrt{2}}[(V_{Sig} + V_{Img})\cos(\omega_{IF}t - 90°) + (V_{Sig} - V_{Img})\sin(\omega_{IF}t - 180°)] \qquad (10.89)$$

$$= \frac{1}{2\sqrt{2}}[(V_{Sig} + V_{Img})\sin(\omega_{IF}t) - (V_{Sig} - V_{Img})\sin(\omega_{IF}t)]$$

$$= \frac{1}{\sqrt{2}}V_{Img}\sin(\omega_{IF}t)$$

and

$$v_{LSB}(t) = \frac{1}{2\sqrt{2}}[(V_{Sig} + V_{Img})\cos(\omega_{IF}t - 180°) + (V_{Sig} - V_{Img})\sin(\omega_{IF}t - 90°)] \qquad (10.90)$$

$$= \frac{1}{2\sqrt{2}}[-(V_{Sig} + V_{Img})\cos(\omega_{IF}t) - (V_{Sig} - V_{Img})\cos(\omega_{IF}t)]$$

$$= -\frac{1}{\sqrt{2}}V_{Sig}\cos(\omega_{IF}t).$$

This proves that the lower sideband signal has been separated from the upper sideband image.

POLYPHASE

A polyphase network has N inputs and N outputs, and is rotationally symmetric with respect to the port indices on each side. Treating the N inputs as one complex signal sampled at equal phase intervals, the polyphase network acts as a filter with an asymmetrical response for positive and negative frequencies. For example, the differential I and Q outputs of the two mixers in Figure 10-55 can be interpreted as a 4-phase signal with phase offsets 0°, 90°, 180°, and 270°. The desired signal is then centered at −IF, and the image at +IF. A simple RC polyphase network can reject the positive frequencies around +IF, while admitting the negative frequencies around −IF. Cascading several RC polyphase stages can greatly reduce amplitude response variations over a wide frequency range. A polyphase filter of the same topology can also be used to generate the 0° and 90° phases of the LO.

Obviously, only perfect components result in perfect separation of the image from the desired signal. An important performance parameter of this mixer is its image rejection, defined as the ratio of the desired signal to the image at the IF output, and usually expressed in dB. It can be estimated as follows:

$$\text{IR [dB]} = 10 \log\left(\frac{1 + A^2 + 2A\cos\theta}{1 + A^2 - 2A\cos\theta}\right) \quad (10.91)$$

where A is the amplitude imbalance expressed as a voltage ratio $A = 10^{(A\,[\text{dB}])/20}$, and θ is the phase imbalance. A is the aggregate amplitude imbalance, computed as the product of amplitude imbalances in all the stages of this mixer (or as a sum when expressed in dB). Similarly, θ is the sum of the phase imbalances in all the stages. Figure 10-56 plots the image rejection as a function of the two imbalances. As we can see, the requirements for good image rejection are fairly strict if image rejection upwards of 40 dB (needed to replace a minimal image-reject filter) is to be achieved. For example, 40 dB image rejection can be attained with 0.1 dB amplitude imbalance and 1° phase imbalance.

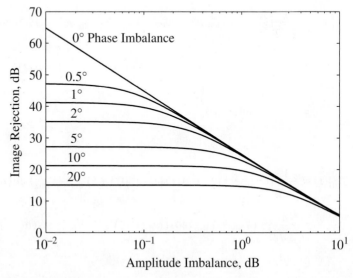

Figure 10-56 Image rejection as a function of amplitude and phase imbalance. Note the double logarithmic scale for amplitude imbalance (dB on a log scale).

Basic Characteristics of Mixers

Although Figure 10-55 shows two mixers, actual implementations usually share some features of the two mixers, such as the RF driver stage, to reduce any amplitude imbalances. Relatively low frequency receivers are able to use a double-frequency oscillator together with divide-by-two circuits to produce the two LO signals 90° out of phase. Issues may arise when realizing the IF quadrature hybrid with passive components because of the low frequency and tight tolerances. Resistor-capacitor phase shift (polyphase) networks together with an active summer can be used instead. Finally, the linearity of this mixer becomes even more important because the potentially strong image signal passes through the mixer stage unfiltered.

Practically Speaking. Simulating practical oscillators using Harmonic Balance analysis

In this section, we return to our oscillator discussion, examining the common-base circuit using Harmonic Balance analysis in the commercial circuit simulator ADS. We consider two possible modes of operation, "negative resistance" and "negative conductance", as well as two different resonator types. Simulation results include waveforms and phase noise.

Figure 10-57(a) shows the base feedback negative resistance circuit, which is simply a variation of the previously studied emitter feedback oscillator, taking the input impedance from the point of view of the capacitor C_2. Building simple small-signal and large signal models in the same manner as in the previous cases, we can determine the input impedance and admittance in the small-signal case

$$Z_{in} = (-g_m + j\omega C_1) \frac{\omega^2 L_1 C_1 - 1}{g_m^2 + \omega^2 C_1^2}$$

$$Y_{in} = \frac{-g_m - j\omega C_1}{\omega^2 L_1 C_1 - 1}$$

where $g_m = I_E/V_T$, and in the deep Class C large-signal case

$$Z_{in(1)} = (-g_{m(1)} + j\omega C_1) \frac{\omega^2 L_1 C_1 - 1}{g_{m(1)}^2 + \omega^2 C_1^2}$$

$$Y_{in(1)} = \frac{-g_{m(1)} - j\omega C_1}{\omega^2 L_1 C_1 - 1} = -\frac{2I_E}{|v_{in}|} - j\frac{1}{\omega L_1 - \frac{1}{\omega C_1}}$$

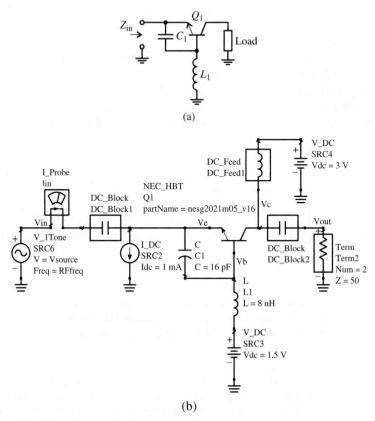

Figure 10-57 Base feedback negative resistance circuit: (a) simplified, (b) with bias and excitation components for input impedance analysis.

where we define the large-signal transconductance at the first harmonic frequency as

$$g_{m(1)} = \frac{2I_E}{|v_{in}|}(\omega^2 L_1 C_1 - 1).$$

We notice that negative resistance is available only above the L_1-C_1 resonance frequency. Observing how the negative resistance changes with input voltage amplitude, we can conclude that this circuit can serve as either negative resistance or negative conductance, depending on the size of g_m compared to ωC_1. If $g_m < \omega C_1$, the circuit is a "negative resistance", where $R_{in(1)}$ starts out at its most negative and becomes more positive as the signal amplitude increases. On the other hand, if $g_m > \omega C_1$, the initial behavior is that of "negative conductance", with $R_{in(1)}$ becoming more negative with increasing $|v_{in}|$ until $g_{m(1)}$ becomes less than ωC_1, at which point $R_{in(1)}$ starts moving towards more positive values. The input conductance $G_{in(1)} = \text{Re}(Y_{in(1)})$ always reduces in magnitude as the input signal increases, suggesting that this circuit is a "negative conductance", but it is the initial behavior or $R_{in(1)}$ that decides whether the circuit is to be used with a parallel or a series resonator according to Kurokawa's condition. This behavior makes this circuit versatile, but also tricky to use. In this section we examine both modes of operation.

If we apply a small enough bias current to the transistor and install a large enough capacitor C_1, we can operate this circuit as a "negative resistance" with a series resonator setting the oscillation frequency. For example, setting $I_E = 1$ mA, $C_1 = 16$ pF and $f = 1$ GHz (same values we used in the earlier discussion of the emitter feedback circuit) results in $g_m = 0.0385$ S $< \omega C_1 = 0.101$ S, assuring that the circuit is a pure "negative resistance". Figure 10-57(b) shows this circuit with the addition of an 8 nH base feedback inductance and idealized bias. The transistor used is the NEC NESG2021M05, a low-power, discrete SiGe HBT with f_T up to 25 GHz, which is supplied with a nonlinear ADS model. We will use this transistor for both the relatively low-frequency 1 GHz oscillator and a mid-frequency 5 GHz oscillator in a subsequent example. Figure 10-58 presents the Harmonic Balance simulated input impedance of the circuit in Figure 10-57(b) together with the analytical approximations. The simple model predicts the circuit behavior relatively well, particularly for the input resistance. Some deviation is seen in the reactance, mainly due to the parasitic reactances present in the transistor.

Given the resonator series resistance R, we can estimate the oscillation amplitude for an oscillator built from this circuit:

$$|v_{in}| \approx \frac{2I_E}{R}\left(\omega L_1 - \frac{1}{\omega C_1}\right)^2$$

$$|i_{in}| \approx \frac{2I_E}{R}\left(\omega L_1 - \frac{1}{\omega C_1}\right)$$

$$P \approx \frac{2I_E^2}{R}\left(\omega L_1 - \frac{1}{\omega C_1}\right)^2.$$

Thus, we can optimize the resonator power by increasing the bias current, reducing the resonator series resistance, or maximizing the residual reactance of L_1 and C_1 in series.

To create an oscillator, we need to install an impedance $Z = -Z_{in(1)} = (4.6 - j45.9)\ \Omega$ at the input terminals of the base feedback negative resistance circuit, where we used the rule of thumb $R = -R_{in}/3$. As previously mentioned, the quality factor of the

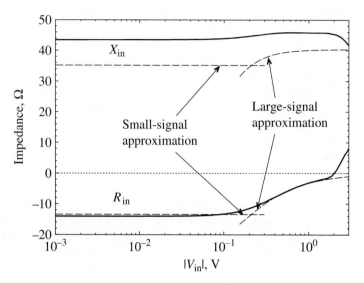

Figure 10-58 Harmonic balance simulation results for the example base feedback negative resistance circuit at 1 GHz.

resulting series resonator should be maximized. Choosing to use lumped components, the inductors typically limit the Q. Assuming inductors with $Q \approx 50$ (typical of wire-wound chip inductors), we need a total series inductance of $QR/\omega = 36.6$ nH, which is the combination of L_1 and the resonator inductance in series. We subdivide this into $L_1 = 8$ nH with 1 Ω series resistance and the resonator inductor $L_2 = 28.6$ nH with 3.6 Ω series resistance. To assure proper impedance at 1 GHz, the resonator capacitor C_2 must be 0.706 pF.

Figure 10-59 presents one practical realization of our 1 GHz oscillator, set up for Harmonic Balance analysis in ADS. The idealized bias components have been replaced with more realistic lumped components. The base voltage is set by the voltage divider R_2-R_3, while the emitter current is controlled by the resistor R_1. The large value inductors (200 nH) and capacitors (200 pF) act as RF chokes and DC blockers, respectively. In a more practical circuit, L_3 and C_3 may be omitted, but that would result in R_1 reducing the Q of the resonator.

Figure 10-60 presents the Harmonic Balance analysis results for the oscillator in Figure 10-59. The circuit oscillates at 1.005 GHz. The magnitude of the first harmonic of the input (emitter) voltage is 0.637 V, compared to 0.71 V obtained from the analytical expression. The voltage waveforms in Figure 10-60(a) highlight the characteristics

Basic Characteristics of Mixers

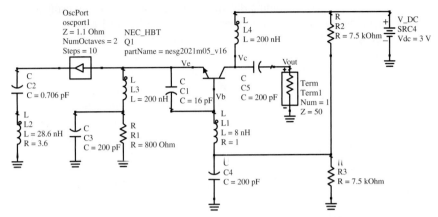

Figure 10-59 1 GHz oscillator utilizing base feedback and a series resonator.

of both the negative resistance circuit and the series resonator. The collector voltage v_C, which is mainly $-i_C R_{\text{load}} + V_C$, is far from sinusoidal. We can see Class C operation (pulses), as well as the effects of parasitic reactances in the transistor, seen as smaller secondary pulses. The emitter voltage waveform contains significant harmonics. This is the result of the pulsed emitter current flowing into a series LC circuit, whose impedance increases with frequency. Here, the parasitic reactances in the transistor actually dampen the higher harmonics of the voltage waveform. The presence of harmonics may or may not be a desirable characteristic of an oscillator circuit.

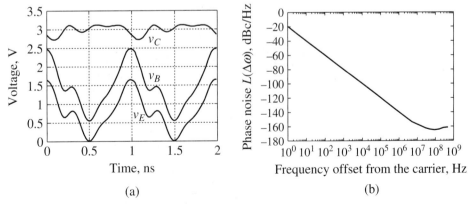

Figure 10-60 Harmonic balance simulation of the 1 GHz oscillator: (a) transistor voltage waveforms, (b) phase noise.

The ADS Harmonic Balance analysis tool is capable of estimating the phase noise of our oscillator, as depicted in Figure 10-60(b). We can recognize the basic shape of the phase noise spectrum given by Leeson's model, generally decreasing by 20 dB/decade of frequency offset from the carrier. At extremely large frequency offsets, a semblance of a noise floor is reached. The −30 dB/decade part of the noise spectrum at low frequency offsets is missing due to the lack of $1/f$ noise models for any of the circuit components.

The output power of this oscillator is −8.6 dBm, slightly higher than the theoretical −10 dBm, which is again due to transistor parasitics.

The second example oscillator operates at 5 GHz with a higher bias current of 3 mA and omits the capacitor C_1. The internal base-emitter capacitance C_{BE} in the transistor takes over the role of C_1. With $C_1 = C_{BE}$ being relatively low, and g_m relatively high due to higher bias current, we can anticipate that this circuit will be of "negative conductance" type. However, we do not expect our analytical expressions to make very accurate quantitative predictions due to transistor parasitics playing a much greater role at this higher frequency. Figure 10-61 shows this circuit with idealized bias components and a 1 nH inductance used for base feedback. Here, we use a 10 Ω load resistance to reduce the collector voltage swing and also reduce the influence of the load on the oscillator. Figure 10-62 shows

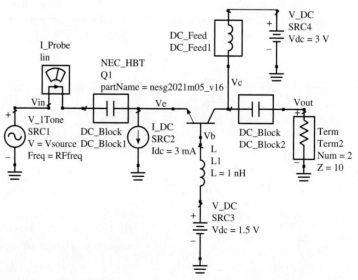

Figure 10-61 Simplified base feedback negative resistance circuit set up for input impedance analysis at 5 GHz.

the Harmonic Balance analysis results for the input impedance and admittance. As anticipated, the circuit produces negative resistance that initially becomes more negative as $|v_{in}|$ increases. Thus, this is a "negative conductance" circuit for use with a parallel resonator. We observe that R_{in} eventually turns around and heads towards positive values. However, the input conductance G_{in} monotonically moves towards more positive values as signal amplitude increases. Unlike in the analytical expression, the input susceptance B_{in} is not constant with signal amplitude due to the parasitic reactances present in the transistor. To ensure that the circuit oscillates at the desired frequency, it is best to use a high-Q resonator, whose susceptance changes rapidly with frequency near resonance. High resonator Q also optimizes oscillator phase noise.

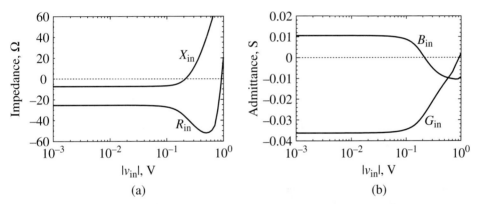

Figure 10-62 Harmonic balance simulation results for the simplified base feedback negative resistance circuit at 5 GHz: (a) input impedance, (b) input admittance.

For the parallel resonator, we can use a circuit consisting of two open-circuited microstrip stubs with lengths $l_+ = \lambda/4 + \Delta l$ and $l_- = \lambda/4 - \Delta l$, where Δl is small. If the transmission lines are lossless, the input admittance of this circuit is

$$Y = j\frac{1}{Z_0}\left[\tan\left(\frac{\omega}{\omega_0}\left(\frac{\pi}{2} + \Delta\theta\right)\right) + \tan\left(\frac{\omega}{\omega_0}\left(\frac{\pi}{2} - \Delta\theta\right)\right)\right]$$

where Z_0 is the microstrip characteristic impedance, ω_0 is the center frequency, and $\Delta\theta$ is the electric length equivalent of Δl. At the center frequency, the two tangent terms cancel, resulting in a parallel resonance.

Any losses in the circuit can be modeled as a shunt conductance G (which may be artificially set by installing a shunt terminating resistor). The quality factor of a parallel resonator is generally defined as

$$Q = \frac{\omega_0}{2G}\frac{dB}{d\omega}\bigg|_{\omega = \omega_0}$$

where B is the susceptance (from $Y = G + jB$). Combining the two expressions, we can approximate the quality factor of this resonator as

$$Q \approx \frac{\pi}{2GZ_0\Delta\theta^2}$$

as long as $\Delta\theta$ is small. An interesting result is available in the literature for the unloaded quality factor of a certain class of transmission line resonators, including simple quarter-wave resonators as well as this resonator:

$$Q_U = \frac{\pi}{\alpha\lambda} = \frac{\beta}{2\alpha}$$

where α is the transmission line attenuation constant in Np/m (1 Np/m = 8.686 dB/m), and λ is the wavelength at the center frequency. If we operate our resonator without an external load resistance for the best possible Q, we can equate the two expressions for Q, and solve for the shunt conductance:

$$G \approx \frac{\alpha\lambda}{2Z_0\Delta\theta^2}.$$

Following the same procedure for a short-circuited quarter-wave resonator, we obtain

$$G_{\lambda/4} = \frac{\alpha\lambda}{4Z_0}.$$

Thus, the advantage of the $\lambda/4 \pm \Delta l$ resonator compared to a simple quarter-wave resonator is that its unloaded shunt conductance is easily adjustable in a useful range using $\Delta\theta$, while the quality factor remains the same.

For the PCB substrate, we choose 0.020 inch thick Rogers 4350, which exhibits a stable dielectric constant of 3.48 and a low loss tangent of 0.0031 (specified at 2.5 GHz). The computed propagation parameters for a 50 Ω microstrip in 1 oz copper on this substrate at 5 GHz are: $\varepsilon_\text{eff} = 2.713$ and attenuation = 3.68 dB/m. Thus, $\lambda = 36.4$ mm, $\alpha = 0.424$ Np/m, and the quality factor of our parallel resonator is $Q_U = 204$.

The small-signal input conductance of the circuit in Figure 10-61 is $G_{in} = -0.0363$ S as seen in Figure 10-62(b). According to the negative conductance rule of thumb, the resonator conductance should be $G = -G_{in}/3 = 0.0121$ S. Inverting the expression for G, we obtain $\Delta\theta = 0.113$ rad $= 6.47°$. Converting to useful PCB length units, the open-circuited stub lengths are $\lambda/4 \pm \Delta l = 358.3 \pm 25.8$ mil. The 50 Ω microstrip width was computed as 43.8 mil.

Figure 10-63 shows one realization of the 5 GHz oscillator using the dual open-circuited stub parallel resonator, set up for Harmonic Balance oscillator analysis in ADS. Here, narrow (high Z_0) $\lambda/4$ lines with large terminating capacitors are used as RFCs for biasing, the base inductance is replaced by the appropriate length of microstrip, and the 50 Ω load is transformed to 10 Ω using a single-stub matching network. The idealized bias voltage and current sources have been replaced by a single 3 V power supply and resistors. Large-value capacitors are used for DC blocking. It is worth noting that this schematic can be further improved by adding microstrip junction patches and device pads.

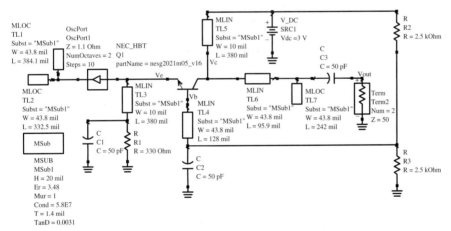

Figure 10-63 5 GHz oscillator utilizing base feedback and a parallel resonator.

Figure 10-64 presents some Harmonic Balance simulation results for the circuit in Figure 10-63. The circuit oscillates at 5.01 GHz, slightly away from the resonator center frequency (since the active circuit has nonzero input susceptance B_{in}). Figure 10-64(a) shows the time domain voltage waveforms at the transistor terminals. The emitter voltage, which is the same as the resonator voltage, has a first harmonic component of 0.511 V in amplitude, which agrees well with the $|v_{in}|$ in Figure 10-62(b) at the point where $G_{in} = -0.0121$ S. Additional harmonics are present in the waveforms, and the transistor operating regime is no longer clearly evident, masked by the transistor parasitic reactances.

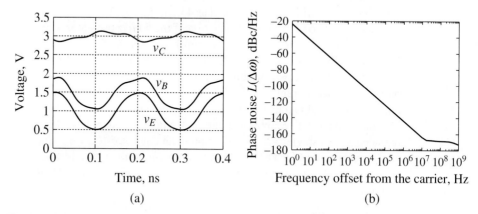

Figure 10-64 Harmonic balance simulation of the 5 GHz oscillator: (a) transistor voltage waveforms, (b) phase noise.

Figure 10-64(b) shows the oscillator phase noise as a function of frequency offset from the carrier. Again, the general shape of Leeson's noise model is recognized, including the −20 dB/decade section and the noise floor, but excluding the −30 dB/decade part, since $1/f$ noise is not modeled. We notice that the phase noise in this oscillator is slightly lower than that in the 1 GHz oscillator (Figure 10-60(b)), even though the frequency has increased fivefold. This improvement is mainly due to the increase in resonator quality factor from 50 to 204 as well as the increase in resonator power from 0.44 mW to 1.6 mW.

The output power of this oscillator at the first harmonic is −1.7 dBm according to the simulation, significantly higher than the −7.4 dBm anticipated from the Class C approximation. This further underscores that, at these high frequencies, the simple analysis is no longer quantitatively useful.

This long example underscores the utility of Harmonic Balance analysis. It also demonstrates the influence of harmonics on the circuit behavior. Moreover, we note how P, Q, and the center frequency influence the phase noise. The design engineer has to pay careful attention to the differences that exist between negative resistance and conductance behaviors.

10.4 Summary

Oscillators and mixers require a nonlinear transfer characteristic and are therefore more difficult to design than standard linear amplifiers. It is not uncommon to encounter circuits that perform as desired, but the design engineer does not understand exactly why they behave this way. Today's extensive reliance on CAD tools has often reduced our thinking to trial-and-error approaches. This certainly applies both to oscillators and mixer RF circuits.

One of the key design requirements of an oscillator is the negative resistance condition as a result of the feedback loop equation, which can be formulated as the Barkhausen criterion

$$H_F(\omega)H_A(\omega) = 1.$$

For instance, the feedback Pi-type network results in a host of different oscillator types, of which we discussed the Hartley, Colpitts, and Clapp designs. At frequencies up to approximately 250 MHz one of the passive feedback elements can be replaced by a quartz crystal whose mechanical vibrations allow substantial improvements in frequency and temperature stability.

For higher frequencies, the S-parameters again become the preferred design procedure. For a two-port oscillator, the oscillation conditions assume importance:

$$k < 1, \Gamma_{in}\Gamma_S = 1, \Gamma_{out}\Gamma_L = 1.$$

A typical approach would start with the test of the stability factor k. Next, from a particular output load within the unstable region, the output reflection coefficient is determined from the knowledge of the input reflection coefficient. Conversely, the design can also be conducted from the input side. To enhance high-frequency performance, a dielectric resonator can be added, whose behavior is that of a parallel resonance circuit with normalized impedance:

$$z_{DR} \approx \frac{R/Z_0}{1 + j2Q_U(\Delta f/f_0)} = 2\beta.$$

Instead of the dielectric resonator, a magnetically induced resonance condition can be established with the help of a YIG element. A Gunn diode finds applications in very high frequency oscillators. To add frequency tuning flexibility, a varactor diode is often employed to adjust the resonant circuit capacitively.

Besides oscillators, mixers are the second group of practical circuits directly exploiting the nonlinear transfer characteristic of active solid-state elements such as diodes and bi- and monopolar transistors. The ability of a mixer to achieve frequency translation finds applications in heterodyne receiver and transmitter circuits. An RF signal ω_{RF} mixed with a local oscillator frequency ω_{LO} results in an output with mixing products of

$$I = \ldots + BV_{RF}V_{LO}(\cos[(\omega_{RF} + \omega_{LO})t] + \cos[(\omega_{RF} - \omega_{LO})t]).$$

Here, the first term signifies upconversion and the second term downconversion. This second-term response can, for instance, be utilized as the required intermediate output signal in a receiver. To isolate the desired signal frequency, extensive filtering is required on the input (image filter) and output (bandpass) sides of the mixer. Single-ended, single-balanced, and double-balanced designs can be constructed by appropriate impedance matching of the source and load to the active device. Additional complications over the amplifier matching network designs arise because of the need to isolate the RF, LO, and IF ports from each other, the most important being LO-to-RF isolation. Balanced mixers, in addition to inherent port isolation, offer improved signal performance through partial cancellation of undesired harmonic responses. However, they require additional complexity, as well as trade-offs in noise figure and power.

Further Reading

B. Gilbert, "A high performance monolithic multiplier using active feedback," *IEEE Journal of Solid State Circuits (JSSC)*, Vol. SC-9, No. 6, 1974.

C. D. Hul and R. G. Meyer, "A systematic approach to the analysis of noise in mixers," *IEEE Trans. on Microwave Theory and Technique (MTT)*, Vol. 40, No. 12, pp. 909–919, 1993.

T. H. Lee, *The Design of CMOS Radio-Frequency Integrated Circuits*, Cambridge Press, 2004.

A. Hajimiri and T. H. Lee, "A General Theory of Phase Noise in Electrical Oscillators," *IEEE Journal of Solid-State Circuits*, Vol. 33, pp. 179–194, 1998.

D. B. Leeson, "A simple model of feedback oscillator noises spectrum," *Proc. IEEE*, Vol. 54, pp. 329–330, 1966.

K. Kurokawa, "Some basic characteristics of broadband negative resistance oscillator circuits", *Bell System Tech. J.*, Vol. 48, pp. 1937–1955, 1969.

K. L. Fong, and R. G. Meyer, "Monolithic RF Active Mixer Design," *IEEE Trans. on Circuits and Systems II: Analog and Digital Signal Processing*, Vol. 46, pp. 231–239, 1999.

G. Watanabe, H. Lau, and J. Schoepf, "Integrated mixer design," *Proc. Second IEEE Asia Pacific Conference on ASICs*, pp.171–174, 2000.

Y. Anand and W. J. Moroney, "Microwave Mixer and Detector Diodes," *Proceedings of IEEE,* Vol. 59, pp. 1182–1190, 1971.

R. J. Gilmore and F. J. Rosenbaum, "An Analytical Approach to Optimum Oscillator Design Using *S*-Parameters," *IEEE Trans. on Microwave Theory and Techniques,* Vol. 31, pp. 633–639, 1983.

G. Gonzalez, *Microwave Transistor Amplifiers, Analysis and Design,* Prentice-Hall, Upper Saddle River, NJ, 1997.

J. B. Gunn, "Effect of Domain and Circuit Properties on Oscillations on GaAs," *IBM Journal of Res. Development,* Vol. 10, pp. 310–320, 1966.

J. M. Manley and H. E. Rowe, "Some General Properties of Nonlinear Elements," *Proceedings of IRE,* Vol. 44, pp. 904–913, 1956.

S. A. Mass, *Microwave Mixers,* Artech House, Dedham, MA, 1986.

M. A. Smith, K. J. Anderson, and A. M. Pavio, "Decade-Band Mixer Covers 3.5 to 35 GHz," *Microwave Journal,* pp. 163–171, Feb. 1986.

G. Vendelin, A. Pavio, and U. L. Rhode, *Microwave Circuit Design Using Linear and Nonlinear Techniques,* John Wiley, New York, 1990.

G. Vendelin, *Design of Amplifiers and Oscillators by the S-Parameter Method,* John Wiley, New York, 1982.

P. C. Wade, "Novel FET Power Oscillators," *Electronics Letters,* September 1978.

Problems

10.1 In Section 10.1.5, the Colpitts oscillator centered around the h-parameter description of the BJT in common-emitter configuration is derived. Follow similar steps and derive the Hartley oscillator. Specifically, find the oscillator frequency in terms of L_1, L_2, C_3, and the h-parameters. Further, establish the ratio of L_2 to L_1.

10.2 A Colpitts oscillator is to be designed for 250 MHz. At the bias point $V_{CE} = 2.7$ V and $I_C = 2$ mA, the following circuit parameters are given at room temperature of $25°$ C: $C_{BC} = 0.2$ fF, $r_{BE} = 3$ kΩ, $r_{CE} = 12$ kΩ, $C_{BE} = 80$ fF. If the inductance is fixed at 47 nH, find values for the capacitances in the feedback loop. Examine whether it is appropriate to use the h-parameters obtained under DC conditions.

10.3 In Section 10.1.6, the quartz element is discussed. Show that solving (10.44) results in the approximate series and parallel resonance conditions of equations (10.45a) and (10.45b). Hint: Use the Taylor Series expansion and retain the first two terms.

10.4 Quartz resonators are typically specified in terms of their series and parallel resonance frequency. For the electric equivalent circuit parameters of $R_q = 50$ Ω, $L_q = 50$ mH, $C_q = 0.4$ pF, and $C_0 = 0.8$ pF, find the series and parallel resonance frequencies based on (10.45a) and (10.45b). Plot the reactance of this quartz resonator over a suitable frequency range.

10.5 A particular crystal oscillator operates with the crystal in the parallel resonance mode. Then, a lossless inductor is added in parallel with the crystal. If the combination of inductor and crystal is required to have the same reactance as the crystal originally did, will the oscillator frequency go up or down? Explain your answer.

10.6 In an oscillator design, the S-parameters of the transistor operating in common-base (CB) mode are often required. Unfortunately, the manufacturer typically supplies the S-parameters for the transistor measured in common-emitter (CE) mode. We therefore have to convert them into CB S-parameters. The usual practice is that the S-parameters are first converted into Y-parameters, then the CE Y-parameters are converted into CB mode, and the result is finally converted into S-parameter representation. Derive CE to CB conversion formulas for the Y-parameter representation.

10.7 A GaAs MESFET chip has the following S-parameters in common-source configuration measured at 4 GHz: $S_{11} = 0.83\angle-67°$, $S_{21} = 2.16\angle119°$, $S_{12} = 0.17\angle61°$, $S_{22} = 0.66\angle-23°$. Using the conversion formulas derived in the previous problem, compute the transistor S-parameters in common-base mode. Determine the stability circles both with and without a positive feedback of $L = 0.5$ nH at 4 GHz.

10.8 In Section 10.2.1, we discussed the case where Γ_L is chosen such that $|\Gamma_{in}|$ becomes greater than unity. Show that $|\Gamma_{in}| > 1$ implies $|\Gamma_{out}| > 1$ and vice versa when the oscillating condition is satisfied.

10.9 In designing oscillators based on the S-parameter approach, it was stated that the following conditions must be satisfied: $k < 1$ and $\Gamma_S\Gamma_{in} = \Gamma_L\Gamma_{out} = 1$. By representing the input impedance as $Z_{in} = R_{in} + jX_{in}$ and output impedance $Z_{out} = R_{out} + jX_{out}$ as well as the source $Z_S = R_S + jX_S$ and load $Z_L = R_L + jX_L$ impedances, show that $R_{in} = -R_S$, $X_{in} = -X_S$, $R_{out} = -R_L$, and $X_{out} = -X_L$. This proves that the S-parameter design is equivalent to the negative resistance design.

10.10 An oscillator has to be designed for 3.5 GHz. The S-parameters of the BJT in common-base configuration are determined to be $S_{11} = 1.1\angle127°$, $S_{12} = 0.86\angle128°$, $S_{21} = 0.94\angle-61°$, and $S_{22} = 0.9\angle-44°$. By adding an inductance to the base, the instability can be enhanced. Determine the inductance for which the instability of the BJT is maximized according to Rollet's stability factor k.

10.11 In Section 10.2.2, the dielectric resonator is introduced and the S-parameters for the angular resonance frequency ω_0 are derived in (10.62). Show that near resonance, (10.62) has to be modified to the form

$$[S] = \begin{bmatrix} \dfrac{\beta}{1+\beta+j2(Q_U\Delta f/f_0)} & \dfrac{1+j2(Q_U\Delta f/f_0)}{1+\beta+j2(Q_U\Delta f/f_0)} \\ \dfrac{1+j2(Q_U\Delta f/f_0)}{1+\beta+j2(Q_U\Delta f/f_0)} & \dfrac{\beta}{1+\beta+j2(Q_U\Delta f/f_0)} \end{bmatrix}.$$

10.12 Since $|\Gamma_{in}| > 1$ and $|\Gamma_{out}| > 1$, they cannot be displayed in a conventional Smith Chart. Extend the Smith Chart in such a way as to be able to display these quantities. What happens with the circles of constant resistance when the reflection coefficients are larger than unity?

10.13 Design a 7.5 GHz oscillator in common-emitter BJT configuration. The S-parameters at $V_{CE} = 5.0$ V and $I_C = 20$ mA are as follows: $S_{11} = 0.87\angle{-40°}$, $S_{12} = 0.25\angle{-32°}$, $S_{21} = 0.6\angle{100°}$, and $S_{22} = 1.21\angle{165°}$. Sketch the circuit, including the DC biasing network ($\beta = 80$).

10.14 A BJT is used in common-base configuration with biasing conditions specified as $V_{CE} = 3$ V and $V_{BE} = 0.9$ V. For this case, the transistor has the following S-parameters at 2.5 GHz: $S_{11} = 1.41\angle{125°}$, $S_{12} = 0.389\angle{130°}$, $S_{21} = 1.5\angle{-63°}$, and $S_{22} = 1.89\angle{-45°}$. Design a series feedback oscillator that satisfies the three conditions (10.50a)-(10.50c).

10.15 The S-parameters of a GaAs FET in common source configuration are measured at 9 GHz and have the following values: $S_{11} = 0.30\angle{-167°}$, $S_{12} = 0.15\angle{21.3°}$, $S_{21} = 1.12\angle{-23.5°}$, and $S_{22} = 0.90\angle{-25.6°}$. Design an oscillator with 9 GHz fundamental frequency and a 50 Ω load impedance. Use microstrip lines for a substrate FR4 with 40 mil thickness ($\varepsilon_r = 3.6$) and determine the widths and lengths of the elements.

10.16 A tunable oscillator involving a varactor diode has to be designed. For the varactor diode, the following data is known: equivalent series resistance of 5 Ω and a capacitance ranging from 15 pF to 35 pF for reverse voltages between 30 V and 2 V. Design a voltage-controlled Clapp-type oscillator with center frequency of 300 MHz and ±10% tuning capability. Assume that the transconductance of the transistor is constant and equal to $g_m = 115$ mS.

10.17 The output power of an oscillator can be approximated by

$$P_{out} = P_{sat}\left[1 - \exp\left(\frac{G_0 P_{in}}{P_{sat}}\right)\right]$$

where P_{sat} is the saturated output power, $G_0 = |S_{21}|^2 > 1$ is the small signal power gain, and P_{in} is the input power. For maximum output power, we obtain

$$d(P_{out} - P_{in}) = 0 \text{ or } \frac{dP_{out}}{dP_{in}} = 1.$$

Show that this leads to the maximum oscillator output power

$$P_{out,\,max} = P_{sat}\left(1 - \frac{1}{G_0} - \frac{\ln G_0}{G_0}\right).$$

For a typical MESFET at 7 GHz, with $G_0 = 7$ dB and $P_{sat} = 2$ W, find the maximum oscillator power.

10.18 The basic downconverting receiver system is shown in Figure 10-38. Draw a similar block diagram describing an upconversion transmitter system and explain its functionality.

10.19 When building BJT and diode-based mixers, intermodulation is an important design consideration. Ideally, over the entire range of RF input signal magnitudes, the mixer should not generate any intermodulation beyond the first harmonic. In reality, however, there may be a significant influence. Follow the same derivation as discussed in Section 10.3.1 and derive the first-, second-, and third-order harmonics, as well as cross terms, for the combined mixer input signal $V = V_{RF}\cos(\omega_{RF}t) + V_{LO}\cos(\omega_{LO}t)$. If the RF signal is 1.9 GHz and the output IF is 2 MHz, determine all frequencies up to the third-order harmonics that are generated by this mixer.

10.20 Design a single-ended BJT mixer as shown in Figure 10-45. Compute values for the resistors R_1 and R_2 such that biasing conditions $V_{CE} = 2.5$ V, $V_{BE} = 0.8$ V, $I_C = 2.5$ mA, and $I_B = 40$ μA are satisfied based on a supply voltage of $V_{CC} = 3.2$ V. RF and IF frequencies are $f_{RF} = 2.5$ GHz and $f_{IF} = 250$ MHz. The BJT is measured at IF to have an output impedance of $Z_{out} = (650 - j2400)\Omega$ for short-circuit input and an input impedance of $Z_{in} = (80 - j136)\Omega$ for short-circuit output at RF frequency.

10.21 For the balanced diode mixer in Figure 10-50 assume the following voltages:

$$v_{RF}(t) = V_{RF}\cos(\omega_{RF}t) \quad \text{and} \quad v_{LO}(t) = [V_{LO} + v_n(t)]\cos(\omega_{LO}t)$$

where the constant amplitudes are such that $V_{RF} \ll V_{LO}$ and where the noise voltage v_n is much smaller than V_{LO}.

(a) Find the currents through the upper diode $i_1(t)$ and lower diode $i_2(t)$ if the transfer characteristic is

$$i_n = C(-1)^{n+1}v_n^2, \quad n = 1, 2$$

where C is a constant, and v_1, v_2 is the respective diode voltage.

(b) Explain how some of the noise cancellation occurs and show that the IF current, after suitable low-pass filtering (behind each diode), can be written as

$$i_{IF} = -2CV_{RF}(V_{LO} + v_n)\sin[(\omega_{RF} - \omega_{LO})t]$$
$$\cong -2CV_{RF}V_{LO}\sin(\omega_{IF}t)$$

A

Useful Physical Quantities and Units

Table A-1 Physical constants

Quantity	Symbol	Units	Value
Permittivity in vacuum	ε_0	F/m	8.85418×10^{-12}
Permeability in vacuum	μ_0	H/m	$4\pi \times 10^{-7}$
Speed of light in vacuum	c	m/s	2.99792×10^8
Boltzmann's constant	k	J/K	1.38066×10^{-23}
Electron charge	e	C	1.60218×10^{-19}
Electron rest mass	m_0	kg	0.91095×10^{-30}
Electon volt	eV	J	1.60218×10^{-19}

Table A-2 Relevant quantities, units, and symbols

Quantity	Symbol	Units	Value
femto	f	—	10^{-15}
pico	p	—	10^{-12}
nano	n	—	10^{-9}
micro	μ	—	10^{-6}
milli	m	—	10^{-3}
kilo	k	—	10^{3}
mega	M	—	10^{6}
giga	G	—	10^{9}
tera	T	—	10^{12}
mil	mil	0.001 inch = 25.4 μm	

International System of Units			
Quantity	Symbol	Units	Dimensions
Electric Charge	C	coulomb	$A \cdot s$
Current	A	ampere	C/s
Voltage	V	volts	J/C
Frequency	Hz	hertz = cycles per second	$1/s$
Electric field	E	V/m	
Magnetic field	H	A/m	
Magnetic flux	Wb	weber	$V \cdot s$
Energy	J	joule	$N \cdot m$
Power	W	watt	J/s
Capacitance	F	farad	C/V
Inductance	H	henry	Wb/A
Resistance	Ω	ohm	V/A
Conductance	S	siemens	A/V
Conductivity	σ	S/m	
Resistivity	ρ	$\Omega \cdot m$	

Table A-3 Relative permittivity and loss tangent for different dielectric materials

Material	ε_r	Loss Tangent			
		f = 1 kHz	f = 1 MHz	f = 100 MHz	f = 3 GHz
Aluminum oxide	9.8	0.00057	0.00033	0.0003	0.001
Barium titanate	37	0.00044	0.0002		0.0023
Porcelain	5	0.0140	0.0075	0.0078	
Silicon dioxide	4.5	0.00075	0.0001	0.0001	0.00006
Araldite CN-501	3.35	0.0024	0.0190	0.0340	0.0270
Epoxy resin RN-48	3.52	0.0038	0.0142	0.0264	0.0210
Foamed polystyrene	1.03	<0.0002	<0.0001	<0.0002	0.0001
Bakelite BM120	3.95	0.0220	0.0280	0.0380	0.0438
Polyethylene	2.3	<0.0002	<0.0002	0.0002	0.00031
Polystyrene	2.5	<0.00005	0.00007	<0.0001	0.00033
Teflon	2.1	<0.0003	<0.0002	<0.0002	0.00015
Sodium chloride	5.9	<0.0001	<0.0002		<0.0005
Water (distilled)	80		0.0400	0.0050	0.1570

Table A-4 American wire gauge chart

Wire Size (AWG)	Diameter in mils	Diameter in millimeters	Area in square mils	Area in square millimeters
1	289.3	7.34822	262934	169.6345
2	257.6	6.54304	208469	134.4959
3	229.4	5.82676	165324	106.6606
4	204.3	5.18922	131125	84.59682
5	181.9	4.62026	103948	67.06296
6	162.0	4.1148	82448.0	53.19212
7	144.3	3.66522	65415.8	42.20364

Table A-4   American wire gauge chart (Continued)

Wire Size (AWG)	Diameter in mils	Diameter in millimeters	Area in square mils	Area in square millimeters
8	128.5	3.2639	51874.8	33.46752
9	114.4	2.90576	41115.2	26.52585
10	101.9	2.58826	32621.1	21.04581
11	90.7	2.30378	25844.2	16.67370
12	80.8	2.05232	20510.3	13.23244
13	72.0	1.8288	16286.0	10.50709
14	64.1	1.62814	12908.2	8.327859
15	57.1	1.45034	10242.9	6.608296
16	50.8	1.29032	8107.32	5.230518
17	45.3	1.15062	6446.83	4.159237
18	40.3	1.02362	5102.22	3.291754
19	35.9	0.91186	4048.92	2.612199
20	32.0	0.8128	3216.99	2.075474
21	28.5	0.7239	2551.76	1.646293
22	25.3	0.64262	2010.90	1.297354
23	22.6	0.57404	1604.60	1.035224
24	20.1	0.51054	1269.23	0.818860
25	17.9	0.45466	1006.60	0.649417
26	15.9	0.40386	794.226	0.512403
27	14.2	0.36068	633.470	0.408690
28	12.6	0.32004	498.759	0.321780
29	11.3	0.28702	401.150	0.258806
30	10.0	0.254	314.159	0.202683
31	8.9	0.22606	248.846	0.160545

Table A-4 American wire gauge chart (Continued)

Wire Size (AWG)	Diameter in mils	Diameter in millimeters	Area in square mils	Area in square millimeters
32	8.0	0.2032	201.062	0.129717
33	7.1	0.18034	158.368	0.102172
34	6.3	0.16002	124.690	0.080445
35	5.6	0.14224	98.5203	0.063561
36	5.0	0.127	78.5398	0.050671
37	4.5	0.1143	63.6173	0.041043
38	4.0	0.1016	50.2654	0.032429
39	3.5	0.0889	38.4845	0.024829
40	3.1	0.07874	30.1907	0.019478

B

Skin Equation for a Cylindrical Conductor

The starting point of the skin effect analysis is Maxwell's equations expressed by the laws of Ampère and Faraday in differential form:

$$\nabla \times \mathbf{H} = \mathbf{J} = \sigma \mathbf{E} \tag{B.1a}$$

$$\nabla \times \mathbf{E} = -\mu\left(\frac{\partial \mathbf{H}}{\partial t}\right) \tag{B.1b}$$

where the displacement current density $\varepsilon(\partial \mathbf{E}/\partial t)$ in (B.1a) is neglected inside a conductor. This is permissible since the electric field in conjunction with the dielectric constant is very small, even for rapidly changing fields, when compared with the conduction current. We evaluate these equations in a cylindrical coordinate system where E_z, E_r, and H_ϕ are the only nonzero components. Carrying out the curl in cylindrical coordinates, results in

$$\frac{1}{r}\frac{\partial}{\partial r}(rH_\phi) = \sigma E_z \tag{B.2a}$$

$$-\frac{\partial H_\phi}{\partial z} = \sigma E_r = 0 \tag{B.2b}$$

$$\frac{\partial E_z}{\partial r} - \frac{\partial E_r}{\partial z} = \mu\frac{\partial H_\phi}{\partial t}. \tag{B.2c}$$

The second equation is zero because H_ϕ does not depend on the z-coordinate. Consequently, E_r is also zero. Differentiating the last equation with respect to r, and then substituting the first into it, yields a second-order differential equation:

$$\frac{\partial^2 E_z}{\partial r^2} + \frac{1}{r}\left(\frac{\partial E_z}{\partial r}\right) - \mu\sigma\left(\frac{\partial E_z}{\partial t}\right) = 0. \tag{B.3}$$

For time-harmonic fields, the time derivative can be replaced by $j\omega$ and combined with $\mu\sigma$ to form the new parameter $p^2 = -j\omega\mu\sigma$. The final form

$$\frac{d^2 E_z}{dr^2} + \frac{1}{r}\left(\frac{dE_z}{dr}\right) + p^2 E_z = 0 \qquad (B.4)$$

is the standard Bessel equation with the solution $E_z = AJ_0(pr)$, where A is a constant and J_0 is the zeroth order Bessel function of first kind. Substituting this solution into the time-harmonic form of (B.2c) gives us

$$j\omega\mu H_\phi = ApJ_0'(pr) \qquad (B.5)$$

with the prime denoting differentiation with respect to the argument. The current is related to the line integral of H_ϕ along the outer perimeter $r = a$ of the conductor: $H_\phi 2\pi a = I$. Thus, we can write

$$H_\phi = A\left(\frac{p}{j\omega\mu}\right)J_0'(pa) = \frac{I}{2\pi a} \qquad (B.6)$$

which allows us to determine the constant A. Substituting A into the solution of the Bessel equation leads to

$$E_z = \frac{j\omega\mu}{2\pi pa}I\left(\frac{J_0(pr)}{J_0'(pa)}\right). \qquad (B.7)$$

An interesting property of Bessel functions is the fact that $J_0'(pa) = -J_1(pa)$, which yields, after algebraic manipulation, the final result

$$E_z = \frac{p}{2\pi\sigma a}I\left(\frac{J_0(pr)}{J_1(pa)}\right). \qquad (B.8)$$

This equation is used in Chapter 1. The validity of (B.8) for the case of zero frequency, or a DC condition, can be proved easily. For low frequency we see that

$$J_0(pr) = 1 - \left(\frac{pr}{2}\right)^2 + \frac{(pr)^4}{(2 \cdot 4)^2} - \frac{(pr)^6}{(2 \cdot 4 \cdot 6)^2} + \ldots \approx 1 \qquad (B.9a)$$

$$J_1(pa) = \frac{pa}{2}\left[1 - \frac{(pa)^2}{2 \cdot 4} + \ldots\right] \approx \frac{pa}{2}. \qquad (B.9b)$$

Substituting (B.9) into (B.8) yields Ohm's law for uniform current density J_z:

$$E_z = \frac{Ip}{2\pi a\sigma}\left(\frac{2}{pa}\right) = \frac{I}{\sigma\pi a^2} \equiv \frac{J_z}{\sigma}. \qquad (B.10)$$

To find the per-unit-length resistance of the wire, we use the power dissipation relation for a resistor

$$P = \frac{1}{2}|I|^2 R \qquad (B.11)$$

Skin Equation for a Cylindrical Conductor

and the volume power loss density in a conductor

$$p = \frac{1}{2} \mathbf{E} \cdot \mathbf{J}^* = \frac{1}{2}\frac{|J|^2}{\sigma}. \tag{B.12}$$

Interpreting (B.11) in per-unit-length terms, the per-unit-length power loss can also be expressed in terms of (B.12):

$$P = \int_A p\, ds = \frac{1}{2}\int_A \frac{|J|^2}{\sigma} ds \tag{B.13}$$

where A represents the cross-sectional area of the conductor. Equating the right hand sides of (B.11) and (B.13), the per-unit-length resistance becomes

$$R = \frac{1}{|I|^2}\int_A \frac{|J|^2}{\sigma} ds \tag{B.14}$$

After substituting the parameters for a round wire, the per-unit-length resistance is

$$R = \frac{|p|^2}{2\pi a^2 \sigma |J_1(pa)|^2}\int_0^a |J_0(pr)|^2 r\, dr \tag{B.15}$$

where the remaining integral must be evaluated numerically.

C

Complex Numbers

This appendix provides a brief summary of several useful concepts and definitions regarding complex numbers and their manipulations as repeatedly used throughout this textbook. Emphasis is placed on the basic definition of a complex number, its use in the magnitude computations, and its meaning in terms of the circle equation.

C.1 Basic Definition

A complex number z, such as the normalized impedance, can be represented in rectangular and polar forms as

$$z = x + jy = |z|e^{j\Theta} \tag{C.1}$$

where the magnitude is given by

$$|z| = \sqrt{z \cdot z^*} = \sqrt{(x+jy) \cdot (x-jy)} = \sqrt{x^2 + y^2} \tag{C.2}$$

and the phase is

$$\Theta = \operatorname{atan2}(y, x) \tag{C.3}$$

where $\operatorname{atan2}$ denotes the inverse tangent for $-\pi < \Theta < \pi$. The star notation denotes the complex conjugate (i.e., $z^* = x - jy$).

C.2 Magnitude Computations

Let us apply the preceding definition to a typical computation involving the magnitude of two complex numbers such as

$$|z + w^*|^2$$

where w is another complex number of the form $w = u + jv$. Substituting w yields

$$|z + w^*|^2 = (z + w^*) \cdot (z^* + w) = |z|^2 + |w|^2 + 2\text{Re}(z \cdot w) \tag{C.4}$$

where we used the fact that the terms $z \cdot w = ux - vy + j(uy + vx)$ and $z^* \cdot w^* = ux - vy - j(uy + vx)$ can be combined to $2\text{Re}(z \cdot w)$. Here, Re(...) represents the real part.

C.3   Circle Equation

Perhaps one of the most useful equations involving complex numbers in RF circuits is the circle equation

$$|z - w| = r \text{ or } |z - w|^2 = r^2 \tag{C.5}$$

which forms the foundation of the Smith Chart. We can verify that this is indeed a circle equation by going through the magnitude computation

$$|z - w|^2 = (z - w) \cdot (z - w)^* = (x - u)^2 + (y - v)^2 = r^2. \tag{C.6}$$

It is seen that u and v are the coordinates of the circle center in the complex z-plane and r is its radius, as depicted in Figure C-1.

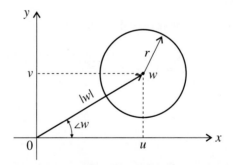

Figure C-1   Circle representation in the complex z-plane.

D

Matrix Conversions

Conversion between **Z**, **Y**, **h**, and **ABCD** representations

	[Z]	[Y]	[h]	[ABCD]
[Z]	$\begin{bmatrix} Z_{11} & Z_{12} \\ Z_{21} & Z_{22} \end{bmatrix}$	$\begin{bmatrix} \dfrac{Z_{22}}{\Delta Z} & -\dfrac{Z_{12}}{\Delta Z} \\ -\dfrac{Z_{21}}{\Delta Z} & \dfrac{Z_{11}}{\Delta Z} \end{bmatrix}$	$\begin{bmatrix} \dfrac{\Delta Z}{Z_{22}} & \dfrac{Z_{12}}{Z_{22}} \\ -\dfrac{Z_{21}}{Z_{22}} & \dfrac{1}{Z_{22}} \end{bmatrix}$	$\begin{bmatrix} \dfrac{Z_{11}}{Z_{21}} & \dfrac{\Delta Z}{Z_{21}} \\ \dfrac{1}{Z_{21}} & \dfrac{Z_{22}}{Z_{21}} \end{bmatrix}$
[Y]	$\begin{bmatrix} \dfrac{Y_{22}}{\Delta Y} & -\dfrac{Y_{12}}{\Delta Y} \\ -\dfrac{Y_{21}}{\Delta Y} & \dfrac{Y_{11}}{\Delta Y} \end{bmatrix}$	$\begin{bmatrix} Y_{11} & Y_{12} \\ Y_{21} & Y_{22} \end{bmatrix}$	$\begin{bmatrix} \dfrac{1}{Y_{11}} & -\dfrac{Y_{12}}{Y_{11}} \\ \dfrac{Y_{21}}{Y_{11}} & \dfrac{\Delta Y}{Y_{11}} \end{bmatrix}$	$\begin{bmatrix} -\dfrac{Y_{22}}{Y_{21}} & -\dfrac{1}{Y_{21}} \\ -\dfrac{\Delta Y}{Y_{21}} & -\dfrac{Y_{11}}{Y_{21}} \end{bmatrix}$
[h]	$\begin{bmatrix} \dfrac{\Delta h}{h_{22}} & \dfrac{h_{12}}{h_{22}} \\ -\dfrac{h_{21}}{h_{22}} & \dfrac{1}{h_{22}} \end{bmatrix}$	$\begin{bmatrix} \dfrac{1}{h_{11}} & -\dfrac{h_{12}}{h_{11}} \\ \dfrac{h_{21}}{h_{11}} & \dfrac{\Delta h}{h_{11}} \end{bmatrix}$	$\begin{bmatrix} h_{11} & h_{12} \\ h_{21} & h_{22} \end{bmatrix}$	$\begin{bmatrix} -\dfrac{\Delta h}{h_{21}} & -\dfrac{h_{11}}{h_{21}} \\ -\dfrac{h_{22}}{h_{21}} & -\dfrac{1}{h_{21}} \end{bmatrix}$
[ABCD]	$\begin{bmatrix} \dfrac{A}{C} & \dfrac{\Delta ABCD}{C} \\ \dfrac{1}{C} & \dfrac{D}{C} \end{bmatrix}$	$\begin{bmatrix} \dfrac{D}{B} & -\dfrac{\Delta ABCD}{B} \\ -\dfrac{1}{B} & \dfrac{A}{B} \end{bmatrix}$	$\begin{bmatrix} \dfrac{B}{D} & \dfrac{\Delta ABCD}{D} \\ -\dfrac{1}{D} & \dfrac{C}{D} \end{bmatrix}$	$\begin{bmatrix} A & B \\ C & D \end{bmatrix}$

$\Delta Z = Z_{11}Z_{22} - Z_{12}Z_{21}$, $\Delta Y = Y_{11}Y_{22} - Y_{12}Y_{21}$,
$\Delta h = h_{11}h_{22} - h_{12}h_{21}$, $\Delta ABCD = AD - BC$

Conversion from S-parameters to Z, Y, h, or ABCD representations

[Z]	$Z_{11} = Z_0 \dfrac{(1+S_{11})(1-S_{22}) + S_{12}S_{21}}{\Psi_1}$ $\quad Z_{12} = Z_0 \dfrac{2S_{12}}{\Psi_1} \quad Z_{21} = Z_0 \dfrac{2S_{21}}{\Psi_1}$ $Z_{22} = Z_0 \dfrac{(1-S_{11})(1+S_{22}) + S_{12}S_{21}}{\Psi_1}$ where $\Psi_1 = (1-S_{11})(1-S_{22}) - S_{12}S_{21}$
[Y]	$Y_{11} = \dfrac{(1-S_{11})(1+S_{22}) + S_{12}S_{21}}{Z_0 \Psi_2}$ $\quad Y_{12} = \dfrac{-2S_{12}}{Z_0 \Psi_2} \quad Y_{21} = \dfrac{-2S_{21}}{Z_0 \Psi_2}$ $Y_{22} = \dfrac{(1+S_{11})(1-S_{22}) + S_{12}S_{21}}{Z_0 \Psi_2}$ where $\Psi_2 = (1+S_{11})(1+S_{22}) - S_{12}S_{21}$
[h]	$h_{11} = Z_0 \dfrac{(1+S_{11})(1+S_{22}) - S_{12}S_{21}}{\Psi_3}$ $\quad h_{12} = \dfrac{2S_{12}}{\Psi_3} \quad h_{21} = \dfrac{-2S_{21}}{\Psi_3}$ $h_{22} = \dfrac{(1-S_{11})(1-S_{22}) - S_{12}S_{21}}{Z_0 \Psi_3}$ where $\Psi_3 = (1-S_{11})(1+S_{22}) + S_{12}S_{21}$
[ABCD]	$A = \dfrac{(1+S_{11})(1-S_{22}) + S_{12}S_{21}}{2S_{21}} \quad B = Z_0 \dfrac{(1+S_{11})(1+S_{22}) - S_{12}S_{21}}{2S_{21}}$ $C = \dfrac{(1-S_{11})(1-S_{22}) - S_{12}S_{21}}{2S_{21}Z_0} \quad D = \dfrac{(1-S_{11})(1+S_{22}) + S_{12}S_{21}}{2S_{21}}$

Conversion from **Z**, **Y**, **h**, and **ABCD** representations to *S*-parameters

[Z]	$S_{11} = \dfrac{(Z_{11} - Z_0)(Z_{22} + Z_0) - Z_{12}Z_{21}}{\Psi_4}$ $S_{12} = \dfrac{2Z_{12}Z_0}{\Psi_4}$ $S_{21} = \dfrac{2Z_{21}Z_0}{\Psi_4}$ $S_{22} = \dfrac{(Z_{11} + Z_0)(Z_{22} - Z_0) - Z_{12}Z_{21}}{\Psi_4}$ where $\Psi_4 = (Z_{11} + Z_0)(Z_{22} + Z_0) - Z_{12}Z_{21}$
[Y]	$S_{11} = \dfrac{(1 - Z_0 Y_{11})(1 + Z_0 Y_{22}) + Y_{12}Y_{21}Z_0^2}{\Psi_5}$ $S_{12} = \dfrac{-2Y_{12}Z_0}{\Psi_5}$ $S_{21} = \dfrac{-2Y_{21}Z_0}{\Psi_5}$ $S_{22} = \dfrac{(1 + Z_0 Y_{11})(1 - Z_0 Y_{22}) + Y_{12}Y_{21}Z_0^2}{\Psi_5}$ where $\Psi_5 = (1 + Z_0 Y_{11})(1 + Z_0 Y_{22}) - Y_{12}Y_{21}Z_0^2$
[h]	$S_{11} = \dfrac{(h_{11}/Z_0 - 1)(h_{22}Z_0 + 1) - h_{12}h_{21}}{\Psi_6}$ $S_{12} = \dfrac{2h_{12}}{\Psi_6}$ $S_{21} = \dfrac{-2h_{21}}{\Psi_6}$ $S_{22} = \dfrac{(h_{11}/Z_0 + 1)(1 - h_{22}Z_0) + h_{12}h_{21}}{\Psi_6}$ where $\Psi_6 = (h_{11}/Z_0 + 1)(h_{22}Z_0 + 1) - h_{12}h_{21}$
[ABCD]	$S_{11} = \dfrac{A + B/Z_0 - CZ_0 - D}{\Psi_7}$ $S_{12} = \dfrac{2(AD - BC)}{\Psi_7}$ $S_{21} = \dfrac{2}{\Psi_7}$ $S_{22} = \dfrac{-A + B/Z_0 - CZ_0 + D}{\Psi_7}$ where $\Psi_7 = A + B/Z_0 + CZ_0 + D$

E

Physical Parameters of Semiconductors

See Table E1 on the following page

Table E-1 Properties of Ge, Si, GaAs, InP, 4H-SiC, GaN and SiGe at 300 K

Properties	Ge	Si	GaAs	InP	4H-SiC	GaN	Si$_{.5}$Ge$_{.5}$
Dielectric constant	16	11.9	13.1	12.5	10	9.5	13.9
Energy gap, eV	0.66	1.12	1.424	1.344	3.23	3.39	0.945
Intrinsic carrier concentration, cm^{-3}	2.40×10^{13}	1.45×10^{10}	1.79×10^{6}	1.30×10^{7}	1.50×10^{-8}	3.00×10^{10}	1.20×10^{13}
Intrinsic resistivity, $\Omega \cdot$ cm	47	2.30×10^{5}	1.00×10^{8}	8.60×10^{7}	1.00×10^{12}	1.00×10^{10}	1.15×10^{5}
Minority carrier lifetime, s	1.00×10^{-3}	2.50×10^{-3}	1.00×10^{-8}	2.00×10^{-9}	1.00×10^{-9}	1.00×10^{-9}	1.75×10^{-3}
Electron mobility (drift), cm^2/(V $\cdot$ s)	3900	1350	8500	4600	1140	1250	7700
Normalized effective mass of the electron	0.55	1.08	0.067	0.073	0.29	0.2	0.92
Hole mobility (drift), cm^2/(V $\cdot$ s)	1900	480	400	150	50	850	1175
Normalized effective mass of the hole	0.37	0.56	0.48	0.64	1	0.8	0.54
Saturated electron velocity, cm/s	6.00×10^{6}	1.00×10^{7}	1.00×10^{7}	1.00×10^{7}	2.00×10^{7}	2.20×10^{7}	1.00×10^{7}
Breakdown electric field, V/cm	1.00×10^{5}	3.00×10^{5}	6.00×10^{5}	5.00×10^{5}	3.50×10^{6}	2.00×10^{6}	2.00×10^{5}
Electron affinity χ, V	4	4.05	4.07	4.38	3.7	4.1	4.025
Specific heat, J/(g $\cdot$ K)	0.31	0.7	0.35	0.31	0.69	0.49	0.505
Thermal conductivity, W/(cm $\cdot$ K)	0.6	1.5	0.46	0.68	3.7	1.3	0.083
Thermal diffusivity, cm^2/s	0.36	0.9	0.24	0.372	1.7	0.43	0.63

F

Long and Short Diode Models

The current flow through a diode under an applied forward bias voltage (see Chapter 6) can be evaluated based on the concentration of the injected **excess charge carriers** in each semiconductor region. Depending on the length of the semiconductor layers, we need to differentiate between a long and short diode model. In the following discussion the current flow is derived for both cases.

With reference to Figure F-1, let us examine the *pn*-junction under forward bias voltage V_A.

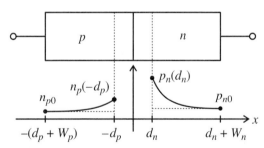

Figure F-1 P_n-junction under forward bias.

Under this applied voltage, the junction is no longer in thermal equilibrium, and minority concentrations are created that exceed the equilibrium condition n_{p0} in the *p*-layer and p_{n0} in the *n*-layer. Indeed, thermodynamic considerations predict the minority concentrations in each layer as

$$p_n(d_n) = p_{n0} e^{V_A/V_T} \quad \text{and} \quad n_p(-d_p) = n_{p0} e^{V_A/V_T}. \tag{F.1}$$

The corresponding excess charge concentrations

$$\Delta p_n = p_n - p_{n0} \quad \text{and} \quad \Delta n_p = n_p - n_{p0} \tag{F.2}$$

begin to diffuse into the semiconductor layers, a process governed by the steady-state diffusion equation. For the *n*-layer, the equation reads

$$\frac{d^2(\Delta p_n)}{dx^2} = \frac{\Delta p_n}{D_p \tau_p} \tag{F.3}$$

where D_p and τ_p are the diffusion constant for holes in the *n*-layer and the excess charge carrier lifetime (on the order of $10^{-7} - 10^{-6}$ s), respectively. It is the so-called **diffusion length**

$$L_p = \sqrt{D_p \tau_p} \text{ and } L_n = \sqrt{D_n \tau_n} \tag{F.4}$$

with respect to the length of each semiconductor layer that determines whether we have to deal with a long or short diode model. The general solution to (F.3) is $\Delta p_n = C_1 e^{x/L_p} + C_2 e^{-x/L_p}$, with two unknown constants to be determined through the boundary conditions on either end of the semiconductor layer. The following two cases are considered:

F.1 Long Diode ($W_n > L_p$, $\Delta p_n \to 0$ as $x \to \infty$)

Since the excess carriers completely decay to zero before reaching the end of the layer, only C_2 has to be specified and $C_1 = 0$. Applying (F.1) as a boundary condition, we can find C_2 and insert it into the general solution, with the result

$$\Delta p_n = p_{n0}(e^{V_A/V_T} - 1)(e^{-(x-d_n)/L_p}) \tag{F.5}$$

In an identical way, we can find for the *p*-layer ($W_p > L_n$, $\Delta n_p \to 0$ as $x \to -\infty$)

$$\Delta n_p = n_{p0}(e^{V_A/V_T} - 1)(e^{(x+d_p)/L_p}). \tag{F.6}$$

F.2 Short Diode ($W_n > L_p$, $\Delta p_n \to 0$ as $x \to d_n + W_n$)

Here, the situation is more complicated since the decay takes place over a finite distance. As as result, both coefficients have to be retained. The additional boundary condition on the right-hand layer now reads $p_n(d_n + W_n) = p_{n0}$. Going through the mathematics eventually leads to

$$\Delta p_n = p_{n0}(e^{V_A/V_T} - 1) \frac{\sinh[(d_n + W_n - x)/L_p]}{\sinh(W_n/L_p)} \tag{F.7}$$

which can be further simplified by approximating the hyperbolic sine function, sinh, by its argument. This is permissible as long as the layer length is less than the diffusion length ($W_n < L_p$). The final result is

$$\Delta p_n = p_{n0}(e^{V_A/V_T} - 1) \frac{d_n + W_n - x}{L_p}. \tag{F.8}$$

Short Diode

Similarly for the p-layer ($W_p < L_n$, $\Delta n_p \to 0$ as $x \to -(d_p + W_p)$)

$$\Delta n_p = n_{p0}(e^{V_A/V_T} - 1)\frac{x - (d_p + W_p)}{L_p}. \tag{F.9}$$

Similarly to (6.14), equations (F.5), (F.6) or (F.8), (F.9) can be used to find the total current through the diode:

$$I = A[J_p(d_n) + J_n(d_p)] = A\left[(-q)D_p\left(\frac{d\Delta p_n}{dx}\right)\bigg|_{d_n} + qD_n\left(\frac{d\Delta n_p}{dx}\right)\bigg|_{-d_p}\right]. \tag{F.10}$$

Inserting (F.5), (F.6) or (F.8), (F.9) into (F.10) finally results in the Shockley equation:

$$I = I_0(e^{V_A/V_T} - 1) \tag{F.11}$$

where the reverse saturation current is for the long diode

$$I_0 = A\left(\frac{qD_p p_{n0}}{L_p} + \frac{qD_n n_{p0}}{L_n}\right) \tag{F.12}$$

and for the short diode

$$I_0 = A\left(\frac{qD_p p_{n0}}{W_n} + \frac{qD_n n_{p0}}{W_p}\right). \tag{F.13}$$

A typical numerical example for a short Si diode involves the following parameters:

$A = 2 \times 10^{-5}$ cm^2, $D_n = 22$ cm^2/s, $D_p = 9$ cm^2/s, $N_A = 1.5 \times 10^{16}$ cm^{-3}, $n_i = 1.5 \times 10^{10}$ cm^{-3}, $N_D = 3 \times 10^{16}$ cm^{-3}, $\tau_p = \tau_n = 10^{-7}$ s, $W_n = W_p = 25$ μm.

With these data we can compute the minority carrier electron and hole concentrations at thermal equilibrium:

$$p_{n0} = n_i^2/N_D = 7.5 \times 10^3 \text{ cm}^{-3}, \quad n_{p0} = n_i^2/N_A = 15 \times 10^3 \text{ cm}^{-3}.$$

Inserting into (F.13) results in a reverse saturation current of 0.5 fA.

G

Couplers

Branchline couplers and power dividers play important roles in RF circuits and measurement arrangements, since they allow the separation and combination of RF signals under fixed phase references. Notably, in the mixer section of Chapter 10 and the measurement protocol of characterizing a device under test in Chapter 4, we see their usefulness. The purpose of this appendix is to discuss some of the couplers and dividers encountered most often in terms of their S-parameters and figures of merit.

G.1 Wilkinson Divider

The transmission line configuration and the microstrip line implementation of this power divider are shown in Figure G-1. The S-parameters for such a three-port network are given by the matrix

$$[\mathbf{S}] = \frac{-1}{\sqrt{2}} \begin{bmatrix} 0 & j & j \\ j & 0 & 0 \\ j & 0 & 0 \end{bmatrix}. \tag{G.1}$$

The figures of merit are the **return loss** at ports 1 and 2

$$\mathrm{RL}_1 \, [\mathrm{dB}] = -20 \, \log|S_{11}| \text{ and } \mathrm{RL}_2 \, [\mathrm{dB}] = -20 \, \log|S_{22}| \tag{G.2}$$

the **coupling** between ports 1 and 2

$$\mathrm{CP}_{12} \, [\mathrm{dB}] = 20 \, \log|S_{21}| \tag{G.3}$$

and **isolation** between ports 2 and 3

$$\mathrm{IL}_{23} \, [\mathrm{dB}] = -20 \, \log|S_{23}|. \tag{G.4}$$

Figure G-2 provides a typical frequency response of RL_1, CP_{12}, and IL_{23} for a center frequency of $f_0 = 1$ GHz.

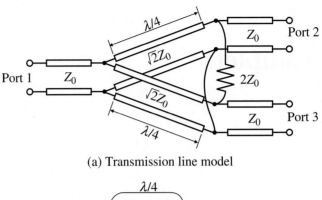

(a) Transmission line model

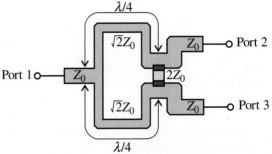

(b) Microstrip line realization

Figure G-1 3 dB Wilkinson power divider.

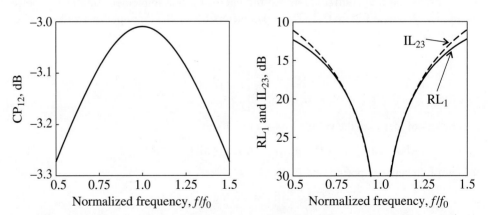

Figure G-2 Frequency response of Wilkinson power divider.

Ideally, return loss and isolation should approach infinity at the center frequency, and the coupling should be as close to –3 dB as possible. We also note that the Wilkinson coupler is not a broadband device. Typical bandwidths do not exceed 20% of the center frequency.

Wilkinson Divider

The derivation of matrix (G.1) is most conveniently carried out by an even and odd mode analysis, as depicted in Figure G-3 for the computation of the S_{12} coefficient. We attach a source V_S to port 2 and terminate the other two ports with a Z_0 load. To make the circuit symmetric, the source V_S at port 2 is divided into a series combination of two $V_S/2$ sources operating in phase. At port 3, two $V_S/2$ sources have a 180° phase shift and their sum is equal to zero. Also, the Z_0 load impedance connected to port 1 is replaced by the parallel combination of two $2Z_0$ impedances.

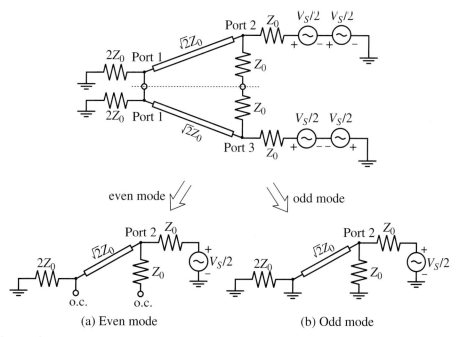

Figure G-3 Even and odd mode representation of Wilkinson divider (o.c.= open circuit).

The reason for choosing the odd and even mode decomposition becomes immediately apparent. Let us consider at first the circuit in Figure G-3(a), which is driven by an even mode, meaning that the drive signals at ports 2 and 3 are in phase. In this case, both ends of the $2Z_0$ cross impedance have the same potential. Thus, there is no current flow and the impedance can be neglected. The input impedance seen at port 2 for this case is the impedance of a $\sqrt{2}Z_0$ quarter-wave transformer terminated with a $2Z_0$ load (i.e., $Z_2 = (\sqrt{2}Z_0)^2/(2Z_0) = Z_0$). Consequently, with an even mode excitation, port 2 is perfectly matched, and the voltage at port 2 is $V_2^e = 0.5(V_S/2) = V_S/4$. The corresponding voltage at port 1 can be found based on our discussion regarding the voltage distribution along a transmission line (see Chapter 2):

$$V_1^e = V^+(1 + \Gamma_0^e) \qquad (G.5)$$

where $\Gamma_0^e = (2Z_0 - \sqrt{2}Z_0)/(2Z_0 + \sqrt{2}Z_0)$ is the even mode reflection coefficient at port 1. Therefore, the even mode voltage at port 1 is

$$V_1^e = V^+(1 + \Gamma_0^e) = jV_2^e \frac{\Gamma_0^e + 1}{\Gamma_0^e - 1} = \frac{-j\sqrt{2}}{4}V_S \qquad (G.6)$$

and where the factor j is due to the $\lambda/4$ transmission line.

For the odd mode excitation, voltages at ports 2 and 3 have opposite polarities and there is a zero potential along the middle of the circuit. This means that the middle is a virtual ground. Since the input impedance seen from port 2 is again Z_0 and port 1 is grounded, we find that $V_1^o = 0$ and $V_2^o = V_S/4$.

The total voltage at ports 1 and 2 is found by adding the even and odd mode voltages. The corresponding S_{12} parameter is then computed as

$$S_{12} = \frac{V_1}{V_2} = \frac{V_1^e + V_1^o}{V_2^e + V_2^o} = -\frac{j}{\sqrt{2}}. \qquad (G.7)$$

An identical analysis for the port 3 to 1 configuration results in $S_{13} = -j/\sqrt{2}$. Furthermore, because the divider is a linear, passive network, we conclude that $S_{21} = S_{12}$ and $S_{31} = S_{13}$. Also, both in the even and odd mode analysis, port 2 is isolated from port 3 by either an open circuit or ground, we find that $S_{23} = S_{32} = 0$. Thus, all off-diagonal terms in (G.1) are verified.

In addition, $S_{22} = S_{33} = 0$ is due to the matching of the odd and even modes. This leaves us only to prove that $S_{11} = 0$. We notice that when port 1 is driven, the current through the $2Z_0$ resistor between ports 2 and 3 is again zero, and has no influence on the circuit. Thus, the impedance Z_1 seen at port 1 is a parallel combination of two Z_0 terminations connected through $\sqrt{2}Z_0$ quarter-wave transformers

$$Z_1 = \frac{1}{2}\frac{(\sqrt{2}Z_0)^2}{Z_0} = Z_0. \qquad (G.8)$$

This proves that port 1 is matched (i.e., $S_{11} = 0$).

G.2 Branch Line Coupler

There are two 3 dB branch line couplers of importance. According to their phase shifts, they are either referred to as 90° (quadrature) or 180° couplers. The S-parameter representation for the 90° coupler is

Branch Line Coupler

$$[S_{90}] = \frac{-1}{\sqrt{2}}\begin{bmatrix} 0 & j & 1 & 0 \\ j & 0 & 0 & 1 \\ 1 & 0 & 0 & j \\ 0 & 1 & j & 0 \end{bmatrix} \quad (G.9)$$

and a circuit schematic is shown in Figure G-4.

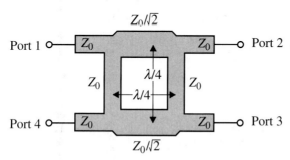

Figure G-4 Microstrip line realization of quadrature hybrid.

Besides return loss, isolation, and coupling definitions given in (G.2)–(G.4), the directivity of a branch coupler is a key parameter, and defined as

$$D_{34} \text{ [dB]} = -20 \log\left|\frac{S_{41}}{S_{31}}\right| \quad (G.10)$$

where D_{34} ideally approaches infinity at f_0.

In our derivation of (G.9), we start with an even and odd mode analysis, as depicted in Figure G-5. We drive the hybrid at port 1 with an RF source V_S and terminate the remaining ports into the characteristic line impedance Z_0. An equivalent circuit results if the source voltage at port 1 is written as the sum of an even (V_{1e}) and odd (V_{1o}) voltages, such that $V_1 = V_S = V_{1e} + V_{1o}$ with $V_{1e} = V_S/2$ and $V_{1o} = V_S/2$. At port 4, we can enforce a zero voltage condition by setting $V_4 = 0 = V_{4e} + V_{4o}$, where $V_{4e} = V_S/2$ and $V_{4o} = -V_S/2$.

The total transmitted voltage at port 2 due to the input voltage at port 1 can be established as

$$V_2 = (T_e + T_o)\frac{V_S}{2} = S_{21}V_S \quad (G.11)$$

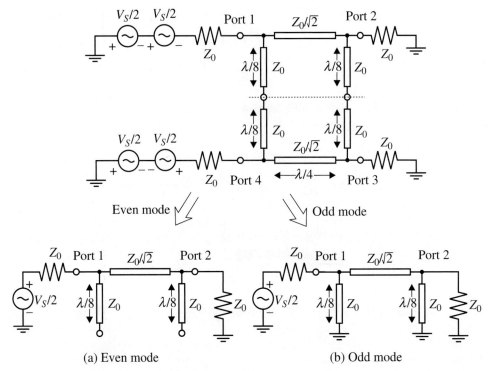

Figure G-5 Building blocks of branch line coupler.

and similarly

$$V_3 = (T_e - T_o)\frac{V_S}{2} = S_{31}V_S \tag{G.12}$$

$$V_4 = (T_e - T_o)\frac{V_S}{2} = S_{41}V_S. \tag{G.13}$$

The reflected signal at port 1 is

$$V_1 = (\Gamma_e + \Gamma_o)\frac{V_S}{2} = S_{11}V_S. \tag{G.14}$$

We must next turn our attention toward finding T_e, T_o, Γ_e, and Γ_o. The transmission line circuits in Figure G-5(a) and (b) can be represented as a three-element model involving either short- or open-circuited stub lines $\lambda/8$ in length.

The even mode and odd mode stub lines have an admittance of

$$Y_e = Y^{oc} = \frac{1}{Z_0}\tan\left(\frac{\pi}{4}\right) \quad \text{and} \quad Y_o = Y^{sc} = \frac{-1}{Z_0}\cot\left(\frac{\pi}{4}\right). \tag{G.15}$$

The three-component circuit in *ABCD* network representation is then

$$\begin{Bmatrix} V_{U1} \\ I_{U1} \end{Bmatrix} = \begin{bmatrix} 1 & 0 \\ jY_{e,o} & 1 \end{bmatrix} \begin{bmatrix} \cos(\beta l) & jY_A^{-1}\sin(\beta l) \\ jY_A\sin(\beta l) & \cos(\beta l) \end{bmatrix} \begin{bmatrix} 1 & 0 \\ jY_{e,o} & 1 \end{bmatrix} \begin{Bmatrix} V_{U2} \\ -I_{U2} \end{Bmatrix}$$

$$= \begin{bmatrix} A & B \\ C & D \end{bmatrix} \begin{Bmatrix} V_{U2} \\ -I_{U2} \end{Bmatrix} \qquad (G.16)$$

where $Y_A = 1/Z_A$ is the characteristic admittance of the $\lambda/4$ line element. Multiplying the three matrices and converting the result into S-parameter form yields, after some algebra, the following nonzero coefficients: $S_{21} = S_{12} = -j(Z_A/Z_0)$, $S_{43} = S_{34} = -j(Z_A/Z_0)$, and $S_{31} = S_{13} = -[1 - (Z_A/Z_0)^2]^{1/2} = S_{42} = S_{24}$. Setting $Z_A = Z_0/\sqrt{2}$ gives the desired matrix listed in (G.9). It is noted that all four ports are matched into Z_0.

The 180° coupler can be constructed by adjusting the lengths of the four transmission line segments and arranging them in a ring configuration, as shown in Figure G-6.

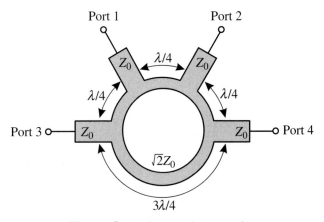

Figure G-6   A 180° ring coupler.

The S-parameter matrix for this configuration, also known as a "rat race," is given by

$$[\mathbf{S}_{180}] = \frac{-j}{\sqrt{2}} \begin{bmatrix} 0 & 1 & 1 & 0 \\ 1 & 0 & 0 & -1 \\ 1 & 0 & 0 & 1 \\ 0 & -1 & 1 & 0 \end{bmatrix}. \qquad (G.17)$$

G.3 Lange Coupler

A popular implementation of the quadrature hybid in microstrip line form is the so-called Lange coupler shown in Figure G-7 for a four-strip configuration. Additional variations involve six- and eight-strip realizations. The interdigital form of the microstrips permits a very compact geometric size and provides for tight coupling.

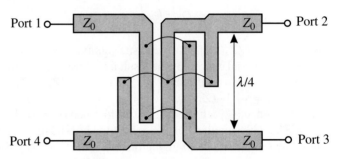

Figure G-7 A 3dB Lange coupler.

Typical coupling values range between –5 and –1 dB. By choosing the length of the microstrip elements appropriately, a very broadband realization of up to 40% bandwidth can be achieved.

Further Reading

P. Karmel, G. Colef, and R. Camisa, *Introduction to Electromagnetic and Microwave Engineering,* John Wiley, New York, 1998.

J. Lange, "Interdigitated Stripline Quadrature Hybrid," *IEEE Trans. on MTT,* Vol. 17, pp. 1150–1151, 1969.

H

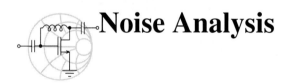Noise Analysis

The intent of this appendix is to provide an overview of the most important noise definitions and concepts as related to the noise figure analysis conducted in Chapter 9.

H.1 Basic Definitions

In a broad sense, noise can be characterized as any undesired signal that interfers with the main signal to be processed. Examples of noisy signals are AC power coupling, crosstalk between circuits, and electromagnetic (EM) radiation, to name but a few sources. Mathematically, we use **random variables** of Gaussian distribution and zero mean to describe noise behavior. Although the mean is zero, the **root mean square** (RMS) value of a noisy voltage signal $v_n(t)$ is not. This can be expressed as

$$V_{n\text{RMS}} = \sqrt{\overline{V_n^2}} = \left(\lim_{T_M \to \infty} \frac{1}{T_M} \int_{t_1}^{t_1 + T_M} [v_n(t)]^2 dt \right)^{1/2} \neq 0 \qquad (\text{H.1})$$

where t_1 is an arbitrary point in time and T_M is the measurement interval.

In 1928, Johnson first observed the fact that a resistor in the absence of any external current flow generates noise due to the random motion of charge carriers in the conductor. The **noise power** in a conductor is quantified as

$$P_n = kT\Delta f = kTB \qquad (\text{H.2})$$

where k is Boltzmann's constant, T is the absolute temperature in K, and $\Delta f = B$ is the **noise bandwidth** of the measurement system. The noise bandwidth is defined as the integration of the instrument's gain $G(f)$ over all frequencies normalized with the respect to the maximum gain $G_{\max}$:

$$B = \frac{1}{G_{\max}} \int_0^\infty G(f) df. \qquad (\text{H.3})$$

We next turn our attention to the noise voltage. Let us consider the simple circuit shown in Figure H-1.

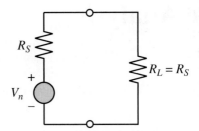

Figure H-1 Noise voltage of a circuit.

According to this circuit, the noise power is treated as if a noise voltage source drives a noiseless resistor R_S. Under matching condition $R_S = R_L$, the noise power of the resistor is given as

$$P_n = \frac{V_{n\text{RMS}}^2}{4R_S} = kTB \tag{H.4}$$

from which the RMS noise voltage is found

$$V_{n\text{RMS}} = \sqrt{4kTBR_S}. \tag{H.5}$$

To keep the notation simple (and since no ambiguity will arise) the subscript RMS is dropped (i.e., $V_{n\text{RMS}} \equiv V_n$). In general, we represent a noisy resistor R as a noise voltage source in series with the noise-free resistor R (Thévenin equivalent circuit) or as a noise current source $I_n = \sqrt{4kTB/R}$ in shunt with a noise-free resistor, as shown in Figure H-2.

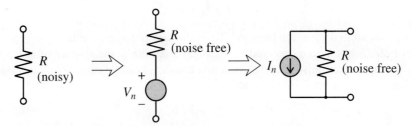

Figure H-2 Equivalent voltage and current models for noisy resistor.

If the the bandwidth is eliminated from (H.5) we can define also-called **spectral noise voltage** and a **spectral noise current**:

$$\overline{V}_n = V_n/\sqrt{B} \quad \text{and} \quad \overline{I}_n = I_n/\sqrt{B} \tag{H.6}$$

whose units are given in $V/\sqrt{Hz}$ and $A/\sqrt{Hz}$.

Frequently, the **spectral density** $S(f)$ is used to quantify the noise content in a unit bandwidth of 1 Hz. For the thermal noise source associated with resistor R, it is given by

$$S(f) = \frac{V_n^2}{B} = 4kTR. \tag{H.7}$$

If $S(f)$ is independent of frequency, we speak of **white noise**. Care is required when noisy elements are added in a circuit. For instance, if two noisy resistors R_1 and R_2 are added, the associated noise sources V_{n1} and V_{n2} cannot be linearly summed. Instead, the resultant noise source V_n is

$$V_n = \sqrt{V_{n1}^2 + V_{n2}^2} \tag{H.8}$$

provided both noise sources are **uncorrelated**. This is equivalent to saying that only power-proportional voltage square quantities can be added because of their random distribution of amplitudes and phases as well as different nonharmonic frequencies.

If the noise sources are **correlated**, a correlation coefficient $C_{n1,n2}$ enters (H.8) such that

$$V_n^2 = V_{n1}^2 + V_{n2}^2 + 2C_{n1,n2}V_{n1}V_{n2} \tag{H.9}$$

where $-1 \leq C_{n1,n2} \leq 1$. It is interesting to observe that if V_{n1} and V_{n2} are 100% correlated ($C_{n1,n2} = 1$), then $V_n^2 = V_{n1}^2 + V_{n2}^2 + 2V_{n1}V_{n2} = (V_{n1} + V_{n2})^2$ and the voltages can again be added, in agreement with Kirchhoff's linear circuit theory.

The thermal noise of a resistor is also referred to as an **internal noise** source, since no external current has to be impressed to observe the noise voltage. However, many noise mechanisms only occur due to externally impressed current flow through the device. They are collectively known as **excess noise**. Chief among them are the $1/f$ noise (also known as flicker noise, semiconductor noise, pink noise) and shot noise. **$1/f$ noise** is most prominent at low frequencies and exhibits, as the name implies, an inverse frequency-dependent spectral distribution. It was first encountered in vacuum tubes as a result of "flickering" noticed on the plates. **Shot noise** is most important in semiconductor devices and can be attributed to the discontinuous current flow across junction potential barriers. As an example, in a semiconductor diode the reverse-bias noise current I_{Sn} is given as

$$I_{Sn} = \sqrt{4qI_SB} \tag{H.10}$$

where I_S is the reverse saturation current and q is the electron charge.

H.2 Noisy Two-Port Networks

The previous analysis can be expanded to two-port networks. Figure H-3 shows a noisy network and the equivalent noise-free network augmented by two current noise sources I_{n1} and I_{n2}.

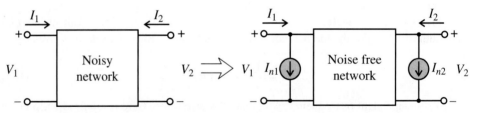

Figure H-3 Noisy two-port network and its equivalent representation.

In *Y*-parameter matrix representation, we can write

$$\begin{Bmatrix} I_1 \\ I_2 \end{Bmatrix} = \begin{bmatrix} Y_{11} & Y_{12} \\ Y_{21} & Y_{22} \end{bmatrix} \begin{Bmatrix} V_1 \\ V_2 \end{Bmatrix} + \begin{Bmatrix} I_{n1} \\ I_{n2} \end{Bmatrix}. \tag{H.11}$$

A more useful representation is obtained by rearranging (H.11) as follows:

$$V_1 = -\frac{Y_{22}}{Y_{21}}V_2 + \frac{1}{Y_{21}}I_2 - \frac{1}{Y_{21}}I_{n2} \tag{H.11a}$$

and

$$I_1 = \frac{Y_{11}Y_{22} - Y_{12}Y_{21}}{Y_{21}}V_2 + \frac{Y_{11}}{Y_{21}}I_2 + I_{n1} - \frac{Y_{11}}{Y_{21}}I_{n2}. \tag{H.11b}$$

Defining the transformed voltage and current noise sources

$$V_n = -\frac{1}{Y_{21}}I_{n2} \text{ and } I_n = I_{n1} - \frac{Y_{11}}{Y_{21}}I_{n2} \tag{H.12}$$

we arrive at the network model shown in Figure H-4.

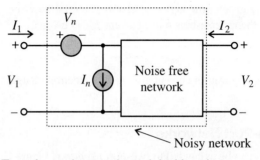

Figure H-4 Transformed network model with noise source at the input.

Example H-1. Noise analysis of a low-frequency BJT amplifier

In Figure H-5, a simplified BJT amplifier is treated as a two-port network consisting of the following parameters: $V_S = 1.5$ mV, $R_S = 50\ \Omega$, $R_{in} = 200\ \Omega$, voltage gain $g_V = 50$, and measurement bandwidth $B = 1$ MHz. The spectral noise voltage and current of the amplifier are given by the manufacturer as $\overline{V}_n = 9\ \text{nV}/\sqrt{\text{Hz}}$ and $\overline{I}_n = 9\text{fA}/\sqrt{\text{Hz}}$. Find the signal-to-noise ratio $\text{SNR} = 20 \log(V_2/V_{n2})$ at the output.

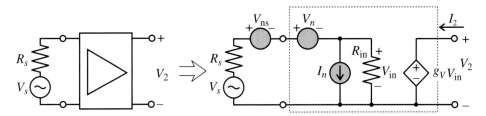

Figure H-5 Amplifier model and network representation with noise source.

Solution. The output voltage V_2 is directly found from $V_2 = g_V R_{in}/(R_{in} + R_S)V_S = 1$ V. The spectral noise sources of the network are next expressed in RMS noise voltage and current:

$$V_n = \overline{V}_n \sqrt{B} = 9\mu\text{V} \quad \text{and} \quad I_n = \overline{I}_n \sqrt{B} = 9\ \text{pA}.$$

The voltage source creates through the voltage divider rule the following noise voltage across R_{in}:

$$\frac{R_{in}}{R_{in} + R_S} V_n = 7.2\ \text{nV}.$$

The noise current source is responsible for the noise voltage of

$$\frac{R_{in} R_S}{R_{in} + R_S} I_n = 0.36\ \text{nV}.$$

Finally, the source resistor contributes the voltage

$$\frac{R_{in}}{R_{in} + R_S} V_{ns} = 728 \text{ nV}$$

where $V_{ns} = \sqrt{4kTBR_S} = 910$ nV, assuming $T = 300°$K.
Therefore, the total noise voltage at the output is

$$V_{n2} = g_V \sqrt{\left(\frac{R_{in}}{R_{in} + R_S} V_n\right)^2 + \left(\frac{R_{in} R_S}{R_{in} + R_S} I_n\right)^2 + \left(\frac{R_{in}}{R_{in} + R_S} V_{ns}\right)^2}$$

$$= 36.4 \text{ μV}.$$

Finally, the signal to noise ratio is

$$\text{SNR} = 20 \log\left(\frac{V_2}{V_{n2}}\right) = 122.8 \text{ dB}.$$

We notice that the noise voltage is dominated by the source.

The example makes clear how the noise voltages are individually computed, added, and amplified to provide the output noise voltage. This is in stark contrast to linear circuit theory.

H.3 Noise Figure for Two-Port Network

The **noise figure** is defined as the ratio between the SNR at the input to the SNR at the output port of a network. Specifically, Figure H-6 depicts the relevant power flow conventions, including the noise representation of the source Z_S.

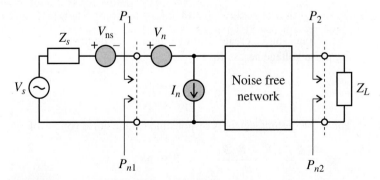

Figure H-6 Generic noise model for noise figure computation.

Noise Figure for Two-Port Network

The noise figure F can be cast into several equivalent representations. The first form involves the ratios of the signal-to-noise power at the input and output ports:

$$F = \frac{P_1/P_{n1}}{P_2/P_{n2}} = \frac{P_{n2}/P_2}{P_{n1}/P_1}. \tag{H.13}$$

Employing the available power gain G_A from Section 9.2.3 to express $P_2 = G_A P_1$ and $P_{n2} = G_A P_{n1} + P_{ni}$, (H.13) is re-expressed as

$$F = 1 + \frac{P_{ni}}{G_A P_{n1}} \tag{H.14}$$

where P_{ni} is the internally generated noise power within the amplifier.

Based on Figure H-6, we see that the signal power P_1 is

$$P_1 = \frac{1}{2}\frac{\mathrm{Re}(Z_{in})}{|Z_S + Z_{in}|^2}V_S^2 \tag{H.15}$$

which is less than the power under source matching ($Z_S = Z_{in}^*$):

$$P_1\big|_{Z_S = Z_{in}^*} = \frac{1}{2}\frac{|V_S|^2}{4\mathrm{Re}(Z_{in})}. \tag{H.16}$$

The thermal noise at the input side is, with $Z_S = R_S + jX_S$:

$$P_{n1} = 4kTR_S B\frac{\mathrm{Re}(Z_{in})}{|Z_S + Z_{in}|^2} = V_{ns}^2\frac{\mathrm{Re}(Z_{in})}{|Z_S + Z_{in}|^2}. \tag{H.17}$$

The power ratio is therefore

$$P_1/P_{n1} = \frac{V_S^2}{V_{ns}^2}. \tag{H.18}$$

The signal power P_2 is simply $P_2 = G_A P_1$, where P_1 is given by (H.15). For the noise power P_{n2}, we state $P_{n2} = G_A P_{n1} + P_{ni}$, where the internally generated noise power P_{ni} takes into account the noise sources associated with the two-port network V_n and I_n. Thus, V_{ns}^2 in (H.17) has to be replaced by all three noise sources: $V_{ns}^2 + V_n^2 + (I_n|Z_s|)^2$. Since the gain applies equally to signal and noise, it cancels and we arrive at

$$P_2/P_{n2} = \frac{V_S^2}{V_{ns}^2 + V_n^2 + (I_n|Z_s|)^2}. \tag{H.19}$$

The noise figure therefore takes the form

$$F = \frac{V_{ns}^2 + V_n^2 + (I_n|Z_s|)^2}{V_{ns}^2} = 1 + \frac{V_n^2 + (I_n|Z_s|)^2}{4kTBR_s}. \tag{H.20}$$

The preceding treatment does not take into account the fact that the same noise mechanisms are usually responsible for both V_n and I_n. Thus, these sources are to a certain degree correlated. This can be incorporated into the noise model by splitting I_n into an uncorrelated I_{nu} and a correlated current I_{nc} contribution, respectively. The correlated current contribution is related to the noise voltage V_n via a complex correlation factor $Y_C = G_C + jB_C$, such that $I_{nc} = Y_C V_n$. Since it is more convenient to deal with noise currents than voltages for our network, we convert the source into an equivalent Norton representation, as seen in Figure H-7.

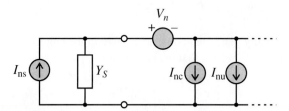

Figure H-7 Noise source modeled at network input.

The total RMS noise current I_{ntot} under short-circuit input conditions can be expressed as

$$I_{ntot}^2 = I_{ns}^2 + V_n^2 |Y_S + Y_C|^2 + I_{nu}^2 \tag{H.21}$$

where $I_{nc} = Y_C V_n$ and $I_n = V_n Y_S$ are combined because of their correlation. We can now rewrite (H.20) as

$$F = \frac{I_{ns}^2 + V_n^2 |Y_S + Y_C|^2 + I_{nu}^2}{I_{ns}^2}. \tag{H.22}$$

Under the assumption that all noise sources are represented by an equivalent thermal noise source, we identify in (H.22)

$$I_{ns}^2 = 4kTBG_S: \text{ noise due to the source } Y_S = G_S + jB_S \tag{H.23}$$

$$I_{nu}^2 = 4kTBG_u: \text{ noise due to the equivalent noise conductance } G_u \tag{H.24}$$

$$V_n^2 = 4kTBR_n: \text{ noise due to the equivalent noise resistance } R_n. \tag{H.25}$$

Inserting (H.23)–(H.25) into (H.22) gives

$$F = 1 + \frac{G_u + R_n |Y_S + Y_C|^2}{G_S} = 1 + \frac{G_u}{G_S} + \frac{R_n}{G_S}[(G_S + G_C)^2 + (B_S + B_C)^2]. \tag{H.26}$$

The circuit designer can minimize (H.26) through an appropriate choice of source admittance Y_S. This process is accomplished by first observing that the imaginary part can be chosen such that $B_S = -B_C$. This eliminates the $(B_S + B_C)^2$ term in (H.26). Next, the remaining expression is minimized with respect to G_S; that is,

$$\left.\frac{dF}{dG_S}\right|_{B_s = -B_c} = \frac{1}{G_{Sopt}^2}\{R_n[2G_{Sopt}(G_{Sopt} + G_C) - (G_{Sopt} + G_C)^2]\} = 0 \qquad (H.27)$$

which yields the explicit optimum value

$$G_{Sopt} = \frac{1}{\sqrt{R_n}}\sqrt{R_n G_C^2 + G_u} \qquad (H.28)$$

The **minimum noise figure** is thus obtained by the optimal source admittance

$$Y_{Sopt} = \left(\frac{1}{\sqrt{R_n}}\sqrt{R_n G_C^2 + G_u}\right) - jB_C. \qquad (H.29)$$

Substituting (H.28) into (H.26) results in the expression

$$F_{min} = 1 + \frac{G_u}{G_{Sopt}} + \frac{R_n}{G_{Sopt}}(G_{Sopt} + G_C)^2. \qquad (H.30)$$

Eliminating G_u in (H.30) by using $G_u = R_n G_{Sopt}^2 - R_n G_C^2$ from (H.28) gives

$$F_{min} = 1 + 2R_n(G_{Sopt} + G_C). \qquad (H.31)$$

This number is typically provided by the device manufacturer. It is dependent on frequency and bias conditions. Equation (H.31) can be incorporated into (H.26) with the result

$$F = F_{min} - 2R_n G_{Sopt} - 2R_n G_C + \frac{G_u}{G_S} + \frac{R_n}{G_S}[(G_S + G_C)^2 + (B_S - B_{Sopt})^2]. \qquad (H.32)$$

Replacing G_u by $G_u = R_n G_{Sopt}^2 - R_n G_C^2$ and rearranging terms provides the final result

$$F = F_{min} + \frac{R_n}{G_S}[(G_S - G_{Sopt})^2 + (B_S - B_{Sopt})^2] = F_{min} + \frac{R_n}{G_S}|Y_S - Y_{Sopt}|^2. \qquad (H.33)$$

This is the starting point of our noise circle analysis in Section 9.5. Based on the characteristic line impedance $Z_0 = 1/Y_0$, (H.33) is often expressed in terms of normalized noise resistance $r_n = R_n/Z_0$, conductance $g_S = G_S/Y_0$, and admittances $y_S = Y_S/Y_0$, $y_{Sopt} = Y_{Sopt}/Y_0$ in the form

$$F = F_{min} + \frac{r_n}{g_S}|y_S - y_{Sopt}|^2. \qquad (H.34)$$

H.4 Noise Figure for Cascaded Multiport Network

The previous noise figure discussion for a single two-port network, with P_{n1} being the input noise and $P_{n2} = G_A P_{n1} + P_{ni}$ being the output noise, can be extended to multiple cascaded networks, as shown in Figure H-8.

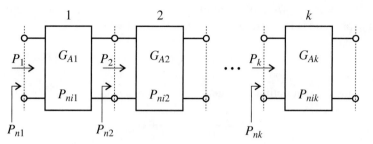

Figure H-8 Cascaded network representation.

In accordance with Figure H-8, we adopt a suitable notation such that G_{Ak} and P_{nik} denote power gain and internal noise generated by amplifier block $k = 1, 2, \ldots$ Thus, for the noise power at the output of the second amplifier section, it is seen that

$$P_{n3} = G_{A2}(G_{A1}P_{n1} + P_{ni1}) + P_{ni2} \tag{H.35}$$

or for the total noise figure F_{tot}, we see

$$F_{\text{tot}} = \frac{P_{n3}}{P_{n1}G_{A1}G_{A2}} = 1 + \frac{P_{ni1}}{P_{n1}G_{A1}} + \frac{P_{ni2}}{P_{n1}G_{A1}G_{A2}}. \tag{H.36}$$

It is customary to retain the same noise figure expression for the individual blocks as derived for the single network; that is,

$$F_1 = 1 + \frac{P_{ni1}}{P_{n1}G_{A1}}, \quad F_2 = 1 + \frac{P_{ni2}}{P_{n2}G_{A2}}, \quad \ldots, \quad F_k = 1 + \frac{P_{nik}}{P_{nk}G_{Ak}}. \tag{H.37}$$

For two networks, this concept leads to the expression

$$F_{\text{tot}} = F_1 + \frac{F_2 - 1}{G_{A1}} \tag{H.38}$$

or for multiple cascaded networks

$$F_{\text{tot}} = F_1 + \frac{F_2 - 1}{G_{A1}} + \frac{F_3 - 1}{G_{A1}G_{A2}} + \ldots + \frac{F_k - 1}{G_{A1}G_{A2}\ldots G_{A(k-1)}} + \ldots \tag{H.39}$$

The preceding considerations have important practical implications. For instance, if two amplifier stages with different gains and noise figures (F_1, G_{A1} and F_2, G_{A2}) are to be cascaded, which sequence of these stages results in the lowest noise figure? To answer this question, let us hypothetically assume amplifier block 1 (F_1, G_{A1}) is followed by amplifier block 2 (F_2, G_{A2}). The total noise figure for this configuration is

$$F_{\text{tot}}(1, 2) = F_1 + \frac{F_2 - 1}{G_{A1}}. \tag{H.40}$$

On the other hand, if block 2 is followed by block 1, we obtain

$$F_{tot}(2, 1) = F_2 + \frac{F_1 - 1}{G_{A2}}. \tag{H.41}$$

Under the assumption that $F_{tot}(1, 2)$ is lower $F_{tot}(2, 1)$, the following inequality must hold:

$$F_1 + \frac{F_2 - 1}{G_{A1}} < F_2 + \frac{F_1 - 1}{G_{A2}}. \tag{H.42}$$

Rewriting (H.42)

$$(F_1 - 1)\left(1 - \frac{1}{G_{A2}}\right) < (F_2 - 1)\left(1 - \frac{1}{G_{A1}}\right) \tag{H.43}$$

allows us to define

$$NM_1 < NM_2 \tag{H.44}$$

where $NM_1 = (F_1 - 1)/(1 - G_{A1}^{-1})$ and $NM_2 = (F_2 - 1)/1(-G_{A2}^{-1})$ are the **noise measures** of amplifiers 1 and 2, respectively. In other words, it is a combination of noise figure and gain that determines the noise measure as a basis of an overall noise performance comparison.

I

Introduction to Matlab

A considerable number of Matlab simulations have been created to enable the reader to reproduce the results presented in the examples. Moreover, it is hoped that these so-called **m-files** will stimulate and encourage the reader to develop code on his or her own relative to the RF topics covered in the ten chapters. This appendix is neither a tutorial of Matlab nor a detailed discussion of the software written in support of this textbook. Rather, it is hoped that sufficient background is provided to understand how Matlab routines are created, and how code can be written to reproduce some of the results and graphs presented in the text. Being a general-purpose mathematical spreadsheet tool, Matlab does not replace specifically developed RF and MW CAD programs, such as MMICAD and ADS, with their powerful circuit analysis, optimization, and even layout utilities. Such dedicated simulation packages cannot be expected to be available to the general reader. For this reason, the authors have attempted to use Matlab as a package that is widely available to students and at very reasonable cost. For more information regarding the use of Matlab, the reader should refer to the following website: *www.mathworks.com*.

This appendix first provides some general background as to how we created the m-files, followed by a brief example of how they are used in the context of a stability analysis, as done in Chapter 9. All m-files can be downloaded from our website: *ece.wpi.edu/RF_Circuit_Design*.

I.1 Background

Matlab is an easy-to-use mathematical spreadsheet with the capabilities for writing custom routines to evaluate the equations discussed in the main text and display the results graphically.

Upon executing Matlab, a window is opened with a command line indicator >>. The appropriate directory can be checked with the command `pwd`, which yields

```
>>pwd
ans =
d:\RF\simulations
```

indicating that the directory is located on d-drive under subdirectories RF\simulations. Changing to a different directory can be initiated through the command cd, and a listing of the files within a directory is done with the commands ls or dir.

By way of an example from Chapter 2, let us consider the following command lines, which can be executed sequentially, each line terminated by pressing **ENTER**.

```
I=5
a=0.005
N=100
M=10
r=(0:N)/N*(M*a)
for k=1:N+1
    if(r(k)<=a)
        H(k)=I*r(k)/(2*pi*a*a)
    else
        H(k)=I/(2*pi*r(k))
    end
end
plot(r*1000,H,'k')
```

In the first line of the program we specify a current through the wire. The second line defines the wire radius. Variables N and M specify the number of points and the maximum distance from the center of the wire at which the magnetic field will be computed. In our case M=10, which means that we will look at distances ranging from 0 to 10 wire radii, and the number of points is set to N=100. The fifth line defines a one-dimentional array of points that determine the actual position from the center of the wire. The command (0:N) creates an array of $N+1$ elements with values of 0,1,2,..., N. After dividing this array by N, the values range between 0 to 1. Next, the array is scaled so that the distance changes from 0 to M*a. An alternative way to define this array would be r=(0:M*a/N:M*a), where the parameter between the two colons defines the step size.

The next line of the code starts a for loop cycle for k ranging from 1 to N+1. For each k, we take the corresponding radius from the array r and check whether it is less or greater than the wire radius. As discussed in Chapter 2, the field inside the wire is linear

$$H = \frac{Ir}{2\pi a^2}$$

with respect to the radial distance, whereas outside the wire we observe

$$H = \frac{I}{2\pi r}.$$

The last line of the code instructs the program to plot a graph of the magnetic field H versus radius r. The graph is shown in black, which is specified in the last parameter of the plot instruction. Some of the possible choices for color include 'k'–black, 'r'–red, 'y'–yellow, 'b'–blue, and 'g'–green. Other useful versions of the plot function include the following:

semilogx—logarithmic scale along *x*-axis, linear scale on *y*-axis
semilogy—logarithmic scale along *y*-axis, linear scale on *x*-axis
loglog—logarithmic scale on both axes
polar—polar plot.

The entire list of commands can be entered in an interactive mode by using the command line. Alternatively, the commands can be placed in a file for batch-mode execution. For example, we can save this program in a file with name field.m; then to execute this program we simply type >>field on the Matlab command line. Note that .m is a file extension reserved for use by Matlab.

I.2 Brief Example of Stability Evaluation

Another useful capability of Matlab is the creation of functions. For example, the following listing is a function that takes an array of *S*-parameter data (s_param) and computes two output parameters: the stability factor k and $|\Delta|$, denoted as K and delta.

```
function [K,delta] = K_factor(s_param)
% Usage:  [K,delta] = K_factor(s_param)
%
% Purpose: returns K factor for a given s-parameter matrix
% if K>1 and delta<1 then circuit is unconditionally stable
% otherwise circuit might be unstable

s11=s_param(1,1);
s12=s_param(1,2);
s21=s_param(2,1);
s22=s_param(2,2);

delta=abs(det(s_param));

K=(1-abs(s11).^2-abs(s22).^2+delta.^2)./(2*abs(s12.*s21));
```

The first line in the listing defines a function K_factor that takes one input parameter s_param and returns two values as a result: K and delta, which are computed inside the function. Unlike program scripts, files containing functions must have the same name as the function name. Therefore, this function is stored in the file K_factor.m.

If the user does not know or forgot how to use the function, one can type

```
help K_factor
```

in the command line of Matlab, and the comments that follow the first line in the function will be displayed.

The program file for creating the S-parameter matrix of a particular transistor and the stability check as well as the display of the stability circles is shown in the next file, entitled `test.m`.

```
% s-parameters for hypothetical transistor
close all;

s11=0.7*exp(j*(-70)/180*pi);
s12=0.2*exp(j*(-10)/180*pi);
s21=5.5*exp(j*(+85)/180*pi);
s22=0.7*exp(j*(-45)/180*pi);

s_param=[s11,s12;s21,s22];

% check stability
[K,delta] = K_factor(s_param)

% create a Smith Chart
smith_chart;

% plot input and output stability circles
input_stability(s_param, 'r');
output_stability(s_param, 'b');

% create PostScript copy of the figure
print -deps 'fig9_8.eps'
```

This file is not a function; it is a collection of commands (program script) and therefore can have any name. In our case, we use the name `test.m`.

We notice that the S-parameters are given in magnitude and phase and stored in an array called s_param. Next, a stability check is performed by passing on the s_param array into the m-file `K_factor.m`, whose task is to find the stability factor and $|\Delta|$ based on equations (9.24) and (9.29). After this, we call three user-defined functions:

- `smith_chart`—creates a figure containing the Z Smith Chart.
- `input_stability`—draws the input stability circle computed from the supplied S-parameters. Circles are drawn in the currently active figure (Smith Chart) and use a specified color (red in our case).
- `output_stability`—draws the output stability circle in the currently active figure.

Simulation Software

The last line of the script creates a file called `fig9_8.eps`, which contains the figure stored in PostScript format. This is the format employed to produce most of the simulation results throughout the book.

I.3 Simulation Software

I.3.1 Overview

The intent of the software is to provide support for the material covered in the textbook. All programs have been developed and tested using Matlab Version 5.2. Although the authors believe that all routines should be compatible with all versions of Matlab, this may not be the case. The software is maintained and regularly updated through our website at ece.wpi.edu/RF_Circuit_Design.

I.3.2 File Organization

All files are organized in the directory structure shown below and the content of each folder is described in the table.

Folder name	Description
RF_matlab	Root directory
ch01-ch10	Selected examples and figures for chapters 1 through 10.
tools	Common files for simulations
amplifiers	Programs for computation of stability factor and simultaneous complex-conjugate matching for the bilateral design
circles	Various circle equations
gain	Constant gain circles
noise	Constant noise circles
quality	Constant Q_n circles
stability	Stability circles
conversion	Conversion routines between different two-port network representations
global	Some useful routines for the computation of the input and output reflection coefficients, VSWR, etc.
networks	Routines for the definition of matching network circuit topologies
smith	Programs related to the construction and plotting of various arcs in the Smith Chart

Additional information for each of the programs can be obtained by executing the command `help <program_name>`, where `<program_name>` is the name of the particular m-file. For example, to obtain help about the program `smith_chart.m`, you execute the command `help smith_chart` in Matlab's main window.

Index

A

ABCD-matrix, 150, 196, 248–249
ABCD network representation, 158–162
    of a T-network, 159–160
    of a transmission line section, 160–162
    of an impedance element, 158–159
Abrupt doping profile, 308
Active biasing networks, 463
    design of, for BJT in common-emitter configuration, 383, 464–465
Active RF component modeling, 361–420
    diode models
        linear diode model, 364–367
        nonlinear diode model, 362–364
    measurements, 397–408
        of AC parameters of bipolar transistors, 398–403
        of DC parameters of bipolar transistors, 397–398
        of field effect transistor parameters, 403–404
        scattering parameter device characterization, 169, 404–408
    transistor models, 367–396
        large-signal BJT models, 367–376, 391
        large-signal FET models, 388–391, 410
        small-signal BJT models, 376–386, 420
        small-signal FET models, 391–394

Active RF components
    bipolar-junction transistor (BJT), 313–330
        brief history of, 313
        construction of, 314
        frequency response, 322, 338, 433
        functionality of, 316–338
        limiting values, 328–329
        temperature behavior, 324–328
        types of, 316
    high electron mobility transistors, 331, 342–347
        computation of HEMT-related electric characteristics, 281, 345–346
        construction of, 314, 343
        frequency response, 322, 324
        functionality of, 331–335
        pseudomorphic HEMTs (pHEMTs), 343
    RF diodes, 298–313
        BARRITT diodes, 313
        Gunn diodes, 313, 608–609, 641
        IMPATT diodes, 310–312, 313
        PIN diodes, 301–306
        Schottky diode, 299–301
        TRAPATT diodes, 313
        tunnel diodes, 312
        varactor diodes, 307–309
    RF field effect transistors, 330–331
        construction of, 330–331
        frequency response, 338–339
        functionality, 343–346
        limiting values, 339
    semiconductors, 278–298

    See also Active RF component modeling; Bipolar-junction transistor (BJT); Field effect transistors, 278
Active RF components, 359
Admittance matrix, 148, 149, 167
Admittance Smith Chart, *See* Y-Smith Chart, 6
Admittance transformation, 123–127
    parametric admittance equation, 123
    Y-Smith Chart, 124
    Z-Smith Chart, 124, 125
    Z*Y*-Smith Chart, 126–127
Admittance/**Y**-matrix, 148
Advanced Design Systems (ADS), 121, 362, 408
American wire gauge (AWG):
    chart, 649
    system, 653
Ampère's law, 52–55, 65–66
    in differential (point) form, 53–54
Amplifiers:
    balanced amplifier design, 441, 444, 533–535
    broadband, 529–539
    characteristics of, 486–487
    class A, 458
    class AB, 458
    class B, 458
    class C, 458
    classes of operation, 458–463, 477–478
    dynamic range, 541–542, 616
    efficiency computation, 460–463

Amplifiers: (*Continued*)
  gain compression, 540, 549
  high-power, 540–542
  intermodulation distortion (IMD), 541, 616
  multistage, 543–550
  power amplifier (PA), 3, 4
    RF source, 550
  power relations, 487–492, 550–552
  RF transistor amplifier design, 485–557
  Cascode, 395
Attenuation behavior of filter, 227
Available power, 90, 488
Available power gain, 489
Avalanche breakdown, 329
Avalanche effect, 310

B

Balanced amplifier design, 533–535
Bandgap energy, 279, 281, 363
Bandpass filters, 260–262
  with coupled line transmission line segments, 264
  frequency transformation, 236, 240
Bandstop filters, 206, 214–221, 254
  frequency transformation, 206, 241–242, 249, 254
  response, 216
Bandwidth, 208, 276, 433
  noise, 437, 486, 603
Bandwidth factor, 254
Barkhausen criterion, 561
BARRITT diodes, 313
Base-width modulation, 377
BFG403W BJT, 537–539
BFG540 BJT, 544
BFQ65 BJT, 592
Biasing networks, 464, 469, 470, 471, 479
  amplifier classes of operation and, 458
  bipolar transistor biasing networks, 463
  field effect transistor biasing networks, 469–478
Bilateral matching, 527
BiCMOS, 348
Binomial low-pass filters, 206
Bipolar transistor biasing networks, 463–469

Bipolar-junction transistor (BJT), 151, 154, 313–329
  bias conditions, setting for, 383–403
  brief history of, 278
  construction of, 314–315, 341
  forward active mode, 316, 318–320, 372
  frequency response, 322–324
  functionality of, 316–322
  GaAs field effect transistors (GaAs FETs), 314
  input/output impedances, determining for, 107–110, 383–386
  internal resistances/current gain, 153, 153–154
  limiting values, 328–329, 339
  low-frequency hybrid network, description of, 151–152
  parameter nomenclature, 318
  parameters, 362
  reverse active mode, 317, 321–322
  saturation mode, 317, 322
  S-parameters, computing for, 383–386
  stabilization of, 501–503
  temperature behavior, 324–328
  types of, 316
BJT, *See* Bipolar-junction transistor (BJT), 463
Black box methodology, 146
Blocking capacitors, 303, 469
Boltzmann's constant, 677
Branches (flowgraph), 178
Branch line coupler, 672–675
Branch line couplers and power dividers, 669
Breakdowns, 329
Broadband amplifiers, 529–550
  balanced amplifier design, 533–535
  design of, using frequency compensated matching networks, 530–533
  negative feedback circuits, 535
  parameters of, 529, 530
Butterworth filters, 206, 224–228
  coefficients for, 227
  compared to linear phase Butterworth and Chebyshev filters, 234–236

C

Calibration procedure (Network Analyzer), 189
Capacitance, 16

Capacitors:
  chip, 25, 27
  high-frequency, 19–21
  RF impedance response of, 19–22
Carbon-composite resistors, 16
Cascading bandpass filter elements, 262
Cascading networks, 157–158
Cauer low-pass filters, 206
Cellular phones, 3
Center frequency, 206, 254, 276
Chain scattering matrix, 175–177
Chain/**ABCD**-matrix, 150
Channel inversion, 341
Channel length modulation parameter, 336
Characteristic line impedance, 67–68
Charge:
  electric, 279, 286
  electron, 290
  excess charge carriers, 665
Chebyshev bandpass filter design, 230, 234, 243–245
Chebyshev filters, 206, 224, 228–236
  coefficients for, 230, 233, 242–243
  compared to Butterworth and linear phase Butterworth filters, 206, 230, 234–236
Chebyshev polynomials, 206, 228–230
Chip capacitors, 27
Chip resistors, 25–26
Circle equation, 526, 658
Clapp oscillator, 581
Class A amplifier, 460, 462
Class AB amplifier, 458
Class B amplifier, 460, 462
Class C amplifier, 458
Coaxial cable, 46, 47, 50, 94–95
Colpitts oscillator, 581, 643
  design, 581–585, 592, 594
Commensurate line, 246
Common base, 395
Common collector, 396
Common drain, 396
Common emitter, 395
Common gate, 395
Common source, 395
Complementary MOS, 340
Complex numbers, 657–658
  basic definition, 657
  circle equation, 658
  magnitude computations, 657, 658
Complex propagation constant, 66, 74–75

Index

Compressed Smith Chart, 111
Computer aided design (CAD), 120–122, 361
Conductance, 128, 394, 426
Conduction angle, 458, 460, 462, 565
Conduction parameter, 390
Conductivity, 11, 280
    thermal, 664
Conformal mapping, 68
Constant current flow in a conductor, magnetic field generated by, 54–55
Constant gain, 504–521
    bilateral design, 512–515
    operating and available power gain circles, 515
    unilateral design, 504–512, 551
        applicability test, 511
        unilateral figure of merit, 511
Constant gain circles, derivation of, 506, 509, 510, 531
Constant VSWR circles, 525–529
    design for given gain/noise figure, 527
Conversion compression, mixers, 616
Conversion gain (CG), mixers, 616
Conversion loss (CL), mixers, 615
Correlated noise sources, 679
Coupled filter, 257–266, 270
    bandpass filter section, 260–262
    cascading bandpass filter elements, 262–264
    design example, 264–266, 600
    odd/even mode excitation, 257–259, 263, 264
Couplers, 669–676
    branch line coupler, 669
    dual directional, 405
    Lange coupler, 676
    Wilkinson power divider, 669–672
    rat-race, 675
Coupling, 515
Coupling coefficient, 599
Coupling factor, 405
Cross-coupled oscillator, 571
Current, 67
Cutoff frequency, 206, 239, 322
Cylindrical conductor skin equation for, 653

D

Device spread, 337
Device under test (DUT), 169
Dielectric constant, 34, 48, 68
Dielectric resonator oscillators, 598
    design of, 598
Diffusion constants, 285
Diffusion current, 285
Diffusion length, 666
Digital subscriber line (DSL), 45
Digital-to-analog converter (DAC), 3
Dimensions, 6–9, 422
Dimensions and units, 6–9
Diode equation, 293
Diode models, 362–367
    linear diode model, 364–367
    nonlinear diode model, 362–364
Diodes:
    Gunn, 313
    IMPATT, 310–312
    leakage, 48, 68, 373, 474
    PIN, 301–305
    RF, 298–313
    Schottky, 299–301
    TRAPATT, 313
    tunnel, 312
    varactor, 307–309
Dipole domains, 608
Directivity factor, 405
Dispersion-free transmission, 75
Dissipation factor d, 219
Distributed circuit theory
    fundamental concepts of, 95–96
Double-balanced active mixer, 627
Double-balanced mixer, 623–624
Double-stub matching networks, 454–458
    design of, 456–458
Downconverted frequency components, 612
Drain saturation voltage, 334
Dual directional coupler, 404, 405–406
Dynamic Ebers-Moll model, 369
Dynamic range:
    amplifiers, 543, 544
    mixers, 616
    spurious-free, 542, 544

E

Early effect, 338, 373
Ebers-Moll BJT model, 362, 367–376, 397, 414
    dynamic Ebers-Moll model, 369
    Injection version (Ebers-Moll), 367
    large-signal, transport vs. injection form of, 371–373
    popularity/acceptance of, 373
Effective dielectric constant, 69, 70, 475
Effective width, 72
Electon volt, 647
Electric circuit models, 413–415
Electric field, 42, 52, 598
Electrical length, 111, 448, 450, 596
Electric charge, 648
Electron affinity, 664
Electron charge, 647
Electron mobility, 664
Electron rest mass, 647
Elliptic low-pass filters, 207
Emission coefficient, 362
Emitter degeneration, 395
Emitter efficiency, 314
Emitter follower, 396
Energy, 2
Energy gap, 279
Equivalent circuit representation, 16, 17, 49–51, 153
Equivalent noise resistance, 521
Equivalent series resistance (ESR), 21
Excess charge carriers, 665
Excess minority carrier lifetime, 302
Excess noise, 679
External inductance, 14
External Q, 210, 220, 223

F

Faraday's law, 52, 55–57, 63
Feedback loop, 180
Feedback oscillator design, 441, 578–581
Field effect transistor biasing networks, 469
Field effect transistors (FETs), 330–342
    construction of, 330
    frequency response, 322–324, 433
    functionality, 331–338
    GaAs field effect transistors (GaAs FETs), 314
    hetero FET, 331
    high electron mobility transistors (HEMTs), 314
    junction FET (JFET), 330
    limiting values, 339
    metal insulator semiconductor FET (MISFET), 330
    metal oxide semiconductor FET (MOSFET), 330
    metal semiconductor FET (MESFET), 330
Filter design
    concepts, 268–270

Filter implementation, 245–257, 276
  Kuroda's identities, 247–249, 251
  unit elements (UEs), 247
Filter Q, 210, 220, 223
Filters:
  bandpass, 205, 214–216
  bandstop, 241–243, 254–257
  binomial, 206
  Butterworth, 225–228
  Chebyshev, 228–230
  coupled, 257
  elliptic, 206
  high-pass, 205, 206, 213–214, 224, 238–239
  linear phase, 228
  low-pass, 205, 206, 210–213, 239
  RF Filter Design, 205–270
Finfet, 347
Fire retardant (FR), 29
Fixed-frequency oscillators, 591–598
  design of, 104, 243
  GaAs FET oscillator, microstrip design of, 595–598
Flat coils, 28, 29
Flow graphs, 178–184
  branches, 178
  nodes, 178
Forbidden regions, 431–439
Forward active mode, 318–320, 369, 372, 375, 376
Forward early voltage, 373
Forward power gain, 173
Forward voltage gain, 173
Frequency, 169
  center, 206
  cutoff, 205, 238
Frequency-compensated matching networks, 530–533
Frequency response, 24
  bipolar-junction transistor (BJT), 322
  field effect transistors (FETs), 338
  high electron mobility transistors (HEMTs), 346
  input reflection coefficient, 508
  matching networks, 166, 351, 421
  RF field effect transistors, 330–339
  Wilkinson power divider, 669
Frequency reuse, 9
Frequency spectrum, 9
Frequency transformation, 236, 236–242
  bandpass filters, 214–216
  bandstop filters, 214–216
  high-pass filters, 213–214

G

GaAs field effect transistors (GaAs FETs), 595, 600
GaAs MESFET, 342, 393
  determination of cutoff frequency of, 393–394
Gain-bandwidth product, 382
Gain compression, 540
Generic RF system, block diagram of, 3, 4, 42
Generic transmission line circuit
  phasor representation of source, 87
Gilbert cell mixer, 627
Global positioning systems (GPS), 2
Global System Mobile (GSM), 616
Gradual-channel approximation, 333
Graphical representation
  parametric, 110–111
Gummel-Poon BJT model, 367–375
Gunn diodes, 313, 641
Gunn element oscillator, 608–609

H

Harmonic balance analysis, 566
Harmonic distortions, 541
Harmonic IMD, mixers, 616
Hartley oscillator, 581
Heterodyne receiver, 609, 645
Hetero FET, 331
Heterojunction bipolar transistors (HBTs), 314
High electron mobility transistors (HEMTs), 314, 342–347
  computation of HEMT-related electric characteristics, 345–346
  construction of, 343
  frequency response, 338
  functionality of, 403
  pseudomorphic HEMTs (pHEMTs), 343
High-frequency capacitors, 19–21, 27, 578
  loss tangent, 19
High-frequency inductors, 22–25, 28
  RF chokes (RFCs), 22
High-frequency oscillator configuration, 587–609
High-frequency resistors, 16–19
  surface mounted devices (SMDs), 16
High-frequency systems, 32–33
High-pass filters, 205, 213–215
  frequency transformation, 236–242

High-power amplifiers, 540–542
High-side injection, 612, 628
High-temperature cofired ceramics (HTCCs), 29
$\mathbf{h}$-matrix, 150–153, 196
Hole mobility, 664
Hybrid circuits, 29
Hybrid/$\mathbf{h}$-matrix, 150–153, 196
Hyper-abrupt profile, 308

I

IEEE frequency spectrum, 9
I-layer, 301
Image filter, 614, 628
Image frequencies mapping, 614
Image impedance, 261
Image reject mixer, 628–640
Image rejection, 630
Impact ionization, 310
IMPATT diodes, 310–312
Impedance:
  characteristic line, 67
  input impedance matching, 91–92
  intrinsic, 7, 8
  normalized impedance equation, 106–112, 439
  waves, 67
Impedance matching using discrete components, 422–431
Impedance matrix, 147–149
Impedance transformation, 112–122
  computer simulations, 120–122
  for general load, 112
  special transformation conditions, 116–117
    open-circuit transformations, 117–119
    short-circuit transformations, 119–120
  standing wave ratio (SWR), 113–115
Impedance/$\mathbf{Z}$-matrix, 147–148
Impulse sensitivity function, 577
Induced voltage, in a stationary wire loop, 56–57
Inductance, 14, 15
Inductors:
  high-frequency, 22–28
  surface-mounted, 28
Input impedance matching, 91–92
Input matching network, 525
Input reflection coefficient, frequency response of, 508

Index

Input resistance, 377, 399
Input stability circle equation, 493
Input VSWR, 525, 528
Insertion loss (IL), 92–94, 207, 221–224, 269
Integrated active mixers, 624–628
Intercept point (IP), 542
Interconnecting networks, 154–162
    ABCD network representations, 158–162
    cascading networks, 157–158
    parallel connection of networks, 156–157
    series connection of networks, 154–156
Intermediate frequency (IF), 610
Intermodulation distortion (IMD):
    amplifiers, 541
    mixers, 617
Internal inductance, 11
Internal noise source, 679
Interstage matching network, 4, 543
Intrinsic carrier concentration, 664
Intrinsic impedance, 7
Intrinsic resistivity, 664
Inverse early voltage, 373
Isolation, 669

J

Junction FET (JFET), 330
Junction grading coefficient, 363

K

Kirchhoff's voltage and current laws, 62–66
Kuroda's identities, 247–249

L

Lambda-quarter transformer, 84
Lange coupler, 676
Large-signal BJT models, 367
Large-signal diode model, 362
Large-signal Ebers-Moll circuit model, 367, 368, 371
Large-signal FET models, 388
Large-signal S-parameters, 540
Lead resistance, 16
Leeson's phase noise model, 576
Linear diode model, 364–367
Linear phase behavior, 226

Linear phase low-pass filters, coefficients for, 227
Load impedance
    reflection coefficient, 104–111
Loaded Q, 210, 220, 223, 433
Load-pull, 540
Local oscillator (LO) frequency, 610
    selection of, 610
Long and short diode models, 665–667
Long diode, 666
Loop gain equation, 561
Loss factor, 222, 226, 236
Loss tangent, 19, 21
Low- to high-frequency circuit operation, 2–40
Low-frequency electronics
    resistors at, 16–19
Low-pass filters, 208, 210–212
    actual attenuation profiles for, 206, 209
    frequency transformation, 236
Low-side injection, 612, 618
Low-temperature cofired ceramics (LTCCs), 29
L-section matching networks, 422–431
    analytical approach to design of, 423
    graphical approach to design of, 427

M

μ stability factor, 497
Magnetic field, 56
Magnetic flux, 56
Magnitude computations (complex numbers), 657–658
Mapping:
    conformal, 68, 104
    image frequencies, 614
Matched load reflection coefficient, 512
Matched source reflection coefficient, 512
Matching and biasing networks, 421–483
Matching networks:
    design approaches, 530
    double-stub matching networks, 454–458
    forbidden regions, 431–441
    frequency-compensated matching networks, 530–533
    frequency response, 431–437
    interstage matching network, 543
    with lumped and distributed components, design of, 447–449

    narrow-band matching network, design of, 439–441
    quality factor, 31, 431–438
    single-stub matching networks, 450–454
    T and Pi matching networks, 442–446
    transfer function of 222
    two-component matching networks, 422–431
MathCad, 136, 168
Mathematica, 136
Matlab, 121, 133, 689–694
    background, 689–691
Matlab simulations, 689
Matrix conversions (networks), 659–661
Maximally flat filters, See Butterworth filters, 224, 225, 254
Maximum available gain, 514
Maximum frequency of oscillation, 324
Maximum power hyperbola, 328
Maximum power transfer, 92, 423, 488
Maximum stable gain, 514
Maxwell's equations, 52
Mesa processing technology, 301
Metal-film resistors, 16
Metal-film resistors, RF impedance response of, 17–19
Metal insulator semiconductor FET (MISFET), 330
Metal oxide semiconductor FET (MOSFET), 330, 339–342
Metal semiconductor FET (MESFET), 330
    drain saturation current in, 335
    GaAs MESFET, determination of cutoff frequency of, 393
    idealized MESFET device structure, 392
    I-V characteristic of, 337–338
    small-signal MESFET model, 391, 392
    SPICE modeling parameters for, 391
m-files, 137
Microstrip filter design, examples of, 249–255
Microstrip line matching networks, 446–458
    from discrete components to microstrip lines, 446–449
Microstrip transmission lines, 68–72
    design of, 47, 68
Microwave (MW), 3
Microwave amplifier, 166

Microwave Monolithic Integrated Circuit (MMIC), 29
Miller effect, 379–382, 395
Minimum detectable signal, 541
Minimum noise figure, 523–524, 685
  design of small-signal amplifier for, 523–524
Minority carrier lifetime, 664
Mixers, 609–640
  basic characteristics of, 609–610
  basic concepts, 610–611
  conversion compression, 616
  conversion gain (CG), 616
  conversion loss (CL), 615
  double-balanced mixer, 623–624
  dynamic range, 616
  frequency domain considerations, 612–614
  harmonic IMD, 616
  intermodulation distortion (IMD), 616
  local oscillator (LO) frequency, 610
    selection of 612–614
  noise figure, 521, 523
  single-balanced mixer, 622–623
  single-ended mixer design, 614–617
    BJT mixer, 618–622
  See also Oscillators
Mixing process, spectral representation of, 612, 613
MMIC (Monolithic Microwave Integrated Circuit), 546
Modulation-doped field effect transistors (MODFETs), *See* High electron mobility transistors, 342
Monopolar devices, 330
Multilayer boards, 48
Multistage amplifiers, 543–546
  transistor choices for design of, 544

N

Narrow-band matching network, design of, 439–441
N-channel, 330, 388
Negative conductance oscillator, 563
Negative feedback circuits, 535
Negative feedback loop broadband amplifier, design of, 537
Negative feedback, defined, 474, 535–538
Negative resistance oscillator, 562–573

Network parameter sets, 163–169
Networks, 145–197
  admittance/**Y**-matrix, 148, 156
  basic definitions, 145, 146–154
  chain/**ABCD**-matrix, 150
  hybrid/**h**-matrix, 150
  impedance/**Z**-matrix, 147
  interconnecting networks, 154–156
  ABCD network representations, 158–159
  cascading networks, 157–158, 176
  parallel connection of networks, 156–157
  series connection of networks, 154–156
  network properties/applications, 163–165
    analysis of microwave amplifier, 166–169
    interrelations between parameter sets, 163–165
    scattering parameters (*S*-parameters), 169–195
Nodal quality factor, 437–439
Nodes, 178, 179
Noise analysis, 677–687
  correlated noise sources, 679
  internal noise source, 679
  noise bandwidth, 677
  noise figure:
    for cascaded multiport network, 685–687
    for two-port network, 196
  noise power, 523, 575
  noisy two-port networks, 679–682
  $1/f$ noise, 330, 679
  random variables, 677
  shot noise, 330, 679
  spectral density, 330, 679
  spectral noise current, 678
  spectral noise voltage, 678
  uncorrelated noise sources, 679
  white noise, 679
Noise factor, 521
Noise figure, 486, 521–523, 682
  for cascaded multiport network, 685–687
  equivalent noise resistance, 521
  minimum, 521
  design of small-signal amplifier for, 551
  mixers, 615
  optimal reflection coefficient, 521

optimal source admittance, 521
  for two-port network, 682–685
Noise figure circles, 521–524
Noise measure, 687
Noise performance, 329–330
Noisy signals, 677
Noisy two-port networks, 679–682
Noninsulated-gate FET, 388
Nonlinear diode model, 362–367
Nonlinear transistor model, 367–376, 388–391
Normalized effective mass of the electron, 664
Normalized impedance equation, 106

O

1 dB compression point, 521
$1/f$ noise, 679
Odd and even mode excitation, 257
Open-circuit transformations, 117–119
Open-circuited transmission line, 82–83
Operating point, 300
Operating power gain, 490, 515–517
  circle derivation, 516
  constant operating power gain circles, amplifier design using, 490
Optimal reflection coefficient, 521
Optimal source admittance, 521
Oscillations, adding positive feedback element to initiate, 562, 589–591
Oscillator circuit, 560–561
Oscillator design, 578–581
Oscillator feedback, 560–561
Oscillator phase noise, 574–578
Oscillators
  basic model, 560–587
  Clapp oscillator, 581
  Colpitts oscillator, 567
    design, 583–585
  design steps, 581–583
  dielectric resonator oscillators, 598–603
    design of, 600–603
  feedback oscillator design, 578–581
  fixed-frequency oscillators, 591–598
  Gunn element oscillator, 608–609
  Hartley oscillator, 581
  negative resistance oscillator, 562–563
  output voltage vs. gain characteristic, 561

Index

701

quartz oscillators, 585
varactor diode oscillator, 604–606
　design of, 605–606
voltage-controlled oscillator, 604–606
YIG (yttrium iron garnet) oscillator, 603–604
　See also Mixers
Oscillators and mixers, 559–642
Output matching network, 525
Output stability circle equation, 492
Output VSWR, 525, 528, 529

P

Parallel and series connections, 127–137
Parallel connection of networks, 156–157
Parallel-plate transmission line:
　circuit parameters for, 57–61
　derivation of equations, 63–66
　line parameters of, 60–61
Parameter extraction, 153
Parametric admittance equation, 123
Parametric reflection coefficient equation, 108–110
Passive networks, 150, 464
　biasing for BJT in common-emitter configuration, 383, 464–468
P-channel, 330
Permeability in vacuum, 647
Permittivity in vacuum, 647
Phase noise, 574
Phase velocity, 8–9, 32
Phasors, 6
Pinch-off voltage, 334
PIN diodes, 301
　computation of transducer loss of, 305
Pi-network, matrix representation of, 148–150, 199
Pi-type matching network, 442–446
　design of, 442–444
Planar printed circuit boards (PCBs), 47
pn-junction, 285
Polyphase network, 630
Polysilicon, 340
Power, 4
Power amplifier (PA), 3
Power considerations
　for transmission line, 88–91
Power gain, 486
　available, 486
　forward, 173

operating, 486
reverse, 173
transducer, 488, 489
unilateral, 486, 489
Power gain circles, 515–521
Power standing wave ratio, 77
Power-added efficiency, 463
Printed circuit boards (PCBs), 25–32, 47–49
Pseudomorphic HEMTs (pHEMTs), 471
Puck (dielectric resonater), 598

Q

Quality factor Q, 25, 209
Quantum well, 342
Quarter-wave transmission line, 83–85
Quartz crystal:
　and piezoelectric effect, 585
　prediction of resonance frequencies of, 586–587
Quartz oscillators, 585–587
Quiescent point, 188

R

Radio frequency (RF), 3
Radio frequency chokes (RFCs), 4
Random variables, 677
Reciprocal network, 168
Rat race coupler, 675
Reflection coefficient:
　input, 76
　matched load, 512
　matched source, 512
　normalized impedance equation, 106–107
　optimal, 521
　parametric reflection coefficient equation, 108–110
　in phasor form, 104–106
　representations, 105–106
　simultaneous conjugate matched, 513
　voltage, 73–74
Resistance, 648
Resistivity, 648
Resistors, 2, 16, 22, 26, 503, 537
　chip, 25–26
　high-frequency, 16–17, 196
　metal film, 17–19
Return loss (RL), 92–93
Reverse active mode, 317, 369
Reverse bias, 305, 679

Reverse collector and emitter saturation currents, 369
Reverse power gain, 174
Reverse recovery, 294
Reverse voltage gain, 174
RF amplifier functions, 487–488
RF amplifier, power relations of, 490
RF behavior of passive components, 10–25
　capacitors, 19–21
　inductors, 22–24
　resistors, 16–19
RF chokes (RFCs), 22
　RF impedance response of, 17–19
RF circuits
　manufacturing processes, 29–30
RF diodes, 298–313
　BARRITT diodes, 313
　Gunn diodes, 313
　IMPATT diodes, 310–312
　PIN diodes, 301–305
　Schottky diode, 299–301
　TRAPATT diodes, 313
　tunnel diodes, 312
　varactor diodes, 307–309
RF filter design, 205, 270
　bandpass filters, 214
　bandstop filters, 205, 214–216, 269, 270
　response, 216–218
　basic resonator/filter configurations, 206–224
　coupled filter, 257–266
　　bandpass filter section, 260–262
　　cascading bandpass filter elements, 262–264
　　design example, 264–266
　　odd/even mode excitation, 257–260
　denormalization of standard low-pass design, 236–245
　　frequency transformation, 236, 244
　　impedance transformation, 242
　filter implementation
　　Kuroda's identities, 245, 247–249
　　unit elements, 247
　filter types/parameters, 206–210, 236
　insertion loss, 94, 221–224
　low-pass filters
　　attenuation profiles for, 208
　microstrip filter design, examples of, 249–255

RF filter design, (Continued)
    special filter realizations, 224–245
        Butterworth filters, 224, 225–236
        Chebyshev filters, 224, 228–236, 270
RF filters, calculation of quality factors for, 223–224, 272
RF impedance response of, 20–22
RF impedance response:
    of capacitors, 20
    of metal film resistors, 17–19
    of RFCs, 23
RF transistor amplifier design, 485
    amplifiers:
        characteristics of, 486
        power relations, 486, 489
    broadband amplifiers, 529, 530
    constant gain, 504, 505, 507
        bilateral design, 512
        operating and available power gain circles, 515
        unilateral design, 504, 510, 511
    constant VSWR circles, 525–529
    high-power amplifiers, 540
    multistage amplifiers, 529
    noise figure circles, 521
    power gain (operating power gain), 490
    stability, 395
        circles, 515
        stabilization methods, 501
        unconditional, 494, 496, 499, 509
    transducer power gain, 488–489
RF/microwave semiconductor industry, 347–352
RF/MW circuits, 195–197
RF/MW system, 478–479
Richardson constant, 299
Richards transformation, 245–247
Ring oscillator, 578
Ripple, 206, 208
Rollett stability factor, 516, 547, 549, 589–591
Root mean square (RMS), 677

S

Safe operating area (SOAR), 328
Saturation mode, 317, 322
Saturation voltage, 316
Scattering parameter device characterization, 404–408
Scattering parameters (S-parameters), 169–195, 196
    chain scattering matrix, 175–196
    conversion between Z-parameters and, 177–178
    definition of, 167–171
    generalization of, 184–188
    meaning of, 172–175
    practical measurements of, 169, 188–193
    signal flow-graph modeling, 178–184, 196
Schottky contact, 295–298
Schottky diode, 299–301
Second-order intermodulation products, 611
Self-biased networks, 463
Self-loop, 180–184
Semiconductors physical properties of, 278–285
Series connection of networks, 154–156
Series connections, 127–133
    of C elements, 129
    of L elements, 129
Shape factor, 208
Short air-core solenoid, 24
Short diode, 666
Short-circuit transformations, 119–120
Short-circuited transmission line, 80–81
    input imedance of, as a function of frequency, 246
Shot noise, 330, 679
Shunt connections, 127–135
    of C elements, 128
    of L elements, 127
Signal flow graph, 187, 278
    building blocks, 181
    input impedance computation of a transmission line based on use of, 187–188
    modeling, 178–184
Signal flow graph modeling
    RF network analysis, 178–184
Simultaneous conjugate match, 512, 513, 551
Simultaneous conjugate matched reflection coefficients, derivation of, 513–514
Single-balanced active mixer, 622–623
Single-balanced mixer, 622–623
Single-ended mixer design, 614–622
    BJT mixer, 622
Single-/multiport networks, 145–197
    admittance/Y-matrix, 148, 156
    basic definitions, 145, 146–154
    chain/$ABCD$-matrix, 150
    hybrid/h-matrix, 150
    impedance/Z-matrix, 147
    interconnecting networks, 154–156
        $ABCD$ network representations, 158–159
        cascading networks, 157–158, 176
        parallel connection of networks, 156–157
        series connection of networks, 154–156
    network properties/applications, 163–165
        analysis of microwave amplifier, 166–169
        interrelations between parameter sets, 163–165
    scattering parameters (S-parameters), 169–195
Single-stub matching networks, 450–454
    with fixed characteritic impedances, design of, 450–452
Skin depth, 11–13, 58–61
Skin Equation for a cylindrical conductor, 653
Skirt selectivity, 207
Small-signal BJT models, 376–386
Small-signal current gain, 168
Small-signal diode model, 376
Small-signal equivalent circuit model, 300
Small-signal FET models, 391–394
Smith Chart, 104, 110, 123
    admittance transformation, 123–125
        parametric admittance equation, 123–125
        Y-Smith Chart, 124–125, 131
        Z-Smith Chart, 123–124
        ZY-Smith Chart, 126–127, 425, 426
    compressed, 111
    design of matching network using, 428, 446
    impedance transformation, 112, 236, 242–243, 437
        computer simulations, 120–122
        for general load, 112